Restoration of Aquatic Systems

Marine Science Series

The CRC Marine Science Series is dedicated to providing state-of-the-art coverage of important topics in marine biology, marine chemistry, marine geology, and physical oceanography. The series includes volumes that focus on the synthesis of recent advances in marine science.

CRC MARINE SCIENCE SERIES

SERIES EDITOR

Michael J. Kennish, Ph.D.

PUBLISHED TITLES

Artificial Reef Evaluation with Application to Natural Marine Habitats, William Seaman, Jr.

The Biology of Sea Turtles, Volume I, Peter L. Lutz and John A. Musick

Chemical Oceanography, Second Edition, Frank J. Millero

Coastal Ecosystem Processes, Daniel M. Alongi

Ecology of Estuaries: Anthropogenic Effects, Michael J. Kennish

Ecology of Marine Bivalves: An Ecosystem Approach, Richard F. Dame

Ecology of Marine Invertebrate Larvae, Larry McEdward

Ecology of Seashores, George A. Knox

Environmental Oceanography, Second Edition, Tom Beer

Estuarine Indicators, Stephen A. Bortone

Estuarine Research, Monitoring, and Resource Protection, Michael J. Kennish

Estuary Restoration and Maintenance: The National Estuary Program, Michael J. Kennish

Eutrophication Processes in Coastal Systems: Origin and Succession of Plankton Blooms and Effects on Secondary Production in Gulf Coast Estuaries, Robert J. Livingston

Handbook of Marine Mineral Deposits, David S. Cronan

Handbook for Restoring Tidal Wetlands, Joy B. Zedler

Intertidal Deposits: River Mouths, Tidal Flats, and Coastal Lagoons, Doeke Eisma

Marine Chemical Ecology, James B. McClintock and Bill J. Baker

Morphodynamics of Inner Continental Shelves, L. Donelson Wright

Ocean Pollution: Effects on Living Resources and Humans, Carl J. Sindermann

Physical Oceanographic Processes of the Great Barrier Reef, Eric Wolanski

The Physiology of Fishes, Second Edition, David H. Evans

Pollution Impacts on Marine Biotic Communities, Michael J. Kennish

Practical Handbook of Estuarine and Marine Pollution, Michael J. Kennish

Practical Handbook of Marine Science, Third Edition, Michael J. Kennish

Seagrasses: Monitoring, Ecology, Physiology, and Management, Stephen A. Bortone

Trophic Organization in Coastal Systems, Robert J. Livingston

Restoration of Aquatic Systems

Robert J. Livingston

Director
Center for Aquatic Research and Resource Management
Florida State University
Tallahassee

Taylor & Francis
Taylor & Francis Group

Boca Raton   London   New York   Singapore

A CRC title, part of the Taylor & Francis imprint, a member of the
Taylor & Francis Group, the academic division of T&F Informa plc.

Published in 2006 by
CRC Press
Taylor & Francis Group
6000 Broken Sound Parkway NW, Suite 300
Boca Raton, FL 33487-2742

© 2006 by Taylor & Francis Group, LLC
CRC Press is an imprint of Taylor & Francis Group

No claim to original U.S. Government works
Printed in the United States of America on acid-free paper
10 9 8 7 6 5 4 3 2 1

International Standard Book Number-10: 0-8493-1966-8 (Hardcover)
International Standard Book Number-13: 978-0-8493-1966-2 (Hardcover)
Library of Congress Card Number 2004064948

Library of Congress Cataloging-in-Publication Data

Livingston, Robert J.
    Restoration of aquatic systems / by Robert J. Livingston.
        p. cm. –- (Marine science series)
    Includes bibliographical references (p.) and index.
    ISBN 0-8493-1966-8
    Restoration ecology--Florida. 2. Aquatic ecology--Florida. 3. Wildlife habitat improvement--Florida.
    I. Title. II. Series.

QH105.F6L582 2005
639.9—dc22   2004064948

Taylor & Francis Group
is the Academic Division of T&F Informa plc.

Visit the Taylor & Francis Web site at
http://www.taylorandfrancis.com

and the CRC Press Web site at
http://www.crcpress.com

"The road to Perdition is paved with good intentions."

—**Unattributed aphorism**

"Approximately 80% of our air pollution stems from hydrocarbons released by vegetation, so let's not go overboard in setting and enforcing tough emission standards from man-made sources."

—**Ronald Reagan**

"No witchcraft, no enemy action had silenced the re-birth of new life in this stricken world. The people had done it themselves."

—**Rachel Carson,** *Silent Spring*

"A vast and destructive algae bloom has floated out of the backcountry and settled near Islamorada. ...Crossing the tall arch to Long Key, you noticed a foul-looking stain stretching gulfward to the horizon. What you saw is Florida Bay dying."

—**Carl Hiaasen,** *Miami Herald,* **1993**

"It sounds like a bad 1950s science fiction movie: ugly green slime spreading through the sea, killing fish and threatening children and animals that swim in the water. But experts say this scenario could become reality in the Baltic, the world's largest brackish water sea..."

—**Nina Garlo, Reuters article, 2003**

"A foolish consistency is the hobgoblin of little minds."

—**Ralph Waldo Emerson**

"Collectively, the restoration effort is billed as the nation's premier watershed restoration program and is billed as a model for estuarine restoration programs worldwide. All the while, decades pass and the Bay's most basic environmental indicators suggest little if any sustained improvement.... As the region's political system continues to grapple with environmental protection, it is important not to adjust environmental goals to reflect 'political realities' or to grow content with hollow 'successes' that fail to reflect tangible environmental improvements."

—**Howard R. Ernst,** *Chesapeake Blues,* **2003**

Preface

The research effort on which this book is based has involved continuous analyses of various river–estuarine and coastal systems in the southeastern United States since 1970. These long-term studies have been carried out using a combination of field descriptive and experimental (lab and field) approaches. The research team for this work has included field collection personnel, chemists, taxonomists, experimental biologists, physical oceanographers, hydrological engineers, statisticians, computer programmers, and modelers. A long-term, interdisciplinary, comparative database has been created that is currently being published in a series of books and peer-reviewed scientific journals. This program has been used to evaluate system-level processes that determine the effects of nutrient loading and nutrient dynamics on phytoplankton/benthic macrophyte productivity and associated food web responses. Efforts have been made to determine how human activities affect these processes. The emphasis of the research has been on seasonal and interannual trophic response to habitat changes and an evaluation of the interrelationships of cultural eutrophication and toxic substances on secondary production. This research is currently being used to develop realistic indices of ecosystem condition and determine restoration methods.

This book includes several features that have been integrated using the ecosystem evaluations to determine the efficacy of habitat rehabilitation. In addition, information regarding the way in which the scientific data have been reported by American news media and the public response to such information have also been analyzed. The long-term database (Table 1) has thus been used as objective background material for an evaluation of the efficacy of the restoration process. The various components of the restoration of damaged aquatic systems are not restricted to the scientific effort. Other factors include economic and political interests, information dissemination, public response, and the complex synergism that goes into controversial environmental issues. This book is thus concerned with just how effective the restoration process becomes as a product of a complex mixture of competing interests. The primary emphasis is on the relationship of scientific research to the rehabilitation of aquatic habitats.

Table 1 Field/Experimental Effort of the Florida State University Research Group from 1970 to 2004

System	Met.	PC	LC	POL	PHYTOPL	SAV	ZOOPL	INFAUNA	INV	FISHES	FW	NL	NR
North Florida Lakes	35	14	14	5	6	2	1	2	2	2	2	1	1
Apalachicola System	85	17	2	2	1	1	1	8	14	14	8	2	nd
Apalachee Bay	50	26	12	2	5	15	3	2	18	19	26	3	1
Perdido River–Bay	50	16	16	3	13	3	3	16	16	16	16	16	1
Pensacola Bay System	30	1.5	1.5	1	1.5	nd	1.5	1.5	1.5	1.5	1.5	1.5	nd
Blackwater–East Bay System	4	1.5	1.5	1	nd	1	nd	1.5	1.5	1.5	1.5	1.5	nd
Escatawpa System	5	2	2	2	nd	nd	nd	2	2	2	nd	nd	nd
Choctawhatchee River–Bay	30	4	4	2	2	1	2	4	4	4	2	2	nd
Mobile River Estuary	3	2	2	2	nd	nd	nd	2	2	2	nd	nd	nd
Nassau–Amelia Estuaries	3	3	3	2	3	3	3	2	1	1	2	nd	1
Sampit River–Winyah Bay System	5	4	4	4	nd	nd	nd	4	2	2	nd	nd	nd

Met. = river flow, rainfall.

PC = salinity, conductivity, temperature, dissolved oxygen, oxygen anomaly, pH, depth, Secchi.

LC = NH_3, NO_2, NO_3, TIN, PON, DON, TON, TN, PO_4, TDP, TIP, POP, DOP, TOP, TP, DOC, POC, TOC, IC, TIC, TC, BOD, silicate, TSS, TDS, DIM, DOM, POM, PIM, NCASI color, turbidity, chlorophyll a, b, c, sulfide.

POL = water/sediment pollutants (pesticides, metals, PAH).

PHYTOPL = whole water and/or net phytoplankton.

SAV = submerged aquatic vegetation.

ZOOPL. = net zooplankton.

INFAUNA = infaunal macroinvertebrates taken with cores and/or ponars.

INV = invertebrates taken with seines (freshwater) and otter trawls (bay).

FISHES = fishes taken with seines (freshwater) and otter trawls (bay).

FW = food web transformations.

NL= nutrient loading.

NR = nutrient limitation experiments.

nd = no data.

About the Author

Robert J. Livingston is currently Professor Emeritus of the Department of Biological Sciences at Florida State University (Tallahassee, Florida, USA). His interests include aquatic ecology, pollution biology, field and laboratory experimentation, and long-term ecosystem-level research in freshwater, estuarine, and marine systems. Past research includes multidisciplinary studies of lakes and a series of drainage systems on the Gulf and East coasts of the United States. Over the past 35 years, Livingston's research group has conducted a series of studies in areas from Maine to Mississippi. Dr. Livingston has directed the programs of 49 graduate students who have carried out research in behavioral and physiological ecology with individual aquatic populations and communities in lakes, rivers, and coastal systems. He is the author of more than 170 scientific papers and five books on the subject of aquatic ecology, and has been the principal investigator for more than 100 projects since 1970.

The primary research program of the author has been carried out largely in the southeastern United States. The areas studied include the Apalachicola drainage system, the Choctawhatchee drainage system, the Perdido drainage system, the Escambia River and Bay system, the Blackwater River estuary, the Escatawpa Pascagula drainage system, the Mobile River estuary, the Winyah Bay system (including the Sampit River), Apalachee Bay (the Econfina and Fenholloway River estuaries) and a series of north Florida lake systems (Table 1). This work has included determinations of the impact of various forms of anthropogenous activities on a range of physical, chemical, and biological processes.

A computer system has been developed to aid in the analysis of the established long-term databases. Together with associates and former graduate students, Livingston is currently in the process of publishing the long-term database. Field research is ongoing in the Perdido system. Retirement is not an option.

Acknowledgments

There were many people who contributed to the research effort over the past 35 years. It is impossible to acknowledge all these people. Literally thousands of undergraduate students, graduate students, technicians, and staff personnel have taken part in the long-term field ecology programs. Various people have been involved in CARRMA research associated with the Center for Aquatic Research and Resource Management at Florida State University for the study period.

The program has had the support and cooperation of various experts. Dr. A.K.S.K. Prasad studied phytoplankton taxonomy and systematics. Robert L. Howell IV, ran the field collections and identified fishes and invertebrates. Glenn C. Woodsum designed the overall database organization and, together with Phillip Homann, ran database management and day-to-day data analysis of the research information. Dr. F. Graham Lewis III, was instrumental in the development of analytical procedures and statistical analyses. Dr. David A. Flemer designed the overall approach for analytical chemistry. Dr. Duane A. Meeter, Dr. Xufeng Niu, and Loretta E. Wolfe performed statistical analyses. The taxonomic consistency of the field collections was carried out with the help of world-class systematists, including Dr. John H. Epler (aquatic insects), Dr. Carter R. Gilbert and Dr. Ralph Yerger (freshwater fishes), and Dr. Clinton J. Dawes (submerged aquatic vegetation).

The long-term field ecological program has benefited from the advice of a distinguished group of scientists. Dr. John Cairns, Jr., Mr. Bori Olla, Dr. E.P. Odum, Dr. J.W. Hedgpeth, Dr. O. Loucks, Dr. K. Dickson, Dr. F.J. Vernberg, and Dr. Ruth Patrick reviewed the early work. Dr. D.M. Anderson, Dr. A.K.S.K. Prasad, Mr. G.C. Woodsum, Dr. M.J. Kennison, Dr. K. Rhew, Dr. M. Kennish, Dr. E. Fernald, Dr. David C. White, and Ms. V. Tschinkel have reviewed various aspects of the recent work. Other reviews have been provided by Dr. C.H. Peterson, Dr. S. Snedaker, Dr. R.W. Virnstein, Dr. D.C. White, Dr. E.D. Estevez, Dr. J.J. Delfino, Dr. F. James, Dr. W. Herrnkind, and Dr. J. Travis. Dr. Eugene P. Odum gave me the good advice to write up our results in a series of books, which turned out to be the only way our work could be adequately treated.

Over the years, we have had a long list of excellent co-investigators who have participated in various projects. Recent collaborative efforts in the field have included the following scientists: Dr. D.C. White (microbiological analyses), Dr. D.A. Birkholz (toxicology, residue analyses), Dr. G.S. Brush (long-core analysis), Dr. L.A. Cifuentes, (nutrient studies, isotope analyses), Dr. K. Rhew (microalgal analyses), Dr. R.L. Iverson (primary productivity), Dr. R.A. Coffin (nutrient studies, isotope analyses), Dr. W.P. Davis (biology of fishes), Dr. M. Franklin (riverine hydrology), Dr. T. Gallagher (hydrological modeling), Dr. W.C. Isphording (marine geology), Dr. C.J. Klein III (aquatic engineering, estuarine modeling), Dr. M.E. Monaco (trophic organization), Dr. R. Thompson (pesticide analyses), Dr. W. Cooper (chemistry), Mr. D. Fiore (chemistry), and Dr. A.W. Niedoroda (physical oceanography).

We have had unusually strong support in the statistical analyses and modeling of our data, a process that is continuing to this day. These statistical determinations and modeling efforts have been carried out by the following people: Dr. T.A. Battista (National Oceanic and Atmospheric Administration), Dr. B. Christensen (University of Florida), Dr. J.D. Christensen (National Oceanic and Atmospheric Administration), Dr. M.E. Monaco (National Oceanic and Atmospheric Administration), Dr. Tom Gallagher (Hydroqual, Inc.), Dr. B. Galperin (University of South Florida), and Dr. W. Huang (Florida State University).

Experimental field and laboratory ecology has been carried out by the following graduate students: Dr. B.M.S. Mahoney (experimental biology), Dr. K. Main (experimental biology), Dr. Frank Jordan (biology of fishes), Dr. Ken Leber (experimental biology), Dr. W.H. Clements (trophic analyses), Dr. C.C. Koenig (fish biology), Dr. P. Sheridan (trophic analyses), Dr. J. Schmidt (aquatic macroinvertebrates), Dr. A.W. Stoner (trophic analyses), Dr. R.A. Laughlin (trophic analyses), Dr. J.L. Luczkovich (experimental biology), Dr. J. Holmquist (experimental biology), Dr. D. Bone (experimental biology), Dr. B. MacFarlane (biology of fishes), Ms. C. Phillips (experimental biology), Mr. M. Kuperberg (submerged aquatic vegetation), Ms. T.A. Hooks (experimental biology), Mr. Joseph Ryan (trophic analyses), Mr. C.J. Boschen (trophic analyses), Mr. G.G. Kobylinski (experimental biology), Mr. C.R. Cripe (experimental biology), Mr. T. Bevis (biology of fishes), Mr. P. Muessig (experimental biology), Ms. S. Drake (experimental biology), Mr. K.L. Heck, Jr. (experimental biology), Ms. H. Greening (aquatic macroinvertebrates), Mr. Duncan Cairns (experimental biology), Ms. K. Brady (larval fishes), Mr. B. McLane (aquatic macroinvertebrates), Ms. B. Shoplock (zooplankton), Ms. J.J. Reardon (phytoplankton, submerged aquatic vegetation), and Ms. J. Schmidt-Gengenbach (experimental biology).

Taxonomy is at the heart of a comprehensive ecosystem program, which is something I learned from my major professors (Dr. Carl Hubbs, Dr. C. Richard Robbins, Dr. Arthur A. Myrberg, and Dr. Charles E. Lane) while attending graduate school. Our attention to systematics and natural history has benefited from the efforts of a long line of graduate students, post-doctoral fellows, and trained technicians who include the following: Dr. A.K.S.K. Prasad and Dr. K. Rhew (phytoplankton, benthic microalgae), Dr. F. Graham Lewis III (aquatic macroinvertebrates), Dr. J. Epler (aquatic macroinvertebrates), Dr. R.D. Kalke (estuarine zooplankton), Dr. G.L. Ray (infaunal macroinvertebrates), Dr. K.R. Smith (oligochaetes), Dr. E. L. Bousfield (amphipods), Dr. F. Jordan (trophic analyses), Dr. C.J. Dawes (submerged aquatic vegetation), Dr. R.W. Yerger and Dr. Carter Gilbert (fishes), Mr. R.L. Howell III (infaunal and epibenthic invertebrates and fishes), Mr. W.R. Karsteter (freshwater macroinvertebrates), and Mr. M. Zimmerman (submerged aquatic vegetation).

A number of phycologists have given their time to the taxonomy and nomenclature of phytoplankton species over the years: Dr. C.W. Reimer (Academy of Natural Sciences of Philadelphia), Dr. G.A. Fryxell (University of Texas at Austin), Dr. G.R. Hasle (University of Oslo, Norway), Dr. P. Hargraves (University of Rhode Island), Professor F. Round (University of Bristol, England), Mr. R. Ross, Ms. P.A. Sims, Dr. E.J. Cox and Dr. D.M. Williams (British Museum of Natural History, London, United Kingdom), Dr. L.K. Medlin and Dr. R.M. Crawford (Alfred-Wegener Institute, Bremerhaven, Germany), Professor T.V. Desikachary (University of Madras, India), Dr. J.A. Nienow (Valdosta State University, Georgia), Dr. P. Silva (University of California at Berkeley), Dr. M.A. Faust (Smithsonian Institution), Dr. R.A. Andersen (Bigelow Laboratory for Ocean Sciences, Maine), Dr. D. Wujeck (Michigan State University), Dr. M. Melkonian (University of Koln, Germany), Dr. C.J. Tomas (Florida Department of Environmental Regulation, St. Petersburg), Professor T.H. Ibaraki (Japan), Dr. Y. Hara (University of Tsukuba, Japan), and Dr. J. Throndsen (University of Oslo, Norway). Technical assistance with transmission and scanning electron microscopes was provided by Ms. K.A. Riddle (Department of Biological Science, Florida State University). Mrs. A. Black and Mr. D. Watson (Histology Division, Department of Biological Science, Florida State University) aided in specimen preparation and serial thin sectioning.

The curators of various International Diatom Herbaria have aided in the loan of type collections and other authentic materials: Academy of Natural Sciences of Philadelphia (ANSP), California Academy of Sciences (CAS) (San Francisco), National Museum of Natural History, Smithsonian Institution (Washington, D.C.), Harvard University Herbaria (Cambridge. Massachusetts), The Natural History Museum (BM) (London, United Kingdom), and the F. Hustedt Collections, Alfred Wegener Institute (Bremerhaven, Germany).

Others have collaborated on various aspects of the field analyses. Mr. W. Meeks, Mr. B. Bookman, Dr. S.E. McGlynn, and Mr. P. Moreton ran the chemistry laboratories. Mr. O. Salcedo, Ms. I. Salcedo, Mr. R. Wilt, Mr. S. Holm, Ms. L. Bird, Ms. L. Doepp, Ms. C. Watts, and Ms. M. Guerrero-Diaz provided other forms of lab assistance. Field support for the various projects was provided by Mr. H. Hendry, Mr. A. Reese, Ms. J. Scheffman, Mr. S. Holm, Mr. M. Hollingsworth, Dr. H. Jelks, Ms. K. Burton, Ms. E. Meeter, Ms. C. Meeter, Ms. S. Solomon, S.S. Vardaman, Ms. S. van Beck, Mr. M. Goldman, Mr. S. Cole, Mr. K. Miller, Mr. C. Felton, Ms. B. Litchfield, Mr. J. Montgomery, Ms. S. Mattson, Mr. P. Rygiel, Mr. J. Duncan, Ms. S. Roberts, Mr. T. Shipp, Ms. A. Fink, Ms. J.B. Livingston, Ms. R.A. Livingston, Ms. J. Huff, Mr. R.M. Livingston, Mr. A.S. Livingston, Mr. T. Space, Mr. M. Wiley, Mr. D. Dickson, Ms. L. Tamburello, and Mr. C. Burbank.

Project administrators included the following people: Dr. T.W. Duke, Dr. R. Schwartz, Dr. M.E. Monaco, Ms. J. Price, Dr. N.P. Thompson, Mr. J.H. Millican, Mr. D. Arceneaux, Mr. W. Tims, Jr., Dr. C.A. Pittinger, Mr. M. Stellencamp, Ms. S.A. Dowdell, Mr. C. Thompson, Mr. K. Moore, Dr. E. Tokar, Dr. D. Tudor, Dr. E. Fernald, Ms. D. Giblon, Ms. S. Dillon, Ms. A. Thistle, Ms. M.W. Livingston, and Ms. C. Wallace.

Robert J. Livingston and Julia B. Livingston designed the book cover.

Contents

section I

Definitions

The restoration of aquatic systems has become an important object of research organizations throughout the world. Most of this effort is directed at minor parts of a given drainage system; popular restoration efforts consist of replanting individual marshes and isolated seagrass beds, and the building of new ponds, or replanting denuded upland areas with trees. Citizen-oriented restoration efforts abound, with a variety of community activities that include solid waste removal and a broad array of water quality monitoring programs. These restoration activities are not addressed here for several reasons. Seldom are such efforts grounded on solid scientific investigations. Often, there is media hype that is devoted more to political considerations than actual restoration. And there is almost always a lack of scientific information with regard to the actual efficacy of a given restoration effort. Instead, this author is mainly interested in system-level restoration activities that purport to use the scientific process to restore entire lakes, rivers, and coastal systems to some former level of productivity. These efforts include some of the most important aquatic systems in the world.

The definition of a given restoration effort should include some reference to the natural habitat, the human activities in the region of interest, and the inclusion of unavoidable impacts due to natural causes. That is, there should be some knowledge of system processes so that realistic goals can be projected for the restoration process. This information should also include an estimate of the natural productivity of the target system, the human health aspects involved, the biodiversity of the original system, the state of the fisheries potential (past and present), and the possibilities for enhancement of aesthetic qualities of the natural ecosystem. That is, the restoration process should be addressed as the formal use of scientific and engineering methods to bring back damaged aquatic systems to some level of natural productivity. This effort should be made within the limits of the natural productivity of the system in question.

chapter 1

The Restoration Paradigm

1.1 Definitions

There are many interpretations of the term "restoration." According to the National Oceanic and Atmospheric Administration (NOAA, 1995), restoration is defined as "the process of reestablishing a self-sustaining habitat that closely resembles a natural condition in terms of structure and function." The Environmental Protection Agency (EPA U.S. Environmental Protection Agency, 2003) defines restoration as "the return of a degraded ecosystem to a close approximation of its remaining natural potential." Another interpretation of 'restoration' relates to the return of a polluted or degraded environment to a successful, self-sustaining ecosystem with both clean water and healthy habitats. If the definition includes a return to the "original natural system," this assumes that the attributes of the original system are known, which is seldom the case. There are varied interpretations of the "success" of individual restoration efforts. This is complicated by endless philosophical discussions on an academic level concerning the actual meaning of the term. To add to the complication of interpretation, there are different levels of restoration efforts that range from attempts to restore minor wetland areas and individual grass beds to the proposed recovery of entire ecosystems. The factors that contribute to "successful" restoration efforts are not restricted to scientific efforts. These include political, economic, legal, and sociological processes that often supersede the science in terms of influence on the outcome. These factors are seldom mentioned when a given restoration effort is developed.

Whatever definition is accepted for a given restoration effort, the process should include some reference to factors that include the following: natural habitat, human uses in the region of interest, and inclusion of unavoidable impacts due to human activities. It should also include an estimate of the natural productivity of the aquatic system in question. Other contributing factors include human health aspects, the biodiversity of the original system, the state of existing fisheries, and possibilities for enhancement of aesthetic qualities. That is, restoration should be addressed as the formal use of scientific and engineering methods to bring back aquatic systems that have been damaged by human activities to some level of natural productivity. Most of all, however, there should be a clear definition of the goals of a given restoration effort. Such goals should be realistic, with the clear realization that a return to the "pristine" state is seldom if ever a possibility.

1.2 Ecosystem Research and Restoration

The ecosystem concept is the basis of resource management issues in aquatic systems. Although the definition of what constitutes ecosystem-level research is continuously debated by aquatic scientists, the purpose of such analyses (i.e., development of a factual basis that is used for system management) remains relatively clear. Because ecosystem research is often used to answer questions that are not necessarily asked at the beginning of the program, such research must remain open-ended so that it can be applied to unanticipated impacts. Ecosystem processes cannot be defined entirely by *post hoc* accumulations of information. In many ways, the application of even comprehensive scientific data to the management of coastal systems is not direct, and requires a multidisciplinary approach that often goes beyond the mere assembly of facts. This application can be successful (Livingston, 2002) but requires diverse and innovative approaches to problem solving.

A successful ecosystem approach requires a very different set of initial research questions than that of the usual hypothesis-testing process. These questions are often limited by unknown factors that will occur in the future, and thus require a database that can eventually address future information needs of complex questions. In ecosystem research, it is thus necessary to use a combination of disciplines that include broad yet inclusive approaches to an understanding of the system in question. This should involve integration of descriptive field monitoring and laboratory/field experimentation with an emphasis on comprehensive, long-term scientific observation of the system processes. Ecosystem-level questions should be based on multiple sequences of temporal interactions at various levels of biological organization. Because of the uniqueness of the combination of state variables that drive individual aquatic systems, the final product of any given analysis should also include a comparison with other systems. The establishment of an effective ecosystem-level research program thus requires sophisticated levels of planning, consistent, long-term funding, and a central organization that includes close coordination of the timing and execution of the various parts of the program.

1.3 Human Impacts on Aquatic Systems

Kennish et al. (2003) have outlined various forms of human impacts on aquatic systems. Habitat degradation and water quality deterioration have resulted from the cumulative impacts of multiple human activities. Various stressors affect aquatic areas with complex system-specific impacts that remain incompletely understood in terms of cause-and-effect mechanisms. Nutrient enrichment, organic carbon loading, chemical contaminants, and various pathogens represent the primary forms of pollution. These factors interact with habitat degradation due to wetland reclamation, dredging, fresh water diversion, and physical alterations due to shoreline development. Additional biological stressors include the introduction of exotic species and over-fishing.

The primary sources of human impacts include rapidly expanding urbanization in major drainage basins, loading of agricultural and industrial wastes, and point and non-point sources of toxic agents such as mercury. Anthropogenous nutrient enrichment due to diverse sources is rapidly becoming a leading factor in the deterioration of various freshwater, estuarine, and marine systems (Kennish et al., 2003). There is growing evidence of the increased incidence and virulence of phytoplankton blooms in aquatic systems around the world (Livingston, 2000, 2002). Bricker et al. (1999), in a comprehensive review of the extent of the eutrophication problem in coastal areas of the conterminous United States, found that 44 estuaries (representing 40% of the total estuarine surface area) suffer "high expressions of eutrophic conditions." The primary symptoms on which the survey was based included decreased light availability, high chlorophyll *a* concentrations, high epiphytic/macroalgal growth rates, changes in algal dominance (diatoms to flagellates, benthic to pelagic dominance), and increased decomposition of organic matter due to the high chlorophyll *a* concentrations and macroalgal growth. Bricker et al. (1999) noted that these effects were most pronounced along the coasts of the Gulf of Mexico and the Middle

Atlantic states. The authors projected that eutrophic conditions would worsen in 86 estuaries by the year 2020.

Howarth et al. (2000) reviewed the effects of excess nutrients on coastal systems. This worldwide phenomenon has led to increased algal biomass, excessive concentrations of sometimes toxic algae in the form of harmful brown and red tides, reduced seagrass and coral reef habitat, altered marine biodiversity, hypoxia/anoxia, and the loss of commercial fisheries. The authors emphasized that a considerable number of coastal systems in the United States have some of the symptoms of excessive nutrification. The report was designed to make recommendations for the implementation of management efforts to reduce nutrient loading. These recommendations included the expansion of monitoring programs, development of ways to reduce non-point sources of nitrogen and phosphorus, an increased federal role in eutrophication issues, development of a classification scheme for management of nutrient over-enrichment, improvement of comprehensive assessments of environmental quality and associated modeling efforts, and expansion of our knowledge concerning eutrophication questions.

Human population increases in drainage basins throughout the world have become an increasingly important source of nutrient loading to aquatic systems. Recent studies indicate that phytoplankton blooms are the single most important factor in eutrophication processes of such systems. It is therefore necessary to understand phytoplankton population and community dynamics if the postulated effects of plankton blooms on the trophic organization of coastal systems are to be understood. Unfortunately, scientific inquiry into this phase of aquatic ecology is still in the early stages of descriptive research. Kennish (1997) has reviewed problems associated with anthropogenous nutrient loading to coastal systems. Hypereutrophication has long been correlated with increased biochemical oxygen demand (BOD), hypoxia, and anoxia in receiving estuaries. Hypoxia is also considered a problem in estuaries and bays throughout Europe, the Far East, the United States, and Australia (Kennish, 1997). Other postulated water quality changes due to excessive nutrient loading include increased turbidity, reduced light penetration, and deterioration of sediment quality. Proliferation of nuisance algal macrophytes and plankton blooms, often associated with losses of productive seagrass beds, are considered major factors in the response of coastal areas to anthropogenous nutrient loading.

McComb (1995) outlined the primary effects of nutrient loading on shallow estuaries and lagoons. A common aspect in the review of cultural eutrophication in diverse estuaries includes association of nutrient loading with the development of phytoplankton blooms, associated fish kills, and the general decline of fisheries (Hodgkiss and Yim, 1995). Nutrient enrichment also adversely affects seagrass beds through changes associated with stimulation of micro- and macroalgae (Duarte, 1995; Hein et al., 1995; Valiela et al., 1997). The emphasis on shallow systems highlights the importance of sediments in the nutrient loading and accumulation process (McComb and Lukatelich, 1995). Sediment resuspension as a product of water column dynamics can be a factor in nutrient storage and loading processes (de Jong, 1995; de Jong and Raaphorst, 1995), although such effects are not always associated with hypereutrophication (Marcomini et al., 1995). The timed release and bioavailability of sediment nutrients, as modified by resuspension, remobilization, and regeneration through biotic activity, remain poorly understood although such processes could have an important effect on the response of benthic microalgae and macrophytes (Thornton et al., 1995a).

The lack of information regarding non-point source nutrient loading due to urban and agricultural runoff is another consideration when reviewing the causes of hypereutrophication (King and Hodgson, 1995). De Jong and Raaphorst (1995) reviewed the positive and negative effects of increased nutrient loading. There can be initial increases of secondary production and fisheries output with increased nutrient loading (Thornton et al.,

1995a). However, prolonged increases often end in plankton blooms and accompanying declines of coastal populations that are often not discovered until there is an advanced state of hypereutrophication. The complex processes associated with the interaction and competition of microphytes and macrophytes often complicate a uniform response of different estuaries to increased anthropogenous nutrient loading. However, in most studies, there is a general lack of detailed data concerning how phytoplankton blooms are initiated, and how changes in the phytoplankton community structure actually affect food web structure and secondary production. Despite a plethora of studies concerning hypereutrophication in aquatic systems, the processes that lead to altered phytoplankton populations and associated food web changes remain largely undefined.

The key to any restoration effort rests on the development of adequate scientific knowledge of the eutrophication process and the determination of how specific changes in anthropogenous nutrient loading affect individual estuaries and coastal systems. Just how adequate science plays a role in the successful restoration of aquatic systems remains complicated, however. There is no straight path from good science to the restoration of lost aquatic resources. The common link between the various processes that contribute to such restoration is the awareness and interest of the public at large. It is here that complications occur that often preclude a successful restoration process. However, the links between what the public understands and how such understanding affects the restoration process are usually discounted or ignored by the managers of many such efforts.

1.4   The Paradox of Actual Risks and Public Concerns

According to Wright and Nebel (2002), there is a considerable difference between the hierarchy of environmental risks as defined by the U.S. EPA and the American public's perception of what constitutes such risks (Table 1.1).

Wright and Nebel (2002) attribute the differences of risk perception by the public (as compared to a tabulation of risks by environmental professionals) to the "strong influence" of the news media in the United States, "which are far better at communicating the outrage

Table 1.1 The Primary Environmental Risks According to the U.S. Environmental Protection Agency vs. Concerns as Viewed by the American Public

The EPA's Top 11 Risks (not ranked)	Public Concerns (in rank order)
Ecological Risks	
Global climate change	1. Active hazardous waste sites
Stratospheric ozone depletion	2. Abandoned hazardous waste sites
Habitat alteration	3. Water pollution from industrial wastes
Species extinctions and loss of biodiversity	4. Occupational exposure to toxic chemicals
	5. Oil spills
Health Risks	6. Destruction of the ozone layer
Criteria air pollutants (e.g., smog)	7. Nuclear power plant accidents
Toxic air pollutants (e.g., benzene)	8. Industrial accidents releasing pollution
Radon	9. Radiation from radioactive wastes
Indoor air pollution	10. Air pollution from factories
Contamination of drinking water	
Occupational exposure to chemicals	
Pesticide application	

Source: From Wright, R.T. and B.J. Nebel. 2002. *Environmental Science: Toward a Sustainable Future.* Prentice Hall, Upper Saddle River, NJ. With permission.

elements of a risk rather than they are at communicating the hazard elements." Unfortunately, the processes of law-making and the creation of public policies regarding the environment depend largely on public perceptions rather than the studied opinions of experts. Serious environmental concerns such as global warming and habitat destruction that lead to reduced biodiversity are consequently ignored by the public, whereas industrial pollution, which has been successfully remediated in many instances, remains high on the public's list of environmental risks.

There is a widespread lack of understanding of scientific ideas and procedures in the public sphere (Jenkins, 2003). In some cases, science has been perceived by the public as inextricably linked with politics, and is therefore accorded a minor role in decision making. Environmental knowledge is often evaluated along personal and local perceptions that are influenced by extraneous events and attitudes. Attitudes can be influenced by diverse factors that range from associations with established political entities to socio-political activism as "environmentalists." Any given environmental problem should thus be viewed within the context of broader factors such as economics, sociology, and politics.

The often-contradictory actions by government and regulatory agencies can be seen in various major environmental questions such as the Great Barrier Reef of Australia. In this instance, this wonder of nature is being destroyed by various factors such as nutrient runoff and over-fishing. These effects have led to actions by the Australian government to correctly restore the reef. Yet, Australian politicians have refused to sign the Kyoto Protocol for reduction of greenhouse gases and global warming although projected temperature changes have already been associated with destructive coral bleaching episodes. Despite the fact that global warming is considered the main threat to coral reefs around the world by most experts in this field, there is an obvious disconnect between the scientific projections and political action. There is little doubt that the interaction between public misconceptions and environmental risk factors and the resulting lack of political action concerning important environmental matters is a complex and little-understood phenomenon.

1.5   Factors for Successful Restoration

The initiation of a given restoration action can have many origins, and there are numerous approaches to the restoration process. In most cases, it is advisable that such actions are based on scientific information, although there is considerable controversy concerning just what kind of scientific data are needed for such decisions. The usual restoration procedure starts with the perception of an environmental problem. The scale of a given restoration action can vary from the recovery of a seagrass bed or marsh to the reconstitution of an entire ecosystem. In any restoration effort, regardless of scope, there should be a well-defined goal. Scientific data are needed to determine the appropriate scale of operations to achieve a realistic goal. It is during this process that many mistakes can be made concerning the scale of the restoration of the program and the determination of just what is expected from the restoration effort.

Restoration efforts can also be applied to the endangerment of a given species, the perceived damage to a particular aquatic habitat, or the economic aspects of a failing fishery. These cases are often complicated by long-term, often cyclic, changes in aquatic ecosystems. This complicates verification of the success of a given restoration effort. When a given impact is associated with a specific form of pollution such as a toxic substance or anthropogenous nutrient loading, scientific verification is often not carried out because of complications associated with natural variation in the field. There are also complications associated with the adaptability of populations. For example, nutrient loading and associated phytoplankton blooms can have major adverse impacts on the productivity of a given aquatic system (Livingston, 2000, 2002). However, phytoplankton are highly adaptable and can change in their response to varying levels of nutrient loading from both anthropogenous and natural sources. A given level of loading in a river or bay can have very different effects on the phytoplankton community in a system with no blooms than one that has already been affected by bloom species. The nonlinear processes associated with biological systems thus complicate both the identification and solution of a given environmental problem. It therefore follows that aquatic ecosystems do not always respond in a predictable way when a given pollutant is removed, and there is no guarantee that the system will revert back to its original form. Thus, the patterns of recovery of a given aquatic system may not be the mirror image of the original response to noted impacts. Natural variability and long-term (interannual) successions can complicate the success of a given restoration action.

In an ideal world, there are specific factors that form a necessary foundation for a successful restoration effort, to include:

- *Scientific research.* Any ecosystem-level restoration effort should be based on adequate scientific data that is used to evaluate the problem(s), establish baseline conditions as potential targets for the restoration effort, and post-action monitoring to evaluate the effectiveness of restoration actions.
- *Regulation and enforcement.* There is considerable evidence that voluntary restoration efforts do not work. This is, in part, associated with the costs associated with system-level efforts to restore lakes, rivers, and coastal areas affected by cumulative impacts of a broad spectrum of human activities. Usually, specific source-related impacts are complicated by multiple effects on poorly understood ecosystem processes so that restoration usually requires specific goals that are best drawn by local, state, and federal regulatory authorities.
- *Public education.* In democratic societies, it is broadly assumed that the entire societal process of government depends on an informed and active body politic. Public education based on scientific data thus becomes an important part of what is often a significant public investment in the outcome of a given restoration effort. It is also assumed that public education in all its forms provides the basis for the mechanism of resource restoration.
- *Economic/political considerations.* The single most important part of any given restoration has to do with costs at all levels of the restoration process. The inextricable relationship of politics and economic activities is often involved in the success or failure of a given restoration effort. The basis for all such efforts rests on the development of a system of laws by a theoretically informed body of officials elected by a theoretically informed public.
- *Legal actions.* The heart of effective regulatory actions and restoration efforts is based on the complex web of municipal, state, and federal laws. This aspect of the restoration effort is often underestimated.
- *News media.* The basis of the governmental processes associated with environmental decisions rests on an informed public. The process of information dissemination falls on the way news is transmitted to the public. This factor resides at the heart of the democratic process, both in terms of political action and the development and application of laws at all levels of society.

There are, of course, differences between the ideal system for restoration and the actual application of science to issues related to public resources. The overall success of restoration efforts in aquatic systems remains mixed at best. Although there have been some successes, the scientific documentation of the most highly profiled restoration efforts indicates that restoration remains elusory. There has been an emphasis on prominent aquatic systems such as the Great Lakes, the Mississippi River system, the Florida Everglades, and coastal systems such as Chesapeake Bay, San Francisco Bay, the Long Island Sound, and the various areas along the Atlantic and Gulf of Mexico that are affected by runoff from alluvial rivers. However, many lesser-known aquatic systems have been quietly destroyed without adequate scientific documentation or public notice. In almost all accounts of restoration efforts, there is a general lack of explanation as to why important aquatic resources were not protected in the first place. Often, the complex economic, sociological, legal, and political factors that contributed to the loss of aquatic resources have been overlooked in the restoration process. The role of the news media in these losses is often ignored in any accounting of a given restoration process.

There are many uncertainties associated with the restoration process, and there are, as yet, relatively few unqualified successes in the broad range of extensive restoration efforts in well-known freshwater and coastal systems. The following chapters are devoted to a discussion of a series of restoration efforts in various aquatic systems whereby available scientific information is compared to the sociological, economic, educational, political, and legal factors that contributed to the success or failure of the restoration effort. There is an emphasis on the role of the news media as a dominant factor in the formation of public opinion concerning restoration efforts. This author attempts to outline the scientific information concerning a series of aquatic systems and the factors that contributed to either the protection or destruction of these systems. The tensions between environmental protection and the restoration movement underway in the United States are examined with detailed examples of what constitutes success or failure in any given process. The relationships concerning the available scientific facts, how such facts are reported to the public, and the level of public understanding with regard to the restoration process are also examined.

North Florida as a Microcosm of the Restoration Paradigm

North Florida is one of the last areas in the United States where low population levels, together with relatively little industrial development, have contributed to some of the least polluted aquatic areas in the world. This includes lakes, springs, rivers, and coastal areas that remain pristine in every sense of the word. Spring-fed lakes are unique in terms of the relationship with the karst geological organization of the aquatic landscapes. Springs abound in this region, and are primary sources of clean, fresh water to the many rivers that eventually drain into an untouched series of estuaries in the Gulf of Mexico. Because of the relatively low levels of population and pollution, the impacts of a growing human population are more easily determined. Long-term research in these areas has thus led to various conclusions regarding the impacts of urbanization, agricultural development, and industrial wastes on aquatic resources of the region.

chapter 2

Cultural Eutrophication of North Florida Lakes

Most lakes, rivers, and coastal areas do not get the publicity, or the scientific attention, that characterizes the better-known systems. Likewise, the cumulative impacts of complex combinations of human activities are rarely determined with adequate scientific information. Without such data, there is little chance that anything approaching full restoration can be achieved. This factor, the scientific approach to restoration, has not been lost on development interests and their counterparts in the news media. The complicated interplay of scientific research, economic/political interests, and the role of the news media in environmental matters remains undermined despite the fact that these forces direct the fate of most aquatic systems in the United States.

Over the past 16 years, we have conducted a series of studies concerning the impact of urban storm water on lakes in north Florida (Figure 2.1). The long-term data were taken using methods outlined in Appendix I. The data were released as a series of public reports (Livingston, 1988a, 1989a, 1992a, 1993a, 1995a,b,c, 1996a, 1997a,b, 1998a, 1999a,b).

2.1 Background of Solution (Sinkhole) Lakes

Solution or sinkhole lakes are relatively common in areas dominated by limestones of north and central Florida. The dissolution of subsurface lime-rock forms a karst topography that, together with ample rainfall, provides the conditions of the infiltrated limestone environment (Northwest Florida Water Management District, 1992). Many karst systems in the southeastern United States are interconnected with springs, underground caverns or caves, and sinkholes so that groundwater is freely interconnected with surface water. The solution lake is thus directly connected to the surficial water table, and is dependent on seasonal and interannual drought–flood cycles. This situation is responsible for specific effects of storm water runoff on water and sediment quality that can be natural and/or anthropogenous (i.e., affected by human activities).

The most important groups of solution lakes in the northern hemisphere occur in Florida (Hutchinson, 1951). Although various studies have been carried out in some northern Florida lakes, there have been virtually no comprehensive ecological analyses of these systems. The area is underlain by the Floridan Aquifer, which is the primary source of the groundwater (Hendry and Sproul, 1966). Recharge of the aquifer comes mostly from rain that moves through the aquifer and is discharged into numerous springs to the south. Solution lakes in north Florida are located primarily in the Tallahassee Red Hills (Leon County, Florida) as part of the Miocene–Pliocene delta plain that is characterized by streams, wetlands drainages, and sub-surface limestone (Swanson, 1991). In the Tallahassee

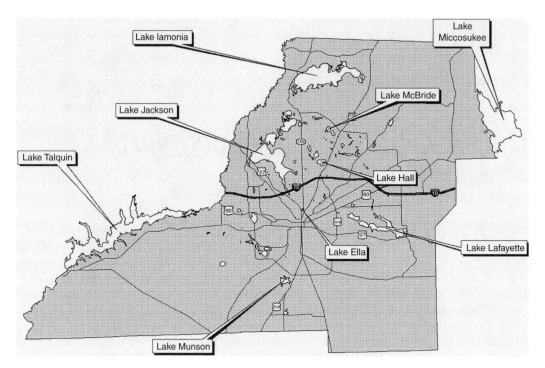

Figure 2.1 Distribution of lake systems in north Florida that were part of the long-term studies by the Florida State University Study Group. Geographic data provided by the Florida Geographic Data Library (FGDL).

Hills, polje-like depressions are produced by sudden developments of sinks in the normal valleys. The part of the valley drained by the sink is then eroded, forming an elongate, closed basin (Hutchinson, 1951). With increasing erosion and deposition, the sinks are plugged, forming elongate basins that remain closed laterally. These solution lakes often have convoluted shorelines, and they experience periodic desiccation during drought periods as a product of the opening of the sink and/or the lowering of the water table. Examples of such lakes include Lakes Jackson and Lafayette. These lakes are thus subject to extremes in water level fluctuation due to the unique combination of precipitation trends and geomorphology of the region.

The lakes of the north Florida region are usually small and relatively shallow (less than 10 m deep), and are controlled by various complex geological, morphological, and meteorological factors. There is considerable variation in the physiography of these lakes. The Lake Jackson Basin, about 25.8 km² (16.1 sq. mi.), includes the littoral zone and flood plain of Lake Jackson, Little Lake Jackson, and Lake Carr (Figure 2.2) as an open-water, clay hill lake system with sinkholes. According to Wagner (1984), Lake Jackson has drained five times in the past 80 years. The steep-sided basin is closed, receiving input from urban storm water in the southeastern and southwestern sections and low-intensity agricultural runoff in the north (Figure 2.3). Two major roads (I-10 and U.S. 27) are part of an extensive commercial growth zone that contributes to the Lake Jackson drainage. Megginnis Arm, Ford's Arm, portions of the western section of the lake, and Little Lake Jackson are most affected by the urban storm water runoff.

During wet periods, groundwater and lake levels are high; and during dry periods, these levels go down. Inflow factors for Lake Jackson include rainfall, surface water runoff,

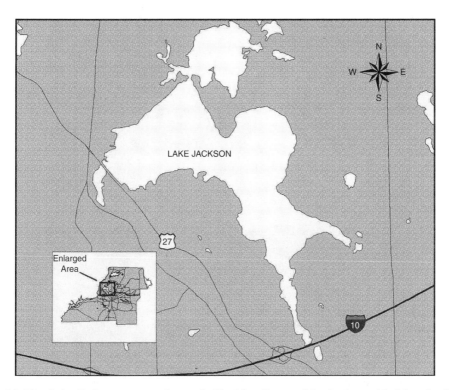

Figure 2.2 The Lake Jackson system in north Florida. Geographic data provided by the Florida Geographic Data Library (FGDL).

and discharge from the Surficial Aquifer. These are also the primary sources of the input of nutrients and toxic substances. Water loss is dominated by evapo-transpiration and leakage either through the bottom or through a loss of water from the sinkholes. According to Wagner (1984), when the level of Lake Jackson reaches 82 ft or less, there is no real inflow, and losses to the groundwater control lake levels. Bottom leakage is insignificant compared to evaporation and transpiration. Loss due to bottom outflow is proportionately higher during prolonged drought. Losses of water through sinks in the lake are considered an important part of the declines in lake levels in recent times (Wagner, 1984). These ecological characteristics make lakes such as Jackson highly susceptible to adverse impacts due to urban storm water flows as the lake is in continuous contact with contaminated surface and groundwaters.

Flushing rates (residence times) are important factors in the eutrophication potential of sinkhole lakes (Richey et al., 1978), and the average residence times of Florida lakes are about an order of magnitude greater than those of comparably sized lakes with rapid hydrological through flow. This indicates water-residence times of 1 to 5 years that are longer by an order of magnitude than those in lakes having rapid surficial runoff. This accounts for the vulnerability of many of the north Florida lakes to eutrophication and acidification (Deevey, 1988). When developing nutrient budgets in such systems, it is necessary to take the above facts into account with sediments, water, and the biota acting as primary nutrient sinks. Increased nutrient loading due to human sources such as sewage plant releases and storm water runoff, together with the relatively long retention times and high efficiency of nutrient recycling, all add to the susceptibility of solution lakes to cultural eutrophication.

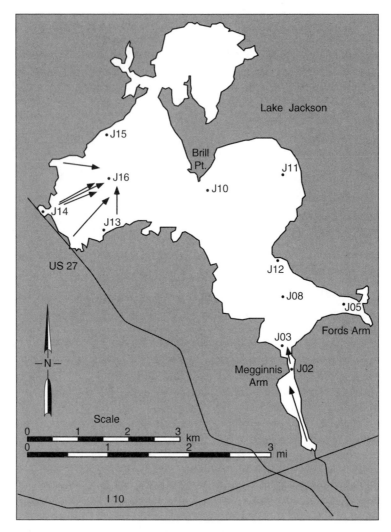

Figure 2.3 The Lake Jackson system in north Florida, showing long-term sampling stations. Arrows indicate main sources of urban runoff. Geographic data provided by the Florida Geographic Data Library (FGDL).

2.2   *Urban Runoff and Solution Lakes*

Although lakes have common driving components (nutrients, water and sediment quality, physical modifying factors, primary producers, predators/prey associations, trophic organization), they behave as unique aggregations of these similar components (Richey et al., 1978). Differences in the response of a given lake system to urban pollutant loading are based on assimilative capacity as determined by physical dimensions and flushing rates. In general, solution lakes are essentially closed systems and, as such, are particularly sensitive to urban storm water runoff. Response to pollutant loading is primarily related to amount, timing, and qualitative composition of surface runoff and surficial groundwater contributions. Loading rates of nutrients, organic compounds, and toxic agents, as qualified by the assimilative capacity of a given lake, are thus crucial to the effects of such substances in systems that are either closed or have limited flushing capabilities. Johnson (1987), in a multivariate analysis of storm water runoff in Leon County, found that significant predictors of runoff volume (in order of importance) are the extent of urban land (impermeable surfaces, reduced wetlands, etc.), the percentage of clay in the soils (permeability), the overall drainage area, and the average slope of the basin. Evapo-transpiration and groundwater leakage also affect the response, but the essential accumulation of nutrients and toxic agents under such circumstances accounts for the high vulnerability of essentially closed solution lakes to inputs of nutrients, organic matter, and toxins.

Various pollutants occur in sediments and animals in receiving areas associated with wastewater treatment plants and storm water runoff (Gossett et al., 1983). Bioaccumulation of pollutants has been associated with the n-octanol/water partition coefficients. Storm water has been associated with high concentrations of hydrocarbon contaminants known as polynucleated aromatic hydrocarbons (PAHs) (Wild et al., 1990a,b). These compounds are introduced into the environment in natural and anthropogenous combustion processes (Menzie et al., 1992). Polynucleated aromatic hydrocarbons are often found in areas affected by the incomplete combustion of organic materials such as coal, oil, natural gas, and wood. Aquatic systems concentrate PAHs through contaminants in the air and/or loading via the drainage basins. The association of urban pollutants and aberrant characteristics of aquatic organisms, including disease, has been well established. The highest frequency of diseased fishes often occurs in so-called "polluted" areas of aquatic systems. McCain et al. (1992) found that sediments and animals taken from areas receiving urban runoff in San Diego Bay were characterized by high levels of aromatic hydrocarbons and their metabolites when compared to areas that did not receive urban runoff. PAH contamination of sediments has been associated with various forms of fish disease, and PAH compounds can cause sufficient stress to cause susceptibility of fish to fatal parasite infestations.

2.3 Lake Ecology Program

The Lakes Program was designed around a series of continuous field collections of data and field/laboratory experiments and analyses. Data were taken from 1988 to 1997 in Lake Jackson and from 1991 to 1997 in a series of other sinkhole lakes in the region (Lakes Lafayette, Hall, Munson, McBride, and Ella and No-Name Pond; Appendix I). Detailed studies of the biological organization of Lake Jackson were carried out concerning phytoplankton, submerged aquatic vegetation, zooplankton, infaunal macroinvertebrates, fishes, and trophic organization. The effects of PAHs on submerged aquatic macrophytes were also analyzed. Storm water quality analyses were carried out in addition to analyses in a series of treatment holding ponds. The primary objective of the project was to analyze the effects of urban storm water on lakes systems at various levels of biological organization, and to evaluate seasonal and interannual changes in background habitat factors relative to the effects of urban storm water. These analyses were supplemented by photographs and by underwater photography. The long-term field-monitoring program was integrated with a series of field and laboratory experimental programs to determine the effects of urban storm water runoff on Lake Jackson.

2.4 Urban Runoff and Lake Jackson

2.4.1 Background

Lake Jackson (Figure 2.2 and Figure 2.3) has been a center of human activity for thousands of years. The cultural peak of Native American occupation around the lake occurred between A.D. 1250 and A.D. 1500. During this time, along the southwestern shore of Megginnis Arm, a series of earthworks were constructed. This complex, composed of farmsteads, hamlets, and six pyramidal, flat-topped, truncated temple mounds, was con-structed and utilized by a Native American culture whose influence and settlements extended across much of the Southeast during this period. Today, Lake Jackson (Figure 2.2) is designated an Outstanding Florida Water and an Aquatic Preserve by the state of Florida. These designations supposedly give legal protection to the lake, although there has been continuous, scientifically documented input of polluted water to the lake from road con-struction and urban development from the early 1970s to the present (Harriss and Turner, 1974; Livingston, 1993a, 1995a, 1997a, 1997b, 1999a).

Until recently, Lake Jackson was famous throughout the country for its bass fishing. Bass grew faster and larger in Lake Jackson than in most other lakes in the country. The lake is a closed system with inputs from three major drainages: (1) Megginnis Creek (draining portions of the southern basin), (2) Ford's Creek (draining portions of the southern basin), and (3) Ox Bottom Creek (draining the northeastern basin) (Figure 2.3). Megginnis Arm Creek drains a major urbanized area characterized by malls, shopping centers, gas stations, a major interstate highway (I-10), and low- to high-density urban/res-idential developments. Ox Bottom Creek is a drainage area entering the northern part of Lake Jackson. Forested areas, light agriculture, and increasing encroachment by housing developments contribute to the storm water runoff in this area. The Ford's Arm basin includes forested uplands, light agriculture, and rapid proliferation of urban housing. The northern extremity of the Jackson basin is managed primarily as an agricultural resource with cattle, timber, and low-intensity farming. Lake Jackson also has various forms of municipal development in the western sections that have led to water quality impacts from roads and various forms of urban development.

A series of studies was carried out concerning the relationship of water quality in Lake Jackson as a consequence of urban sediment and nutrient loading. Harriss and Turner (1974) in a 3-year analysis of water quality measurements and phytoplankton productivity, noted frequent oxygen sags in Megginnis Arm and Ford's Arm. Water quality was char-acterized by fair to poor water quality conditions with urban storm water runoff associated with low Secchi readings, high turbidity and conductivity, and high pH. Conductivity increased in Megginnis Arm over the period of study from about 40 to 100 μmhos cm^{-1}. Phosphorus and nitrogen concentrations were usually highest during winter periods in Megginnis Arm and Ford's Arm. Heavy metals (Pb) and dissolved phosphorus were traced to commercial parking areas in the Megginnis Arm watershed.

Affected lake areas had the highest phytoplankton productivity, with nannoplankton as the primary form. Megginnis Arm was characterized by low phytoplankton diversity and blue-green algae. Ecologically healthy northern and mid-lake areas were characterized by green algae, dinoflagellates, or chrysophytes. Studies by the Florida Game and Fresh Water Fish Commission (July 1975 to June 1976) indicated that the most common macro-phytes in Lake Jackson included water hyssop (*Bacopa caroliniana*), American lotus (*Nelumbo lutea*), spikerush (*Eleocharis baldwinii*), sagittaria (*Sagittaria stagnorum*), and maidencane (*Panicum hemitomon*). Introduced Hydrilla (*Hydrilla verticillata*) was starting to increase at this time (Babcock, 1976). Dominant infaunal macroinvertebrates included scuds (amphipods), oligochaete worms, and midge larvae (Chironomids). Phantom midge larvae, common in eutrophic waters, were found in Megginnis Arm, whereas the amphi-pods were largely absent in this area of the lake. Fletcher (1990) found that numbers of chironomid larvae were directly associated with dissolved oxygen (DO) levels in Lake Jackson. Mason (1977) found that water quality was significantly degraded in the southern parts of the lake (particularly Megginnis Arm) due to loading from the newly constructed I-10 highway and other portions of the urbanized basin through the lake.

Wanielista (1976), Wanielista et al. (1984), and Wanielista and Yousef (1985), using sediment elutriate tests in Megginnis Arm, found high concentrations of turbidity, dissolved phosphorus, ammonia, nitrate, and organic nitrogen. High levels of oils and greases occurred at abandoned boat launching ramps. Class III standards were violated for pH, turbidity, alkalinity, zinc, iron, and especially lead in the elutriate tests. Concentrations of organic matter, nutrients, and heavy metals were considerably higher in the surface sediments relative to deeper sediment layers. Oils and greases were also high in the sediments of Megginnis Arm, especially in the central portion of the Arm. An artificial marsh system was constructed to filter the storm water runoff and to reduce the loading of suspended materials entering the lake at the southern, most urbanized end (Northwest Florida Water Management District, unpublished report). This control system was altered almost continuously since its inception (Schmidt-Gengenbach, 1991). There was evidence that the artificial marsh had not been fully effective (Tuovila et. al., 1987; Alam, 1988). Despite various efforts to improve the water quality of the Megginnis drainage area, the condition of the lake continued to worsen with respect to various forms of hypereutrophication and levels of pollutants during the late 1980s (Wanielista, 1976; Tuovila et. al., 1987; Alam, 1988).

Byrne (1980) carried out a study of the effects of petroleum hydrocarbon concentrations on Lake Jackson. The implications of the results are qualified by the relatively obsolete chemical analyses used by the principal investigator. However, Byrne (1980) found that, by 1978–1979, there were marked increases in petroleum hydrocarbon concentrations in Lake Jackson sediments. These increases were associated with the expansion of urbanized areas around the lake. The principal source of the petroleum hydrocarbons was storm water runoff from urban areas. Some 90% of the 4380 kg of total hydrocarbons transported to Lake Jackson during 1978–1979 were of petroleum origin. Total hydrocarbons were most concentrated in sediments of Megginnis Arm. The primary inputs of such products were from storm water runoff and base flow from the surrounding watershed, along with dust fall, rainfall, and the decomposition of aquatic and terrestrial plant matter. Asphalt, composed of multipolymers of aromatic rings linked by aliphatic and/or naphthenic chains, were a source due to bleeding of petroleum products adsorbed on the asphaltic surfaces. Temperature-driven dissolution of organic molecules (i.e., summer bleeding) followed by storm water incidents accounted for the movement of petrochemical products via oil impregnation into and released from the asphalt. Upon flushing with rainwater, layers of the film were solubilized into a continuous flow phase (Byrne, 1980).

The construction of highway I-10 (see Figure 2.3) by the Florida Department of Transportation in the southern drainage basins of Lake Jackson in the early 1970s was associated with extensive erosion problems. Massive amounts of sediments washed down the relatively steep slopes of the Okeheepkee Road sub-basin, eventually ending up in southern Lake Jackson. Following the construction of a series of intensive commercial developments at the head of the Okeheepkee sub-basin during the mid-1980s, there were increased erosion problems. Again, sediments and degraded water washed through the Okeheepkee drainage into Lake Jackson. Against local opposition, a holding (i.e., collecting) pond was constructed by local officials to capture some of this runoff. Instead of improving the situation, the pond simply concentrated the polluted water and redistributed it into a series of surface and groundwater flows that led to the contamination of local residences. This situation continues to this day, with polluted water entering Lake Jackson during prolonged rainstorms.

The Indian Mounds Creek system is another major tributary to the Megginnis Arm drainage in Lake Jackson (Figure 2.3). This creek was artificially redirected in recent times (1950s) (D. Benton, personal communication, 1993). Continuous observations of the Indian

Creek system indicate that it has been severely affected by storm water runoff from roads (U.S. 27) and shopping malls at the headwaters of the creek. There is a series of hyper-eutrophicated ponds along the upper drainage; polluted runoff from these ponds eventually ends up in Lake Jackson. In addition to storm water pollution, sewage spills have damaged the Indian Mounds system.

The headwaters of the Lake Hall drainage basin (see Figure 2.1) consist of malls, roads, and housing developments. Storm water from the mall drains through a series of ponds directly into Lake Hall. Until recently, the outlet for this pond was damaged, and storm water ran almost continuously into the lake from the mall area. Recently, to the east, Thomasville Road has undergone major expansion with runoff from the road running directly into eastern sections of Lake Hall. Over the past few years, a series of major developments have been established in the Ford's Arm drainage basin that extend from Lake Hall westward over Meridian Road and into Lake Jackson. This development has been accompanied by increasing levels of flooding and entry of polluted water through Ford's Arm into the lake.

Over the past 15 years, there has been increased urban development in the western sub-basins of Lake Jackson (see Figure 2.3). Currently, nutrient loading has led to the filling of Little Lake Jackson with vegetation and associated sediments. Accelerated aquatic plant growth contributes to the impairment of lake habitat, altered sediment quality, increased filling with excess (unassimilated) organic matter, and associated water quality deterioration due to the decomposition of such matter (Livingston and Swanson, 1993). Adverse biological effects are the result of cumulative impacts of the eutrophication process that, through altered aquatic plant assemblages, leads to simplified food webs and reduced fisheries potential. With time, areas of western Lake Jackson, affected by runoff from lakeside urban development and runoff from Little Lake Jackson, have shown increasing signs of deterioration (as outlined above).

The primary source of polluted urban water to Lake Jackson is Megginnis Arm (see Figure 2.3). By 1986–1989, municipal development in the southern sub-basins of the lake was accelerated. During this period, Hydrilla became dominant in receiving areas of eastern Lake Jackson. The Northwest Florida Water Management District completed a small holding pond for the Megginnis Arm basin. Despite efforts to improve water quality of the Megginnis drainage during the late 1980s, lake water quality continued to worsen (Wanielista, 1984; Tuovila et al., 1987; Alam, 1988; Livingston, 1988a). Polluted storm water continued to flow through Megginnis Arm during the 1990s whenever it rained. Today, Megginnis Arm Creek drains a major urbanized area with malls, shopping centers, gas stations, a major interstate highway (I-10), and high-density urban/residential developments. Despite construction of an additional holding pond and a freshwater marsh system, the Megginnis Arm continues to be a major source of polluted urban storm water to southern Lake Jackson with massive runoff and nutrient loading to the lake after prolonged rainfall conditions.

2.4.2   Long-Term Cycles of Rainfall and Storm Water Runoff

The ecological condition of a given lake must be viewed within the context of long-term changes of rainfall and lake water levels (Figure 2.4). The relationship between lake stage and rainfall is complex. The increased lake stages during 1994 reflected preceding rainfall peaks as noted above. Increased rainfall was often noted during the summer months. Peak rainfall occurred during a series of storms spring–summer 1994. This was followed by a drought during 1995 and early 1996. Rainfall peaks again occurred during the summer of 1996. This peak was followed by decreasing rainfall during the summer and fall of 1997. During 1998, there was a drought, which was reflected in reduced lake stages. By 1999,

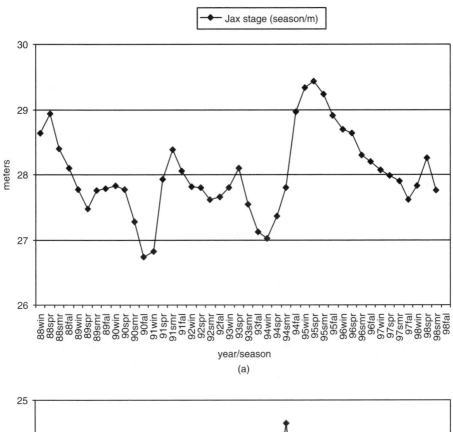

(a)

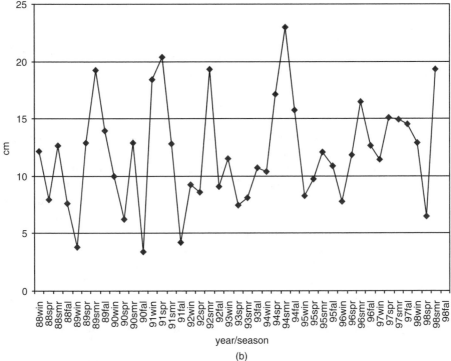

(b)

Figure 2.4 (a) Lake stage (m) and (b) rainfall (cm) in Lake Jackson from winter 1988 to fall 1998. Data provided by the Northwest Florida Water Management District.

major parts of Lake Jackson disappeared into the sinkholes, leading to the drying out of most of the lake during the prolonged drought of 1998–2001.

2.4.3   Water Quality Changes

Station locations in Lake Jackson are given in Figure 2.3. Long-term changes in the water chemistry of Lake Jackson are given in Figure 2.5. Statistical tests of significance were determined by methods noted in Appendix II. Secchi depth readings were reduced in time, and such depths were significantly ($P < 0.05$) different at Stations 3, 5, 8, 10, 14, 15, and 16 between 1988 and 1991 and between 1996 and 1998. These data indicate a gradual loss of light penetration in the lake with increased storm water loading through the southern entry points. With time, the bottom was no longer sighted during all seasons of the year. Conductivity was significantly higher during all seasons at Stations J03, J05, J08, and J14. There was a general trend of increasing conductivity through the sampling period. Ammonia concentrations were also significantly higher during the last 3 years of sampling (compared to the first year) at Lake Jackson Stations J03, J05, J08, and J14, whereas orthophosphorus concentrations were significantly lower at these stations. The general increases in the total inorganic nitrogen/total inorganic phosphorus (TIN/TIP) ratios over the 10-year sampling period during all seasons appeared to reflect these nutrient trends. Chlorophyll *a* concentrations were significantly higher at Stations J03, J05, J08, J10, and J14 during the last 3 years of sampling; these increases were especially pronounced during spring and summer periods at Stations J03, J05, J08, J10, and J16. Overall, the long-term trends indicated increased phytoplankton activity with time in Lake Jackson, with orthophosphate indicated as a limiting nutrient. The increased ammonia levels could have been related to increased blue-green algae blooms (see below).

2.4.4   Sediment Changes

Sediment nutrient data for Lake Jackson are given in Figure 2.6. Sediment nutrient concentrations of phosphorus (P) and nitrogen (N) were highest at Stations J03, J05, J08, J11, and J14 relative to more northerly parts of Lake Jackson (Station J10). Peak concentrations were noted at these stations during the period 1993 to 1994. Sediment nitrogen tended to decline slightly through 1996, whereas sediment phosphorus appeared to decline during 1994, reaching much lower concentrations during 1995 to 1996. These declines in sediment phosphorus followed water column trends of reduced orthophosphate and total phosphorus (TP). The sediment nutrient declines occurred during a series of intensive blue-green algae blooms in 1994 and 1995 (see below). The data suggest that blue-green algae, which are able to fix nitrogen, may have effects on water and sediment quality due to the release of ammonia. At the same time, increased algal biomass was associated with reduced orthophosphate concentrations in the water. Reductions in sediment phosphorus could be associated with these trends. The blooms could be supplied with sediment phosphorus during periods of reduced orthophosphate in the water. Thus, the loading of sediments with phosphorus and subsequent release of this nutrient during bloom periods could represent an important link to the proliferation of blue-green algae in Lake Jackson. The temporal progressions of water and sediment chemistry, with storm water runoff incursions timed to drought–flood cycles, appeared to be linked to microalgal trends in complex ways.

The data indicated that long-term changes of water and sediment quality in Lake Jackson could not be interpreted without an understanding of the changes in the aquatic plant distributions in the lake as a response to anthropogenous nutrient loading. There are continuous feedback cycles associated with seasonal and interannual changes of storm

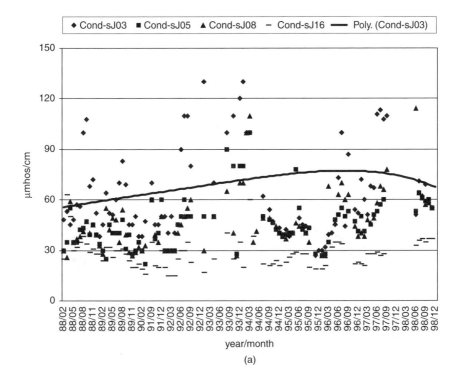

(a)

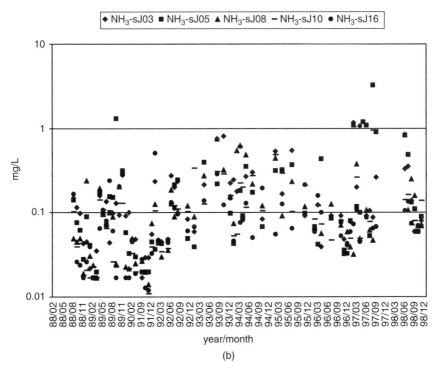

(b)

Figure 2.5 Water quality features of Lake Jackson taken monthly from February 1988 to December 1998: (A) surface conductivity (μmhos.cm^{-1}); (B) surface ammonia (mg L^{-1}); (C) surface orthophosphate (mg L^{-1}); and (D) surface chlorophyll a (μg L^{-1}).

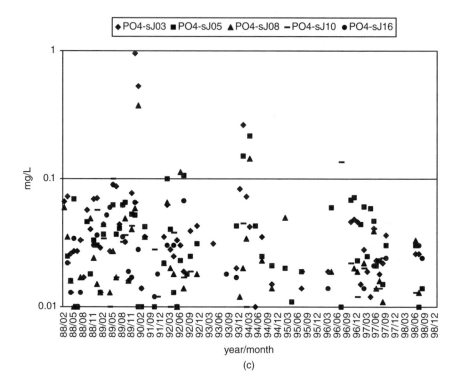

(c)

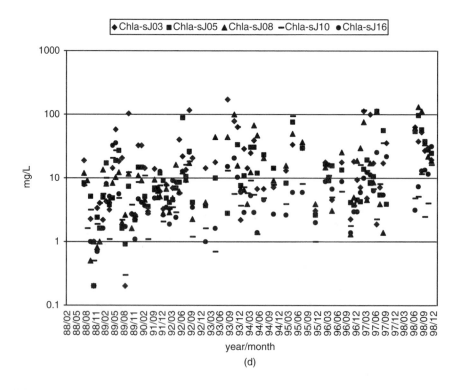

(d)

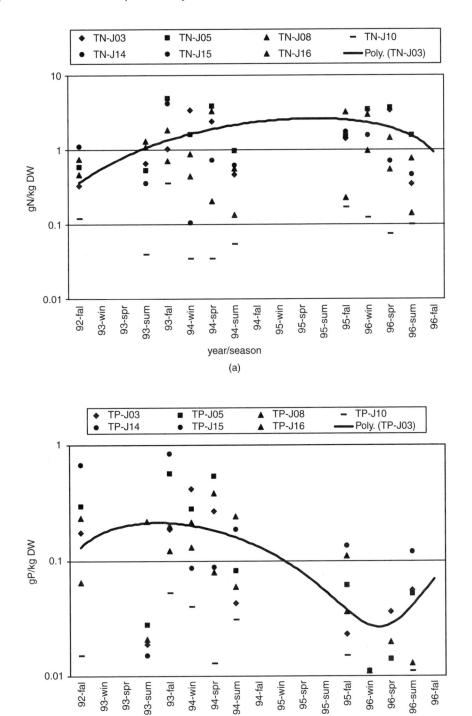

Figure 2.6 (A) Sediment total nitrogen (TN) and (B) total phosphorus (TP) in Lake Jackson from fall 1992 to fall 1996.

water runoff; water and sediment quality conditions thus interact with aquatic plant distributions in space and time. The plants integrate the varying nutrient loading and water/sediment conditions through a continuous pattern of integrated changes in the submerged aquatic vegetation and the phytoplankton. Feedback processes are involved in the plant/water/sediment interactions that are both seasonal and interannual through changes in the qualitative and quantitative composition of the aquatic plant associations.

2.5   Submerged Aquatic Vegetation

There is a long history of changes of submerged aquatic vegetation (SAV) in Lake Jackson. In 1954, when urban development in the Jackson basin was just beginning, there was little evidence of surface vegetation in the lake as most was composed of species with relatively short blades. By 1970, there was increasing municipal development of the Megginnis Arm and Ford's Arm sub-basins with associated (polluted) runoff. By 1980, there was enhanced growth of SAV in southern parts of the lake. By 1986, there was considerable municipal development in the southern sub-basins, filling of Megginnis and Ford's Arms with *Hydrilla verticillata*, which had been in Lake Jackson for over a decade, expanded its distribution with spectacular overgrowth of the native SAV. Blue-green algae blooms were also noted. By 1987, an herbicide called fluridone (SONAR®) was tested against the Hydrilla with some success. By this time, a storm water treatment system at the head of Megginnis Arm was developed although it was not considered large enough to handle the entire storm water load discharging from the upland basin that was, by now, primarily paved over and developed. By spring 1987, there had been a series of sewage spills in the southern Lake Jackson that compounded the storm water runoff problem. Hydrilla proliferation was accompanied by increased emergent vegetation in Lake Jackson.

By fall 1992, Hydrilla was the dominant form of submergent vegetation in areas extending from Stations J03, and J08 northward to Stations J11 and J15 (Livingston, 1995a). The submergent species *Ceratophyllum demersum* was dominant at Stations J12, J14, and J16. The green alga *Spirogyra* sp. was dominant at Station J05. The area around Station J10 was characterized by *Vallisneria americana* beds as a remnant of what once had been an extensive distribution before the extension of Hydrilla. At Station 13, *Myriopyllum hetero-phyllum* was the dominant species. The highest total biomass was found at Stations J10 and J12. High concentrations of Hydrilla were noted in Megginnis Arm (J03), Ford's Arm (J05), and southern portions of the lake (J08). The least amount of vegetation was found in the western sections (J13, J16, and J14). High dominance and low species richness of submergent vegetation occurred in areas of urban storm water entry. The highest species richness was found at Station J11, an area characterized by remnant good water quality in 1992.

During the four years following the Hydrilla outbreak, the water quality system maintained by the Northwest Florida Water Management District at the head of Megginnis Arm was enlarged. Polluted sediments were dredged out of Megginnis Arm, and a second holding pond was constructed at the head of the arm. Plant control efforts using herbicides were continued. The average annual Fluridone treatment in Lake Jackson was 122 acres from 1987 to 1992, about 4% per year. The plant control program was continued through 1997 with applications in 1987, 1988, 1990, 1992, 1993, 1994, and 1996. In all, over $700,000 was expended on the control of Hydrilla in Lake Jackson. Over the treatment period, this introduced species expanded its distribution throughout the entire lake. During November 1993, Hydrilla occupied 100% of the water column as a surficial mat in the southern part of the lake. The Hydrilla monoculture was rooted in a deep flocculate hydrosoil containing detrital deposits. From December 1993 to January 1994, the plant community completely disappeared in areas off Megginnis Arm.

During May 1994, a phytoplankton bloom extended throughout the entire eastern section of Lake Jackson. By July 1994, the bloom had dissipated and small amounts of sparsely distributed *Ceratophyllum demersum* could be found within the station area (Bevis, 1995). Thus, Hydrilla had almost completely disappeared by the time of the herbicide treatment in March 1994. Station J05 in southern Lake Jackson was characterized by high concentrations of the blue-green algae *Lyngbya* sp. The rise of *Lyngbya* in eastern Lake Jackson was coincident with the reduction and virtual loss of Hydrilla. By winter 1994, blue-green algae (*Microcystis aeruginosa*) covered the macrophyte community of northeastern sections of the lake that had been treated with fluridone. This treatment appeared to be associated with the noted increase of *Microcystis* at the bottom of this area of Lake Jackson. A blanket of gelatinous algae, up to 0.5m thick, was distributed across the entire northeastern section of the lake (Bevis, 1995). In this way, the entire bottom of eastern sections of Lake Jackson was dominated by *Microcystis aeruginosa* in the north and *Lyngbya* to the south.

The proliferation of *Microcystis* in Lake Jackson during 1994 and 1995 marked a distinct shift in the plant distributions of the lake. Colonies of this blue-green alga were enclosed in mucilage. *Microcystis* contains gas vacuoles that can be used to make the colony buoyant, which can add to or cause algal blooms in the water column. Periodically, during the next 4 years, mats of this species were noted in different parts of the eastern section of Lake Jackson. Its presence was accompanied by reduction or elimination of other plants and animals. According to Prescott (1980), lakes can be almost completely overgrown by members of the genus *Microcystis*, and the dense growths of some species "may lead directly to the death of fish through suffocation or by poisoning, and the toxin produced by some species causes the death of cattle and birds." Gorham (1964) noted that *Microcystis* could be responsible for acute poisoning of different animals. Bacteria associated with these algae also produce toxins having a combined toxic effect with the algae.

The upper lake south of Brill Point was characterized by a mixed bed of emergent vegetation composed of *Panicum hemotomon, Nymphaea odorata*, and *Brasenia schrebrri* (Livingston, 1995a). These beds were also dominated by *Sagittaria stagnorum* and several other species, including *Eleocharis baldwinii, Bacopa caroliniana*, and *Utricularia* spp. In southern western lake areas (Station J14), which received urban storm water from roads and Little Lake Jackson, the sediments were flocculated and saturated with methane (Livingston, 1995a). This region was dominated by *Ceratophyllum demersum*, with lesser amounts of *Hydrilla verticillata* and the blue-green algae *Lyngbya*. Farther north is Station J15, an area that was viewed for many years as an undisturbed reference station. However, during 1994 to 1996, high chlorophyll levels and periodic high conductivity readings suggested that this area was being affected by hypereutrophication in southern parts of the lake. During the early years of sampling, this area was represented by an extensive, highly diverse, mixed bed of macrophytes dominated by *Cabomba caroliniana*. Also present were *Ceratophyllum demersum, Bacopa caroliniana, Myriopyllum heterophyllum, Utricularia* spp., and *Eleocharis baldwinii*. Sediments in this area were characterized by a dense organic hardpan topped with 5.0 to 10.0 cm. sand covered with a thin layer of floc. Over the next 2 to 3 years, the area was invaded by *H. verticillata* that totally eliminated most of the other SAV species. To the east (Station J16) (the deepest area of Lake Jackson) was dominated during the early years by *Ceratophyllum demersum*. However, during 1995 to 1997, the area was invaded by *H. verticillata*.

Water quality and SAV changes from 1988 to 1999 indicated progressively worse conditions of hypereutrophication in Lake Jackson. Conductivity increases in the Megginnis Arm drainage and throughout the eastern portions of Lake Jackson showed that storm water had an increasing effect on Lake Jackson with time. Ammonia increases were evident throughout the lake during the later periods of blue-green algae dominance. A major change in the nutrient dynamics of Lake Jackson was noted over the observed time period, and limiting factors may have been altered as nutrient loading to the lake was enhanced by storm water runoff. The primary increases in phytoplankton blooms occurred in the eastern arm of the lake (from the Megginnis Arm to Brill Point) and in areas surrounding southern entry points of storm water in western areas of the lake. It should be emphasized, however, that all portions of Lake Jackson experienced increased surface chlorophyll levels with time.

2.6   Blue-Green Algae Blooms

Blue-green algae (cyanobacteria or Cyanophyceae) are usually found as dominants in polluted ponds, lakes, reservoirs, and rivers. Various blue-green species produce toxins that have been shown to adversely affect aquatic organisms and humans on a worldwide basis (Codd et al., 1995). Blue-green algal blooms have been noted for a long time (Prescott, 1962). Increased numbers of blue-green species can form floating crusts and scums that can be highly toxic to plants and animals that come in contact with them. A secondary effect of blue-green algae blooms is the accompanying low levels of DO in the water column, especially at night. Other effects include changes in the ecological characteristics of the water body affected by the outbreaks of these algae.

Blue-green algae such as *Anabaena flos-aquae* are capable of nitrogen fixation (via heterocyst formation). Nitrogen fixation may be greater at lower DO concentrations (Stewart, 1974). Nitrogen fixation is usually light dependent. Nitrogen is present in relatively high concentrations in lakes, and diffuses more rapidly than either nitrate or ammonium ions (Stewart, 1974). Rates of nitrogen fixation in freshwater systems are positively correlated with concentrations of dissolved organic nitrogen. This means that species such as *A. flos-aquae* are capable of producing organic nitrogen without input from dissolved inorganic nitrogen sources. This adds another dimension to the occurrence of blue-green algae blooms in Lake Jackson. Heterocyst-possessing blue-green algae such as *Anabaena planctonica* and *Aphanizomenon flos-aquae* have a nutritional advantage over other forms of lake algae, especially under conditions of nutrient limitation. In addition to the above advantages, species such as *Anabaena flos-aquae*, *A. planctonica*, and *Aphanizomenon flos-aquae* have another special cell type known as akinetes (spores) that can be developed during periods of adverse habitat conditions. When fully developed, the akinetes sink to the bottom and germinate to form new filaments when the environmental conditions become advantageous.

Codd et al. (1995) presented a review of the history of blue-green infestations of aquatic systems. Toxic compounds such as the microcystins, produced by species of the genera *Microcystis*, *Anabaena*, *Oscillatoria*, and *Nostoc,* have been documented as ecologically disruptive by-products of blue-green algal blooms. The widespread genus *Microcystis* contains many species that produce potent toxins. These blue-green algae move up and down in the water column and often float to the surface. The toxins are in the cells unless the algae die, which then allows the release of the toxins to water. These toxins cause both direct and indirect effects on the aquatic food webs in infected lakes. The species *Microcystis aeruginosa* has been implicated in the release of toxic agents. Skulberg et al. (1994) indicated that the genus *Microcystis* is associated with two species that produce toxic blooms in Norway. These blooms were associated with waters enriched by plant nutrients from agricultural and municipal developments. Komarek (1991) found that *Microcystis* species are an important component of blooms and toxicity in hypereutrophic waters. According to Brank and Senna (1994), blooms of *Microcystis aeruginosa* are particularly prevalent in lakes with high levels of organic pollution. Such blooms are associated with the onset of water column stratification, increased temperature, and increased solar radiation. The undesirable species are stimulated by high nutrient concentrations, especially nitrogen.

The turning point for Lake Jackson came during spring 1995. In April 1995, there was a major phytoplankton bloom that extended from Megginnis Arm to Brill Point. Essentially, the entire lake in this region was filled with algae to an extent never before observed. The bloom extended throughout the entire water column but appeared more concentrated in the top meter of water. The intensity of the bloom was indicated by the color of the water (a deep green). The coverage (more than 70% of Lake Jackson) and the intensity of the bloom lasted well into summer 1995. At the time, water samples were taken for quantitative and qualitative analysis for microalgae. During the bloom period, pH was particularly high and benthic DO was very low at stations affected by the bloom. The DO was low at Station J13 (Figure 2.3), which is not unusual for this area of the lake as it has had relatively high chlorophyll levels in the past. The pH was high at Station J10 but the DO at depth was also relatively high, which is the primary exception to the above generalizations. There was still a functional grass bed at Station J10. Station J14, another storm water entry point for Lake Jackson, had relatively routine pH and DO levels. Little Lake Jackson (Station J14A) had relatively low DO at depth, high water color (especially at depth), and high chlorophyll *a* throughout the water column.

Secchi depths were uniformly low from Megginnis Arm to Brill Point during the spring 1995 blooms. Surface chlorophyll *a* was high throughout the lake with the exception of the extreme northwestern sections of Lake Jackson (Stations J15 and J16). The highest surface chlorophyll *a* data were found at Station J10. This shift in productivity to the northern portions of the lake, previously the least polluted areas (i.e., farthest from the storm water sources), was evidence of a movement of the lake pollution to the north. The high chlorophyll *a* concentrations, low Secchi readings, high pH levels, and low bottom DO concentrations at stations directly affected by the phytoplankton blooms represented a direct link of the phytoplankton with benthic water quality conditions.

The qualitative and quantitative distribution of species populations of microalgae in Lake Jackson during spring 1995 followed the chlorophyll *a* distributions noted above. Extremely high concentrations of the blue-green species *Anabaena flos-aquae* were noted at Station J03 (Figure 2.3). Smaller numbers of this species were noted at the other stations in the lake. Heterocysts and spore phases were noted at all stations where high numbers of trichomes occurred. The spring of 1996 was relatively cool. Concentrated algal blooms in Lake Jackson were first noted in eastern portions of the lake during late May and early June 1996. Chlorophyll levels in the eastern portions of the lake ranged from 57 to 93 µg L^{-1} at Stations J03, J05, J08, and J11. In many cases, these chlorophyll concentrations were higher at the bottom than the top. During this period, low DO (less than 2.0 mg L^{-1}) was noted at depth at Stations J03, J05, J08, J13, J14, F09, J16 (F10), and F04. As noted above, the blue-green algal species *Lyngbya* was noted at the bottom of various stations throughout the northern parts of the lake (corresponding to areas having low DO in Lake Jackson). The benthic proliferation of *Microcystis aeruginosa* in northeastern parts of the lake was apparent during these periods. The relatively cold spring delayed the spring blooms until late May. The species *Aphanizomenon flos-aquae*, a blue-green alga, was dominant at Stations J03 and J08 during June 1996. *Anabaena planctonica* was found as a dominant at Station J05 in June 1996.

Peak abundance of the primary bloom species in Lake Jackson was usually seasonal, with dominants such as *Microcystis aeruginosa* occurring during fall months and others such as *Anabaena planctonica* occurring during winter–spring months (see Figure 2.7). The distribution of *Anabaena flos-aqua* (Figure 2.8) indicates dominance during 1997, whereas *Anabaena planctonica* (Figure 2.9) was prevalent during 1996 with increased dominance

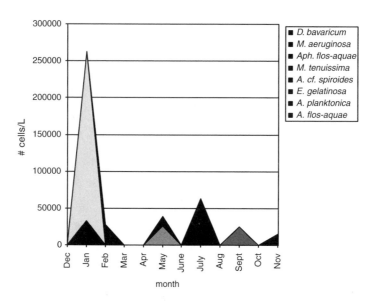

Figure 2.7 Cell numbers L^{-1} of dominant bloom species taken in Lake Jackson averaged by month from monthly collections from April 1995 to November 1998. Species analyzed included *Anabaena planctonica, Aphanizomenon flos-aqua, Microcystis aeruginosa, Anabaena flos-aqua, Dinobryon bavaricum, Merismopedia tenuissima, Elakatothrix gelatinosa, Anabaena cf. Spiroides,* and *Dinobryon bavaricum.* Phytoplankton analyses were made by A.K.S.K. Prasad and include information taken from Reardon (1999).

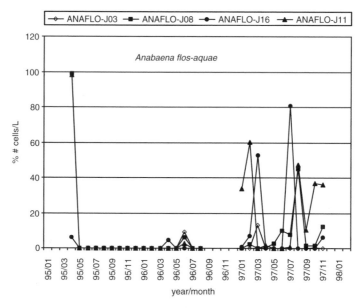

Figure 2.8 Percent of total numbers L^{-1} of *Anabaena flos-aqua* taken in Lake Jackson by month from April 1995 to November 1998. Phytoplankton analyses were made by A. K. S. K. Prasad and include information taken by Reardon (1999).

during 1997. *Microcystis aeruginosa* was present mainly during 1997 (Figure 2.10) and was present throughout Lake Jackson during various times although the main peaks of this species occurred in the fall. During peak dominance of *Microcystis aeruginosa* and *A. flos-aqua*

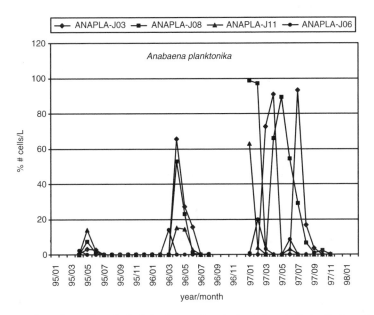

Figure 2.9 Percent of total numbers L^{-1} of *Anabaena planctonica* taken in Lake Jackson by month from April 1995 to November 1998. Phytoplankton analyses were made by A.K.S.K. Prasad and include information taken from Reardon (1999).

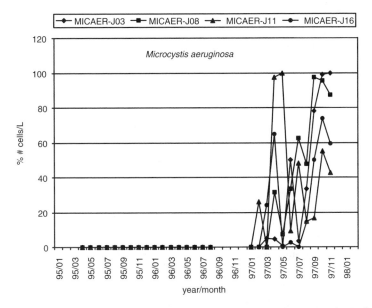

Figure 2.10 Percent of total numbers L^{-1} of *Microcystis aeruginosa* taken in Lake Jackson by month from April 1995 to November 1998. Phytoplankton analyses were made by A.K.S.K. Prasad and include information taken from Reardon (1999).

in 1997, there was a marked reduction in phytoplankton species richness compared to previous years (Figure 2.11). These reductions were strongly correlated with the occurrence of *Microcystis aeruginosa* in the lake.

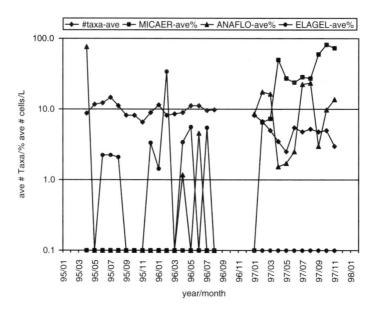

Figure 2.11 Percent of total numbers L^{-1} of *Microcystis aeruginosa*, *Anabaena flos-aqua*, and *Elakatothrix gelatinosa* (averaged over stations in Lake Jackson) compared to the average species richness monthly from April 1995 to November 1997. Phytoplankton analyses were made by A.K.S.K. Prasad and include information taken from Reardon (1999).

Reardon and Livingston (unpublished data) found major blooms of the dominant blue-green algae (*Microcystis aeruginosa*, *Anabaena flos-aquae*, *A. planktonica*) during fall 1997. There were temporal successions as well as spatial differences in the dominance relationships. These blooms were accompanied by a precipitous decline in numbers of phytoplankton species, which were generally lower in areas affected by the primary blooms. In a PCA-regression analysis of the data, there were significant associations between *A. planktonica* and high TIN, high ammonia, high nitrate, high total nitrogen (TN), and high conductivity. *Microcystis aeruginosa* was closely associated (negatively) with nitrate, total organic nitrogen (TON), and TN. *Anabaena flos-aquae* was significantly associated (negatively) with nitrate, TON, and TN. Phytoplankton numbers were significantly associated with high TIN, high ammonia, high nitrate, high total nitrogen, and high conductivity. The data thus show that blue-green algae blooms were associated with various forms of nutrients (Reardon, 1999). Zooplankton numbers (Shoplock, 1999) were significantly (negatively) associated with oxygen anomaly, and the chlorophylls (*a*, *b*, *c*), and (positively) with high TIN, high ammonia, high nitrate, high TN, and high conductivity.

During 1998, Lake Jackson reached another climactic state relative to the blue-green algae blooms and associated habitat deterioration in the form of flocculent sediments (Figure 2.12). Most remaining rooted vegetation deteriorated with the spread of *Lyngbya* in eastern Lake Jackson. Increased storm water runoff from Ford's Arm contributed to the proliferation of this species. During this period there were major blooms of blue-green algae throughout all parts of Lake Jackson. During 1998, Hydrilla appeared to be almost totally eliminated by the blue-green algae in eastern parts of the lake. By fall 1998, the entire lake was taken over by blue-green algae (Figure 2.12). During this period, Secchi depths averaged between 0.5 and 0.7 m throughout the lake. Essentially, hypereutrophication was evident everywhere in the lake, thus completing a process begun in the early 1970s. By 1999, during a prolonged drought, most of the lake drained through existing sinkholes. Following the drying out of the lake, a multimillion-dollar effort was undertaken to remove the polluted sediments.

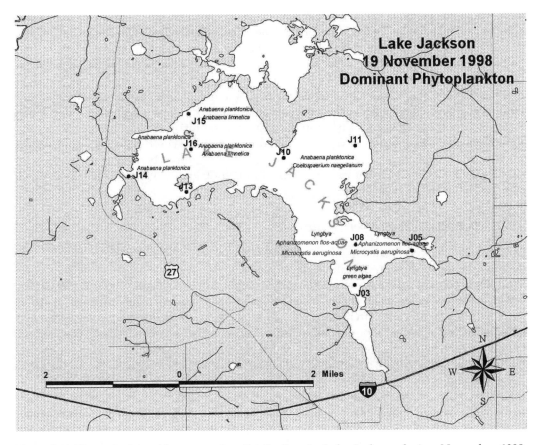

Figure 2.12 Phytoplankton bloom species distribution in Lake Jackson during November 1998. Geographic data provided by the Florida Geographic Data Library (FGDL).

2.7   Biological Response to Blooms

Eutrophication processes in Lake Jackson were strongly influenced by nutrient loading into the southern arms of the lake. The lake was adversely affected by decades of urban development in the Jackson basin, but increased developments near the lake during the period 1988–1998 exacerbated the problem. Conductivity increases in the Megginnis Arm drainage and throughout the eastern portions of Lake Jackson showed that storm water had a cumulative effect on Lake Jackson over time. These effects were not continuous, but were affected by the pattern of rainfall and runoff. Increased frequency and virulence of phytoplankton blooms eventually expanded from the main storm water entry (Megginnis Arm) to areas throughout Lake Jackson. All parts of Lake Jackson experienced increased surface chlorophyll *a* levels with time. The growing levels of ammonia in bottom portions of the lake coincided with the bloom occurrences, and indicated a deterioration of benthic conditions in Lake Jackson that was associated with the increased frequency and extent of the blue-green algae blooms. Reduced orthophosphate in the water and sediment TP with time indicated nutrient limitation by phosphorus as blooms became more predominant.

Analysis of the long-term habitat changes indicated that there was a cumulative adverse impact on Lake Jackson that was directly related to a biological response to long-term storm water runoff. The spatial/temporal distribution of SAV and emergent vegetation (EV) were related to sediment and water quality changes caused by influxes of polluted storm water. The balance between SAV/EV and water column phytoplankton determines the trophic organization of the lake. These plants provide habitat and organic production for the important benthic food webs of Lake Jackson. In turn, the microphyte and macrophyte associations affect habitat conditions that include ammonia distribution in the lake and factors such as DO, pH, light availability, and sediment quality. Habitat deterioration, in the form of flocculent sediments and replacement of rooted vegetation with blue-green algae, was the direct result of repeated nutrient loadings through storm water events. Basic changes of the aquatic plant communities in Lake Jackson were thus translated into major trends of water and sediment quality. These changes, coupled with the loss of phytoplankton species richness and deterioration of benthic macrophytes, led to the major losses of habitat quality throughout the lake. By 1998, the entire lake was affected by various forms of plankton blooms (Figure 2.12).

2.7.1   Infaunal Macroinvertebrates

Infaunal macroinvertebrates live in or on the sediments. These animals are an important part of lake food webs, translating microbial production into animal biomass that forms the basis of lake fisheries resources. An analysis of the infauna of Lake Jackson was carried out by Schmidt-Gengenbach (1991). Two test sites (Megginnis Arm, reference site in upper lake) showed significant differences in several chemical and physical parameters, especially sediment composition. The Megginnis Arm site had higher silt, clay, and organic components, whereas the reference site had a higher sand fraction. The Megginnis Arm site had high levels of lead, chromium, and zinc relative to the reference sites. Conductivity, turbidity, and color were higher at the Megginnis Arm site, and the fluctuations were often much more extreme than at the reference site. Surface DO was significantly lower at the Megginnis Arm site for most of the year, but became higher than that of the reference site in February 1989, and remained so throughout the end of the study. Bottom DO followed a very similar pattern, except that the difference between the Megginnis Arm and reference sites was even greater in the summer months, the values for the Megginnis Arm site reaching close to zero.

There was a significant difference in numbers of individuals and species richness of infaunal macroinvertebrates. Over the course of the 13 sampling months, the total number of individuals collected at the reference site was approximately 3.5 times that of the Megginnis Arm site. Species richness was higher at the reference site. Fifteen more taxa (thirteen more identifiable species) were taken at the reference site than at Megginnis Arm. Species composition was also different between the two stations. A percent similarity index that takes into account the percentage of the total number of individuals that each taxon represents at each station showed a similarity of only 30%.

A plot of log abundance vs. rank indicated that infauna at the reference site were dominated by two highly abundant species, whereas the community at the Megginnis Arm site had a more even distribution. In a comparison of the top five dominant species for the culled data, the two stations had two naidid species in common: *Stylaria lacustris* and *Dero digitata*. These naidids were dominant at the Megginnis Arm site, representing 31% of the 28 fauna, whereas they only represented 9.6% of the reference site, which was dominated by the amphipod *Hyalella azteca* (31.4%) and another naidid, *Pristina synclites* (20%). The remaining three dominants at the Megginnis Arm site included two chironomids, *Dicrotendipes modestus* and *Einfeldia natchitocheae*, and the flatworm *Dugesia tigrina*. The remaining dominant at the reference site was the copepod, *Mesocyclops edax*.

Microcosm experiments were carried out with sediments taken from the field stations according to protocols developed by Livingston and Ray (1989). Sediments from Megginnis Arm had elevated levels of copper, lead, zinc, and chromium compared to control sediments; Megginnis treatments caused decreased infaunal numbers during the first week. Numbers in the toxic treatment recovered to control and sediment control levels by week 3, but and did not undergo an increase by week 5. Three of the taxa that were reduced included *Polypedilum halterale*, immature tubificids with capillary chaetae, and nematodes. The chironomid *P. halterale* showed a strong adverse toxic treatment effect throughout the experiment. The browser/scavenger/grazer-infaunal/mobile burrower guild dropped sharply in the toxic treatment in week 1, staying even with the other two treatments at week 3, and remaining lower in week 5. This group included six naidid species, unidentified Naididae, one tubificid species, both immature tubificid groupings, one chironomid, and unidentified Glossoscolecidae and Lumbriculidae. The microcosm experiment showed that the sediments in Megginnis Arm had an adverse impact on organisms taken from the reference site.

Field data showed considerable differences between the benthic macroinvertebrate communities in the Megginnis Arm and those in the reference sites that indicated habitat differences between the sites (i.e., higher organic, silt, and clay fractions, higher color and turbidity, and lower dissolved oxygen at Megginnis Arm). Tubificids and nematodes comprised one half of all the organisms at the reference site, whereas chironomids were present in higher relative densities in the Megginnis Arm. The high number of the amphipod *Hyalella azteca* at the reference site was one of the most striking differences between the two stations. Pennak (1978) describes this species as being common in unpolluted, clear waters, which would explain the virtual absence of *Hyalella* at the Megginnis Arm station. The bivalves *Musculium partinmeium* and *Musculium* sp. were present in significantly higher numbers at the reference site. Species that are known to be pollutant tolerant, such as *Limnodrilus hofmeisteri* or *Tubifex tubifex*, were dominant at the Megginnis Arm site.

2.7.2 Fishes

2.7.2.1 Fish Diseases

The Leon County Lakes Study started inadvertently during a routine Florida State University (FSU) class field demonstration of fish collection techniques in Megginnis Arm during March 1988. Electrofishing catches were marked by extremely high percentages of diseased fishes. Some species had lesions on the body walls or tumors in the dorso-anterior regions of the body. Some had bloody, open sores on ventral surfaces, massive infections of internal and external parasites, fin rot and deteriorating gill filaments, and abnormalities of the internal organs (Livingston, 1988a). The initial collections were followed by an organized study wherein a series of fish collections was taken at various fixed stations during March and April 1988 and March 1989. Three electroshock samples were taken at each station over set periods during the day. Each fish was examined for any sign of disease and/or parasite infestation. Color photographs were taken of fishes showing signs of disease and/or parasite infestation along with the usual field notes. Larger fishes were opened for signs of disease and/or infection of the internal organs. Pictures were taken with a metric ruler, station, date, and species designation in the field of view. In most cases, the heads of the affected fishes were taken for analysis with searches for nodules and/or any other types of abnormalities in the interior of the fish. Fish species names and standard lengths were recorded in the field for all fishes taken.

Hemorrhagic lesions, cysts, fine erosion, and external/internal parasites were the dominant forms of abnormalities noted in the fishes taken in various southern portions of Lake Jackson. A comparison of the total numbers of fishes taken with the numbers that were diseased/infected in March 1988 indicated that all of the fishes in Megginnis Arm were diseased and/or infected. Some of these fishes were in such bad condition that they died immediately after being caught. High percentages of the diseased/infected fishes were taken at Stations J01, J02, and J03. A small number of diseased fishes were taken at Station J05 (Ford's Arm). No diseased fishes were taken in areas not affected directly by urban runoff. Virtually all the diseased fishes taken in Lake Jackson in March 1988 were located in the southern arms of the lake (Megginnis Arm, Ford's Arm). During April 1988, the highest percentages and numbers of diseased/infected fishes were again noted in Megginnis Arm, although, during this sampling, there were diseased fishes found in Ford's Arm, in areas between the two arms, and at Station J12. Areas in the northern portions of the lake were characterized by relatively low numbers of fishes affected by infection or disease.

Data taken by Byrne (1980) concerning the distribution of hydrocarbons (oil derivatives) in Lake Jackson sediments were used along with our sediment metals data in a series of statistical tests with the fish data. The gradients of diseased fishes taken during March and April 1988 closely followed those of the hydrocarbons reported by Byrne (1980). The highest correlations of the diseased/infected fish indices were noted with the various metal and hydrocarbon indices. Fish numbers and species richness were also highly correlated with these indices. Regressions of the data taken during the March 1988 field collections indicated that the only statistically significant associations of diseased/infected numbers were with copper ($R^2 = 0.59$, $P = 0.05$) and the aromatic hydrocarbons ($R^2 = 0.55$, $P = 0.05$). During the April 1988 analysis, these regressions were significant, with total hydrocarbons ($R^2 = 0.64$, $P = 0.05$) and the aromatic hydrocarbons ($R^2 = 0.86$, $P = 0.05$). The results of these analyses indicate that, of all the different indicators of the impact of storm water run off, the aromatic hydrocarbons represent the factor that was most closely associated with the diseased/infected fish indicators. However, the Byrne data were taken before the availability of sophisticated analytical techniques for determination of PAH concentrations in sediments, and the above correlations do not represent proof of effects of PAHs.

2.7.2.2   Fish Distribution

Field analyses were carried out to determine if fish assemblages taken in areas chronically perturbed by storm water input were significantly different from fish assemblages collected in less affected areas of Lake Jackson (Bevis, 1995). Fish assemblages in areas of the lake receiving different amounts of storm water runoff were examined quarterly from November 1993 through July 1994 using pop nets and electrofishing. This study overlapped a treatment (Fluridone) for control of the exotic *Hydrilla verticillata* in southern and eastern regions of the lake during March 1994. Bevis (1995) found that the bluefin killifish, *Lucania goodei*, which is tolerant of a wide range of environmental conditions, was more abundant and represented a greater percentage of the fish assemblages taken in areas that were stressed by storm water intrusion (eastern and southern regions). Fish assemblages taken in areas that were less affected by runoff (northern and western parts of the lake) were dominated by the bluespotted sunfish (*Enneacanthus gloriosus*). Bevis (1995) found that more adult largemouth bass were collected in the northwestern (i.e., unpolluted) parts of the lake. These fish also were in better health than adult bass taken in the hypereutrophicated southeastern sections.

The regional difference in fish dominance patterns reflected general differences in habitat associated with pollution from storm water runoff. Bevis (1995) found that as urbanization in the various Lake Jackson sub-basins increased, the physical, chemical, and biological habitats of the receiving areas were altered. Although periodic storm water runoff eventually affected the entire lake, habitats within and proximal to receiving areas were the most disturbed. The distribution and abundance of the native plant communities were disrupted and significantly altered by storm water runoff due to long-term exposure to polluted runoff and competition with exploitative, invasive aquatic plants. These trends were exacerbated by periodic chemical plant control because aquatic plants played a significant role in the life histories of many small and juvenile freshwater fishes. Fish assemblages in the lake thus were controlled by long-term water quality changes that reflected the cumulative effects of anthropogenous activities in the southern watersheds of the Lake Jackson basin.

2.7.2.3 Fish Trophic Response to Algal Blooms

In a study of the relationship between eutrophication and fish feeding, Boschen (1996) analyzed the stomach contents of four numerically dominant fish species (*Etheostoma fusiforme, Enneacanthus gloriosus, Lucania goodei,* and *Lepomis macrochirus*) in Lake Jackson. These studies were conducted to determine possible regional differences in dietary characteristics of fishes in eastern areas affected by storm water runoff vs. unpolluted (western) lake areas. Analyses of fishes collected from fall 1993 to July 1994 indicated systematic differences in the trophic organization of fishes taken in lake areas affected by storm water runoff compared to unaffected areas. Boschen (1996) found that there was consumption of less profitable prey types and sizes in polluted (hypereutrophic) areas. This included reduced food consumption and a greater incidence of empty stomachs in fishes exposed to storm water runoff. Stomach fullness comparisons were based on length (SL) – dry weight (mg) regression equations generated for each species: *E. fusiforme* (R^2 = 0.987, $P \le 0.0001$), *E. gloriosus* (R^2 = 0.995, $P \le 0.0001$), *L. goodei* (R^2 = 0.990, $P \le 0.0001$), and *L. macrochirus* (R^2 = 0.996, $P \le 0.0001$). Regional trends showed a greater mean percent stomach fullness in the western region in every species.

Boschen (1996) found significant regional (P = 0.05) differences in diet parameters that were related to observed changes due to eutrophication and to effects of herbicide treatments for Hydrilla control. Percent empty stomachs were significantly (P = 0.05) higher in samples collected in areas affected by nutrient loading and were also enhanced by application of the herbicide Sonar (Fluridone). Two-way ANOVA results for *E. fusiforme* and *E. gloriosus* indicated significant region (P = 0.0215 and 0.0002, respectively), date (P = 0.0005 and 0.0001), and region by date interaction effects (P = 0.0645 and 0.0107). *L. goodei* and *L. macrochirus* exhibited only significant region (P = 0.0086 and 0.0085, respectively) and region by date interaction effects (P = 0.0175 and 0.0981). Comparisons before and after herbicide application exhibited a greater percentage of empty stomachs in the eastern section of Lake Jackson in each species, and regional differences were greatest after the Fluridone application. Displacement of epiphytic invertebrates following macrophyte losses, together with increased zooplankton biomass and turbidity associated with the phytoplankton blooms, probably accounted for the exaggerated decline of fish feeding in the eastern region of Lake Jackson. Fluridone may also adversely affect sensitive prey species.

The Boshen findings agreed with previous studies that indicated that eutrophication-induced habitat modifications leading to food supply alterations (loss of large, profitable prey) adversely affected foraging success and ultimately influenced fish community succession. Prey communities themselves were not studied although known eutrophication-induced factors, including variable prey community composition and dominance patterns, prey patchiness and abundance, and reduced size structure, were thought to cause these regional differences in Lake Jackson (Schmidt-Gengenbach, 1991). Similar work involving yellow perch populations in Lake Erie demonstrated that food supply alterations (loss of large, preferred prey species) due to cultural eutrophication affected consumption rates and diet quality that resulted in reduced growth rates (Hayward and Margraf, 1987). These results coincided with other findings (Leach et al., 1977) that cultural eutrophication leads to a decline of fishes due to unfavorable environmental conditions.

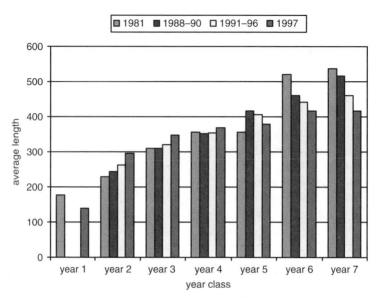

Figure 2.13 Average lengths of largemouth bass in Lake Jackson averaged over year classes (1981, 1988–1990. 1991–1996, and 1997). Data based on a report by Kirk (1999).

2.7.2.4 Long-Term Trends of Largemouth Bass Size

Kirk (1999) examined the records of the Florida Game and Freshwater Fish Commission with regard to long-term changes of the size of largemouth bass in Lake Jackson. His results (Figure 2.13) indicated that, during the period from 1988 through 1997, there was a progressive decline in the size of this species at the larger size classes. Although size trends can be influenced by a number of factors that include fishing pressure and natural changes of habitat conditions, the trend of bass size in Lake Jackson was consistent with the results of Bevis (1995) and Boschen (1996) that showed that progressive changes in the Lake Jackson habitat associated with cultural eutrophication were associated with adverse effects on feeding characteristics of fishes.

2.8 Lake Jackson Restoration Efforts

During the drought of 1999–2002, Lake Jackson drained into sinkholes, and major areas of the lake dried out. A restoration program, funded by the Leon County Commission ($4.4 million), state agencies ($900,000), and the Florida Legislature ($2.65 million), was initiated to clear some of the contaminated sediments from the lake. There was, however, no attempt to improve the water quality of storm water runoff going into the lake. Despite removal of some of the polluted sediments, the same areas of the southern parts of the Lake Jackson basin were still contributing polluted storm water runoff to Lake Jackson during heavy rainstorms.

   An analysis of lake water quality was carried out during 2002–2003 (Livingston, unpublished data), and the results were compared to data taken during the period of maximum pollution of the lake (1997). The data (Figure 2.14) indicated that surface conductivity remains high in the Megginnis drainage, and is even higher than that noted during the major blooms of 1997 (Figure 2.14a). Bottom DO in Megginnis Arm and lower Lake Jackson was extremely low during warm months at levels comparable to those in 1997 (Figure 2.14c). The chlorophyll *a* data indicated major blooms in these parts of Lake Jackson during various seasons of the year (Figure 2.14b and d). This finding was consistent

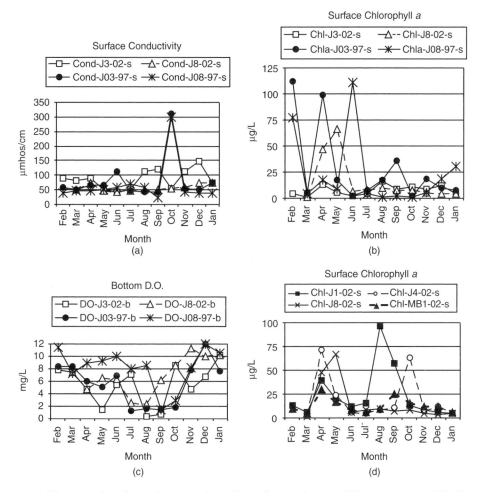

Figure 2.14 Water quality data taken in Lake Jackson from February 2002 to January 2003. Data are compared to similar results taken in Lake Jackson from February 1997 to January 1998: (a) surface conductivity at Stations J03 and J08; (b) surface chlorophyll *a* at Stations J03 and J08; (c) bottom dissolved oxygen at Stations J03 and J08; and (d) surface chlorophyll *a* at Stations J01, J04, and J08 compared to Station MB02 in Lake McBride.

with phytoplankton collections that showed a return of various bloom species (including *Microcystis aeruginosa*) to the lake (Livingston, unpublished data). Without a more comprehensive water treatment system to take out the nutrients and other pollutants from storm water runoff from urban parts of the Jackson basin, the water has again been contaminated, and the cycle of deterioration described above is now being repeated.

2.9   Urban Runoff and North Florida Lakes

The long-term Lake Program included monitoring the water quality and biological features of a series of lakes and ponds in urban areas of the region. The results of these studies were released to the public in the form of reports (Livingston, 1988a, 1989a, 1992a, 1993a, 1995a,b,c, 1996a,b, 1997a,b, 1998a, 1999a,b). The lakes surveyed were mainly sinkhole lakes not unlike Lake Jackson.

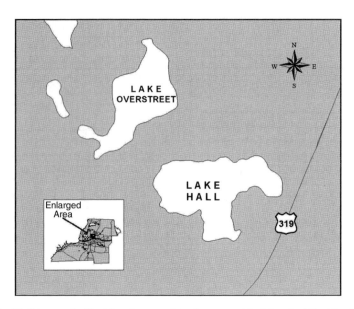

Figure 2.15 Lake Hall in north Florida. Geographic data provided by the Florida Geographic Data Library (FGDL).

2.9.1. Lake Hall

Lake Hall (Figure 2.15) is located at the far eastern corner of the Ford's Arm basin. The Lake Hall drainage basin is affected by intensive residential development with associated roads and commercial areas. The headwaters of the Lake Hall basin is a shopping mall. Storm water from the mall drains directly into Lake Hall. Overall, Lake Hall is receiving storm water from development in the upland basin, and has shown a tendency for increasing levels of eutrophication (Livingston, 1997a). According to McGlynn (unpublished data, 1998), a *Microcystis* (blue-green alga) bloom was noted from July 1998 through October 1998. Results of the long-term sampling effort (Livingston, 1999a) showed increased levels of chlorophyll *a* and ammonia that characterized the *Microcystis* bloom during summer 1998. Increased TN and TP supported these findings.

2.9.2 Lake Lafayette Basin

The Lafayette drainage basin (Figure 2.16) is a complex system with a massive area that includes an extensive series of floodplains and wetlands. Lake Lafayette is comprised of a series of four separate water bodies with partitioning due to anthropogenous activities. The western-most part of the lake (Upper Lafayette) is dry during drought periods, but includes a permanent sinkhole that periodically drains the system. Polluted storm water runoff from urban Tallahassee moves into Upper Lafayette and into the groundwater system via this sinkhole. Piney Z, bounded by control structures, is continuously inundated and can be characterized as a highly eutrophicated open water system. The Alford Arm is a marshy entry point for drainage from the urbanized northeastern sections of the Lafayette Basin. It is periodically inundated by storm water runoff from the north. Lower Lafayette is a shallow pond/cypress system that is dominated by floating and emergent vegetation of various types. The wood stork population in this part of the lake is of considerable importance as the wood stork is an endangered species.

The northern limits of the Lafayette basin originate in Lake McBride (Figure 2.17). The McBride system drains south and east through a series of housing developments

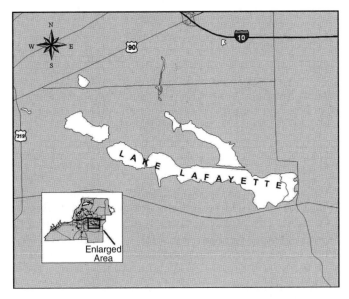

Figure 2.16 Lake Lafayette in north Florida. Geographic data provided by the Florida Geographic Data Library (FGDL).

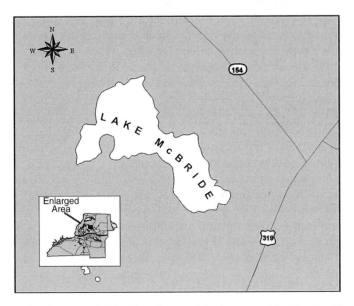

Figure 2.17 Lake McBride in north Florida. Geographic data provided by the Florida Geographic Data Library (FGDL).

ending in the Alford Arm of the lake proper. The upper part of the McBride drainage includes a series of ponded wetlands. This lake and its associated creek remain one of the least polluted parts of the Lafayette basin. The McBride drainage moves through forested areas, ponds, and wetlands into a series of polluted holding ponds in the Killearn area (Lakes Killarney and Kanturk) (Cassidy, 1992). The Lafayette drainage area also includes the low-lying area of Chaires/Capitola to the east and the heavily urbanized (and industrial) portions of Leon County (Weems Road drainage) to the west. This western portion of the drainage basin includes various malls, residential developments, industrial installations, road (and railroad) systems, and light agricultural usage. A ditched high-water channel connects the lower portions of Lake Lafayette with the St. Marks River drainage system.

Lake McBride is a relatively shallow, moderately sized lake (67.2 hectares) in the northeastern section of Leon County, Florida. This lake drains into a series of sinkhole ponds and waterways that include No-Name Pond. Analyses (Livingston, 1992a, 1995a) indicated that nutrient levels in Lake McBride were not high compared to other lake systems in Leon County, and that the lake was in an oligotrophic to mesotrophic state. During the early years of sampling, the McBride drainage was considered to be in a relatively undeveloped stage of human alteration with very good water quality. However, development above No-Name Pond was drained through a holding pond into No-Name Pond during spring 1995. The treatment pond was not built to retain the runoff, and most of the water collected after a rain event ran through a sand filter, down a relatively steep embankment, and into the pond. The treatment pond overflowed after various rainstorms that occurred from summer 1995 until the drought of 1999–2000. Major parts of No-Name Pond and the adjoining swamp developed a thick growth of pond weed (*Lemna minor*) during summer 1995. In some areas, the pond weed grew all the way to the bottom at depths of 1 to 1.5 m, with the lower parts of the growth periodically dying off, causing noxious water quality conditions.

By October 1996, the pondweed blooms in No-Name Pond had intensified and expanded with high color and turbidity. During summer 1997, pond conditions had worsened with thick blooms of pondweed covering the entire swamp and pond area of No-Name Pond. By September 29, 1997, No-Name Pond had progressed to a hyper-eutrophic succession stage of aquatic vegetation characterized by dense growths of water fern, variable leafed milfoil, pickeralweed, and alligator weed. The pond was so filled with vegetation that it was almost impassable by boat. By November 1997, pondweed growths had begun in the downstream drainage area. During the fall and winter of 1997, the new housing development on western portions of Lake McBride was causing major flows of polluted runoff into Lake McBride and the upper portion of McBride Creek in addition to the continued input of the of holding pond. By summer 1998, the massive vegetation had taken most of the DO out of the water, leading to prolonged and massive fish kills in No-Name Pond over the summer months.

The polluted No-Name Pond remains in this state to this day, and is an example of conditions that typify many of the drainages in the urban lake region. Water quality in these areas is characterized by high levels of watercolor and turbidity with relatively high chlorophyll concentrations. Ammonia and orthophosphate levels in areas such as No-Name Pond are very high as a direct result of the storm water input. The pond acts as a sink for the various nutrients and particulates that enter creeks that drain developed areas. There has been a gradual deterioration of water quality, including increases in color, turbidity, nitrate, orthophosphate, TON, TN, total organic phosphorus (TOP), and TP. The accumulation of organic nutrients due to increased aquatic vegetation was an integral part of the process that eventually led to major declines in DO and massive fish kills. The long-term deterioration of water quality in No-Name Pond was thus due to a combination of episodic input of polluted water from the holding pond and the effects of nutrient loading on aquatic plant associations where the assimilative capacity of the pond was eventually exceeded. This combination of effects, short and long term, defined the cumulative impact of storm water runoff on the habitat quality of No-Name Pond. The combination of fall droughts following extreme events of nutrient loading and proliferation of noxious aquatic plants led to hypoxia and fish kills.

The Upper Lafayette system is dominated by flows from the highly developed basin associated with extensive road development. The Weems Road Pond collects runoff from a broad area, with the massive input of polluted water going through a small holding pond off Weems Road and through a wetland into Upper Lake Lafayette. This water eventually ends up in the Upper Lafayette sinkhole. Surface DO is frequently high in the sinkhole, whereas bottom DO is often low. Periodic occurrences of dense blue-green algae blooms in the sinkhole have been associated with major fish kills in 1996 and 1997. Piney Z, the artificial impoundment, was dominated by blooms of three different species of blue-green algae (*Coelosphaerium naegelianum* Unger, *Anabaena planctonica* Brunnthaler, and *Gomphoshaeria aponina* Kuetzing). However, there has been little action by regulatory authorities concerning the pollution of this public resource from surrounding developments.

Lower Lake Lafayette is bordered by the Leon County Landfill and the Talquin Electric Sewage Treatment Plant (STP) on the southeast side, with increasing pressure from residential and commercial development from the Alford Arm drainage. Holding ponds for the STP drain directly into Lake Lafayette. Part of the lower lake is characterized by high conductivity, seasonally low DO concentrations, and minimal Secchi depths along with high phytoplankton productivity as indicated by chlorophyll *a*, with levels exceeding 50 to 100 μgl^{-1}. A package sewage treatment plant located on the southeastern banks of Lake Lafayette contributed to the adverse water quality conditions in this part of the lake. Overall, the Lake Lafayette drainage basin represents a system that has received little attention in terms of planning and management, with just about every form of urban encroachment being implemented without significant attention to the adverse environmental effects on the basin.

2.9.3 Lake Munson

Lake Munson (Figure 2.18) was once a natural lake but currently it is a shallow impoundment located south of the city of Tallahassee (Maristany et al., 1988). The Lake Munson drainage system is primarily composed of a blackwater basin. The lake receives flows from various urban storm water sources. Since its inception, Lake Munson has been affected by storm water runoff from the city of Tallahassee and treated sewage effluents from municipal wastewater treatment facilities. Between 60 and 80% of the storm water runoff from the city of Tallahassee flows into the lake. Lake Munson has historically

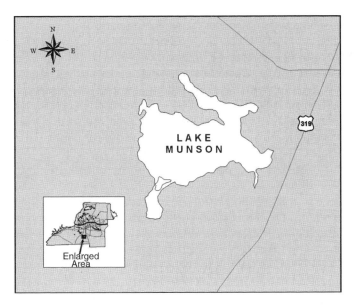

Figure 2.18 Lake Munson in north Florida. Geographic data provided by the Florida Geographic Data Library (FGDL).

received discharges from three sewage treatment plants: the Dale Mabry Plant (until 1978), the T.P. Smith plant (until 1980), and the Lake Bradford Plant (until 1984). Water exits Lake Munson via a control structure and runs to a sinkhole about 5 km to the south. This sinkhole is located about 9 km north of Wakulla Springs.

Lake Munson is dominated by dense peripheral growths of pond cypress (*Taxodium ascendens*) with the exception of the northwest corner of the lake. Recently, it has become infested with a dense growth of Hydrilla. Various water quality problems include high nutrient levels, algal blooms, hypoxic conditions, and associated fish kills. The peak input levels of treated sewage occurred during the period from 1978 to 1979. In 1982, Lake Munson was hypereutrophic. When effluent discharges were eliminated in 1984, there were associated improvements in lake water quality. By 1988, the lake was considered eutrophic. Currently, storm water runoff is the primary source of pollution to the lake. Watercolor and turbidity are uniformly high in Lake Munson. Secchi disk readings are low and conductivity readings are high. Overall, levels of these factors showed no trends after leveling out in 1992. Periods of surface DO super saturation are also evident in Lake Munson.

The data indicated that, with the exception of Lake McBride, most of the subject lakes surveyed in the region were polluted by urban runoff and followed a pattern of ecological deterioration similar to that of Lake Jackson.

2.10 *Holding Pond Ecology*

Holding ponds are being constructed throughout the United States as part of the effort to mitigate the effects of storm water runoff from developed areas. There are two basic functions of storm water holding ponds: (1) flood control and (2) water quality treatment. Treatment processes depend on variables such as the physical dimensions of the holding pond relative to the upland drainage basin that empties into the pond, the nature of urban development in the upland basin, and the spatial/temporal aspects of rainfall relative to the level of the surficial water table. Holding capacity and treatment processes are closely associated with the level of the surficial water table. Water table levels vary both seasonally and interannually. During winter–spring periods of heavy rainfall years, with maximally high water tables and prolonged rainfall over relatively long periods, the functions of the holding pond are stressed, especially with respect to ponds designed for water quality control.

Water quality control necessitates that storm water be held long enough for treatment to occur. Storm water runoff is affected by varying concentrations of silt particles that are suspended in the water. Nutrients such as orthophosphorus are attached to such particles as are other forms of pollutants such as heavy metals and PAHs. Inorganic nitrogen (ammonia, nitrate, nitrite) is usually carried in the dissolved form. Ordinary sand filters, commonly used as water quality instruments, are of no use for removal of these substances. Thus, with each storm, the turbid, nutrient-enriched storm water is transferred through the holding pond to the receiving waters where nutrients and pollutants accumulate in the water and the sediments of receiving areas such as lakes and ponds.

The construction of I-10 by the Florida Department of Transportation in the southern drainage basins of Lake Jackson in the early 1970s (Figure 2.13) was associated with extensive erosion problems. Massive amounts of sediments were washed down the relatively steep slopes of the Okeheepkee Road sub-basin into the Megginnis Arm of Lake Jackson (Livingston, 1988a). Following the construction of a series of intensive commercial developments at the head of the Okeheepkee sub-basin, there was a continuation of erosion problems with the loading of sediments and degraded water that adversely affected the Okeheepkee drainage. A collecting pond (Woodmont Pond) was constructed that concentrated the polluted water and redistributed it into a series of surface and groundwater flows that led to the contamination of downstream water bodies. The polluted runoff eventually ended up in Lake Jackson.

Upper Lake Lafayette is dominated by flows from the highly developed basin associated with the Capitol Circle NE, which includes residential, commercial, and industrial development. The Weems Road Pond collects runoff from a broad area with the massive input of polluted water going into Upper Lafayette and eventually into the Upper Lafayette sinkhole. The Weems Road Pond is totally inadequate as a treatment system due to the massive amounts of storm water draining into the system. This pond has virtually no effect on the quality of the runoff that washes into the groundwater of the Lafayette basin. Low DO and toxic effects of ammonia have led to continuing fish kills in the sinkhole. Recently, there have been kills of turtles and birds. High concentrations of PAHs have been found in the sinkhole sediments. In Lower Lake Lafayette, a package STP on southeastern Lake Lafayette drains into the Lower Lake Lafayette through groundwater connections. Chloride concentrations are high in receiving parts of the lake. High concentrations of various nutrients (including ammonia) and certain other toxic agents have been found in the vicinity of the lake adjacent to the STP.

There are a number of issues involved in the use of holding ponds in the restoration of storm water runoff. There should be a determination of whether or not the holding pond will add to water quality control before the pond is constructed. Differences exist between pre- and post-construction phases with regard to impacts and management responses. After it has been decided that a holding pond is necessary, there should be a complete review of how the pond is to be constructed. This should include determinations of the size, depth, shape, and other physical characteristics of the pond. Biological maintenance features should be utilized for such ponds (i.e., protection of natural aquatic vegetation in receiving areas, the construction of artificial wetlands). The creation of a storm water holding pond should take into consideration the surrounding land use and proposed development activities during and after such construction. The inclusion of tight restrictions concerning land-clearing activities should be part of a pre-construction review. Holding ponds should not be used as an automatic pretext for the current practices of clearing lands adjacent to sensitive aquatic drainages and resources. The assimilative capacity of the receiving system should be an important consideration with respect to pond construction.

Based on the scientific data concerning the use of holding ponds in the north Florida region, there are serious questions regarding the role of holding ponds in improving the quality of storm water runoff that enters natural surface water and groundwater systems. Often, construction of such ponds relies on too little scientific information. Most of the questions regarding how water quality is going to be improved by the holding pond usually remain unanswered because the primary function of pond construction remains flood control. This means that polluted water is collected and, with subsequent rainfall and high water tables, the low-quality water moves through the pond and into the receiving water body. This condition is particularly noted during periods of low evapotranspiration, prolonged high precipitation, and high surficial water table. Previous observations (Livingston, unpublished data) indicate that the surficial water table can come to within 0.6 m of the surface during prolonged rainfall in the study areas. It is also likely that, in a karst geological system, polluted water collected by the pond will make its way through the groundwater.

The technology for water quality treatment for storm water is well known and available. This technology includes combinations of wet and dry retention and detention ponds with various forms of physicochemical and biological treatment sub-systems. The use of such systems depends on local ecological conditions that include the assimilative capacity of the receiving aquatic systems. There is no evidence that such technology is used in a routine fashion anywhere in Florida or, for the matter, the United States. It appears more likely that public officials will try to "restore" affected water bodies after they are polluted rather than use preventive controls before the pollution occurs. The costs are thus shifted from development interests to the public. There is also no guarantee that the restoration effort will be successful.

The obvious deterioration of aquatic resources in the north Florida area is thus not due to a lack of understanding of the scientific and engineering aspects of the urban runoff question. Rather, it is a product of socioeconomic factors related to the use and misuse of scientific information concerning the effects of urban runoff, and the possible methods of restoration of lakes damaged by urban runoff. There is an obvious lack of regulation of water quality by state and federal agencies as the aquatic resources of a given area are destroyed.

2.11   Press Coverage and Public Response

Between the 1970s and the mid-1990s, the local newspaper (the *Tallahassee Democrat*) employed a series of distinguished environmental reporters whose coverage of the lake research was accurate, objective, and comprehensive. After the discovery of sick and dying fishes and poor water quality in Lake Jackson, various scientific reports on the findings were made public (Livingston, 1988a).

> "Lake Jackson is undergoing serious problems due to the input of nutrients and toxic substances by storm water runoff. Multiple sources are indicated. The nutrients have stimulated an enormous increase in the plant biomass in the lake which, in turn, has caused a loss of dissolved oxygen and a destruction of important habitat during the warm months of the year."

> "A considerable number of fishes have been taken showing various forms of disease and internal/external parasite infestation: external tumors, bloody, open areas on the ventral surfaces, damaged gills and fins, bacterial infections, extensive parasitic infestations (internal and external), and abnormalities of the internal organs. Such disease is a significant sign that the lake has been seriously affected by toxic substances."

> "Despite designations as an international bass fishing area, a major economic asset to the region, a Florida Outstanding Water, and an Aquatic Preserve, in addition to being ranked as the second most important water body in the region by the Northwest Florida Water Management District, Lake Jackson is currently being used as a finishing pond for storm water runoff in the basin. This situation has developed over the past 15 years despite repeated warnings to the local and state agencies responsible for the protection of this resource."

The response by local and state officials was swift and well-publicized:

> "We question his (Livingston's) findings."

> **—Official, Florida Game and Freshwater Fish Commission;**
> *Florida Flambeau,* **October 1988**

"Wattendorf disputes Livingston's figures. He said his division has handled more than 1600 sunfish and none showed obvious signs of disease."

—*Florida Flambeau*, **October 1988**

"As far as our statutory responsibilities are concerned, we've definitely done our part."

—**Official, Florida Department of Natural Resources;**
Tallahassee Democrat, **October 1988**

"...The basic condition of Lake Jackson is good."

—**Official, Florida Game and Freshwater Fish Commission;**
Florida Flambeau, **November 1988**

"I want a very holistic management strategy, I don't want just a research project... we want to see action taken; we do not want to study the death of lakes."

—**Official, Florida Department of Environmental Regulation;**
Tallahassee Democrat, **November 1988**

"We are concerned about the status of Lake Jackson, but do not think exaggerated statements will benefit the situation."

—**Official, Florida Game and Freshwater Fish Commission;**
Tallahassee Democrat, **November 1989**

Another report (Livingston, 1993a) on continuing scientific analyses of Lake Jackson was released to the public. Again, these revelations were met with official consternation that included sharp attacks on the quality of the lake research and the integrity of the principal investigator.

"The data indicate that the areas of storm water runoff (i.e., Megginnis Arm, Ford's Arm) were characterized by lower numbers of individuals and species of infauna than other parts of the lake. The highest such in areas of the lake that were remote from the areas of storm water runoff."

"Chlorophyll peaks occurred during summer periods of high rainfall showing a direct connection of such productivity with storm water both in terms of spatial and temporal variability. In bottom areas of Lake Jackson, mean dissolved oxygen was lowest during late summer and fall months with the lowest concentrations found at stations J02, J03, J12, and J13."

"Overall, the temperature and D. O. data indicate that Lake Jackson is being adversely affected by urban storm water in major portions of the eastern and western portions of the lake with some sections of the northern part of the lake (distant from storm water entry points) still showing ... healthy D. O. conditions."

Again, these revelations were met with official consternation:

> "We believe the fish population of Lake Jackson is in extremely good shape. We feel there is no apparent disease problem."

> **—Official, Florida Game and Freshwater Fish Commission;**
> ***Tallahassee Democrat,* July 1993**

> "The Game and Freshwater Fish Commission has routinely conducted fish sampling in Lake Jackson. Their sampling has not indicated an increase in diseased fish over background levels...." "...these results (are) an overestimation of oxygen depletion problem areas. It is not uncommon for eutrophic lakes to have summertime DO readings below 4 ppm." "The report has many inconsistencies. ...the report contradicts data that is being obtained in other agencies' current studies."

> **—Official, Florida Game and Freshwater Fish Commission; 1993**

There were a series of highly critical reports by local and state officials who criticized the report and made what could be called personal attacks on the character of the principal investigator. The Lake Jackson controversy led to the appointment of a Science Advisory Committee (SAC) by the Leon County Commission in 1995. A well-respected group of scientists and managers were chosen by the commission with the charge to review the results of the lake studies with an emphasis on the controversial Lake Jackson data. The following are excerpts from the SAC findings:

> "Over the last few months, the Leon County Science Advisory Committee (SAC) has reviewed, at the request of the Board of County Commissioners, a multitude of scientific studies concerning the ecological status of Lake Jackson. ...The SAC ... found the overall conclusions of the reports (i.e., Livingston studies) acceptable and consistent with other studies reviewed. ... Second, the inflammatory language used in the comments of the reviewers sent in by you is unacceptable in a scientific review; it clearly demonstrates a bias on the part of the reviewer, and, as such, cast serious doubt on the objectiveness of the review. ... The SAC urges you to re-evaluate both your motives and your analyses in developing the reviews that were sent to us and the Commission."

> **—Letter to FDEP by SAC, 12 March 1996**

> "The SAC finds that (1) based on the Livingston and other studies reviewed by the SAC, Lake Jackson is in an increasing state of ecological decline; (2) the majority of the problems facing the lake can be attributed to non-point pollution (storm water runoff) entering the lake; (3) the current condition of the lake is also the result of the failure to implement a comprehensive lake management program for the entire watershed; and (4) increased development in lake basins is adversely affecting the ecological integrity of Lake Jackson as well as most of the lakes in the County. ...In light of these findings, the SAC recommends ... an immediate moratorium on any new development (excluding one house on existing lots) should be imposed on the Lake Jackson basin."

> **—Report; The Ecological Condition of Lake Jackson,**
> **Leon County, Florida; 1996, SAC**

"While the problems in Lake Jackson may be more severe, many other lakes are experiencing similar impacts from increasing development."

**—Report; The Ecological Condition of Lake Jackson,
Leon County, Florida; 1996, SAC**

Public officials rejected the suggestion of a moratorium on development in the Lake Jackson basin. There was considerable unhappiness with the SAC as well as the continuing research on the lakes in the region.

Other reports concerning the regional lake situation were made public:

"The admission that Game Commission studies and management of Lake Jackson 'almost exclusively for largemouth bass under the guise that managing the top predator will in turn manage the rest' is inconsistent with just about everything we know concerning management of natural aquatic resources. The lake should be managed for habitat and other variables that are consistent with the welfare of the lake as a whole."

—Livingston, 1996

Again, officials were quick to respond:

"Florida Game and Freshwater Fish Commission officials say Lake Jackson is in good shape. In fact, they say it's doing better than it has in years. ... 'The fish look much healthier because they have food to eat,' said Jeff Nordhaus, a fishery biologist with the commission. ...Nordhaus said 'Green water doesn't scare us too bad...Overall, the way Jackson looks now, we're tickled with it.'"

—*Tallahassee Democrat*, May 17, 1996

Meanwhile, county commissioners had enough of scientific analyses of the lakes, and they fired the head of the Leon County Department of Permitting while attacking their own scientific advisors. The SAC was put on notice:

"The (Science Advisory) Committee was born in early 1995. Commissioners wanted some way to evaluate scientific reports that Lake Jackson was mortally wounded by development runoff. Committee members later concurred with much of Florida State University aquatic biologist Skip Livingston's findings even urging the county to ban development near the popular fishery. ...Fed up with the committee's suggestions, some commissioners now want the group silenced. 'I think some people on that committee have their own agenda,' said Commissioner Bruce Host last week... 'I think it's time for that committee to go away.' ... 'We as a committee decided that if they were going to appoint us to look at water quality issues, we are going to look at that and how to handle that in Leon County,' said Graham Lewis. ... 'If we are not addressing the problem that we need to address, we can fix that in a heartbeat,' (Helge) Swanson said. ...County commissioners in 1996 fired Swanson from his environmental permitting post for failing to follow management procedures. Critics of the county administration, however, said the firing was really over Swanson's opposition to the permit for the Talquin sprayfield near Lake Iamonia. "

—*Tallahassee Democrat*, December 18, 1997

In 1997, the Leon County Commission (LCC) decided to submit the lake monitoring program run by the Florida State University (FSU) team to a bidding process. Although the university bid for the work was the lowest, the LCC voted to "reconsider" (i.e., rescind) approval of the FSU bid for the water quality monitoring. It turned out that the LCC had responded to a letter from an influential developer that the FSU lake studies should be suspended. The following are direct quotes from the Leon County commissioners involved in this change:

> "I'm not sure that where the folk who are taking the water samples might not be under a rookery on purpose to shade the findings, and I would respectfully request that we award this to the next bidder."

> **—County Commissioner, LCC meeting, April 15, 1998**

> "… the most recent study that he's completed and the contract for that study states that all reports and findings resulting from this agreement must be presented to the Board of County Commissioners prior to the distribution to the public. Now what I have found out is that that did not occur. … this board has never been presented the findings from that contract."

> **—County Commissioner, LCC meeting, April 15, 1998**

> "I'm not a farmer, I don't have chickens, but if I did, I wouldn't want guarding the henhouse someone who was chicken-friendly."

> **—County Commissioner, *Tallahassee Democrat*, April 15, 1998**

From September 1997 through June 1998, the LCC suspended the lakes monitoring program. In April 1998, the LCC withdrew the lake contract from the low bidder (Center for Aquatic Research and Resource Management, Florida State University) and awarded the grant to the next bidder for approximately $10,000.00 more. County officials also altered the station locations and numbers of sampling stations. Of the original 34 stations that were used, the monitoring effort was reduced by 17 stations. This action seriously downgraded the effectiveness of the already poorly financed Leon Lakes program.

Meanwhile, during this period, the *Tallahassee Democrat*, a regional newspaper, was taken over by a conservative publisher who cleared out news personnel deemed sympathetic to the environment. This coincided with similar actions by the LCC, who fired county personnel who were identified as environmentally sensitive. A new regime was instituted whereby any stories concerning sensitive issues to development interests in the region were suppressed or replaced by false or misleading stories, a situation that exists to this day. Even letters to the editor were not printed when complaints were made concerning the firings and intimidation of state environmental personnel by elected county officials.

By 1998, there was a new regime with respect to the reportage of regional environmental issues. Accordingly, the news media repeatedly touted the sediment removal program as the solution to the problems of Lake Jackson.

> "Government officials say the more than $8 million spent by Leon County and the state to remove 2 million cubic yards of muck from the southern part of the lake will help Lake Jackson regain its reputation as a home for trophy-size

largemouth bass, though it will take a couple of years before this year's crop matures."

—*Tallahassee Democrat,* **August 26, 2001**

Meanwhile, led by a media blitz headed by the *Tallahassee Democrat,* a sales tax initiative called Blueprint 2000 was passed by voters in November 2000 after enthusiastic acceptance of this so-called "holistic" program of road-building and storm water projects by a team of development interests and local "environmentalists." It did not take long for this plan to unravel due to cost increases and the avoidance of any real restoration of Lake Jackson and other polluted lakes in the region. To date, virtually nothing has been done to mitigate storm water runoff to Lake Jackson while road-building and associated development has accelerated in the Jackson basin. Promises of bike paths and nature walkways satisfied so-called "environmentalists" while lake water quality throughout the region worsened.

Consultants gave optimistic reviews of the Lake Jackson situation, and public officials rejoiced in the newly discovered clean health of lakes in the area. Even a walk through the Tallahassee regional airport was rewarded with the assurance by the mayor of the pristine nature of the lakes. Not to be outdone, the *Tallahassee Democrat* published a series of articles on the recovery of Lake Jackson thanks to the sediment removal program.

"...some other water quality researchers say they expect the lake to be better than it was before.... the county's water quality contractor tested water quality at Megginnis Arm on Wednesday and said he found low conductivity, an indicator of pollution....he expects it is going to be healthier than before it went dry and the muck was removed...(an) FSU Oceanography professor ... says the muck removal and storm water treatment ponds will help the lake in the future."

—**Tallahassee Democrat, June 18, 2001**

All of the publicity regarding the miraculous recovery of Lake Jackson ran counter to the actual scientific data. A sampling program executed during 2002–2003 indicated that the optimistic predictions concerning Lake Jackson water quality were not correct (Figure 2.19). If anything, specific conductance was as high or higher in Megginnis Arm than it was during previous years of admittedly high levels of pollution loading to the lake.

Other variables, such as DO and chlorophyll *a* (Figure 2.14) indicated poor water quality in the lake compared to an unpolluted Lake McBride during recent months. Surface water samples from three sites (J01, J02, and J02A) in Lake Jackson on July 27, 2000, revealed that gas-vesiculate blue-green algae, particularly, *Anabaena spiroides* and irregular gelatinous colonies of *Microcystis aeruginosa*, were dominant (Livingston, unpublished data). Some of the other blue-green algae that were present included *Anabaena circinalis, Anabaena planktonica, Coelosphaerium Naegelianum, Gomphoshaeria aponina,* and *Euglena* spp.

Meanwhile, the news media, led by the incomparable *Tallahassee Democrat,* continued to suppress scientific information. Stories proclaimed high water quality in regional lakes.

"Government officials say the more than $8 million spent by Leon County and the state to remove 2 million cubic yards of muck from the southern part of the lake will help Lake Jackson regain its reputation as a home for trophy-size largemouth bass... though it will be a couple of years before this crop matures."

—*Tallahassee Democrat,* **August 26, 2001**

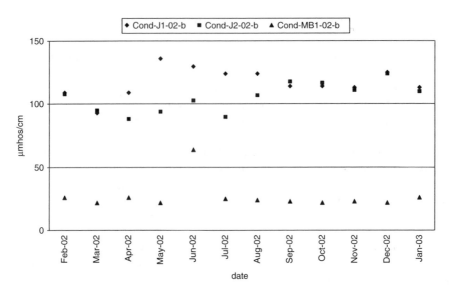

Figure 2.19 Conductivity in the Megginnis Arm of Lake Jackson (J1, J2) compared to a reference Lake McBride (MB1) from February 2002 to January 2003.

"City about to begin cleaning up ponds… It's hard to tell where the grass ends and the water begins on Chapman Pond in Myers Park, but residents won't have to look at the thick layer of algae anymore, once the city cleans it up."

—*Tallahassee Democrat*, **June 20, 2003**

"Leon County has a lot of quality going for it. The area's lakes are generally good…the president of the northern chapter of the Florida Lake Management Society said local lakes aren't pristine but are faring well."

—*Tallahassee Democrat*, **June 20, 2003**

The fact is that, since the final FSU report (Livingston, 1997a), no scientific report concerning the current status of the regional lakes has been made public, and most of the news concerning the lakes has been upbeat and without any semblance of reality. Lake water quality in the region has not improved. Taking Lake Jackson as an example, the loading of nutrients continues to adversely affect the lake. This includes a return of massive growths of Hydrilla to southern parts of the lake with continued phytoplankton blooms that go unreported to the public.

2.12   The Failure of Restoration

There have been no effective state or federal regulatory actions concerning the continuing pollution of the regional freshwater systems. Proposed targets for nutrient loading to various freshwater bodies by the Florida Department of Environmental Regulation have been largely ignored by local authorities. The multimillion-dollar Lake Jackson restoration program represents a very expensive failure, and the public does not have any idea of the continuing pollution of this supposedly protected body of water.

   The situation of the north Florida lakes can be extended to many other parts of Florida and the United States with regard to the misrepresentation of environmental news concerning the health of public resources. The role of the news media has been to control public awareness by obfuscation of scientific data and deliberately misrepresenting existing factual information. This effort involves the usual tools of the press: omission, misinformation, and obfuscation of scientific data. The objective of this journalistic *tour de force* is to control the news in a way that precludes up-front infrastructure costs to developmental interests who are busy cutting down forests to make way for the malls and superstores that have proliferated in the region. This effort has been complemented by the recent "discovery" that recent tax increases, formerly billed as comprehensive planning, have not covered the costs of road-building in the area. Consequently, with the exception of some bike paths and greenways to satisfy the "environmentalists," there will be no money for treating the polluted storm water associated with the massive building efforts. Although this is not a unique or even unusual story, it goes totally unreported by the media who carefully guard their own financial interests in the urban sprawl that is the final result of this process.

   The regional problem of urban runoff and the lack of public awareness or concern for water quality in north Florida water bodies can be illustrated by the history of Lake Ella (Figure 2.20), a small, formerly natural pond that has been developed into a major recreational area where people take their children and dogs for various activities. Actually, Lake Ella has been turned into a storm water treatment pond in the Lake Jackson drainage (Livingston, 1997a). There are indications of sewage input in addition to serving as a receptacle for polluted storm water. Lake Ella is continuously treated with alum to precipitate toxic agents and nutrients into the sediments. Surface and bottom conductivity is continuously high as a response to the collection of storm water runoff. DO is seasonally low and there is occasionally a complete collapse of the DO regime due to herbicide treatment of the considerable aquatic vegetation in the lake. The high biochemical oxygen

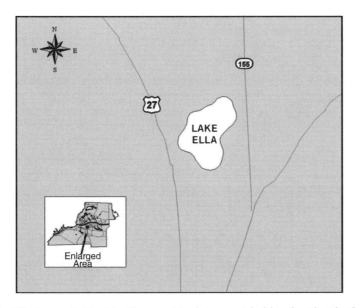

Figure 2.20 Lake Ella in north Florida. Geographic data provided by the Florida Geographic Data Library (FGDL).

demand accompanying the decomposition of the dead and dying vegetation is the usual cause of severe hypoxia and anoxia (Livingston, 1997a).

Lake Ella serves as one of the most popular recreational areas in the region. Dogs are walked and children play along the banks of Lake Ella as carloads of citizens participate in this grand outdoor experience. However, one anecdote sums up the sad fate of Lake Ella. As is true of various treatment ponds, there is a water fountain in the middle of the lake. This public effort of aeration is more symbolic than real, as is most everything associated with the Lake Ella Recreation Area. However, every now and then, the fountain gets clogged and the water fountain shuts down. One day, while sampling the lake, our field personnel noted that a municipal worker was cleaning out the clogged fountain. It seems that every now and then, according to the worker, the water shower is clogged with condoms. Indeed, Lake Ella serves as a metaphor for the situation of lake water quality in the north Florida region, and remains the painted lady of the lakes that have been continuously damaged by urban developers. And so it goes.

chapter 3

Industrial Pollution: Pulp Mills

The manufacture of pulp and paper represents a major industry in the southern United States. Because of the high volume of the effluent from pulp mills (20 to 50 million gallons per day) and the discharge of various agents that have an impact on water and sediment quality, the paper industry represents a formidable challenge in terms of restoration. The discharge of various components leads to varying levels of impacts, depending on the level of treatment and the assimilative capacity of the receiving system. These factors include high concentrations of total dissolved solids (mainly sodium, chloride, and sulfide) that contribute to the high specific conductance of the effluent. Conductivity contributes to the elimination of primary freshwater species due to osmotic imbalances. Enhanced loading of dissolved organic carbon (DOC) increases Biochemical Oxygen Demand (BOD) in the receiving areas that, in turn, causes hypoxic and even anoxic conditions. Various organic compounds that include tannins, lignins, and fulvates are components of the DOC. These compounds also contribute to the high levels of color in pulp effluents that, in turn, adversely affect light penetration in receiving water bodies.

Pulp effluents contain high levels of nutrients (mainly ammonia and orthophosphate) that contribute to adverse impacts on microphyte and macrophyte assemblages in receiving areas. Toxic agents in pulp effluents include ammonia, certain metals, and chlorinated compounds (that can include dioxin). Various methods are available to mitigate many of the toxic components of pulp effluents. Publications of the impacts of pulp effluents in the study areas are available (Livingston, 1975a, 1980a, 1981, 1982a, 1984a, 1985a,b, 1987a,b, 1993b, 1997a, 1999b, 2000, 2002; Livingston et al., 1998a).

3.1 Study Area

The Gulf coastal zone of north Florida extends from the alluvial Perdido River–Bay system in eastern Alabama and the western Florida Panhandle to the black water rivers of Apalachee Bay in the Big Bend area (Figure 3.1A). The drainage systems in this region have been part of the long-term studies of our research group (Livingston, 2000, 2002). The panhandle landscape is the result of stream and river flows and wave action that has acted on the land surface over the past 10 to 15 million years. Beach ridges, spits, cliffs, barrier islands, swales, sloughs, dunes, lagoons, and estuaries along a relatively flat upland configuration characterize the northeast Gulf region. Western bay systems are associated with alluvial rivers having relatively restricted coastal plain areas. The Apalachicola and Apalachee Bay basins (Figure 3.1) are part of broad coastal plains that include extensive marsh areas. Barrier islands start in Apalachicola Bay, extending west to the Pensacola and Perdido Bay systems. On the eastern end of the Panhandle coast (Apalachee Bay), there is no barrier island development, and coastal swamps and marshlands dominate

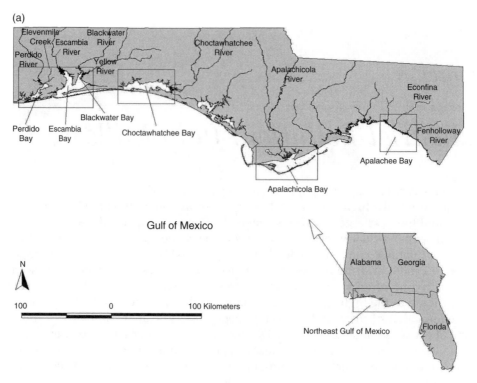

Figure 3.1a The Florida Panhandle, showing the primary river basins and distribution of associated river–estuaries and coastal systems. This figure is a composite of information provided by the National Oceanic and Atmospheric Administration, the Florida Department of Environmental Protection, and the ESRI Corporation.

the coast. This area is characterized by shallow, sloping margins, the lack of wave action, and an inadequate supply of sand (Tanner, 1960) for barrier island development.

Upland watersheds for coastal regions of the northeast Gulf are located in Alabama, Florida, and Georgia, in an area approximating 135,000 km² (Figure 3.1a). Associated estuarine/coastal systems are characterized by habitats that are mainly controlled by the upland freshwater drainage basins. Intersecting habitats of the coastal zone include salt-water marshes, sandy beaches, tidal creeks, intertidal flats, oyster reefs, seagrass beds, sub-tidal unvegetated soft bottoms, and various transitional areas. The salinity regimes of these areas are variously affected by major river systems or, as is the case in Apalachee Bay, by a series of small rivers and groundwater flows. Nine of the twelve major rivers and five of the seven major tributaries of Florida occur in this region. The major alluvial rivers of the northwest Florida Panhandle (Perdido, Escambia, Choctawhatchee, Apalachicola) have their headwaters in Georgia and Alabama (Figure 3.1a). A series of smaller streams along the panhandle coast include the Blackwater and Yellow Rivers of the Pensacola Bay system, the Chipola River (part of the Apalachicola drainage), and the Ochlockonee River on Apalachee Bay. Farther down the coast, a series of small streams (St. Marks, Aucilla, Econfina, and Fenholloway) with drainage basins in Florida, flow into Apalachee Bay.

The Florida Panhandle has a range of human populations from low densities in the eastern drainages to the more populous areas to the west (Figure 3.1b). The Apalachicola and Apalachee Bay drainage basins are among the least populated coastal areas in the United States. Coastal urban areas include Pensacola (Escambia Bay–Pensacola Bay),

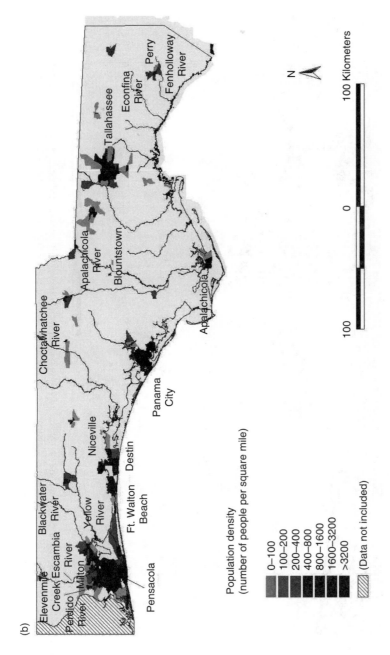

FIGURE 3.1b Population density in the Northwest Florida area of study. This figure is a composite of information provided by the National Oceanic and Atmospheric Administration (NOAA), the Florida Department of Environmental Protection (FDEP), and the ESRI Corporation.

FIGURE 3.1c Distribution of point sources of discharges to coastal waters in the northwest Florida study area. This figure is a composite of information provided by the National Oceanic and Atmospheric Administration (NOAA), the Florida Department of Environmental Protection (FDEP), and the ESRI Corporation.

Destin/Fort Walton Beach/Niceville (western Choctawhatchee Bay), and Panama City (St. Andrews Bay) (Figure 3.1b).

The distribution of sewage treatment facilities, hazardous waste sites, and NPDES permit sites follows closely the distribution of human populations along the northeast Gulf coast (Figure 3.1c). Outstanding sources of pollution loading in the Perdido Bay system include a pulp mill in the upper bay, and agricultural and urban runoff in the lower bay. The highest concentration of point and non-point sources of pollution occurs in the Pensacola Bay system. In recent years, there have been major increases in urban storm water runoff in western Choctawhatchee Bay. Apalachicola Bay and Apalachee Bay remain relatively free of such discharges, and are among the least polluted coastal systems in the conterminous United States. In the Apalachicola system, the single most important pollution source is a sewage treatment plant in Apalachicola that discharges into a creek north of the city. In Apalachee Bay, the primary source of pollution is a pulp mill on the Fenholloway River, with discharges near the inland city of Perry, Florida.

There are significant differences in the development of emergent and submergent vegetation in the subject coastal systems of the northeast Gulf. Perdido Bay and the Pensacola system have moderate concentrations of marshes, whereas the Choctawhatchee system is practically devoid of emergent vegetation (EV). The Apalachicola and Apalachee Bay systems have extensive and well-developed marsh systems. The big alluvial river-bay systems have limited development of submerged aquatic vegetation (SAV), whereas Apalachee Bay is characterized by one of the highest concentrations of seagrass beds in the Northern Hemisphere. The Econfina and Fenholloway estuaries are distinguished from the alluvial river–bay systems to the west by the relatively low flow/watershed and flow/open water ratios, and by the well-developed marsh areas relative to the flow rates of the contributing rivers to Apalachee Bay. The shallowness of Apalachee Bay, together with the relatively low freshwater flows and considerable development of fringing coastal wetlands, contribute to the seagrass beds as the dominant offshore habitat along the Big Bend area of north Florida.

The Perdido Bay system (Figure 3.2) lies in an area of submergence on the north flank of the active Gulf Coast geosyncline, and is a shallow to moderately deep (average depth; 2.2 m) inshore water body oriented along a northeast–southwest axis. A study outline is given in Appendix I. The bay can be divided into four distinct geographic regions: (1) lower Perdido River, (2) upper Perdido Bay (north of the Route 98 bridge), (3) lower Perdido Bay, and (4) the Perdido Pass complex. The bay system has a length of 53.4 km and an average width of 4.2 km. The primary source of freshwater input to the estuary is the Perdido River system that flows southward about 96.5 km, draining an area of about 2937 km^2 (Livingston, 1998b, 2000, 2002). The Elevenmile Creek system (including Elevenmile Creek; Figure 3.2) is a small drainage basin (about 70 km^2) that receives input from a small municipal waste system and a paper mill. The Bayou Marcus Creek drains a residential area of western Pensacola with input from urban storm water runoff. A sewage treatment plant (STP) (recently diverted to adjoining marshes) occupies the area north of the Bayou Marcus Creek. The Gulf Intracoastal Waterway (GIWW) runs through the lower end of Perdido Bay about 5.6 km northeast of Perdido Pass. The U.S. Army Corps of Engineers maintains the Perdido Pass channel at about 4 m as part of the GIWW (U.S. Army Corps of Engineers, 1976). As in the Choctawhatchee Bay system, there is a shelf that can extend up to 400 m in width around the periphery of the Perdido estuary. This shelf usually does not exceed 1 m in depth. The upper bay is shallow, and depth tends to increase southward. The deepest parts of the estuary are located at the mouth of the Perdido River and in the lower bay.

Prior to the opening of Perdido Pass in the early 1900s, Perdido Bay was a largely freshwater system, covered with freshwater plants (Brush, 1991). Access to the Gulf was

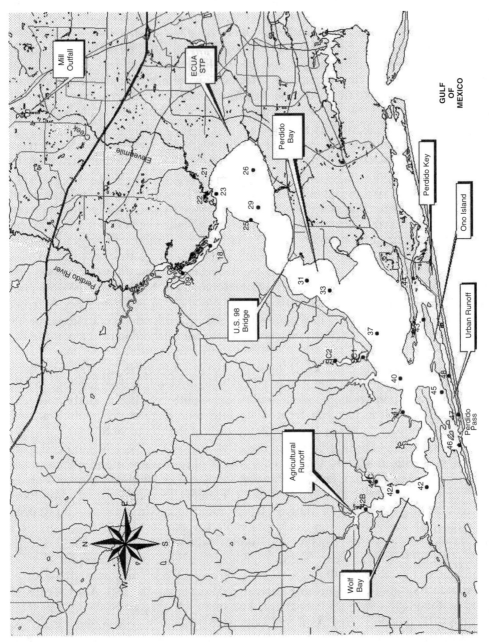

Figure 3.2 Perdido drainage systems and near-shore parts of the Gulf of Mexico with distributions of permanent sampling stations used in the long-term studies of the area. The Florida Geographic Data Library (FGDL) provided geographic data.

restricted by the shallow, shifting body of freshwater. At the time of an early survey (1767), the pass had a depth of approximately 2 m. During an outbreak of malaria in the early 1900s, the mouth of the bay was opened to the Gulf. This transformed the bay into the saltwater system it is today. This action resulted in the creation of the Ono Island/Old River complex. The history of Perdido Bay is thus comparable to that of the Choctawhatchee Bay system (Livingston, 1986a,b, 1989b) that was also opened at the mouth by another group of citizens with shovels in 1929. Since the turn of the century, there has been a steady increase in the human population in the Perdido basin. Problems of sewage treatment and urban storm water runoff remain unresolved in various residential areas around the bay to the present day. Agricultural runoff, based largely in Alabama, contributes to nutrient loading in Wolf Bay at the lower end of the basin (Figure 3.2).

Detailed, long-term analyses of the Perdido drainage system have been carried out for the past 16 years (Livingston, 1992b, 1994, 1995d, 1997b, 1998b, 2000, 2002; Livingston et al., 1998a) that include long-term, bay-wide synoptic analyses at fixed stations (Figure 3.2). According to a U.S. Fish and Wildlife Service Report (1990), SAV was largely concentrated in the lower bay. Historically, SAV has decreased by more than half from 1940–1941 to 1979. Dredging of the Gulf Intracoastal Waterway (GIWW) in the 1930s and continuous pass enlargement and open water spoil disposal have been postulated as factors in the decline of SAV in the lower bay (Bortone, 1991). SAV development has been restricted to Grassy Point on the west side of the upper bay (Figure 3.4) with *Vallisneria americana* as the dominant species. Field/laboratory experiments (Livingston, 1992b) indicated light as the chief limiting factor with grass development not extending deeper than 0.8 to 1.0 m. Based on the success of previous grass bed transplant experiments, Davis et al. (1999) concluded that *V. americana* beds in upper Perdido Bay were recruitment-limited rather than constrained by water quality, toxic substances, light inhibition, or unsuitable substrate. The Perdido Bay system has been adversely affected by pulp mill effluents and overloads from the sewage treatment plant in the upper bay, the dredged opening to the Gulf, and urbanization, and agricultural runoff in the lower bay.

The region along the upper Gulf coast of peninsular Florida from the Ochlockonee River to the Suwannee River has a series of drainage basins that include the Aucilla, Econfina, and Fenholloway Rivers (Figure 3.3). These streams drain into Apalachee Bay, which occupies a broad, shallow shelf along the Gulf coast. The smaller basins are wholly within the coastal plain as part of a poorly drained region that is composed of springs, lakes, ponds, freshwater swamps, and coastal marshes. The Econfina and Fenholloway river estuaries both originate in the San Pedro Swamp (Figure 3.3). A study outline is given in Appendix I. This basin has been affected, in terms of water flow characteristics, by long-term physical modifications through forestry activities. However, most of Apalachee Bay remains in a relatively natural state due to the almost complete lack of human development in the primary drainage basins.

The dominant habitat feature of Apalachee Bay is an extensive series of seagrass beds that extends from Florida Bay in the south to Ochlockonee Bay in the north (Iverson and Bittaker, 1986). The one area of significant anthropogenous effect in an otherwise pristine system is the Fenholloway River estuary where pulp mill discharges have caused adverse effects due to high DOC and water color, high BOD, low dissolved oxygen (DO), and high nutrient loading (ammonia and orthophosphate) (Livingston 1980a, 1982a, 1985a,b, 1988a; Livingston et al., 1998b). The Econfina River remains one of the most natural black water streams along the coast, and has been used as a reference area for studies in the Fenholloway system since 1971. Both the Econfina and Fenholloway drainages (Figure 3.5) share a common meteorological regime. River flow characteristics in the two drainages are comparable in rate and seasonal variation.

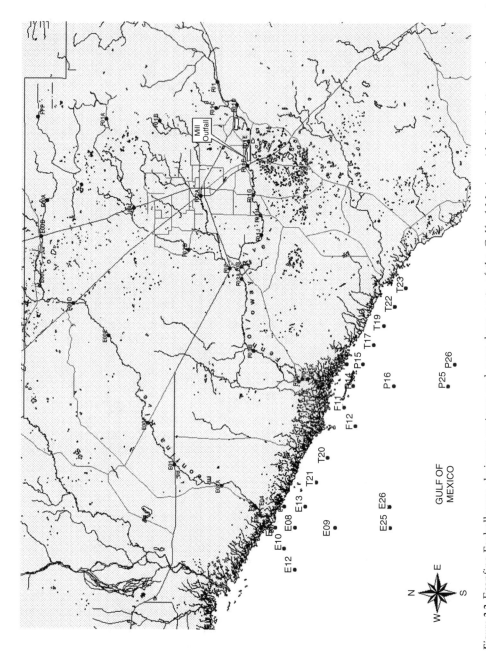

Figure 3.3 Econfina-Fenholloway drainage systems and near-shore parts of the Gulf of Mexico with distributions of permanent sampling stations used in the long-term studies of the area. The Florida Geographic Data Library (FGDL) provided geographic data.

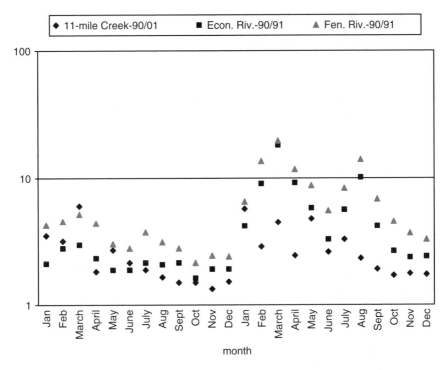

Figure 3.4 River flows of the Econfina and Fenholloway Rivers and Elevenmile Creek (monthly, January 1990–December 1991).

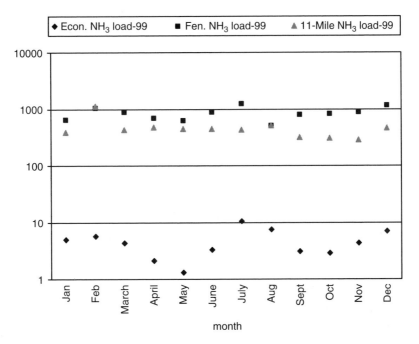

Figure 3.5 Ammonia loading in the Econfina and Fenholloway Rivers and Elevenmile Creek (monthly, January–December 1999).†

Comparative analyses were conducted of various features of the rivers affected by kraft pulp mills (Fenholloway River, Elevenmile Creek) and the reference blackwater stream, the Econfina River. Methods used for the comparison of monthly data (river flows, nutrient loading, water quality) were developed to determine significant differences between matching Amelia and Nassau sites (polluted and unpolluted) over the 12-month study periods (Livingston, 2001a). A detailed explanation of the statistics used is given in Appendix II. For independent, random samples from normally distributed populations, the parametric *t*-test was used to compare the sample means. For cases where one or both of the data sets violated the assumption of normality, a data transformation was made to bring the data into normality. Tests were also developed to compare two serially correlated populations of numbers taken at subject stations by calculating differences of the observations and plotting the autocorrelations (months) of the differences. If differences were not serially correlated, we applied the Wilcoxon sign-rank test to compare (0.05 confidence level) the two sets of numbers.

3.2   River Flows, Nutrient Loading, and Water Quality Changes

River flow rates of the three rivers were comparable (Figure 3.4). However, ammonia loading in the Fenholloway River and Elevenmile Creek (Figure 3.5) were significantly (P < 0.05) higher than that in the Econfina River. Ammonia loading was higher in the Fenholloway River than that in Elevenmile Creek. In both streams affected by pulp mill effluents, ammonia loading was generally high throughout the season, whereas loading in the reference stream generally followed seasonal changes in river flow. Orthophosphate loading (Figure 3.6) was also significantly (P < 0.05) higher in the rivers affected by pulp effluents compared to the Econfina system. However, such loading was significantly

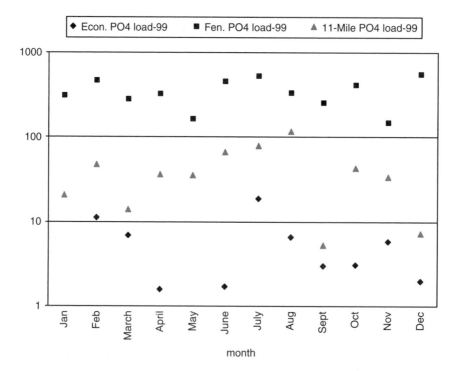

Figure 3.6 Orthophosphate loading in the Econfina and Fenholloway Rivers and Elevenmile Creek (monthly, January–December 1999).

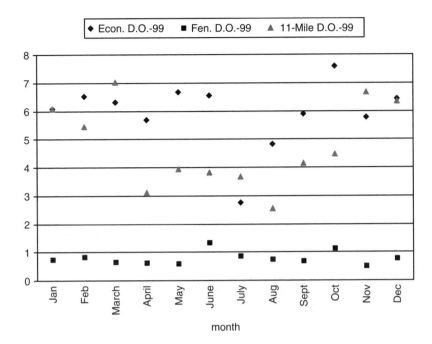

Figure 3.7 DO concentrations in the Econfina and Fenholloway Rivers and Elevenmile Creek (monthly, January–December 1999). Data were taken from Station P13 in Elevenmile Creek, Station E001 in the Econfina River, and Station F01 in the Fenholloway River.

($P < 0.05$) lower in Elevenmile Creek than that in the Fenholloway River. This was due in part to differences in treatment facilities of the mills associated with these rivers. Overall, however, the loading rates of the primary nutrients in streams affected by pulp mill releases were 2 to 3 orders of magnitude greater than such loadings in a natural stream. Orthophosphate concentrations in Elevenmile Creek have been successfully reduced, however, using alum treatments during periods of high loading of this nutrient. Recently, the pulp mill on Elevenmile Creek has been successful in reducing both ammonia and orthophosphate using various methods that include within-mill changes in the treatment system. This restoration effort will be described in depth in a subsequent chapter.

A comparison of DO concentrations in the subject rivers is shown in Figure 3.7. Dissolved oxygen in the Fenholloway River was uniformly low during all months of the year, and was significantly lower ($P < 0.05$) than the DO levels in the Econfina River. In Elevenmile Creek, the DO was uniformly higher than that in the Fenholloway River, but was generally below 4 mg L^{-1} from April through August. The DO concentrations in Elevenmile Creek, although generally lower than those in the Econfina system, were not significantly ($P < 0.05$) different from those in the reference stream. The complex relationships of DO in Elevenmile Creek will be treated more completely below.

Specific conductance was significantly ($P < 0.05$) higher in both contaminated streams than the reference river (Figure 3.8). Total dissolved solids in pulp mill effluents are difficult to restore, and, as noted, both the Fenholloway River and Elevenmile Creek have conductivity levels that usually exceed water quality standards for freshwater systems. In Florida, the requirement for specific conductance is 1275 µmhos cm^{-1}.

Watercolor in the Fenholloway River was significantly higher in the Fenholloway system than color concentrations in the Econfina River (Figure 3.9). However, watercolor in Elevenmile Creek was not significantly different from the reference site as a result of a treatment system that was initiated in the early 1990s. Watercolor and light penetration

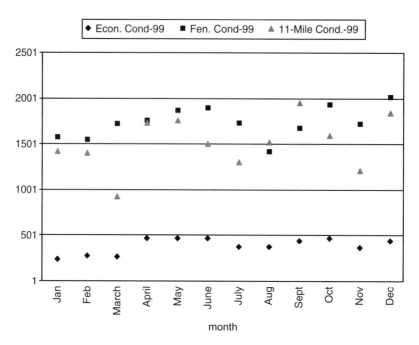

Figure 3.8 Specific conductance in the Econfina and Fenholloway Rivers and Elevenmile Creek (monthly, January–December 1999).

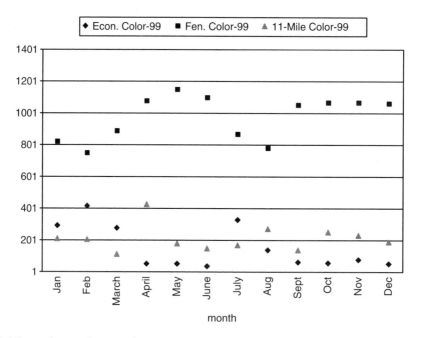

Figure 3.9 Watercolor in the Econfina and Fenholloway Rivers and Elevenmile Creek (monthly, January–December 1999).

in the Perdido Bay system were not affected by mill effluents subsequent to the successful treatment by the paper mill, and microcosm and field experimental analyses showed that submerged aquatic vegetation (*Vallisneria americana*) was not adversely affected by color levels associated with mill effluents (Livingston, 1992b; Davis et al., 1999).

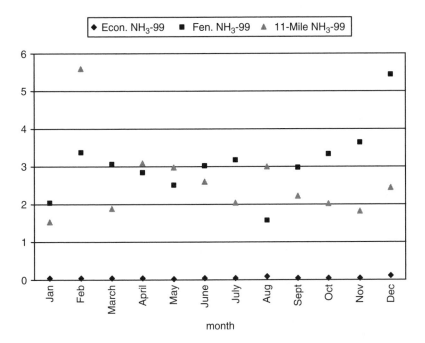

Figure 3.10 Ammonia concentrations in the Econfina and Fenholloway Rivers and Elevenmile Creek (monthly, January–December 1999).

Ammonia concentrations in both of the rivers affected by mill effluents were significantly higher than those in the reference river (Figure 3.10). However, orthophosphate concentrations were reduced in Elevenmile Creek relative to the significantly ($P < 0.05$) higher concentrations of this nutrient in the Fenholloway River (Figure 3.11). The dynamics of orthophosphate chemistry in aquatic systems (proclivity of attachment to fine particulates) relative to those of ammonia (largely dissolved in water) provide the basis for successful treatment of orthophosphate that can reduce loading of this nutrient to background levels. However, this interpretation of orthophosphate levels in areas receiving nutrients from pulp mills will be qualified by changes in phytoplankton bloom populations that will be explained in more detail below.

The water quality data indicate some of the problems associated with the release of pulp mill effluents contaminated with an assortment of chemicals that significantly alter water quality in relatively low flow receiving areas with limited assimilative capacities for such loading. These data also show that some of these factors, such as watercolor and orthophosphate, can be mitigated by treatment and changes in water treatment systems.

3.3   Biological Responses in Freshwater Receiving Areas

Detailed studies were carried out concerning the biological response of the Fenholloway River and Elevenmile Creek to releases of high volumes of pulp mill effluents relative to natural conditions in an unpolluted black water system, the Econfina River (Livingston 1992b, 1993b, 2000, 2002).

3.3.1   Mill Effects on Freshwater Biota: Fenholloway River

3.3.1.1   Periphyton

Periphyton distribution indicated stress in the upper Econfina River and in portions of the Fenholloway River above the mill outfall (Livingston, 1993b). The highest cumulative

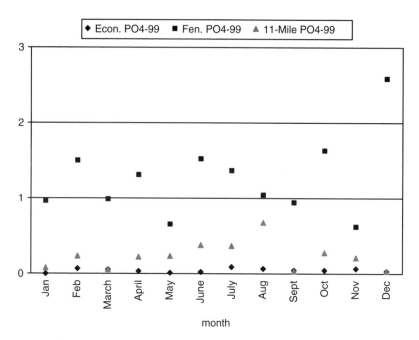

Figure 3.11 Orthophosphate concentrations in the Econfina and Fenholloway Rivers and Elevenmile Creek (monthly, January–December 1999).

numbers of periphyton were noted at Econfina stations E00D, E00F, and E01 (Figure 3.3). The periphyton data indicated stress in areas of the Fenholloway River that were associated with the mill outfall. In the Fenholloway system, the highest numbers were found in Spring Creek and at the mill outfall. The lowest annual totals were found in the middle and lower reaches of the Fenholloway River. Seasonal distributions were most evenly spread out at stations E01 and F01F (Figure 3.3). The periphyton species richness of the Fenholloway River was considerably lower than that of the Econfina River or Spring Creek. These data are consistent with the qualitative data that showed stress in the Fenholloway River below the outfall station. The various diversity and evenness indices followed this pattern closely, which would indicate that both the species richness component and the evenness components of the periphyton reflected adverse effects due to mill effluents. There were thus indications that there was enrichment at the outfall station with subsequent deterioration of the periphyton numbers farther downstream in the Fenholloway River.

3.3.1.2   Hester–Dendy Macroinvertebrates

Macroinvertebrate data from the Fenholloway River indicated that conditions caused by the mill effluent (high summer temperatures, low DO, high conductivity, initially high ammonia, high nutrients, and high dissolved organic carbon (DOC) were associated with the dominance of pollution-tolerant species such as the chironomid insects (Livingston, 1993b). These organisms are well adapted for hypoxic conditions and increased stress due to the high conductivity and periodically high ammonia. Other indicator species such as the crustacean *Asellus* and the gastropod mollusk *Physella* sp. were also found in areas affected by mill effluents. There was recovery along the lower Fenholloway River with a return to species noted in the various unpolluted reference sites. The lowest numbers of species were found at the mill outfall station as an area of high impact. Low numbers of species taken during the first 6 months could also have been associated with the relatively high ammonia levels during this period. Increased species richness farther downstream in the Fenholloway River indicated recovery with distance from the mill outfall.

3.3.1.3 Suction Dredge Macroinvertebrates

Due to the high volumes of water discharged and the loading of organic matter (particulate organic carbon [POC], dissolved organic carbon) and nutrients to the receiving system, a reduction in DO (below the ambient condition) is often a secondary impact of pulp mill effluents (Livingston, 1975a, unpublished data). In freshwater systems, the impact of mill discharge is often dependent on the assimilative capacity of the receiving water body. Variables to consider in the impact of the discharge on a freshwater system depend on water temperature, current/turbulence characteristics, habitat type, and existing biological conditions. Pearson (1980) and Pearson and Rosenberg (1978) have reported on the effects of disturbances due to pulp effluents on marine biota. These authors showed that combinations of changes in sediment characteristics (eH, organic content) and hypoxic conditions in the overlying waters can cause changes in the infaunal macroinvertebrates that are reflected in altered species composition, biomass, and community characteristics.

In the upper parts of the Econfina and Fenholloway Rivers, there was no clear trend of differences in the various community indices derived from the suction dredge macroinvertebrates (Livingston, 1993b). Numbers of animals and species richness were generally low in both rivers. Data from the mill outfall station (F01F; Figure 3.7) indicated relatively high numbers, low species richness, and low species diversity. These trends occurred throughout the sampling period and were noted farther downstream (F03, F04, F05, F05A) with some differences. In the estuarine areas (E04, F05, E05, F05A), there were often higher numbers in the Econfina system or no real difference between the two river biotas. Species richness and diversity indices, however, were often lower in the upper portions of the Fenholloway estuary. In the lower estuaries (E06, F06), there were generally higher numbers of individuals and species in the Econfina system.

Statistical analyses of the data indicated no significant trends of numbers of organisms with the various physicochemical factors. However, macroinvertebrate species richness was significantly correlated with surface and bottom DO concentrations ($R^2 = 0.13$, P = 0.0001). The data indicated that low DO in the Fenholloway River-estuary was an important determinant of the macroinvertebrate community structure compared with the reference system. Because conductivity and ammonia, the other known forms of stress in the Fenholloway River, showed longitudinal trends in this system, the close and continuous association of hypoxia with biotic trends was significant.

Overall, the river–estuarine suction dredge data indicated that species richness and diversity can reach a "natural" level at DO concentrations approximating 3 mg L^{-1} or more. In a system characterized by low levels of DO and high organic carbon loading, a few species can adapt to such conditions and are therefore released in terms of population increases so that low species richness and diversity are accompanied by high numbers in the Fenholloway River estuary. There were indications that reduced species richness and diversity in the Fenholloway were due to other factors in addition to hypoxic conditions. The data indicated that 24-hour means between 3 and 4 mg L^{-1} with minima between 2 and 3 mg L^{-1} are recommended targets for maintaining viable macroinvertebrate assemblages.

3.3.1.4 Fishes

Freshwater fishes were low in numbers of individuals and species richness in the Fenholloway River compared to the Econfina River (Table 3.1). Summer–fall periods of high temperature and low DO were typified by generally low numbers of fish species in the Fenholloway River. The outfall area was not always characterized by hypoxia, and was

Table 3.1 Fishes Taken in the Econfina and Fenholloway Rivers (monthly from 1992 to 1993)

Station	Yymm	TotalN	Nbrtaxa	GAMAFFn	GAMAFF%	station	Yymm	TotalN	Nbrtaxa	GAMAFFn	GAMAFF%
E00D	92/04	8	5	3	37.5	F01D	92/09	0	0	0	0
E00D	92/05	7	2	0	0	F01D	92/10	1	1	1	100
E00D	92/06	10	2	1	10	F01D	93/02	0	0	0	0
E00D	92/08	56	4	37	66	F01D	92/08	nd	nd	nd	nd
E00D	92/11	194	12	146	75.2	F01D	92/11	nd	nd	nd	nd
E00D	92/12	50	6	28	56	F01D	92/12	nd	nd	nd	nd
E00D	93/02	71	2	70	98.5	F01D	93/02	nd	nd	nd	nd
E00F	92/04	0	0	0	0	F01F	92/04	0	0	0	0
E00F	92/05	1	1	1	100	F01F	92/05	24	3	19	79.1
E00F	92/06	0	0	0	0	F01F	92/06	38	3	36	94.7
E00F	92/08	13	2	2	15.3	F01F	92/08	6	3	4	66.6
E00F	92/09	2	2	1	50	F01F	92/09	2	1	0	0
E00F	92/10	0	0	0	0	F01F	92/10	7	3	5	71.4
E00F	92/11	0	0	0	0	F01F	92/11	3	1	0	0
E00F	92/12	0	0	0	0	F01F	92/12	3	1	0	0
E00F	93/01	1	1	0	0	F01F	93/01	25	4	22	88
E00F	93/02	49	2	48	97.9	F01F	93/02	22	1	22	100
E00F	93/03	6	4	0	0	F01F	93/03	nd	nd	nd	nd
E01	92/04	148	8	133	89.8	F02	92/04	7	1	7	100
E01	92/05	27	2	24	88.8	F02	92/05	0	0	0	0

E01	92/06	70	2	68	97.1	F02	92/06	0	0	0	0
E01	92/08	176	8	54	30.6	F02	92/08	19	3	6	31.5
E01	92/09	31	5	2	6.4	F02	92/09	7	2	0	0
E01	92/10	36	3	30	83.3	F02	92/10	5	1	0	0
E01	92/11	17	3	13	76.4	F02	92/11	0	0	0	0
E01	92/12	6	3	1	16.6	F02	92/12	9	5	1	11.1
E01	93/01	7	3	5	71.4	F02	93/01	7	2	2	28.5
E01	93/02	74	6	67	90.5	F02	93/02	2	1	2	100
E01	93/03	127	4	124	97.6	F02	93/03	2	1	2	100
	Average	37.9	2.9	20.8	40.1		Average				
E01A	92/04	0	0	0	0	F03	92/04	nd	nd	nd	nd
E01A	92/05	4	1	0	0	F03	92/05	2	2	0	0
E01A	92/06	4	3	2	50	F03	92/06	4	1	4	100
E01A	92/08	38	5	12	31.5	F03	92/08	0	0	0	0
E01A	92/09	17	3	3	17.6	F03	92/09	1	1	1	100
E01A	92/10	9	3	7	77.7	F03	92/10	nd	nd	nd	nd
E01A	92/11	9	4	3	33.3	F03	92/11	12	1	12	100
E01A	92/12	9	3	4	44.4	F03	92/12	31	1	31	100
E01A	93/01	10	4	5	50	F03	93/01	32	6	25	78.1
E01A	93/02	8	1	8	100	F03	93/02	nd	nd	nd	nd
E01A	93/03	1	1	0	0	F03	93/03	nd	nd	nd	nd
							Average	8.2	1.5	6.1	43.9

Note: nd = no data.

enriched by the mill effluents: high numbers and high species richness of fishes during late spring and early summer were found here (Livingston, 1993b). However, based on the cumulative species richness index, the Econfina River stations had the highest total numbers of fish species. The progressive deterioration along the Fenholloway River was probably due to the distribution of hypoxia in this system, with increasingly low DO downstream from the outfall. Stations along the lower parts of the Fenholloway River were hypoxic during most months of the year. The initially high ammonia levels and the high conductivity from the mill effluents also were implicated in the generally reduced numbers and low species richness in the Fenholloway River.

It should be noted that the mosquitofish (GAMAFF: *Gambusia affinis*) was a major dominant in the Fenholloway River. This live-bearing topminnow is able to live under hypoxic conditions through a series of adaptations.

A comparison of marsh fishes taken in the Econfina and Fenholloway systems is given in Table 3.2 (Livingston, 1975a). There was an overall reduction of both numbers and species richness in the marsh fish associations of the Fenholloway system relative to the Econfina reference area.

3.3.1.5 Phytoplankton

The blue-green species *Merismopedia tenuissima* did not occur in Perdido Bay until the enhanced orthophosphate and ammonia loading in 1996 (Figure 3.12). The primary concentrations of this alga occurred mainly in Elevenmile Creek (Station P22) although some concentrations of *Merismopedia* were noted at the mouth of the creek (Station P23). *Merismopedia tenuissima* occurred all year long in Elevenmile Creek, with peak numbers during December and January. This species appeared to have increased numbers during periods of low creek flow. The drought stimulated this species to bloom levels with high nutrient concentrations as an added stimulus for the blooms. High salinity and silica and low orthophosphate and DO were significantly associated with *Merismopedia tenuissima* numbers of cells L^{-1} (Livingston, 2002). Numbers in the creek peaked during 2002 and there was a reduction of *Merismopedia* during 2003 and 2004. Increased numbers of *Merismopedia* coincided with the drought of 1999 and 2002 that broke during 2003. This pattern was also evident at Station P23 in the bay.

Livingston (1992b) indicated that infaunal macroinvertebrates in Elevenmile Creek were dominated by tubificid (oligochaete) worms such as *Limnodrilus hoffmeisteri*, naidid worms such as *Dero* spp., and chironomid insects (*Chironomus decorus* group, *Goelldichironomus* spp.). These species are well adapted to the high levels of organic matter, hypoxic conditions, and rigorous physical habitat of the creek. Peak numbers of the dominants occurred in the upper creek near the outfall where the concentrations of organic matter in the water and sediments were highest. The numbers of infaunal organisms tended to fall off quickly farther downstream. There was a gradual decrease of infaunal species richness along the creek from the upper stations to mid portions of the stream. At the end of the creek, numbers of individuals and species were extremely low, indicating stressful conditions in the deeper parts of the estuarine parts of the creek. These effects were associated with low DO. Long-term trends (Figure 3.13 and Figure 3.14) indicated a precipitous decrease of infaunal numbers and species richness from 1996 to 1999 that corresponded with increased numbers of *Merismopedia tenuissima*. During late 2003 and 2004, there was a general increase in both indices that again corresponded to decreases of the blue-green blooms. Trends in the bay (Station P23) were somewhat different. There was a general decrease in infaunal numbers and species richness from 1992 to 1999 that followed trends of phytoplankton blooms in the bay (see below). The influence of the blue-green blooms did not appear to extend into the bay.

Table 3.2 Fishes Taken in the Econfina and Fenholloway Saltwater Marshes (monthly from 1974 to 1976)

Date	TotalN-E4a	Nbrtaxa-E4a	Date	TotalN-F5b	Nbrtaxa-F5b	Date	TotalN-E6a	Nbrtaxa-E6a	Date	TotalN-F5b	Nbrtaxa-F5b
74/01	17	3	74/01	0	0	74/01	15	6	74/01	0	0
74/02	107	4	74/02	0	0	74/02	229	10	74/02	0	0
74/03	0	0	74/03	1	1	74/03	36	3	74/03	1	1
74/04	4	4	74/04	5	3	74/04	nd	nd	74/04	5	3
74/05	19	3	74/05	8	2	74/05	11	2	74/05	8	2
74/06	125	5	74/06	8	1	74/06	21	4	74/06	8	1
74/07	27	8	74/07	0	0	74/07	26	6	74/07	0	0
74/08	18	3	74/08	6	3	74/08	11	2	74/08	6	3
74/09	nd		74/09	4	2	74/09	nd	nd	74/09	4	2
74/10	27	6	74/10	11	1	74/10	15	3	74/10	11	1
74/11	22	6	74/11	3	2	74/11	82	7	74/11	3	2
74/12	8	5	74/12	6	4	74/12	0	0	74/12	6	4
75/01	6	3	75/01	nd	nd	75/01	56	4	75/01	nd	nd
75/02	nd	nd	75/02	nd	nd	75/02	nd	nd	75/02	nd	nd
75/03	25	4	75/03	nd	nd	75/03	5	2	75/03	nd	nd
75/04	2	2	75/04	1	1	75/04	46	5	75/04	1	1
75/05	15	5	75/05	21	6	75/05	36	8	75/05	21	6
75/06	26	4	75/06	5	2	75/06	7	2	75/06	5	2
75/07	nd	nd	75/07	4	2	75/07	32	3	75/07	4	2
75/08	nd	nd	75/08	4	3	75/08	10	3	75/08	4	3
75/09	nd	nd	75/09	nd	nd	75/09	3	3	75/09	nd	nd
75/10	44	7	75/10	4	2	75/10	41	4	75/10	4	2
75/11	12	4	75/11	1	1	75/11	nd	nd	75/11	1	1
75/12	1	1	75/12	2	2	75/12	1	1	75/12	2	2
76/01	44	2	76/01	0	0	76/01	nd	nd	76/01	0	0

Note: nd = no data.

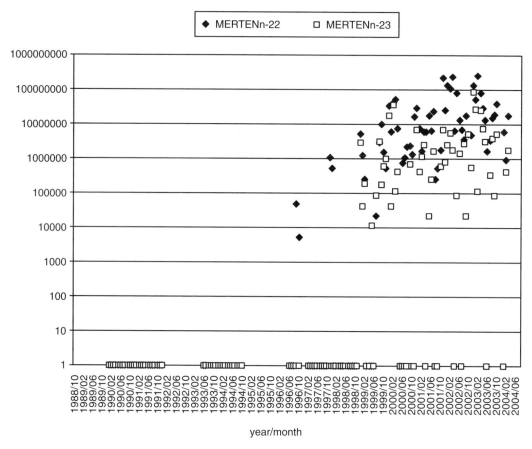

Figure 3.12 Numbers of cells per liter of the blue-green alga *Merismopedia tenuissima* at Stations P22 and P23 taken monthly over the 15-year study period.

Results of the principal components analysis (PCA)-regression analyses of Elevenmile Creek (Figure 3.15) indicated that *Merismopedia tenuissima* numbers were negatively associated with nutrient loading and nutrient ratio indices in Elevenmile Creek (Station 22), but were positively associated with drought conditions. This somewhat counterintuitive result indicated that this bloom species responded to habitat conditions associated with the drought rather than nutrient loading *per se*. Stabilization of creek flows together with high concentrations of orthophosphate and ammonia was apparently important for the initiation of the *Merismopedia* blooms that started to dissipate after a return of increased rainfall and creek flows. *Merismopedia* was positively associated with salinity and negatively associated with DO and orthophosphate concentrations, thus adding credence to the above hypothesis. *Merismopedia* numbers were positively associated with total phytoplankton abundance and species richness. However, this species was negatively associated with a full range of biological variables, indicating adverse impacts on the biota of the creek. The adverse impacts of this blue-green alga appeared to be restricted to Elevenmile Creek.

Livingston (1992b) found that Elevenmile Creek was characterized by high conductivity (which eliminated various primary freshwater fishes and sensitive macroinvertebrates), high levels of free ammonia (which were associated by the U.S. EPA criteria with chronic toxicity), and high concentrations of organic carbon (which was stimulatory to

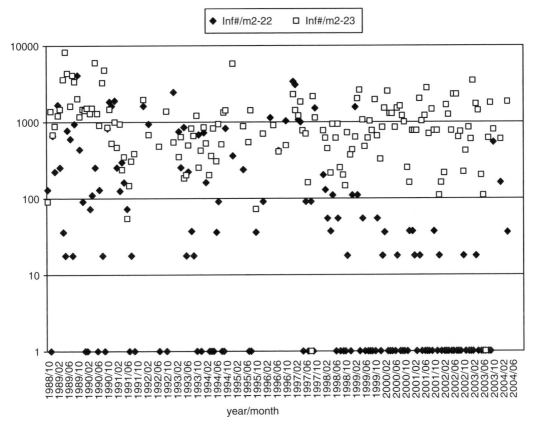

Figure 3.13 Numbers (per square meter) of infaunal macroinvertebrates at Stations P22 and P23 taken monthly over the study period.

resistant chironomids and naidid oligochaetes that were found in high numbers in the upper creek). Upper Elevenmile Creek was characterized by an almost complete lack of primary freshwater fishes (cyprinids, cyprinodonts, percids, and atherinids). On the other hand, secondary freshwater fishes (capable of withstanding high conductivity) such as the lepomids (*Lepomis macrochirus, L. microlophus, Micropterus salmoides)* and various cat-fishes (*Ameiurus natalis, Ictalurus* sp.) were present in considerable numbers at the outfall station, along with estuarine types such as *Mugil cephalus* and fishes that were resistant to organic loading (*Dorosoma cepedianum*). These fishes were subject to periodic (summer) disease, but were drawn by the rich source of macroinvertebrate food that was a direct response to the release of high levels of organic carbon. The incidence of high levels of free ammonia usually preceded or was contemporaneous with the observations of dis-eased fishes at the outfall station.

Farther downstream in Elevenmile Creek, the fish biota changed, with low numbers of fishes and high dominance by secondary freshwater types such as the mosquitofish (*Gambusia affinis, G. holbrooki*). The distribution of fishes in Elevenmile Creek was consistent with the various microhabitat conditions at each station and the changing water quality along the length of the stream. The trophic organization of Elevenmile Creek was thus altered as a direct result of the organic loading at the outfall site.

These data indicate that the effects of pulp mill effluents on small creeks with limited assimilative capacities show a range of adverse impacts that are affected by drought–flood conditions and secondary effects associated with phytoplankton blooms.

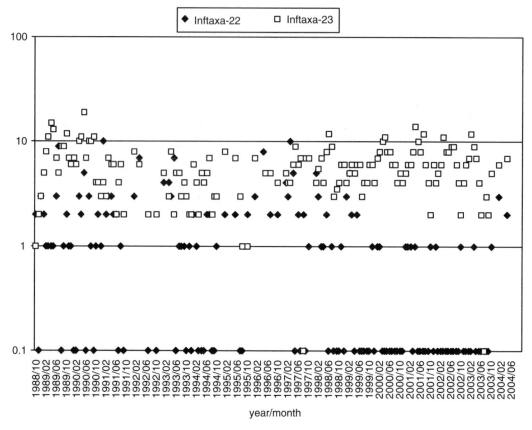

Figure 3.14 Species richness of infaunal macroinvertebrates at Stations P22 and P23 taken monthly over the study period.

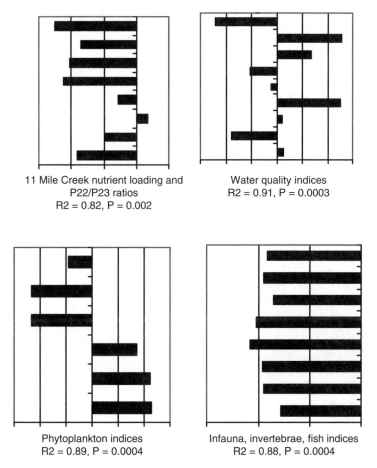

Figure 3.15 PCA-regression analysis of *Merismopedia tenuissima* numbers per liter against nutrient loading and nutrient ratios, water quality indices, phytoplankton community indices, and infaunal, invertebrate and fish indices taken at Station P22 (Elevenmile Creek) monthly over the study period (1988–2004). Only significant associations are shown for eigenvector values.

chapter 4

Pulp Mill Effluents
and Apalachee Bay

4.1   Study Area

Apalachee Bay (Figure 3.3) is located along the Florida Panhandle known as the Big Bend area. The bay is characterized by the absence of barrier islands as a result of the shallow, sloping margins; the lack of wave action; and an inadequate supply of sand (Tanner, 1960). Apalachee Bay is lined by a series of drainage basins that include the Aucilla, Econfina, and Fenholloway Rivers (see Figure 3.3). The smaller basins are wholly within the coastal plain as part of a poorly drained region characterized by springs, lakes, ponds, freshwater swamps, and coastal marshes. The Econfina and Fenholloway River estuaries both originate in the San Pedro Swamp, a basin that has been affected, in terms of water flow characteristics, by long-term physical modifications through forestry activities. The small streams along the Big Bend area are associated with relatively limited estuarine areas. The coastal wetlands are highly developed in the saltwater marshes of the Apalachee Bay region, and Apalachee Bay estuaries have very high wetlands/open water ratios compared to the alluvial estuaries (Livingston, 2000).

The extensive seagrass beds of the region, stretching from north Florida to the Florida Keys, are the dominant ecological feature of the Apalachee Bay system in terms of primary production and habitat distribution. Accordingly, the inshore marine habitat of Apalachee Bay serves as a physically stressed but highly productive nursery for developing stages of offshore forms, many of which form the basis of the sports and commercial species in this area. Phytoplankton productivity is relatively low in the largely oligotrophic offshore parts of Apalachee Bay, The highly complex but extremely productive food webs of the inshore marine systems of this area are thus largely associated with the extensive development of seagrass beds (Livingston, 2002).

The Apalachee Bay Study extends from 1970 to 2004 (see Appendix I). The Fenholloway River estuary represents the one area of significant anthropogenous effects in an otherwise pristine system. Pulp mill discharges in the Fenholloway River have caused adverse effects due to high dissolved organic carbon (DOC) and watercolor, high Biochemical Oxygen Demand (BOD), low dissolved oxygen (DO), and high nutrient loading (ammonia and orthophosphate) (Chapter 2; Livingston 1980a, 1982a, 1984a, 1985a, 1988, Livingston et al., 1998a). The Econfina River (see Figure 3.3) remains one of the most pristine blackwater streams in the United Sates, and provides an excellent reference area for studies in the Fenholloway River and Elevenmile Creek systems. Both the Econfina and Fenholloway drainages, sharing a common origin (San Pedro Swamp), have similar dimensions and comparable flow rates (Livingston, 2002). The reference Econfina system is in relatively pristine condition, with virtually no known anthropogenous impacts (Livingston, 2000).

4.2   Impact Analyses

The study area, with a full complement of sampling stations, is shown in Figure 3.3. Comprehensive field analyses, along with a series of graduate student programs, were carried out from 1970 through 1984 (Appendix I). A field and laboratory program was performed from 1991 through 1993, and was continued through a subsequent monitoring program from 1993 through June 2004. A detailed evaluation of a color removal program by the pulp mill in 1998 was carried out from fall 1998 through June 2004.

In addition to the primary variables affecting light transmission, the study also showed that nutrient loading and sediment characteristics in the Fenholloway near-shore area could affect seagrass growth. After a reduction of color in the treated wastewater by 50% in October 1998, it remained uncertain how nutrient loading (i.e., ammonia, orthophosphate) from the Foley mill would affect phytoplankton growth and possible bloom formation (Livingston, 2000, 2002). It was also necessary to understand whether sediment quality played a role in seagrass growth. The direct loading of nitrogen and phosphorus compounds to an oligotrophic system dominated by seagrasses was considered problematic in the sense that phytoplankton productivity changes are connected to the seagrass problem both in terms of altered light transmission and associated effects on water and sediment quality. There was a possibility that reduced color and increased light penetration in the Fenholloway estuary could result in the production of damaging plankton blooms similar to those observed in the Perdido Bay and Pensacola Bay systems (Livingston, 2000, 2002). Improvements in watercolor thus presented some concern in terms of the nutrient loading issue and possible near-shore phytoplankton response.

Shallow offshore habitats of Apalachee Bay along the northeast Gulf coast of Florida are dominated by seagrasses such as *Thalassia testudinum, Syringodium filiforme,* and *Halodule wrightii* (Iverson and Bittaker, 1986; Zimmerman and Livingston, 1976a,b, 1979). These seagrasses remain relatively undisturbed by human activity, with the exception of one area off the Fenholloway drainage where the release of pulp mill effluents has adversely affected the near-shore beds. Livingston et al. (1998a) showed that pulp mill effluents in the Fenholloway River were associated with increased loading of DOC, watercolor, and nutrients to offshore areas relative to the unpolluted (Econfina River) reference system. This loading resulted in changes of water quality factors and light transmission characteristics in the polluted system.

Sediments in offshore Fenholloway areas were characterized by increased silt/clay fractions and altered particle size relative to reference sites. Field analyses showed that the best predictors of submerged aquatic vegetation (SAV) distribution in the Fenholloway offshore areas were photic depths, qualitative aspects of wavelength distributions (light transmission data), and water quality factors such as color, DOC, and chlorophyll *a* that were inversely related to seagrass biomass. Experimental analyses (Livingston et al., 1998a) using mesocosms of dominant seagrasses (*Thalassia testudinum, Syringodium filiforme, and Halodule wrightii)* indicated that direct contact with pulp mill effluents resulted in adverse growth responses at concentrations as low as 1 to 2%. Mesocosm experiments with light conditions that reproduced actual field conditions showed that light conditions in offshore areas affected by pulp mill effluents adversely affected the growth indices of *S. filiforme* and *H. wrightii*. Field transfer experiments showed that the growth indices of the three dominant seagrass species were adversely affected by water quality and sediment quality in the Fenholloway offshore receiving area (Livingston et al., 1998a).

Apalachee Bay is an oligotrophic system with relatively low phytoplankton productivity compared to the alluvial systems to the west (Livingston, 1993b, 2000, 2002). Nutrient input is restricted to small stream loading. Nutrient limitation experiments in the Econfina and Fenholloway systems indicated a complex set of nutrient–phytoplankton interactions (Livingston, 1993b, unpublished data). Inshore areas of both systems were generally limited by nitrogen, whereas areas farther offshore were increasingly phosphorus-limited. Relatively low TIN/TIP ratios in inshore areas during the experimental period also suggested nitrogen-limitation. Farther offshore, higher TIN/TIP ratios indicated increased phosphorus-limitation. Estuarine phytoplankton populations in the Econfina system were more abundant than those in comparable areas of the Fenholloway system (Livingston, 2002; unpublished data). Estuarine and inshore parts of the Fenholloway system were low in phytoplankton species richness and diversity compared to the Econfina cognate stations. The highest numbers of phytoplankton were found in inshore portions of the Econfina and Fenholloway systems. Fenholloway offshore areas had higher numbers of phytoplankton than comparable Econfina areas. The suppression of phytoplankton species richness and diversity in the inshore Fenholloway system was most likely due to reduced light transmission, although other effects such as inhibition by high concentrations of ammonia (Livingston, unpublished data) could not be ruled out as an additional modifying factor.

Various studies (Hooks et al., 1975; Greening, 1980; Greening and Livingston, 1982; Dugan and Livingston, 1982) indicated that invertebrates taken in the Econfina offshore system were generally more abundant than those found in the Fenholloway system, and that macrophyte abundance was directly associated with invertebrate numbers. Livingston (1975a) reported reduced fish species richness and diversity in the offshore Fenholloway system. These differences corresponded to previously described distributions of SAV and macroinvertebrates. The effects on fishes were also attributed to physiological impacts (respiration, growth, food conversion efficiency; Stoner 1976) and/or behavioral responses such as avoidance of the mill effluents (Lewis and Livingston, 1977; Livingston et al., 1976a), although the habitat changes due to the loss of seagrass beds, together with changes in trophic organization due to altered nutrient loading in the Fenholloway system, were the primary features associated with the loss of secondary production in areas associated with pulp mill effluent disposal.

Changes in the Fenholloway seagrass beds were associated with altered food web patterns in terms of species shifts and feeding alterations relative to reference areas (Livingston, 1975a, 1980a, 2002). Feeding habits of the dominant fish species were different in the offshore Fenholloway system than those found in similar areas of the Econfina drainage. These differences were traced to basic changes in habitat due to reduced seagrass beds in the Fenholloway system (Livingston, 2002; Clements and Livingston, 1983, 1984). Reduced SAV, together with increased phytoplankton productivity due to nutrients from the paper mill, appeared to alter the habitat conditions and fish trophic organization in the Fenholloway offshore area relative to the reference system. These effects were also associated with the complex effects of sediment quality on population distribution and community organization.

Improvement of water quality in 1974 due to construction of treatment ponds by the pulp mill was associated with some recovery of SAV and with recovery of the fish trophic organization in outer areas of the Fenholloway system (Livingston, 1982a, 1984a, 2002). During the earlier period of untreated mill effluents, plankton-feeding fishes were dominant in areas affected by discharges, replacing grass bed species. The water quality improvement was accompanied by trophic shifts that followed habitat changes in the outer portions of the Fenholloway estuary. Areas affected by pulp mill effluents were characterized by altered food webs compared to the reference areas; these changes were directly influenced by shifts of primary producers and altered benthic habitats as seagrasses were turned into unvegetated soft-sediment areas. In offshore Fenholloway areas, these food web shifts were also affected by increased phytoplankton production as dilution enhanced light transmission in areas of increased nutrient concentrations. The benthic food web species were thus replaced in part by planktivorous fish species in the Fenholloway system due to habitat-related food web changes (Livingston, 2002).

Different studies showed the interrelationships of seagrass beds with the offshore biological organization (Livingston, 1984a). This included the effects of pulp mill effluents on behavioral responses (Main, 1983); physiological responses of pinfish (*Lagodon rhomboides*) (Stoner and Livingston, 1984); avoidance reactions of fishes (Lewis and Livingston, 1977, Livingston, 1982a); changes in organism distribution (Hooks et al., 1975; Dugan and Livingston, 1982; Stoner, 1979a,b, 1980a,b,c); ontogenetic (developmental) processes (Livingston, 1980a); trophic organization (Greening and Livingston, 1982; Clements and Livingston, 1983, 1984; Leber, 1983, 1985; Livingston, 1982a, 1984a; Stoner, 1980a,b, Stoner and Livingston, 1984; Stoner et al., 1982); community composition (Livingston, 1975a; Leber, 1983; Lewis and Stoner, 1983; Livingston, 1988); and long-term changes of water quality trends (Livingston, 1975a, 1982a). These studies indicated that offshore Gulf areas were characterized by extremely complex biological associations that were influenced by both physicochemical phenomena (such as pollution and natural background phenomena) and natural biological processes. Relatively small changes in critical habitat features had measurable impacts on the biological organization of the Apalachee Bay system.

The U.S. Environmental Protection Agency (EPA) (2000) completed a nutrient study of the Econfina and Fenholloway systems. Objectives of the study included assessment of phytoplankton blooms in the Fenholloway near-shore areas, determination of relationships of color and phytoplankton biomass in the two study areas, analysis of nutrient loading and color contributions from point sources, determination of light profiles, and analysis of dissolved oxygen concentrations in the study areas. Results of the EPA study indicated that chlorophyll *a* concentrations were low in both estuaries, with maximal concentrations noted at 10 µg L^{-1} in the Fenholloway estuary. Results from the Marine Algal Growth Potential Test (AGPT) were higher in the Fenholloway system, but were greater than 10 mg L^{-1} only in June 1999. Nutrient addition experiments indicated that most of the samples in both estuaries were nitrogen-limited. The EPA study indicated that there were net releases of nutrients from the Fenholloway estuarine sediments, whereas there were net fluxes to the sediments in the Econfina system.

4.3   Water Quality

A review of the long-term changes in the Apalachee Bay area (Livingston, 2004a) indicated complex changes in the Fenholloway system due to altered discharges of color combined with high nutrient loading from the pulp mill. These changes were dependent on long-term drought–flood cycles that dominated the estuarine water quality in both the Econfina and Fenholloway drainages.

4.3.1   Rainfall and River Flow

The importance of riverine input to coastal processes has been reviewed by Livingston (2000, 2002). Analysis of Econfina and Fenholloway River flow trends (Livingston, 2004a) indicated long-term peak flows occurred during 1973, 1984, and 1997 at intervals between 11 and 13 years. Long-term flow rates in the Fenholloway system were characterized by a series of drought periods (1974–1975, 1977–1978, 1981–1982, 1984–1985, 1989–1990, 1992–1993, and 1999–2002). The study during 1992–1993 was carried out during a moderate

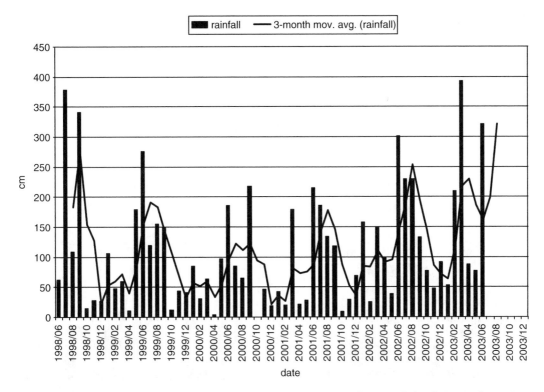

Figure 4.1 Rainfall in the Econfina–Fenholloway drainage system by month (with 3-month moving average) from 1988 to 2003.

drought, whereas the most recent study period (October 1998 to June 2004) represented a period major (century) drought from 1999–2002 (Figure 4.1). The drought broke during 2002–2003, and the changes in rainfall and flows to the Gulf in the study areas had a major effect on water quality, which was then reflected in the biological responses of the mill-affected area relative to the reference site.

4.3.2 Temperature and Salinity

A comparison of seasonal temperature differences between the periods 1971–1980 and 1992–2004 (Livingston, 2004a) indicated that, during the latter period of study, winter temperatures did not go below 13°C, whereas winter temperatures routinely went below this level during the 1970s. This trend was not evident during spring periods. During summer periods from 1992 to 2004, water temperature did not go below 26°C; during the 1970s, water temperature routinely went below such levels. During fall periods, this pattern of relatively higher temperature during the latter study periods was also evident. Data taken from representative areas in the Econfina and Fenholloway systems during drought periods indicated that average temperatures increased between 1992–1993 and 1999–2000 in both areas of study. The greatest ranges of such changes in temperature occurred during fall–winter periods. The lowest temperatures in both systems were noted during drought periods previous to the latest drought. Average salinities were higher during the 1999–2000 period in the Econfina system and during the 1974–1975 period in the Fenholloway system. The lowest salinities in both systems usually occurred during August, with the highest salinities during fall months. These analyses suggest that both temperature and salinity were highest during the latest drought (1999–2002).

4.3.3   Dissolved Oxygen

Livingston (2004a) noted hypoxic conditions in the Fenholloway River during most months compared to relatively high concentrations of DO in the Econfina River (see Chapter 2). During warm months, hypoxic conditions intensify in the Fenholloway estuary and DO levels go below 4 mg L^{-1} (Figure 4.2). Surface DO usually remains higher in the

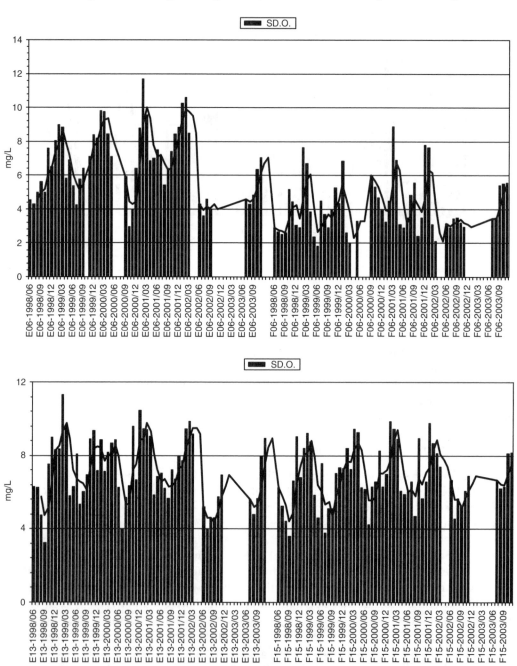

Figure 4.2 Dissolved oxygen (DO) in the Econfina and Fenholloway estuaries and offshore areas from 1988 to 2003.

Econfina estuary, with levels usually remaining at or near 4 mg L⁻¹. Offshore areas of the Econfina and Fenholloway systems remain well oxygenated during most times of the year, with levels remaining above 4 mg L⁻¹ (Figure 4.2). DO is not a limiting factor in offshore parts of the Econfina and Fenholloway drainages.

4.3.4   Watercolor and Light Transmission

Previous studies (Livingston et al., 1998a) indicated that watercolor and DOC were high in the Fenholloway estuary and near-shore area (Stations F06, F09, F10, F11, and F14) relative to the Econfina system. Both factors tended to become equivalent between Stations F10/E08 and F16/E09. Comparative color levels were found somewhere around 3 km offshore. Light transmission trends tended to follow these distributions. Extinction coefficients were higher in the Fenholloway estuary and inshore portions of the Fenholloway system than in the Econfina cognates (Livingston et al., 1998a). The spectroradiometric data indicated that light penetration was most affected by the mill effluent in the Fenholloway estuary and along inshore parts of the Fenholloway Gulf coast within 3 km of the shore and along shore for distances ±4 to 5 km northwest or southeast of the mouth of the Fenholloway River.

A Use Attainability Analysis (Livingston et al., 1998a) indicated that approximately 28 km² of seagrasses offshore of the Fenholloway River had been adversely affected by mill effluents when compared to the relatively natural Aucilla and Econfina near-shore areas. The study recommended that color reduction should be based on annual averages found in what was considered a potential seagrass habitat about 1.5 mi. offshore of the Fenholloway River (Station F10: Figure 3.3). To accomplish a return of seagrasses to this area, water quality models developed by HydroQual, Inc., showed that a 50% reduction of the treated wastewater color from the mill would be required for a return to a median color level of 36 platinum-cobalt units (PCU). This represented a 38% reduction of color at Station F10 (Livingston, 2000, 2002). The response of the Fenholloway system to the color-reduction program was determined as part of the studies of 1998–2004. The target of 36 PCU at Station F10 was evaluated relative to the response of the seagrasses (1999–2002) and associated fishes and invertebrates (1999–2004).

Watercolor was significantly reduced after the first water quality treatment effort in 1973–1974 (Livingston, 2004a). However, by early 1992, watercolor was again relatively high. The treatment effort of 1998 appeared to reduce color in the Fenholloway River and in offshore areas to levels comparable to those in the 1970s after the first treatment. The effect of the drought of 1999–2002 was also evident in the relatively low color levels in the Econfina River and offshore areas. Statistical comparisons of surface color at Stations E08 and F10 over the 5-year study period from 1998 to 2004 indicated that there were significantly higher color levels at Station F10 than at Station E08 during each of the study years (Figure 4.3; Livingston, 2004a). Watercolor at Station F10 exceeded the model prediction of 36 PCU (HydroQual, Inc., unpublished data) for 74% of the noted color levels, whereas watercolor exceedances at Econfina Station E08 totaled about 14% of such levels. There was a noted decrease of color levels in the Fenholloway system during the 1998–2004 study period that probably reflected water quality changes due to both the drought and the color removal program. This decrease should be viewed within the context of the long-term changes of watercolor in both systems over the 33-year period of study. In any case, the color targets for Station F10 were not met.

Spectrophotometric data (Figure 4.4) indicated that photic depths (lowest levels of light needed by submerged aquatic vegetation) were generally higher in the Econfina system during all periods. There was a general increase of photic depths at inshore Econfina stations during peak drought months (2000–2001). There was increased light

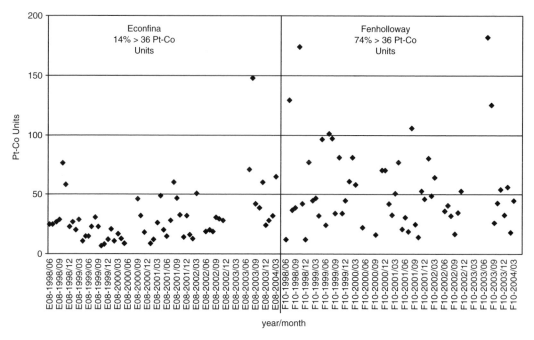

Figure 4.3 Comparison of watercolor at Stations E08 and F10, monthly from June 1988 to April 2004. Data show percent (%) of analyses exceeding 36 PCU in the respective study areas.

penetration with distance from shore in the reference area. Farther offshore, there was less evidence of a drought effect in the Econfina system. There was little evidence of a drought effect in the Fenholloway offshore system. There was a pronounced reduction in photosynthetically active radiation (PAR) extinction coefficients at the inshore Fenholloway stations (F06, F09) that was significant in terms of light penetration changes in areas of high nutrient concentration gradients. Fenholloway estuarine areas did not show improvement with time, and were characterized by losses of light at lower wavelengths, including the 400–500 nm levels of chlorophyll *a* absorption. At station F10, the general increase of photic depths during the drought paralleled similar increases at cognate Station E08, indicating that the drought was probably part of the improvement of light penetration in both systems. However, photic depths here were not deep enough to support SAV except at the outer stations (F12, F16).

It is likely that the observed trend of light penetration in the Fenholloway system was the result of a combination of effects related to the drought and the color removal program. As an introduction to the phytoplankton and SAV reviews, a statistical analysis was made of the various factors associated with light transmission at Stations E08-F10 in the Econfina and Fenholloway systems (Table 4.1). Station F10 was considered a representative area that was monitored for achievement of color reductions to 36 PCU during the recent color reduction program by the pulp mill. Data were averaged over the primary growth season for SAV in the study area. Surface color was significantly higher in the Fenholloway system during most years. Color during 2003 was higher in both systems than all previous periods except 1972. Turbidity was also significantly higher in the Fenholloway system during most years. There were no significant differences in the Secchi depth readings of these stations. However, photic depths were significantly lower at Station F10 during the years 2001, 2002, and 2003. The trends of these photic depths reflected the effects of the drought

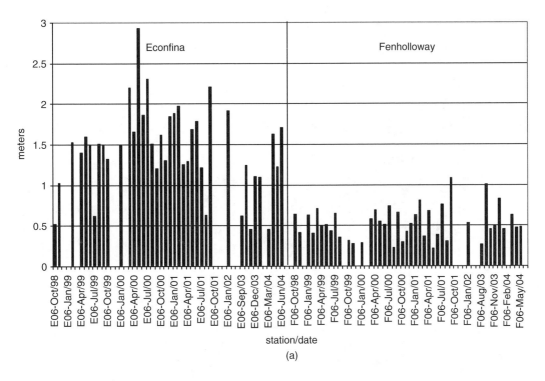

(a)

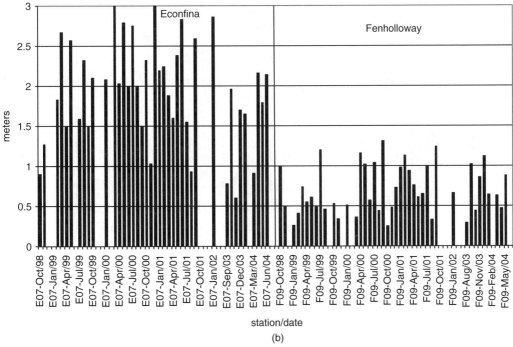

(b)

Figure 4.4 Photic depths (m) in the Econfina and Fenholloway systems from 1998 to 2004.

and the changes in light transmission with a resumption of rainfall and river flows in 2003. The significant differences in photic depths indicate that Secchi depths do not adequately represent light penetration in the study area.

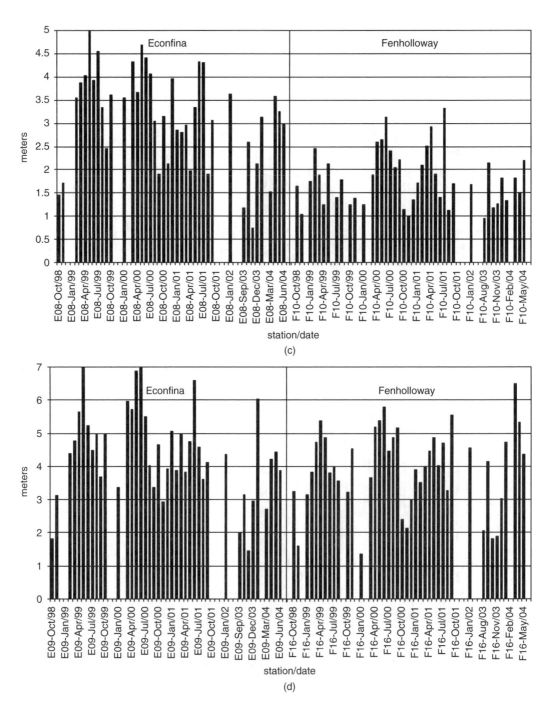

FIGURE 4.4 (continued)

4.4 Sediment Quality

The offshore sediments were high in sand-size particles with relatively low complements of silt and clay (Livingston, 2004a). The distribution of particle size was relatively uniform across stations in the Econfina and Fenholloway offshore areas. Percent organics were also

Table 4.1 Comparison of NCASI Color, Turbidity, Seccchi Depths, Photic Depths, and Biomass of Submerged Aquatic Vegetation (SAV) at Stations E08 and F10 Averaged from June to November, 1972 to 2003

Factor Cognate Sta.	June–November				
	Surface Color (PCU)	Surface Turb. (NTU)	Secchi Depth (m)	Photic Depth (m)	SAV Biomass (g m^{-2})
E08-F10-1972	11.1/118 WS (P < 0.02)	nd	1.7/0.6 WS (P < 0.05)		988/2 WS (P < 0.05)
E08-F10-1974	51.6/96.7 WS (P < 0.05)	nd	1.4/1.0 WN		702/26 WS (P < 0.02)
E08-F10-1976	32.5/67.5 WS (P < 0.01)	3.8/5.3 WS (P < 0.05)	1.5/1.4 WN		508/38 WS (P < 0.05)
E08-F10-1978	37.3/72.5 WN	4.1/4.5 WN	4.1/4.5 WN		499/23 WS (P < 0.05)
E08-F10-1992	44.6/65.7 WN	1.7/2.6 WN	1.7/2.6 WN		31/5 WS (P < 0.02)
E08-F10-1999	17.8/61.8 WS (P < 0.02)	0.18/0.34 WS (P < 0.02)	1.8/1.2 WN		186/10 WS (P < 0.02)
E08-F10-2000	nd	nd	1.7/1.4 WN	3.5/1.8 WS (P < 0.02)	152/7 WS (P < 0.02)
E08-F10-2001	32.8/41.3 WS (P < 0.05)	0.66/1.25 WS (P < 0.05)	1.6/1.5 WN	3.2/2.0 WS (P < 0.05)	162/17 WS (P < 0.02)
E08-F10-2002	24.5/41.3 WS (P < 0.05)	0.51/1.1 WS (P < 0.05)	1.5/1.6 WN	3.1/1.6 WS (P < 0.05)	nd
E08-F10-2003	72/106 WS (P < 0.02)	2.1/2.5 WN	2.1/2.5 WN	1.6/0.9 WS (P < 0.05)	nd

Note: Data were analyzed with *t*-tests and Wilcoxon Means tests for differences (see Appendix II).

nd = no data.

WN = not sig.

WS = sig.

relatively low in the offshore sediments. This is an indication of the highly active areas in the shallow offshore areas of the open Gulf due to wind and tidal effects.

4.4.1   Toxic Agents

The chlorinated organic compounds found in bleached chemical pulp and paper mill effluents represent a complex mixture of more than 200 low-molecular-weight compounds and high-molecular-weight chlorinated lignin compounds (Reeve and Earl, 1988). There are several multi-step methods to analyze these compounds in order to provide estimates of AOCl/AOX (absorbable organically bound chlorine [Cl] or halogen [X]), TOCl or TOX (total organically bound chlorine or halogen), or extractable EOCl or EOX (extractable, organically bound chlorine or halogen). The EOCl fraction represents the highly lipophilic,

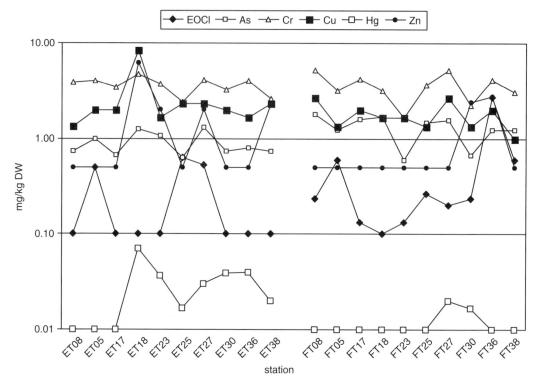

Figure 4.5 Sediment metals and extractable organically bound chlorine (EOCl) taken during the 1999 study in the Econfina and Fenholloway offshore areas.

potentially bioaccumulable compounds that are characterized by having more than a 1000-fold greater concentration in 1-octanol than in water (at equilibrium) (Reeve and Earl, 1988).

The distributions of toxic substances in sediments taken from offshore areas of the Econfina and Fenholloway drainages in 1999 are shown in Figure 4.5. Birkholz (1999) noted that Econfina and Fenholloway sediments "had trace levels of chlorine" that were "extremely low." Metal concentrations, with the exception of three samples (zinc, copper, and mercury), were also uniformly low. With the exception of the lead data, the results of the most recent study (1999) were compatible with the 1992–1993 study. Application of the NOAA effects range-low (ER-L) and effects range-medium (ER-M) guidelines indicated that metal concentrations were not even close to effects ranges for animals.

4.4.2 Nutrients

The importance of sediments in the nutrient loading and accumulation process was noted by McComb and Lukatelich (1995). Sediment resuspension as a product of water column dynamics can be a factor in nutrient storage and loading processes (de Jong and Raaphorst, 1995; de Jong, 1995). The timed release and bioavailability of sediment nutrients, as modified by resuspension, remobilization, and regeneration through biotic activity, remain poorly understood although such processes could have an important effect on the response of benthic microalgae and macrophytes (Thornton et al., 1995a). Sediment nutrient data (Figure 4.6) showed that total sediment phosphorus in the Econfina system was most highly concentrated in the inshore estuary, with moderately high concentrations noted at Station ET-35. A similar pattern of sediment TP was noted in the Fenholloway system,

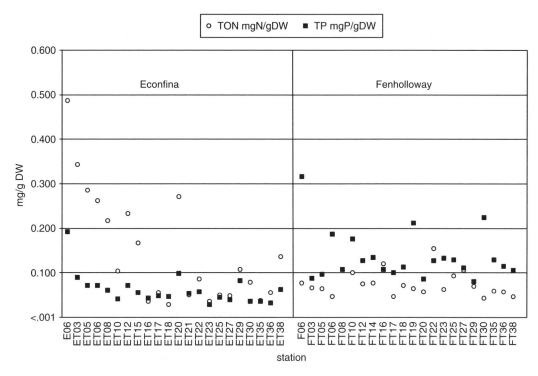

Figure 4.6 Sediment total organic nitrogen (TON) and total phosphorus (TP) taken during the 1999 study in the Econfina and Fenholloway offshore areas.

with somewhat higher concentrations also noted at Stations FT10 and FT06 (F11). Total nitrogen was highest in the inshore estuary of the Econfina system. In the offshore Fenholloway system (Stations FT10, FT16, FT22, FT27), TON was high. The differences in sediment TON were not statistically significant (P = 0.05). Total phosphorus was uniformly higher in the Fenholloway system (Figure 4.6); these differences were statistically significant. Thus, the most obvious differences between the sediments taken in the Econfina and Fenholloway systems were the uniformly higher concentrations of phosphorus in the Fenholloway coastal areas and higher concentrations of TON in parts of the Econfina system.

4.5 Nutrients: Loading, Limitation, and Concentration

4.5.1 Nutrient Loading

Nutrient loading is an important process in the determination of the ecological response of a coastal system to freshwater inflow (Flemer et al., 1997). The primary effects of nutrient loading on shallow estuaries and lagoons have been outlined by McComb (1995). A nutrient loading analysis was carried out. Nutrient loading models, based on a ratio estimator developed by Dolan et al. (1981), were run to determine nutrient loading into Apalachee Bay via the Econfina and Fenholloway Rivers (Appendices A through D).

Results of the quarterly effort are shown in Figure 4.7. Due to nutrient loading from the pulp mill, the Fenholloway River was a primary source of ammonia and orthophosphate to the Apalachee Bay system during most of the study period. Loading of both nutrients in the Fenholloway system were generally about 2 orders of magnitude higher than the reference Econfina system. During the most recent drought, Econfina loading of

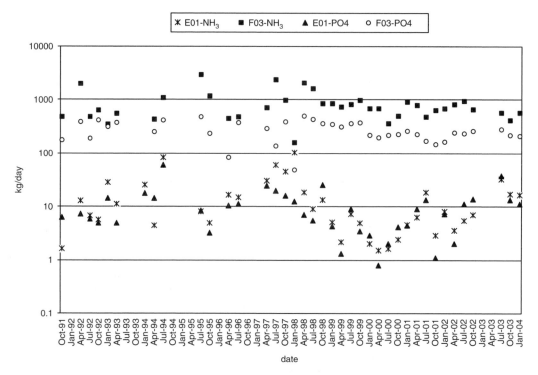

Figure 4.7 Quarterly loading of ammonia (kg day^{-1}) and orthophosphate (kg day^{-1}) from April 1991 through March 2004. The U.S. Geological Survey (Tallahassee, Florida; M. Franklin, personal communication) provided river flow data.

ammonia and orthophosphate was relatively low, whereas the continuous loading of these nutrients from the Fenholloway was maintained at relatively high levels. There was a slight decrease in the Fenholloway loading over the past 5 years, with possible effects of the drought leading to reduced loading from 1999 to 2002. However, loading of ammonia and orthophosphate remained high in the Fenholloway River relative to the Econfina, and there these differences were statistically significant ($P < 0.05$).

4.5.2   Nutrient Limitation

Livingston (unpublished data; 1993b) carried out a series of nutrient limitation experiments in the Econfina and Fenholloway systems. Methods used were similar to those described by Flemer et al. (1997). The estuarine portion of the Econfina system (E06) was characterized by either N-limitation or N+P-limitation during most months. Similar results were found in the Fenholloway estuary (F06), with the exception that phosphorus was found as the chief limiting factor during May, August, and October 1992 and spring 1993. Both systems had the highest levels of chlorophyll during warm months and N+P treatments were the dominant form of chlorophyll stimulation in the inshore areas of both systems. Nitrogen-limitation was dominant in inshore Econfina areas, whereas the cognate Fenholloway area was characterized by phosphorus limitation during spring and fall months. During other periods, nitrogen was still predominant as a limiting nutrient although the highest summer chlorophyll peaks were associated with the N+P treatments. Offshore Econfina and Fenholloway areas were dominated by P-limitation throughout the experimental period.

4.5.3   Nutrient Concentration Gradients

Livingston (2004a) noted that various nutrient concentrations were highest in the Fenhol-loway estuary from Station F06-Station F09. In the Fenholloway system, the highest ammo-nia and orthophosphate levels were in the Fenholloway River estuary, which appeared to be a major source of ammonia to offshore areas. This pattern was not evident in the Econfina system. At Station F09, there were increased concentrations of ammonia during the 1999–2002 period, whereas orthophosphate concentrations decreased. These changes were accompanied by increased chlorophyll *a* concentrations at this station, which could account for the decreasing orthophosphate trends as phytoplankton increased in abun-dance in the Fenholloway estuary (see below). Concentrations of both nutrients were generally lower in the Econfina system, with some pronounced reductions of both ortho-phosphate and chlorophyll during the drought. Farther out in the bay (Station F10), these trends were not as obvious, with some increase in chlorophyll and decreased orthophos-phate. Ammonia levels were somewhat increased in this area during the latest sampling period. At Station F11, there were general decreases in both orthophosphate and chloro-phyll, whereas ammonia concentrations went up during early 2000 and went down again during subsequent periods.

4.6   Phytoplankton and Zooplankton

4.6.1   Introduction

The discharge of industrial wastes may lead to the replacement of indigenous phytoplank-ton populations with species that are better suited to cope with the stress imposed by pollution. Biotic interrelationships within the community can be modified, and a substan-tial alteration of the biotic community often results (Rainville et al., 1975). Stockner and Costella (1976) showed that algae can adapt to relatively high concentrations of pulp mill effluent.

4.6.2   Chlorophyll a Trends

Livingston (2000, 2002) found that chlorophyll *a* concentrations were generally correlated with phytoplankton biomass but were not a reliable indicator of plankton blooms. High chlorophyll *a* concentrations do indicate blooms but low concentrations do not necessarily preclude bloom activity. Livingston et al. (1998b) found that chlorophyll *a* concentrations were relatively low in the Econfina and Fenholloway systems but were consistently higher in the Econfina estuary than the Fenholloway estuary during 1992–1993. This situation was reversed just offshore where the Fenholloway chlorophyll readings were often higher than those at the cognate Econfina stations. The mean isopleth of 4 µg L^{-1} chlorophyll *a* in the Fenholloway offshore system reached an area 3 to 5 km offshore, whereas the 4 µg L^{-1} isopleth in the Econfina system occurred up to 1 km offshore (Livingston et al., 1998b). The highest concentrations of chlorophyll *a* in the Econfina and Fenholloway systems tended to occur during summer–fall months (Figure 4.8). There was a general increase in chlorophyll *a* in the Fenholloway system during 1998–2003 compared to the Econfina system. There was a general trend of relatively high chlorophyll *a* concentrations during 2000–2001 and 2003–2004. Chlorophyll concentrations in both systems were lowest during the drought. During the 2003 period, chlorophyll *a* concentrations were significantly lower in the Fenholloway estuary and significantly higher in most of the Fenholloway offshore areas. The Fenholloway chlorophyll levels during 2003 were higher than those in Econfina

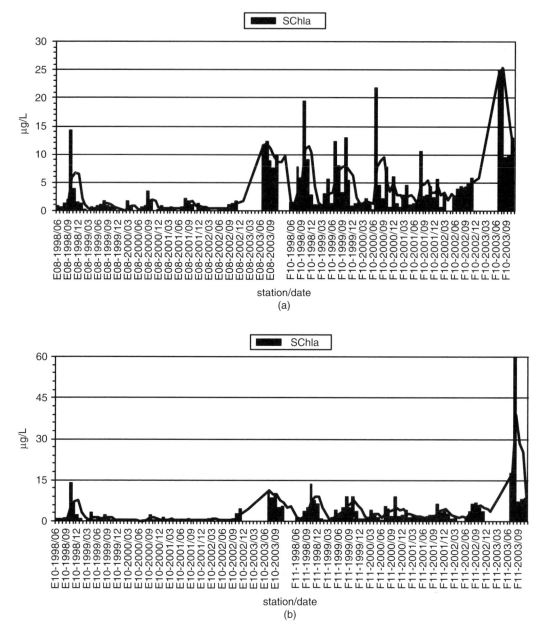

Figure 4.8 Chlorophyll *a* concentrations in the Econfina and Fenholloway areas from 1998 to 2003.

cognate areas, and these concentrations were higher than in any preceding years. The data indicated a basic change in phytoplankton activity in the Fenholloway system during 2003.

4.6.3 *Phytoplankton Distribution (1992–1993)*

During 1992–1993, a total of 264 taxa of phytoplankton were identified, representing the following groups: Bacillariophycae (diatoms) (203), Dinophyceae (dinoflagellates) (26), Cyanophyceae (blue-green algae or cyanobacteria) (10), Cryptophyceae (cryptophytes) (2), Chlorophyceae (green algae) (2), Chrysophyceae (golden-brown algae) (2), Prasinophyceae (3), Prymnesiophyceae (haptophytes) (3), and Euglenophyceae (euglenoids) (4). Most

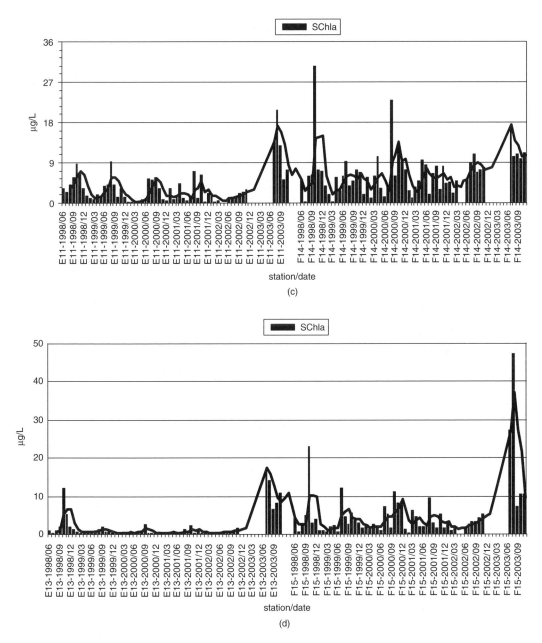

Figure 4.8 (continued)

identified nannoflagellates included small cryptophytes with close affinity to Komma and Falcomonad species. Ultra-plankton appeared to be related to coccoid blue-green algae described by Waterbury et al. (1979) and Johnson and Sieburth (1979).

Most of the dominant phytoplankton populations in the Econfina and Fenholloway estuaries and near-shore waters are euryhaline. Salinity was generally not statistically associated with phytoplankton numbers during the 1992–1993 study. In both systems, numbers of phytoplankton cells were positively associated with temperature (Econfina $R^2 = 0.2$, P = 0.006; Fenholloway $R^2 = 0.23$, P = 0.0007). Secchi disk readings and DOC were significantly (negatively) associated with numbers of cells in the Econfina system (Secchi $R^2 = 0.52$, P = 0.002; DOC $R^2 = 0.62$, P = 0.03). Phytoplankton numbers tended to

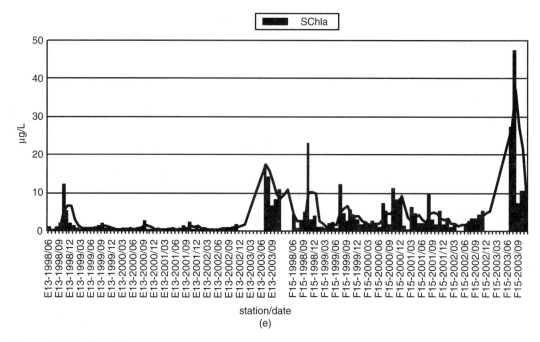

Figure 4.8 (continued)

decrease with increased ammonia in the Fenholloway system, whereas there was a weak positive relationship between these two factors in the Econfina system. Overall, with the exception of temperature and certain light features, phytoplankton numerical abundance was not strongly associated with any of the environmental features of the two systems.

Phytoplankton species richness in the Fenholloway system was weakly (negatively) associated with DOC ($R^2 = 0.07$, P = 0.03), ratio turbidity ($R^2 = 0.06$, P = 0.04), ammonia ($R^2 = 0.16$, P = 0.0006), orthophosphate ($R^2 = 0.16$, P = 0.0003), and silicate ($R^2 = 0.19$, P = 0.0004). Weakly positive (statistically significant) associations in the Econfina system were found between species richness and temperature ($R^2 = 0.22$, P = 0.0001) and DOC ($R^2 = 0.26$, P = 0.0001). The weak R^2 values reduced the significance of these associations although the general trends indicated that phytoplankton species richness decreased with reduced light transmission, high nutrients, and high dissolved organic carbon in the Fenholloway system. Major differences in the phytoplankton assemblages were associated with the differing light transmission characteristics of the two systems. Phytoplankton numbers and species richness varied inversely with distance from shore in both areas, with light and nutrient alterations directly implicated in the differences noted between the respective phytoplankton biotas.

The data indicated that, during 1992–1993, high nutrient loading, combined with increased light penetration, accounted for differences in the phytoplankton distributions in the Econfina and Fenholloway systems (Livingston et al., 1998a). Light transmission characteristics were important, and the dominant wavelength at most of the stations in both systems was between 575 and 580 nm. Both Gulf drainages were affected by river-derived dissolved substances (*Gelbstoff*); such substances were associated with the loss of light at lower wavelengths. This *Gelbstoff* effect was stronger in the Fenholloway estuary and near-shore waters (F06, F09, F11, F14) than cognate stations in the Econfina system. Phytoplankton numbers in both systems peaked in near-shore waters characterized by increasing light penetration and high nutrients. High nutrient loading from the Fenholloway River was most apparent in these near-shore waters. Farther offshore, there was a general decrease in phytoplankton cell numbers and species richness that was directly related to reduced nutrient concentrations. Light was limiting in the Fenholloway estuary; in inshore areas, light became less a factor and increased nutrient loading took over as the primary limiting agent relative to the reference system. Farther offshore, nutrients became limiting in both systems, with corresponding decreases in phytoplankton numbers and species.

4.6.4   Color Removal and Bloom Generation

A comparison was made of the whole water phytoplankton communities in 1992, 1999, and 2000. Data were analyzed from June to October for each of these years because summer–fall represented the peak bloom period. During 1992–1993, the inshore (estuarine) whole water phytoplankton in the Fenholloway system (F06, F09, F11) showed significant reductions in phytoplankton cumulative species richness compared to the Econfina areas (Table 4.2). Inshore Econfina areas were dominated by nannoflagellates, cryptophytes, nannococcoids, and diatom species such as *Navicula* spp. during 1992–1993. Overall, the nannoflagellates and cryptophytes were generally higher in the Econfina estuarine and near-shore waters than in similar areas of the Fenholloway system, whereas the nanno-coccoids were evenly distributed in the two study areas. The diatom *Chaetoceros fragilis* Meun was dominant in offshore areas of the Econfina system. Inshore stations in the Fenholloway system were well represented by species such as *Minutocellus scriptus*, Hasle, Stosch, and Syvertsen, *Cryptomonas* spp., *Cylindrotheca closterium* (Ehrenberg) Reimann and Lewin, and various forms of nannoflagellates. These data followed other indications that the estuarine Fenholloway phytoplankton assemblages were relatively depauperate com-pared to the Econfina estuary.

During 1999, whole water phytoplankton species richness increased in both systems relative to 1992–1993 (Table 4.2). This index remained low at inshore Fenholloway stations. During this period, phytoplankton numbers went up in both systems, with particularly high numbers at Stations F10 and F11. Species diversity and evenness in both systems was low in inshore areas in 1999. By 2000, Econfina phytoplankton numbers were down; this trend continued during the drought. Species richness in the Econfina remained high during 2000. In the Fenholloway system during 2000, inshore numbers remained high. Species richness, diversity, and evenness tended to be low where such high numbers occurred (F09, F10, F14). These community changes reflected the incidence of phytoplank-ton blooms in the Fenholloway system during 1999–2000.

Peak whole water phytoplankton numbers in the Fenholloway system (Table 4.2) were associated with phytoplankton blooms (number cells $L^{-1} > 10^6$) of *Leptocylindrus danicus* (1,831,582 cells L^{-1}, Station F09, September 1999; 4,390,902 cells L^{-1}, Station F10, September 1999); *Johannesbaptistia pellucida* (2,111,220 cells L^{-1}, Station F11, September 1999); and *Cerataulina pelagica* (1,625,299 cells L^{-1}, Station F14, October 1999). The diatom *L. danicus* has a worldwide distribution in a variety of tropical coastal environments, and has adverse impacts on various fish species (see above). *L. danicus* was a key pioneer bloom species in the Perdido Bay system during the initial blooms of 1994. *J. pellucida* is a filamentous blue-green alga with a mucous sheath. It has worldwide distribution and is not considered a noxious species (A.K.S.K. Prasad, personal communication). *C. pelagica* is a filamentous diatom that is a known bloom species that closely resembles *L. danicus* in body type and natural history (A.K.S.K. Prasad, personal communication). These were the first blooms noted in either of the Apalachee study areas.

During 2000, there were blooms during July in the Fenholloway estuary and in the offshore Fenholloway system (Station F15) (Table 4.2). The primary bloom species was *Lepto-cylindrus minimus*. There was also high dominance of *Johannesbaptistia pellucida, Rhizosolenia*

Table 4.2 Numerically Dominant Phytoplankton Species Taken with 25-μm Nets and Whole Water Samples in the Econfina and Fenholloway Systems from April 1992–March 1993 and April–December 1999, and Whole Water Phytoplankton Taken from June to October 2000, June–November 2001, June–November 2002, and August–November 2003

Whole Water Phytoplankton: #cells L⁻¹
Econfina (April 1992–March 1993)

Taxon	E06	E07	E08	E09	E10	E11	TOTAL	%
Unidentified nannoflagellates	53,169	52,856	75,818	52,448	74,278	53,643	362,212	19.57
Unidentified cryptophytes	21,340	54,642	63,058	23,810	122,674	58,941	344,464	18.61
Unidentified nannococcoids	20,147	17,059	34,011	13,753	18,963	26,670	130,602	7.06
Cyclotella sp. 17 (5–10μ)	0	0	0	0	91,390	0	91,390	4.94
Thalassionema nitzschioides	3,191	4,950	32,104	2,914	13,357	5,933	62,450	3.37
Pyramimonas spp.	8,464	10,172	16,559	2,298	6,845	13,169	57,506	3.11
Navicula spp.	20,147	8,567	4,919	5,245	6,475	7,144	52,497	2.84
Unidentified pennate diatoms	13,459	14,531	4,904	2,881	8,140	6,721	50,635	2.74
Chaetoceros spp.	694	4,829	15,606	2,864	12,284	6,872	43,148	2.33
Cylindrotheca closterium	12,266	10,823	4,950	2,015	3,719	8,840	42,611	2.3

Whole Water Phytoplankton: #cells L⁻¹
Fenholloway (April 1992–March 1993)

Taxon	F06	F09	F10	F16	F11	F14	TOTAL	%
Unidentified nannoflagellates	35,893	47,377	57,017	53,327	56,407	46,737	296,757	19.52
Skeletonema costatum	555	1,816	116,883	4,657	52,836	95,804	272,552	17.93
Unidentified cryptophytes	21,337	23,249	82,436	32,402	29,452	34,382	223,259	14.69
Unidentified nannococcoids	14,934	18,012	24,883	12,697	21,997	34,016	126,539	8.32
Navicula spp.	18,383	16,711	12,007	8,492	8,159	11,156	74,906	4.93
Pyramimonas spp.	1,748	4,541	11,526	5,387	10,656	23,710	57,568	3.79
Nitzschia section *pseudonitzschia* spp.	0	394	3,811	1,807	42,735	133	48,879	3.22
Cylindrotheca closterium	2,778	16,166	8,584	2,567	8,862	9,024	47,981	3.16
Minutocellus scriptus	3,968	5,086	1,517	466	1,628	8,858	21,523	1.42
Nitzschia spp.	4,259	3,845	2,294	1,693	2,442	5,511	20,043	1.32

25-μm Net Phytoplankton: #cells L⁻¹
Econfina (April 1992–March 1993)

Taxon	E06	E07	E09	TOTAL	%
Unidentified nannoflagellates	300	166	183	649	12.74
Cylindrotheca closterium	245	94	66	405	7.95
Navicula spp.	142	127	76	345	6.77
Skeletonema costatum (narrow)	101	155	0	256	5.02
Unidentified nannococcoids	145	32	59	235	4.62
Asterionella japonica	4	163	68	234	4.6
Amphora spp.	49	73	91	212	4.16
Rhizosolenia stolterfothii	4	81	88	173	3.39
Chaetoceros diversus	25	31	75	131	2.57
Thalassionema nitzschioides	33	23	75	131	2.57

25-μm Net Phytoplankton: #cells L⁻¹
Fenholloway (April 1992–March 1993)

Taxon	F06	F09	F16	TOTAL	%
Skeletonema costatum	453	10,388	14	10,854	71.35
Unidentified nannoflagellates	265	236	183	684	4.5

Table 4.2 (continued) Numerically Dominant Phytoplankton Species Taken with 25-μm Nets and Whole Water Samples in the Econfina and Fenholloway Systems from April 1992–March 1993 and April–December 1999, and Whole Water Phytoplankton Taken from June to October 2000, June–November 2001, June–November 2002, and August–November 2003

Taxon	F06	F09	F16	TOTAL	%
Cylindrotheca closterium	150	211	96	456	3
Navicula spp.	176	115	93	384	2.53
Nitzschia section pseudonitzschia spp.	0	0	257	257	1.69
Chaetoceros diversus	3	15	206	224	1.48
Thalassionema nitzschioides	10	17	148	174	1.15
Chaetoceros compressus	21	0	134	155	1.02
Unidentified nannococcoids	85	36	31	152	1
Chaetoceros simplex	0	0	103	104	0.68

Whole Water Phytoplankton: #cells L^{-1}
Econfina (April 1999–December-1999)

Taxon	E06	E07	E08	E09	E10	E11	TOTAL	%
Undetermined cryptophyte	320,614	184,092	63,585	81,119	96,228	299,027	1,044,665	30.54
Cyclotella choctawhatcheeana	2,072	42,718	17,428	23,707	141,212	386,243	613,380	17.93
Undetermined nannoflagellates	86,181	79,417	61,567	58,018	40,659	75,433	401,275	11.73
Johannesbaptistia pellucida	2,035	39,590	58,018	102,704	2,701	1,406	206,454	6.04
Asterionellopsis glacialis	37	296	2,590	138,760	74	93	141,850	4.15
Dactyliosolen fragilissimus	815	12,412	5,076	8,391	315	77,793	104,802	3.06
Chaetoceros laciniosus	2,186	70,665	20,343	815	0	334	94,343	2.76
Undetermined nannococcoids	2,629	5,924	22,501	7,812	46,650	3,367	88,883	2.6
Gymnodinium spp.	4,514	10,676	9,087	16,711	9,090	28,196	78,274	2.29
Cyclotella cf. *atomus*	5,665	6,106	10,476	8,936	15,695	23,757	70,635	2.06

Whole Water Phytoplankton: #cells L^{-1}
Fenholloway (April 1999–December 1999)

Taxon	F06	F09	F10	F16	F11	F14	TOTAL	%
Undetermined cryptophyte	358,064	270,651	138,850	70,457	229,703	195,946	1,263,671	26.33
Leptocylindrus danicus	18,755	206,469	493,799	10,028	15,985	1,554	746,590	15.56
Undetermined nannoflagellates	92,645	98,513	79,576	48,093	121,982	100,025	540,834	11.27
Cyclotella cf. *atomus*	59,715	54,184	41,313	352	126,789	80,402	362,755	7.56
Johannesbaptistia pellucida	2,206	6,697	39,886	3,293	279,721	20,805	352,608	7.35
Cerataulina pelagica	625	1,258	10,510	11,785	7,451	181,736	213,365	4.45
Undetermined nannococcoids	499	1,518	2,868	4,497	162,792	481	172,655	3.6
Cerataulina pelagica (veg.)	0	0	0	0	119,963	0	119,963	2.5

Table 4.2 (continued) Numerically Dominant Phytoplankton Species Taken with 25-μm Nets and Whole Water Samples in the Econfina and Fenholloway Systems from April 1992–March 1993 and April–December 1999, and Whole Water Phytoplankton Taken from June to October 2000, June–November 2001, June–November 2002, and August–November 2003

Taxon	F06	F09	F10	F16	F11	F14	TOTAL	%
Dactyliosolen fragilissimus	1,915	3,145	23,348	1,647	74,009	10,361	114,425	2.38
Rhizosolenia setigera	17,402	7,552	66,382	13,988	4,289	4,367	113,980	2.37

25-μm Net Phytoplankton: #cells L⁻¹
Econfina (April 1999–October 1999)

Taxon	E06	E07	E08	E09	TOTAL	%
Asterionellopsis glacialis	34	72	1,392	79,216	80,714	33.84
Johannesbaptistia pellucida	1,607	7,512	21,770	40,634	71,523	29.98
Chaetoceros laciniosus	413	24,921	17,888	556	43,778	18.35
Chaetoceros lauderi	3	9,374	1,041	58	10,476	4.39
Bacteriastrum hyalinum	33	2,799	2,699	108	5,639	2.36
Chaetoceros compressus	71	2,219	2,130	449	4,868	2.04
Chaetoceros cf. *subtilis*	411	2,224	0	0	2,635	1.1
Rhizosolenia setigera	42	928	1,480	158	2,607	1.09
Pseudonitzschia spp.	37	938	940	342	2,257	0.95
Peridinium quinquecorne	1,066	28	0	3	1,097	0.46

25-μm Net Phytoplankton: #cells L⁻¹
Fenholloway (April 1999–October 1999)

Taxon	F06	F09	F10	F16	TOTAL	%
Leptocylindrus danicus	2,608	16,490	38,363	1,877	59,338	27.93
Rhizosolenia setigera	2,658	2,629	39,170	2,683	47,140	22.19
Falcula hyalina	16,081	15,966	9	4	32,060	15.09
Chaetoceros laciniosus	43	33	538	20,690	21,305	10.03
Chaetoceros lauderi	10	133	2,346	8,347	10,837	5.1
Chaetoceros compressus	54	11	8,745	573	9,384	4.42
Johannesbaptistia pellucida	0	28	1,604	5,925	7,557	3.56
Cerataulina pelagica	142	307	987	4,521	5,956	2.8
Thalassionema nitzschioides	90	173	2,073	72	2,409	1.13
Ceratium hircus	70	42	1,101	177	1,390	0.65
					212,452	

Whole Water Phytoplankton: #cells L⁻¹
Econfina (June 2000–October 2000)

Taxon	E06	E07	E08	E09	E10
Undetermined cryptophyte	123,543	1,038,628	254,579	178,489	275,058

Table 4.2 (continued) Numerically Dominant Phytoplankton Species Taken with 25-μm Nets and Whole Water Samples in the Econfina and Fenholloway Systems from April 1992–March 1993 and April–December 1999, and Whole Water Phytoplankton Taken from June to October 2000, June–November 2001, June–November 2002, and August–November 2003

Taxon	F06	F09	F10	F16		TOTAL	%
Johannesbaptistia pellucida	0	38,961	85,415	99,734	91,576		
Cyclotella cf. *atomus*	38,961	104,229	34,299	46,787	24,809		
Undetermined nannoflagellates	24,975	119,215	44,789	43,124	40,960		
Undetermined nannococcoids	0	24,309	67,100	22,645	112,555		
Gymnodinium spp.	666	48,785	50,784	31,137	36,797		
Cocconeis spp.	4,995	55,445	32,303	9,159	38,296		
Pseudonitzschia spp.	666	12,987	73,260	12,988	24,144		
Proboscia alata	0	999	7,993	62,438	1,499		
Chaetoceros spp.	0	19,481	28,306	11,823	21,813		

Taxon	E11	E12	E13	TOTAL	%
Undetermined cryptophyte	682,984	262,071	297,037	3,112,388	38.6
Johannesbaptistia pellucida	3,497	114,219	233,600	667,001	9.6
Cyclotella cf. *atomus*	247,420	24,310	33,634	554,448	7.3
Undetermined nannoflagellates	89,078	42,458	57,777	462,375	6.4
Undetermined nannococcoids	33,633	139,861	55,279	455,381	6.4
Gymnodinium spp.	43,291	35,300	82,253	329,013	4.3
Cocconeis spp.	45,455	15,320	8,993	209,965	2.9
Pseudonitzschia spp.	2,166	56,112	11,656	193,978	2.7
Proboscia alata	1,166	500	79,088	153,681	2.4
Chaetoceros spp.	6,494	27,307	13,488	128,712	1.8

Whole Water Phytoplankton: #cells L^{-1}
Fenholloway (June 2000–October 2000)

Taxon	F06	F09	F10	F16	F11
Leptocylindrus minimus	0	2,137,860	2,601,396	360,306	102,731
Undetermined cryptophyte	78,921	594,072	175,492	384,283	170,830
Proboscia alata	102,231	364,302	549,950	282,384	170,164
Johannesbaptistia pellucida	0	215,118	165,501	190,976	201,798
Thalassiosira minima proschikinae complx	0	321,345	0	0	0
Chaetoceros spp.	333	79,254	101,566	71,263	141,859
Rhizosolenia setigera	1,332	42,291	68,765	50,117	52,615
Undetermined nannoflagellates	9,324	79,587	45,788	56,777	43,125
Dactyliosolen fragilissimus	666	666	99,567	666	1,333
Gymnodinium spp.	5,661	5,661	22,645	51,949	49,618

Table 4.2 (continued) Numerically Dominant Phytoplankton Species Taken with 25-μm Nets and Whole Water Samples in the Econfina and Fenholloway Systems from April 1992–March 1993 and April–December 1999, and Whole Water Phytoplankton Taken from June to October 2000, June–November 2001, June–November 2002, and August–November 2003

Taxon	F14	F12	F15	TOTAL	%
Leptocylindrus minimus	234,932	1,288,544	0	6,725,769	37.4
Undetermined cryptophyte	590,909	175,159	138,030	2,307,696	12.8
Proboscia alata	232,767	458,208	41,292	2,201,298	12.3
Johannesbaptistia pellucida	55,112	199,634	164,670	1,192,808	6.6
Thalassiosira minima proschikinae complx	565,934	0	0	887,279	4.9
Chaetoceros spp.	42,291	80,254	7,659	524,478	2.9
Rhizosolenia setigera	29,804	202,964	1,332	449,220	2.5
Undetermined nannoflagellates	114,719	43,791	50,285	443,395	2.5
Dactyliosolen fragilissimus	102,731	103,563	5,495	314,687	1.8
Gymnodinium spp.	21,313	60,940	31,636	249,421	1.4

Whole Water Phytoplankton: #cells L^{-1}
Econfina (June 2001–October 2001)

Taxon	E06	E07	E08	E09	E10
Undetermined cryptophyte	158,742	107,005	64,076	58,221	74,982
Undetermined nannoflagellates	18,848	13,931	12,544	10,074	11,766
Dactyliosolen fragilissimus	6,660	8,548	6,327	4,440	500
Gymnodinium spp.	4,996	11,017	29,500	32,634	31,775
Pyramimonas spp.	6,694	5,524	4,996	1,526	2,331
Navicula sp. 54 small (F09 8/01)	67	56	—	—	—
Thalassiosira minima proschikinae complx	—	333	—	—	—
Undetermined centric diatoms	26,341	2,499	1,831	1,249	2,027
Cyclotella cf. *atomus*	166	112	56	250	196
Cylindrotheca closterium	6,926	4,273	277	501	418

Taxon	E11	E12	E13	TOTAL	%
Undetermined cryptophyte	162,811	56,500	60,608	682,337	
Undetermined nannoflagellates	15,236	11,184	10,962	93,583	
Dactyliosolen fragilissimus	126,291	83	93,518	152,849	
Gymnodinium spp.	16,790	21,063	24,893	147,775	
Pyramimonas spp.	5,523	1,694	3,691	28,288	
Navicula sp. 54 small (F09 8/01)	—	—	—	123	
Thalassiosira minima proschikinae complx	—	—	—	333	
Undetermined centric diatoms	2,027	1,527	2,277	37,501	
Cyclotella cf. *atomus*	139	278	112	1,197	
Cylindrotheca closterium	4,247	556	167	17,198	

Whole Water Phytoplankton: #cells L^{-1}
Fenholloway (June 2001–October 2001)

Taxon	F06	F09	F10	FII	F12
Undetermined cryptophyte	135,809	118,115	135,420	100,844	81,170
Undetermined nannoflagellates	38,351	30,004	21,702	18,621	10,684

Table 4.2 (continued) Numerically Dominant Phytoplankton Species Taken with 25-μm Nets and Whole Water Samples in the Econfina and Fenholloway Systems from April 1992–March 1993 and April–December 1999, and Whole Water Phytoplankton Taken from June to October 2000, June–November 2001, June–November 2002, and August–November 2003

Taxon	F06	F09	F10	FII	F12
Dactyliosolen fragilissimus	—	67	7,659	3,968	7,964
Gymnodinium spp.	9,990	3,796	10,990	12,738	22,950
Pyramimonas spp.	24,309	16,085	10,268	5,829	3,887
Navicula sp. 54 small (F09 8/01)	14,763	56,576	18,788	4,940	1,805
Thalassiosira minima proschikinae complx	63,270	27,072	1,554	—	389
Undetermined centric diatoms	5,384	9,523	4,107	3,081	2,136
Cyclotella cf. *atomus*	33,189	25,242	611	1,082	806
Cylindrotheca closterium	11,766	12,788	4,496	3,720	1,473
Johannesbaptistia pellucida	—	—	20,979	3,386	1,055

Taxon	F14	F15	F16	TOTAL
Undetermined cryptophyte	159,574	106,811	66,906	904,649
Undetermined nannoflagellates	28,105	22,339	13,460	183,266
Dactyliosolen fragilissimus	7,592	7,937	889	36,076
Gymnodinium spp.	6,727	13,154	20,620	100,965
Pyramimonas spp.	25,475	16,790	3,470	106,113
Navicula sp. 54 small (F09 8/01)	15,251	3,802	28	115,953
Thalassiosira minima proschikinae complex	333	7,076	56	99,750
Undetermined centric diatoms	4,497	2,137	1,083	31,948
Cyclotella cf. *atomus*	8,259	167	361	69,717
Cylindrotheca closterium	11,056	3,526	751	49,576
Johannesbaptistia pellucida	—	19,037	500	19,537

Whole Water Phytoplankton: #cells L⁻¹
Econfina (June 2002–October 2002)

Taxon	E06	E07	E09	E10	E11
Undetermined cryptophyte	45,733	100,370	27,391	19,066	16,380
Cerataulina pelagica	—	—	—	—	—
Gymnodinium spp.	14,042	11,744	22,707	17,012	13,987
Thalassiosira cedarkeyensis	—	167	42	—	—
Undetermined nannoflagellates	9,713	14,259	4,670	4,394	3,105
Navicula spp.	1,166	605	480	293	480
Undetermined coccoid yellow-green algae	—	70,430	—	—	—
Rhizosolenia setigera	—	—	—	83	—
Pyramimonas spp.	611	1,500	272	209	83
Nitzschia spp.	722	1,187	626	1,084	584

Taxon	E11	E12	E13	TOTAL
Undetermined cryptophyte	65,308	21,906	21,962	318,116
Cerataulina pelagica	—	—	—	0
Gymnodinium spp.	19,816	15,643	22,790	137,741
Thalassiosira cedarkeyensis	292	83	125	709
Undetermined nannoflagellates	11,476	3,688	4,918	56,223
Navicula spp.	876	542	792	5,234
Rhizosolenia setigera	—	—	—	70,430
Pyramimonas spp.	1,690	356	272	83
Nitzschia spp.	1,645	751	1,084	4,993

Table 4.2 *(continued)* Numerically Dominant Phytoplankton Species Taken with 25-μm Nets and Whole Water Samples in the Econfina and Fenholloway Systems from April 1992–March 1993 and April–December 1999, and Whole Water Phytoplankton Taken from June to October 2000, June–November 2001, June–November 2002, and August–November 2003

	Whole Water Phytoplankton: #cells L⁻¹ Fenholloway (June 2002–October 2002)				
Taxon	F06	F09	F10	F11	F12
Undetermined cryptophyte	62,504	56,303	111,063	86,177	63,437
Cerataulina pelagica	—	84	44,606	102,593	84,583
Gymnodinium spp.	1,168	999	39,543	21,616	33,301
Thalassiosira cedarkeyensis	123,975	81,729	3,835	20,761	2,289
Undetermined nannoflagellates	25,998	16,909	12,706	20,794	6,869
Navicula spp.	77,149	75,649	417	1,125	792
Undetermined coccoid yellow-green algae	—	—	—	—	—
Rhizosolenia setigera	167	1,665	13,208	5,162	2,123
Pyramimonas spp.	—	999	5,580	11,252	2,622
Nitzschia spp.	18,497	5,499	668	1,251	542

Taxon	F14	F15	F16	TOTAL
Undetermined cryptophyte	84,244	56,169	60,441	580,338
Cerataulina pelagica	102,106	56,901	—	390,873
Gymnodinium spp.	15,157	40,670	22,172	174,626
Thalassiosira cedarkeyensis	28,178	1,541	42	262,350
Undetermined nannoflagellates	15,077	8,830	8,580	115,763
Navicula spp.	2,376	896	272	158,676
Undetermined coccoid yellow-green algae	—	—	—	0
Rhizosolenia setigera	6,496	12,780	229	41,830
Pyramimonas spp.	9,373	3,456	668	33,950
Nitzschia spp.	1,043	313	750	28,563

	Whole Water Phytoplankton: #cells L Econfina/Fenholloway (August 2003)						
Taxon	E09	E11	E12	E13	F11	F15	TOTAL
Pseudonitzschia pseudodelicatissima	433,190	39,996	53,932	121,099	695,925	894,443	2,238,585
Chattonella sp.	0	0	0	0	491,400	851,787	1,343,187
Bacteriastrum hyalinum	68,882	0	856,240	2,222	0	0	927,344
Johannesbaptistia pellucida	31,108	0	0	307,747	307,800	0	646,655
Undetermined cryptophyte	53,328	82,214	23,352	93,324	73,575	54,653	380,446
Cyclotella atomus var. *nov.*	0	222,200	0	49,995	14,175	0	286,370
Undetermined nannoflagellates	29,997	95,546	13,900	51,106	29,700	41,323	261,572
Thalassiosira cedarkeyensis	14,443	112,211	5,004	114,433	4,725	6,665	257,481
Thalassiosira spp.	0	26,664	2,224	0	5,400	59,985	94,273
Merismopedia spp.	0	0	0	0	0	93,310	93,310

setigera, and dinoflagellates (*Gymnodinium* spp.). These blooms were not as diverse as in the previous year. The trends of both species of *Leptocylindrus* were somewhat different, with high numbers of *L. danicus* in estuarine areas during 1999 and more generalized distributions during 2000. *Leptocylindrus minimus* only appeared in inshore Fenholloway waters during 2000. The species *Skeletonema costatum*, a primary dominant during 1992, virtually disappeared from the Fenholloway system during the bloom periods. There was a more general distribution of the bloom species *Prorocentrum minimus* in the Fenholloway during 2000. Changes in the Econfina phytoplankton community during 1999–2000 could be ascribed to the naturally altered habitat conditions associated with the drought (i.e., increased light penetration, salinity, water temperature, and reduced nutrient loading). Reduced diversity in the Econfina system during 1999 could be ascribed to increased dominance by various species and general increases in overall numbers of phytoplankton during this period.

The data thus showed that during the period of increased light penetration, there were increased phytoplankton numbers in both systems. However, with time during the drought, numerical abundance of phytoplankton went down. Increased dominance by the bloom species *Leptocylindrus danicus* noted at Stations F09 and F10 in 1999 and the related species *L. minimus* during 2000 suggest the possibility that increased light penetration coupled with continued high ammonia and orthophosphate loading led to an overlap of deeper photic depths coupled with high nutrient gradients, a situation similar to that of the 1994 initiation of *L. danicus* blooms in Perdido Bay under similar habitat conditions (drought + increased nutrient loading; Livingston, 2000, 2002).

Net phytoplankton data (Table 4.2) supported the hypothesis that there was a basic shift of phytoplankton species composition and abundance in the Fenholloway system during 1999. Numbers, diversity, and evenness indices of the Fenholloway net phytoplankton during 1992 were relatively low compared to the Econfina. Species richness was generally lower in the Fenholloway during both sampling periods, with particularly low numbers in the estuary. On the other hand, numbers of phytoplankton were higher in the Fenholloway than the Econfina during 1999, with comparable reductions of Shannon diversity and evenness.

The phytoplankton data taken during 2001 (June–October; Table 4.2) indicated that there were no blooms during this period. In June, cryptophytes and nannoflagellates predominated, with bloom species such as *Rhizosolenia setigera* taken in the Fenholloway system in low numbers. During July, bloom species such as *R. setigera* and *Leptocylindrus danicus* were taken in the Fenholloway system in relatively low numbers. In August, cryptophytes and *Gymnodinium* spp. predominated in both systems. *Navicula* sp. was dominant in the Fenholloway system, but not in very high numbers. During September, *Navicula* sp. and *Cyclotella* cf. *atomus* were dominant in the Fenholloway system but were noted in relatively low numbers. In October, *Dactyliosolen fragilissimus* was a dominant in the Econfina, but not in the Fenholloway where bloom species such as *Rhizosolenia imbricata*, *Merismopedia* spp., and *Navicula* sp. were present. However, these species were taken in relatively low numbers. Overall, there were qualitative differences of the phytoplankton biota between the Econfina and Fenholloway systems, but the bloom species that were present in the Fenholloway system during 1999–2000 were taken in relatively low numbers. These conditions were noted during the major drought that occurred during this period.

During summer–fall 2002, no blooms were noted in the Fenholloway system (Table 4.2). Phytoplankton numbers were comparable to those in the previous year. Cryptophytes were once again the most abundant group in both systems. The lack of blooms in the Fenholloway system during 2001–2002 could be related to low rainfall (and associated low river flows). It is possible that river pressure was insufficient to carry the nutrients from the mill into offshore Fenholloway areas characterized by higher light penetration. Because of the unusually high chlorophyll *a* concentrations in offshore areas of the Fenholloway drainage during summer 2003, phytoplankton samples were taken from August to November 2003. Results of a spot check of samples taken at Stations F11 and F15 in August 2003 are noted in Table 4.2. Because of the methods used for collection of the August 2003 samples, it is likely that the data underestimated the actual cell counts during this month. During August 2003, the dominant phytoplankton in the Fenholloway system were the bloom species *Pseudonitzschia pseudodelicatissima* and *Chattonella* sp. The former was present in both the Econfina and Fenholloway systems. The raphidophyte *Chattonella* sp., however, was noted only in the Fenholloway area. Based on the chlorophyll *a* distributions during this time period (see above), it is likely that these species were present in bloom concentrations at intermediate offshore areas in the Fenholloway system during July–August 2003. In any case, the relatively high rainfall and river flow conditions during summer 2003 were associated with major concentrations of these bloom species in the Fenholloway system.

The fish-killing mechanism of raphidophyte blooms is still poorly understood. Both physical clogging of gills by mucus excretion as well as gill damage by hemolytic substances such as polyunsaturated fatty acids may be involved (Shimada et al., 1983; Chang et al., 1990). There is accumulating evidence that the production of superoxide radicals represents the primary mechanism of fish mortality. Imai et al. (1997) reported details concerning life-cycle and bloom dynamics of *Chattonella*, a genus with two known fish-killing species (*C. antqua* and *C. marina*). Nutrients and competitors (mainly diatoms) appear to affect the development of *Chattonella* populations. Anything interfering with diatom proliferation could give *Chattonella* an advantage. During a severe phytoplankton outbreak in the Seta Sea in 1972, a raphidophyte red tide killed 14 million cultured yellow tail fish. Effluent controls were then initiated to reduce the organic carbon loading and the discharge of phosphates from household detergents. Following a time lag of 4 years, the frequency of red tide events in the Seta Sea then decreased by about twofold to a more stable level (Hallegraeff, 1995). A similar pattern of long-term loading of coastal waters was evident for the North Sea in Europe (Smayda, 1989, 1990).

The annual trends of phytoplankton numbers and species richness are shown in Figure 4.9. Estuarine areas (E06, E07; F06, F09) did not have any particular temporal pattern in terms of numbers of cells. However, the offshore stations in both systems were characterized by increased numbers of phytoplankton cells during 1999 and 2003. Reduced numbers in both areas characterized drought years. Numbers of phytoplankton cells were usually higher at intermediate offshore stations in the Fenholloway system (F10, F11, F14). Species richness followed a different pattern, with reduced species richness at intermediate stations during 1999 and 2003. These trends are consistent with the hypothesis that, during years of high rainfall, there are increased observations of phytoplankton blooms in near-offshore Fenholloway areas relative to cognate stations in the Econfina system. These blooms are generally associated with reduced species richness. During drought conditions, nutrient gradients overlap high color and reduced light transmission, thus precluding habitat conditions conducive to plankton blooms. High rainfall and river flows move the nutrients into water where color effects are diluted and light transparency is enhanced, leading to blooms in areas of high nutrient loading from the mill. This pattern is consistent with the finding from 1992–1993. The interannual trend from generally innocuous bloom species to more toxic raphidophytes and possible red tide plankton is consistent with findings in the long-term analyses of nutrient loading from a pulp mill in the Perdido Bay system (Livingston, 2000, 2002). The discovery of the red tide dinoflagellate *Karenia brevis* at Station F16 in September 2003 was the first time this species was noted in the study area.

4.6.5   Comparison of Perdido Bay and Apalachee Bay

Long-term (16-year) studies have been carried out in the Perdido drainage system to determine the effects of pulp mill discharges of ammonia and orthophosphate on Perdido

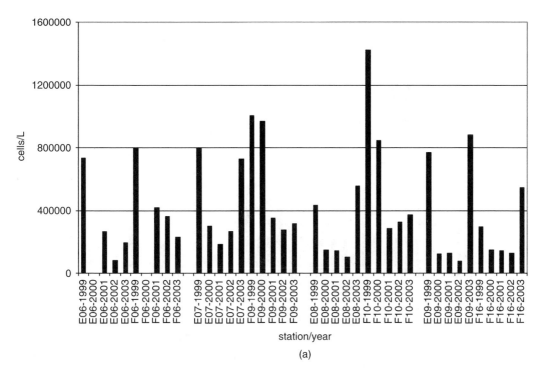

(a)

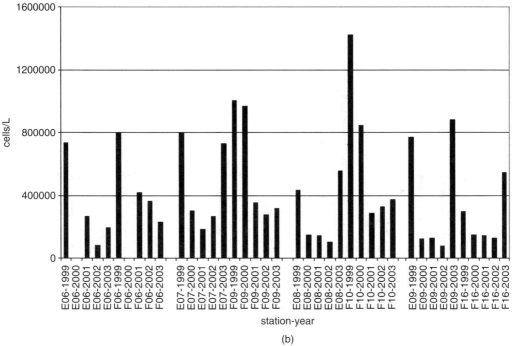

(b)

Figure 4.9 Comparison of annual averages of phytoplankton numbers of cells L^{-1} and species richness at stations in the Econfina and Fenholloway systems from 1999 to 2003.

Bay. A detailed analysis of the nutrient loading characteristics of the Perdido drainage system indicated that, during early years of analysis (1988–1991), orthophosphate and ammonia loading from the pulp mill enhanced secondary production in the immediate

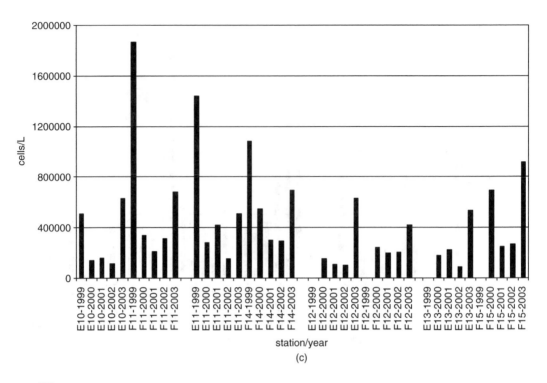

station/year

(c)

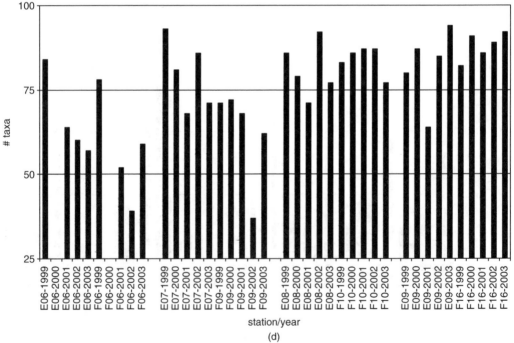

station/year

(d)

Figure 4.9 (continued)

receiving area of the upper estuary. There were no serious phytoplankton blooms during this period, and the phytoplankton community structure remained highly diverse. The chrysophytes, and to a lesser degree the chlorophytes, tended to be abundant and were associated with a balanced food web and relatively high secondary production in upper bay areas that received nutrient loading from the pulp mill. However, during a winter–spring drought in 1993–1994, the pulp mill increased orthophosphate loading to the bay. The combination of drought conditions and increased orthophosphate loading was associated with the initiation of a series of phytoplankton blooms that were dominated by various diatom species.

During 1995–1998, there were periodic increases of ammonia loading to upper Perdido Bay by the pulp mill. From late summer 1997 through spring 1999, the pulp mill reduced its orthophosphate loading, and there were reductions of the high relative dominance of certain bloom species, especially during winter–spring periods. However, by spring 1999, the mill again resumed high loading of orthophosphate to the bay, and this was accompanied by bay-wide spring and summer blooms, increased dominance of bloom species, and associated reductions of phytoplankton species richness. There were also qualitative and quantitative interannual dominance shifts within the bloom sequences over the period from 1993 through 1999 whereby larger-celled raphidophyte and dinoflagellate species replaced diatom species that had been predominant during the initial blooms. High nutrient loading tended to favor cryptophytes, cyanophytes, dinoflagellates, and raphidophytes. During periods of high winter orthophosphate loading, the dinoflagellates were numerically dominant, whereas summer ammonia loading was associated with increased dominance by raphidophytes.

During 2000, there were major increases in whole water phytoplankton abundance in the Fenholloway estuary (F9, F10, F11, F12, F16) compared to the Econfina system. Dominance was higher in the Fenholloway system and species richness generally lower at Stations F06, F09, and F14. Whole water phytoplankton species diversity and evenness were generally lower in the Fenholloway system. Peak phytoplankton numbers in the Fenholloway system were augmented by blooms (numbers $cell^{-1} > 10^6$) of *Leptocylindrus danicus* (1,831,582 cells L^{-1}, Station F09, September 1999; 4,390,902 cells L^1, Station F10, September 1999); *Johannesbaptistia pellucida* (2,111,220 cells L^{-1}, Station F11, September 1999); and *Cerataulina pelagica* (1,625,299 cells L^{-1}, Station F14, October 1999) (Table 4.1). The diatom *L. danicus* has a worldwide distribution in a variety of tropical coastal environments. It is often a major constituent of spring phytoplankton outbursts as well as fall growing periods (Marshall, 1988). This species has been associated with red tide events in Japan. *L. danicus* is believed to be harmful to caged sea trout (*Cynoscion regalis*), Atlantic salmon (*Salmo salar*), and smolt of coho salmon (*Oncothuynchus kisutch*). *L. danicus* was a key pioneer bloom species in the Perdido Bay system during the initial blooms of 1994.

The data thus showed that there were significant differences in the whole water phytoplankton species representation found in the Fenholloway system in 1999 and 2000 relative to data taken in both systems during 1992–1993 and the Econfina data taken in 1999. During January–February 2000, the primary dominants in the Fenholloway system were not bloom species. However, subdominants included *Cyclotella choctawhatcheeana* and *L. danicus*. Increased dominance by the bloom species *L. danicus* was noted at Stations F09 and F10 in 1999 and the related species *L. minimus* during 2000. Dominance by this species in the Perdido system was followed by the *L. danicus* blooms in January–February 1994. The high dominance of various bloom species during 1999–2000 gave support to the hypothesis that the Fenholloway system followed the pattern found in the Perdido estuary (Livingston, 2000). During fall 1993, *Falcula hyalina* was dominant in the Perdido system. The Fenholloway phytoplankton data are thus consistent with the hypothesis that high nutrient loading (ammonia and orthophosphate), habitat conditions, and periodic rainfall fluctuations to contribute to basic changes of the phytoplankton associations in the Fenholloway offshore system.

4.6.6   Zooplankton Distribution

Zooplankton numbers were considerably higher at the estuarine and Gulf Econfina stations when compared with the Fenholloway cognates (Livingston, 1993b). The lowest cumulative species numbers were found in the estuaries of the respective study areas. Somewhat higher numbers of species were found in the Econfina estuary than in the Fenholloway estuary. There was a general increase of species numbers at inshore stations of both systems, although the disparity between the two systems was maximal here. The overall highest numbers of species in both systems were found at the offshore stations. There was a higher number of species taken at the Econfina offshore station. In general, there was evidence of a suppression of a healthy zooplankton community in the Fenholloway inshore and near-shore areas when compared to the zooplankton communities found at the Econfina cognate areas. The alteration of primary (phytoplankton) production in the Fenholloway system could have been a major factor in the changes noted in the zooplankton assemblages.

4.7   Submerged Aquatic Vegetation

The most important component of Apalachee Bay is composed of SAV that provides both habitat and productivity for the complex food webs of the shallow offshore system (Livingston, 1975a). The significant elimination of seagrasses in the offshore Fenholloway system represents the single most important impact of the local pulp mill on the Gulf area. Although various factors affect SAV distribution in Apalachee Bay (Livingston et al., 1998a), watercolor as it affects light transmission was the primary factor in the impacts of mill effluents on offshore seagrass beds.

4.7.1   SAV Distribution in Space and Time

A comparison of changes in SAV distributions in offshore areas of the Econfina and Fenholloway drainages between 1992–1993 and 1999–2002 is shown in Figure 4.10. Inshore stations in both systems (E07, F09) had relatively little SAV during 1992–1993; however, there was a considerable increase of biomass in the Econfina system and, to a lesser degree, in the Fenholloway area during the 1999–2002 sampling period. These increases were pronounced in some Econfina areas (E08, E10). Stations F10 and F11 showed increases in SAV biomass richness compared to the 1992–1993 sampling period, although there was still a considerable difference between the Fenholloway SAV data and data from the reference site. These trends were reflected in data from the other stations. There was general equivalence of SAV biomass at Stations E12/F12 and E09/F16 during the 1999–2002 sampling period. These data illustrate the extreme interannual variation of SAV biomass in the study area. The data also show that there was no recovery of SAV biomass in areas of the Fenholloway offshore system that were affected by reduced light penetration in the past. These data are a further indication that the nominal reductions in watercolor during the drought in the Fenholloway system were not enough to bring back the benthic vegetation at a density comparable to that in the reference area.

A comparison of changes in SAV species richness in offshore areas of the Econfina and Fenholloway drainages from 1992–1993 and 1999–2002 is shown in Figure 4.11. Inshore stations in both systems (E07, F09) had relatively low SAV species richness during 1992–1993; however, there was a considerable increase of species richness in the Econfina system and, to a lesser degree, in the Fenholloway area during the 1999–2002 sampling period. These increases in SAV taxa were evident in some Econfina areas (E08, E10). Stations F10 and F11 showed increases in both SAV biomass and SAV species richness

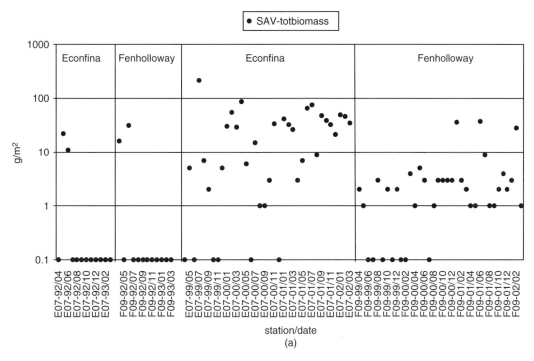

Figure 4.10 Comparison of biomass of SAV in the Econfina and Fenholloway offshore systems between the periods 1992–1993 and 1999–2002.

compared to the 1992–1993 sampling period, although there was still a considerable difference between the Fenholloway SAV data and data from the reference sites. These trends were reflected in data from the other stations, with some recovery at Station F15 during the latter sampling period. There was general equivalence of SAV biomass and species richness at Stations E12/F12 and E09/F16 during the 1999–2002 sampling period. This indicates that the color reduction program could have had a positive impact on the Fenholloway system in addition to the effects of the drought. However, inshore Fenholloway areas (F06, F09, F14) did not show any real improvement over this latter study period; and despite some increase in species richness at Station F10, there was relatively little improvement in SAV development in this area compared to the Econfina reference site.

During 1992–1993, the Econfina system was dominated by *Syringodium filiforme, Thalassia testudinum,* and *Halodule wrightii,* whereas the Fenholloway system was dominated by *H. wrightii* and *S. filiforme* (Livingston, 2004a). Monthly biomass and species richness levels were generally lower in the Fenholloway system, and cumulative species richness in the Fenholloway system was almost half that in the Econfina system. The differences in SAV composition in both systems during 1999–2000 were pronounced. There was a relatively important increase of *Spyridia filamentosa* in both systems. In the Fenholloway area, former dominance by *H. wrightii* was supplanted by *S. filiforme,* and there was a major increase of *Halophila englemannii* in the Fenholloway system. There were indications of a beginning of a recovery of SAV community indices in the Fenholloway system. However, with the exception of Station F12, *T. testudinum* and *S. filiforme* were still not well established in the Fenholloway system by the end of the study period. The seagrass *H. wrightii* was well established in inshore Fenholloway areas, however. Another dominant, *Caulerpa prolifera,* was almost absent from the Fenholloway system.

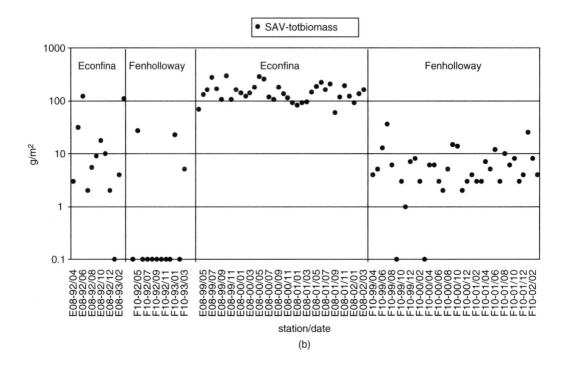

(b)

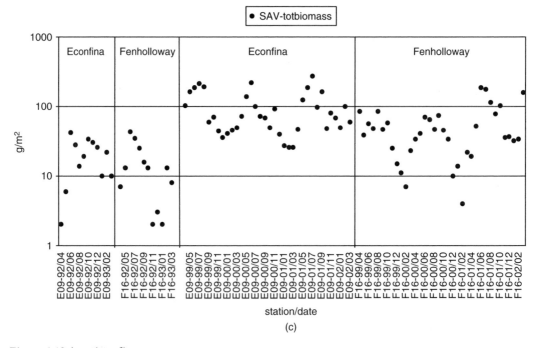

(c)

Figure 4.10 (continued)

There were some important changes in the SAV ecology of the Apalachee Bay area during the 1998–2002 study period. For the first time in over 30 years of sampling, active flowering was noted in four dominant seagrass species (Table 4.3). The flowers and fruiting bodies were noted mainly in offshore stations in both systems during spring and summer months. This form of reproduction makes the expansion of seagrass beds in Apalachee

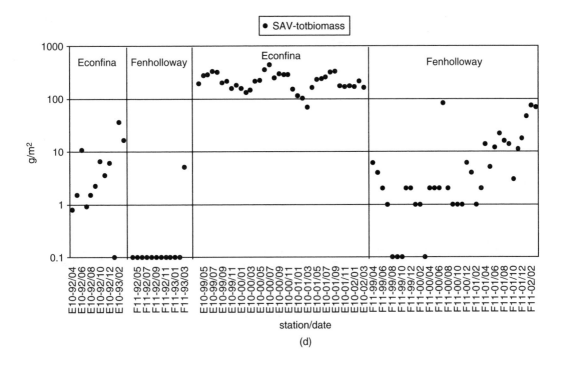

(d)

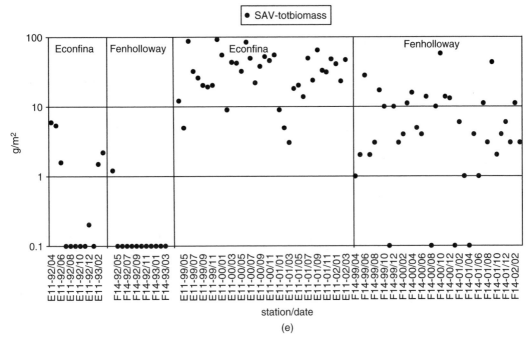

(e)

Figure 4.10 (continued)

Bay more feasible. The flowering of these species also coincides with the substantial increases in water temperature during the study period. In addition, there was a major invasion of the Fenholloway system by *Halophila englemannii,* a seagrass species that is usually found in deeper water areas of the Gulf. The distribution of this species was most abundant at offshore stations to the east (F15, FT29, F16), with some increases at Stations

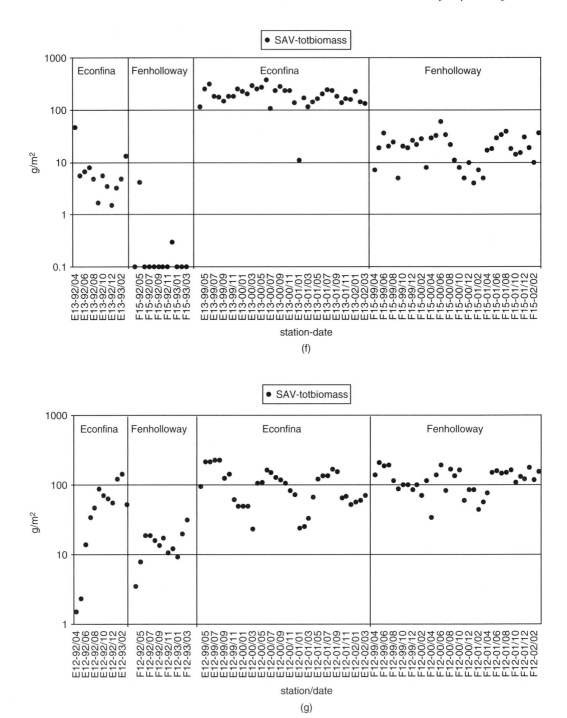

Figure 4.10 (continued)

E11 and E13 in the Econfina system. The temporal distribution at Station FT29 was mainly during summer months, with a trend toward interannual increases with peaks in July and August 2001. There was a general increase in overall biomass at Station FT29, with decreasing diversity indices as biomass increased.

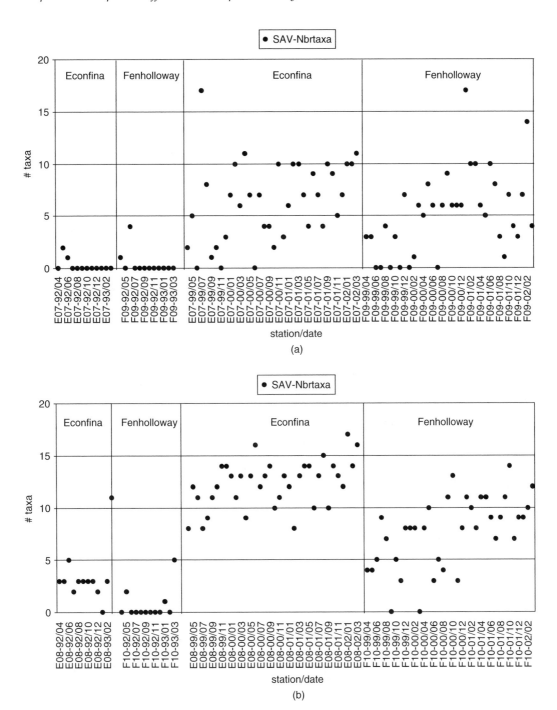

Figure 4.11 Comparison of species richness of submerged aquatic vegetation in the Econfina and Fenholloway offshore systems between the periods 1992–1993 and 1999–2002.

4.8 Invertebrates

In the Econfina system, invertebrate dominance relationships were similar during the three sampling periods from 1999 to 2002 (Livingston, 2004a). In the Fenholloway system, dominance shifted form *Tozeuma carolinensis* to *Hippolye zostericola*. Secondary dominants

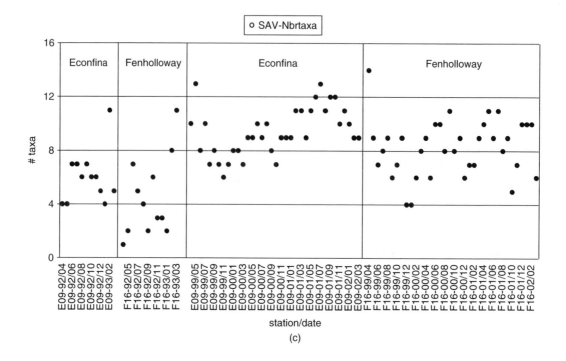

station/date

(c)

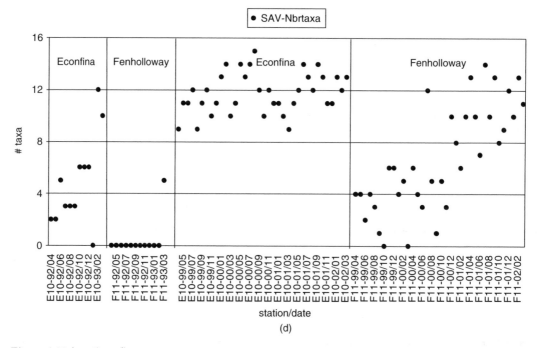

station/date

(d)

Figure 4.11 (continued)

were also different during later sampling periods. A comparison of the trends of invertebrate community indices in the Econfina and Fenholloway offshore systems (Figure 4.12)
indicated significant reductions in the numbers of individuals and species richness in all
Fenholloway areas except Stations F12 and F16. These offshore areas in the Fenholloway
system were the only areas characterized by the presence of seagrass beds, a result of

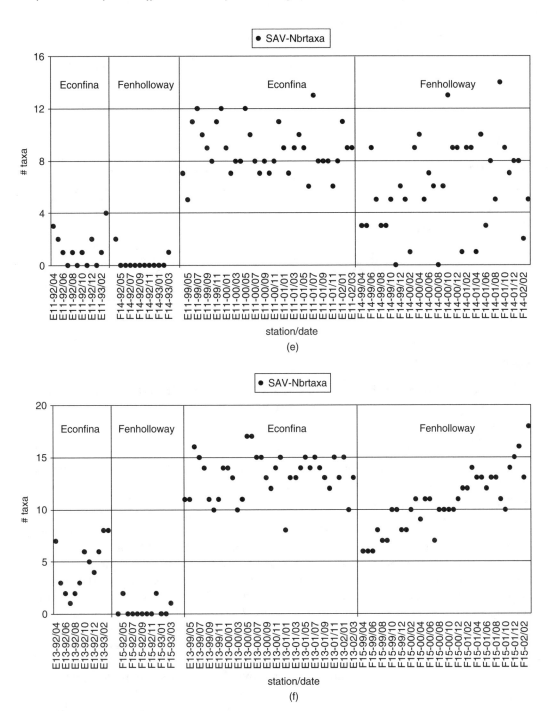

Figure 4.11 (continued)

water quality changes associated with dilution of the effluent as it moved offshore. There was thus a trend toward reduced invertebrate numbers and species richness in areas of the Fenholloway system that were characterized by debilitated seagrass beds. Species richness and diversity at Station F15 tended to increase with time (Livingston, 2004a) as a response to increased SAV species richness in this area. However, overall, this analysis indicated that the continued absence of SAV biomass and species richness was accompa-

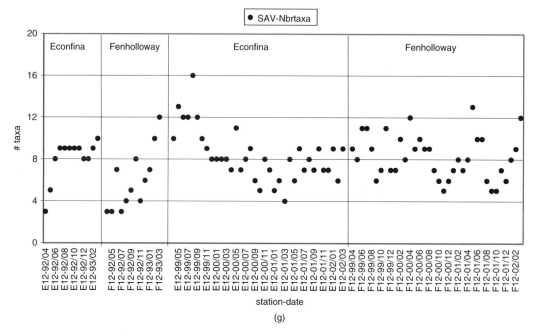

Figure 4.11 (continued)

nied by reduced numbers of invertebrates and reduced invertebrate species richness in the Fenholloway system relative to reference areas.

4.9   Fishes

A comparison of fish abundance taken in the study areas during the 1992–1993 and 1999–2000 sampling periods (Figure 4.13) showed that the greatly increased SAV biomass during the latter sampling effort was associated with increased fish numbers in both systems. These changes were accompanied by shifts of species dominance in both systems. In the Econfina area, increased numbers during the latter sampling period were accompanied by lower cumulative species richness at Stations E08, E09, E10, E12, and E13. This indicated possible competitive interactions and inhibition with higher dominance. The top dominants shifted from *Orthopristis chrysoptera*, *Bairdiella chrysura*, and *Syngnathus floridae* in 1992–1993 to *Micrognathus criniger*, *Diplodus holbrooki*, and *Paraclinus fasciatus* during the latter sampling period. *Lagodon rhomboides* remained as a main dominant during both sampling periods. These shifts could have been associated with changes in food web interactions that could, in turn, have been related to changes in the seagrass beds.

There was a general trend of increased numbers of fishes during summer periods that tended to follow, inversely, the rainfall patterns over the 5-year study period (Figure 4.1). During the period 2003–2004, fish numbers dropped precipitously at most stations in both systems. In the Fenholloway system, there were low fish numbers even at Stations F12 and F16, areas where fish abundance had been high in previous years. There was no evidence of recovery of fishes in the Fenholloway offshore area, and the pattern of fish distribution tended to follow that of the seagrasses.

A similar pattern was noted concerning the distribution of fish species richness in the Econfina and Fenholloway Rivers (Figure 4.14). In general, fish species richness increased during summer–fall months. There was a pattern of increased fish species richness during

Table 4.3 Incidents of Flowering of Seagrass Species in the Econfina and Fenholloway Systems over the 1999–2002 Sampling Period (also shown are the dry weights of the flowering plants taken by station by date)

Station	Date	Species	Dry Wt. (gm⁻²)	Station	Date	Species	Dry Wt. (gm⁻²)
F16	Jun-00	*Halophila englemannii*	nd	F12	Apr-99	*Syringodium filiforme*	0.85
F15	Jun-00	*Halophila englemannii*	2	F11	Jul-00	*Syringodium filiforme*	nd
F10	Jun-00	*Halophila englemannii*	2.85	E13	Jul-00	*Syringodium filiforme*	nd
FT29	Jul-00	*Halophila englemannii*	nd	E12	Jul-00	*Syringodium filiforme*	nd
E13	Jul-00	*Halophila englemannii*	nd	E10	Jul-00	*Syringodium filiforme*	nd
E08	Jul-00	*Halophila englemannii*	nd	E08	Jul-00	*Syringodium filiforme*	nd
FT29	May-01	*Halophila englemannii*	0.25	E10	Aug-00	*Syringodium filiforme*	nd
F15	May-01	*Halophila englemannii*	0.4	F12	May-01	*Syringodium filiforme*	0.45
F11	May-01	*Halophila englemannii*	0.2	E13	May-01	*Syringodium filiforme*	1.6
FT29	Jun-01	*Halophila englemannii*	0.25	E12	May-01	*Syringodium filiforme*	0.15
F16	Jun-01	*Halophila englemannii*	0.15	E10	May-01	*Syringodium filiforme*	0.9
F15	Jun-01	*Halophila englemannii*	0.25	E09	May-01	*Syringodium filiforme*	0.75
F11	Jun-01	*Halophila englemannii*	0.05	E08	May-01	*Syringodium filiforme*	3.5
F10	Jun-01	*Halophila englemannii*	0.1	F16	Jun-01	*Syringodium filiforme*	0.4
E08	Jun-01	*Halophila englemannii*	0.15	E13	Jun-01	*Syringodium filiforme*	3.5
E07	Jun-01	*Halophila englemannii*	0.1	E12	Jun-01	*Syringodium filiforme*	0.05
F11	Jul-01	*Halophila englemannii*	0.1	E10	Jun-01	*Syringodium filiforme*	0.9
E13	Jul-01	*Halophila englemannii*	0.05	E09	Jun-01	*Syringodium filiforme*	0.3
E10	Jul-01	*Halophila englemannii*	0.05	E08	Jun-01	*Syringodium filiforme*	2.3
E08	Jul-01	*Halophila englemannii*	0.05	E13	Jul-01	*Syringodium filiforme*	0.15
F15	Jan-02	*Halophila englemannii*	0.5	E10	Jul-01	*Syringodium filiforme*	1.4
F15	Feb-02	*Halophila englemannii*	0.1	E09	Jul-01	*Syringodium filiforme*	1.9
F11	Feb-02	*Halophila englemannii*	0.2	E08	Jul-01	*Syringodium filiforme*	0.3
FT29	Mar-02	*Halophila englemannii*	0.2	E08	Jan-02	*Syringodium filiforme*	0.1

Table 4.3 (continued) Incidents of Flowering of Seagrass Species in the Econfina and Fenholloway Systems over the 1999–2002 Sampling Period (also shown are the dry weights of the flowering plants taken by station by date)

Station	Date	Species	Dry Wt. (gm⁻²)	Station	Date	Species	Dry Wt. (gm⁻²)
F15	Mar-02	*Halophila englemannii*	0.25	E13	Feb-02	*Syringodium filiforme*	0.2
F11	Mar-02	*Halophila englemannii*	0.3	E10	Feb-02	*Syringodium filiforme*	0.2
F10	Mar-02	*Halophila englemannii*	0.1	E09	Feb-02	*Syringodium filiforme*	0.9
E11	Mar-02	*Halophila englemannii*	0.05	E08	Feb-02	*Syringodium filiforme*	0.9
E12	May-99	*Thalassia testudinum*	nd	F16	Mar-02	*Syringodium filiforme*	2.05
E09	Jun-01	*Thalassia testudinum*	0.55	F12	Mar-02	*Syringodium filiforme*	0.6
E12	Jun-00	*Thalassia testudinum*	nd	E09	Mar-02	*Syringodium filiforme*	0.1
E12	Jun-01	*Thalassia testudinum*	0.1	E08	Mar-02	*Syringodium filiforme*	0.6
E12	Jul-01	*Thalassia testudinum*	0.1	E11	Apr-01	*Ruppia maritima*	5.7
				E11	May-01	*Ruppia maritima*	2.05
				FT29	Jun-01	*Ruppia maritima*	0.2

the drought period. This was particularly true in the Econfina system. During the 2003–2004 period, fish species richness was particularly low, especially at Fenholloway stations that included areas formerly characterized by high levels of this index.

4.10   Summary of Findings

Statistical analyses were made concerning comparisons of various physicochemical and biological indices at stations in the Econfina and Fenholloway systems over the 33-year study period. Methods used for the comparison of monthly data (water quality, biological factors) were developed to determine significant differences between matching Econfina and Fenholloway sites (unpolluted and polluted) over 9 to 12-month study periods (Livingston et al., 1998a; Livingston, 2000), and are given in Appendix II. Tables were constructed of the means where the statistical test could be run without serial autocorrelations.

4.10.1   Water Quality

Water temperature and specific conductance in the Fenholloway River were significantly higher than that in the Econfina system from 1992 to 2004 (Table 4.4). Differences of color at river Stations E01-F03 were significant from 1992–1993 to 2003–2004. During 2003–2004, the averaged color levels were also significantly higher at Stations F06, F09, and F10. Statistical equivalence of this factor was noted at offshore stations (E09, F16). Differences of DO at river Stations E01-F03 were significant from 1992–1993 to 2003–2004. The seasonally averaged DO at Station F03 was relatively low throughout the extended study period. These reductions were significantly lower in the Fenholloway estuary at Stations

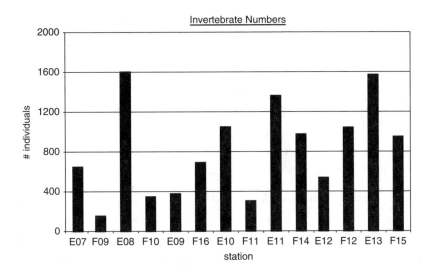

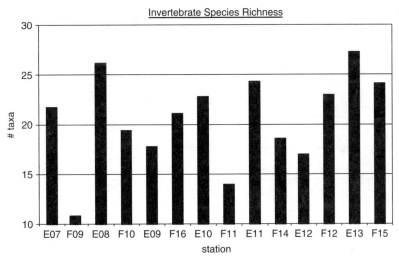

Figure 4.12 Comparison of invertebrate numbers and species richness by cognate stations in the Econfina and Fenholloway offshore systems. Data were averaged over all months from 1999 to 2002.

F06 and F09 during 2003–2004. There were no significant differences of DO taken at offshore stations (E08-F10, E09-F16). There were significantly ($P < 0.05$) higher concentrations of $NO_2 + NO_3$, NH_3, and PO_4 in the Fenholloway River (Station F03) from 1992–1993 through 2003–2004. These differences extended to estuarine stations (E06, F06) during the 2003–2004 period. There were no significant differences between nutrient concentration averages in areas farther offshore (F10, F16).

4.10.2 Chlorophyll a

There was reduced chlorophyll *a* in the Fenholloway estuary during the early years of sampling, with increased chlorophyll *a* at stations just offshore (F10) or near-shore (F11, F14) (Table 4.5). There was also a general trend in relatively higher chlorophyll concentrations during 2000–2001 and 2003–2004. The levels during the final year of sampling were higher than in any preceding years, and indicated a basic change in phytoplankton

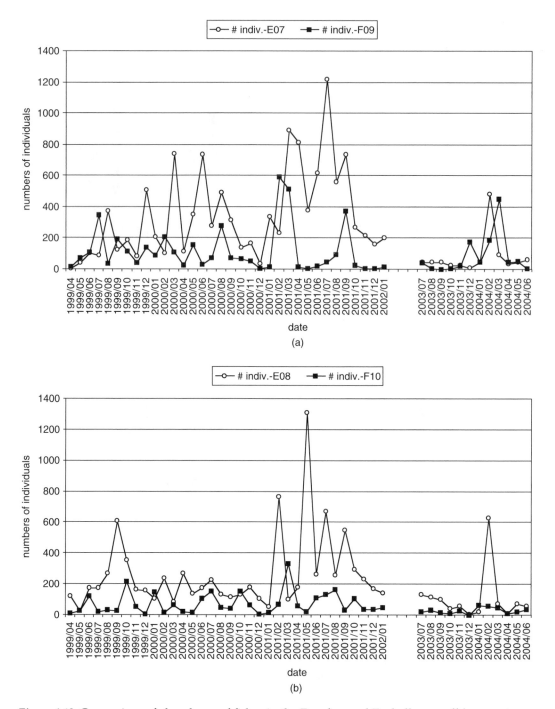

Figure 4.13 Comparison of abundance of fishes in the Econfina and Fenholloway offshore systems from April 1999 through June 2004.

activity that was consistent with the observed changes in the planktonic bloom organization in the Fenholloway system. During the 2003–2004 period, chlorophyll *a* was significantly lower in the Fenholloway estuary and significantly higher in most of the Fenholloway offshore areas.

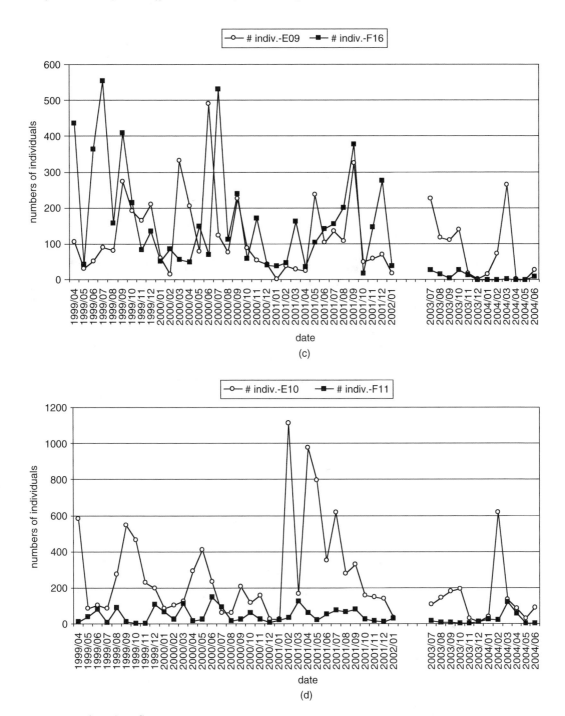

Figure 4.13 (continued)

4.10.3 Phytoplankton

The overall spatial/temporal distribution of phytoplankton in the study area (Table 4.6) indicates that the primary bloom species are located in the Fenholloway offshore system. *Leptocylindrus minimus* and *L. danicus* are located mainly in offshore areas. *Pseudonitzschia*

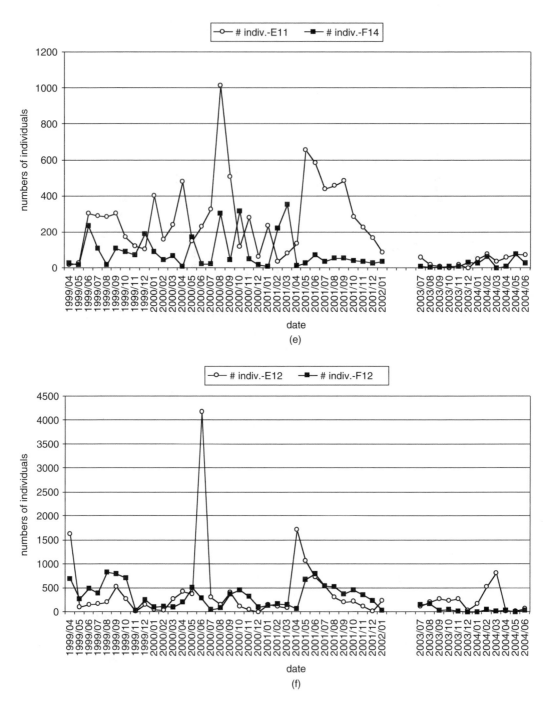

Figure 4.13 (continued)

pseudodelicatissima was located at Stations F11 and F15. *Rhizosolenia setigera* was distributed throughout the Fenholloway system, and was also found in high numbers at Station E12 in the Econfina area. The raphidophyte *Chattonella* sp. was found mainly at Stations F11 and F15. The red tide organism, *Karenia brevis*, was found for the first time at Station F16. With the exception of the offshore stations (F12, F16), the average numbers of cells L^{-1} were considerably higher in the Fenholloway system.

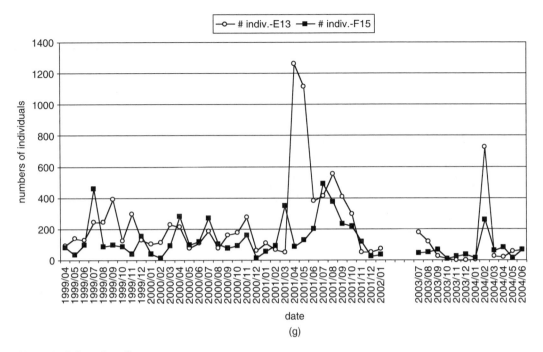

(g)

Figure 4.13 (continued)

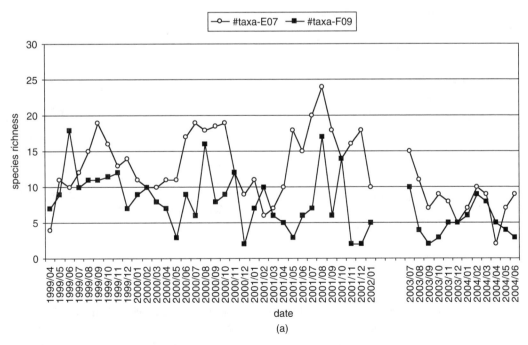

(a)

Figure 4.14 Comparison of species richness of fishes in the Econfina and Fenholloway offshore systems from April 1999 through June 2004.

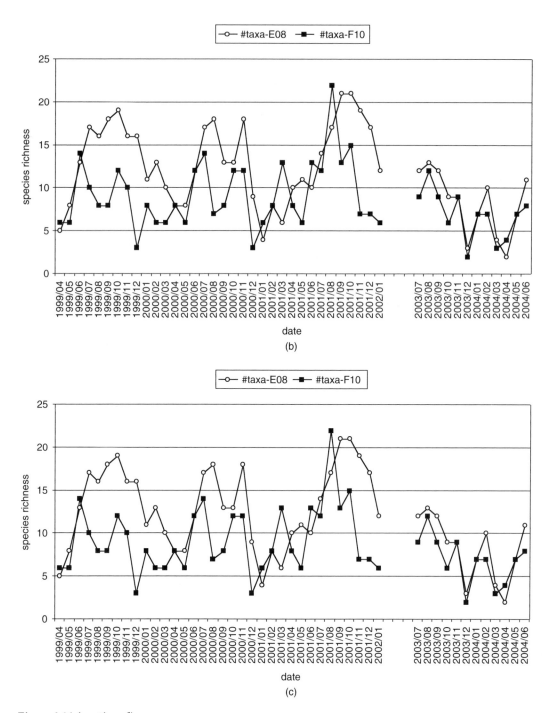

Figure 4.14 *(continued)*

Cell numbers were significantly depressed in the Fenholloway estuary relative to the Econfina estuary in 1992 (Table 4.7). This trend was reversed during more recent years (1999, 2000, 2001). Phytoplankton species richness and diversity were significantly reduced in the Fenholloway estuary during the entire sampling period. These trends were probably related to river flow rates and light penetration trends, along with the continued high

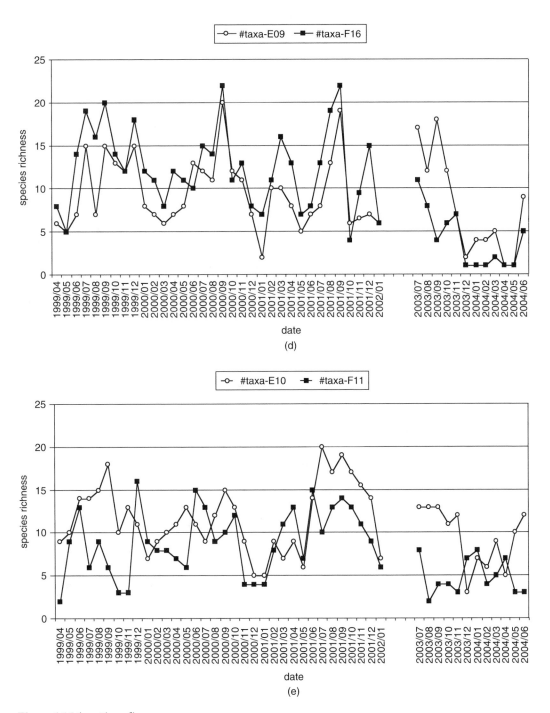

Figure 4.14 (continued)

nutrient loading to the Fenholloway estuary. In near-shore areas (F10, F11, F14), the trend of reduced numbers of cells in 1992 changed to significantly increased numbers of cells with accompanying reductions in phytoplankton species richness during the period 1999–2001.

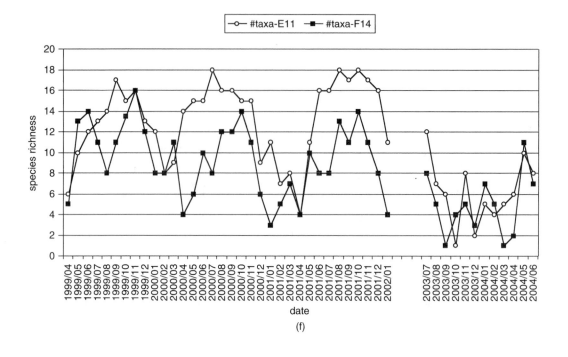

(f)

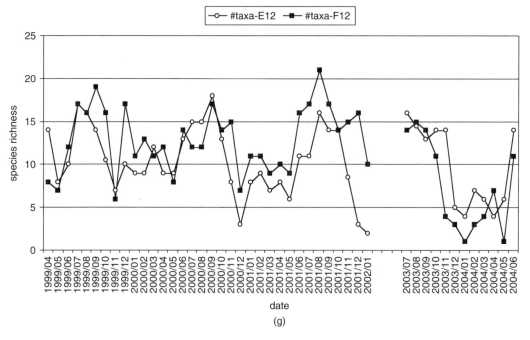

(g)

Figure 4.14 (continued)

4.10.4   *Submerged Aquatic Vegetation*

A summary of the SAV data (Table 4.8) indicated peak SAV concentrations at intermediate Econfina stations (E08, E10, E13) as a response to increased light penetration and increased nutrients. Offshore areas were nutrient limited. With the exception of Station F12, there

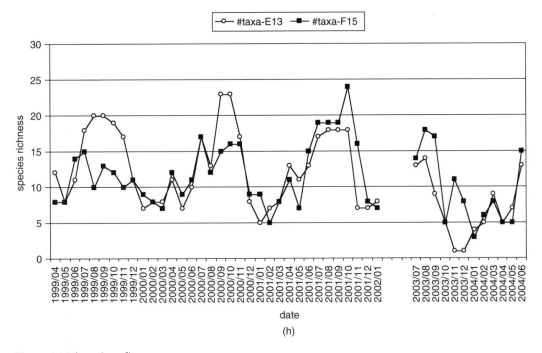

Figure 4.14 (continued)

were major decreases of SAV in the entire Fenholloway study area over the 3-year period of observation. *Syringodium filiforme* was a major dominant in offshore Econfina areas. *Halophila englemanii* moved into the Fenholloway areas during the study, with high concentrations at Stations F15 and F16. *Thalassia testudinum* was concentrated around Station E12.

A statistical analysis was conducted of long-term trends of SAV in the study areas (Table 4.9). With the exception of offshore areas (E12, F12; E09, F16), SAV biomass was significantly lower in the Fenholloway system than in the Econfina system. This includes periods from 1972–1973 through the intensive analyses of 1992–1993 to the most recent analyses (1999–2002). There was considerable interannual variation of SAV biomass over the study period, but there was no significant improvement of this index over the recent 3-year period of analysis. SAV species richness showed a similar trend, with significant reductions of this index in offshore Fenholloway areas relative to reference stations. The primary difference between species richness and biomass was an increase of SAV numbers of taxa at Station F15 during the period 2000–2002 and at Station F11 during 2001–2002. Overall, however, there has not been an appreciable recovery of SAV biomass and species richness since the early 1970s.

4.10.5 Invertebrates

The invertebrates (Table 4.10) in the study area were dominated by various crustacean species in the form of caridean shrimp. With the exception of Stations F12 and F16, there were lower numbers of invertebrates in the Fenholloway system relative to Econfina cognates.

A review of the trends of invertebrate community indices in the Econfina and Fenholloway offshore systems from 1972 (Table 4.11) indicated significant reductions in numbers and species richness of epibenthic invertebrates at Stations F09, F10, F11, and F14

Table 4.4 Comparisons of Water Quality Factors at Stations in the Econfina and Fenholloway Systems

Factor Cognate Station Pair	Color (PCU)	DO (mg L⁻¹)	Cond. (µmhos c(m⁻¹))	Chlor.a (µγΛ⁻¹)	Temp. (°C)	NO₂/NO₃ (mg L⁻¹)	NH₃ (mg L⁻¹)	PO₄ (mg L⁻¹)
E01/F03-(6/92–4/93) River	302/1252 WS	3.4/2.2 WS	68/3993 WS	0.51/0.53 WN	15.1/24.1 WS	0.05/0.79 WS	0.06/1.64 WS	0.05/0.99 WS
E01/F03-(4/98–3/99) River	207/955 WS	5.2/1.1 WS	392/1504 WS	0.05/0.12 WN	21.1/24.2 WS	0.01/0.39 WS	0.05/2.67 WS	0.04/0.87 WS
E01/F03-(4/99–3/00) River	88/1005 WS	6.1/0.79 WS	431/1804 WS	1.29/0.49 WN	20.3/23.5 WS	0.24/0.86 WS	0.07/3.21 WS	0.08/1.24 WS
E01/F03-(4/00–3/01) River	57/956 WS	6.1/0.75 WS	390/1702 WS	1.1/0.6 WN	19.9/22.1 WS	0.023/0.073 WN	0.04/2.38 WS	0.05/0.45 WS
E01/F03-(4/01–3/02) River	167/944 WS	6.1/0.85 WS	434/1710 WS	0.76/0.69 WN	20.1/23.5 WS	0.02/0.14 WS	0.09/3.10 WS	0.06/0.80 WS
E01/F03-(6/02–11/02) River	302/902 WS	nd	nd	1.7/3.9 WS	nd	0.03/0.06 WS	0.05/3.02 WS	0.08/0.88 WS
E01/F03-(7/03–3/04) River	574/794 WS	5.7/1.7 WS	194/1021 WS	5.7/1.7 WS	18.8/21.2 WS	0.02/0.10 WS	0.07/1.36 WS	0.06/0.46 WS
E06/F06-(7/03–3/04) (River Mouth)	405/488 WS	6.7/4.7 WS	—	1.83/2.3 WN	18.9/21.5 WS	0.03/0.33 WS	0.051/0.163 WN	0.30/0.21 WS
E07/F09-(7/03–3/04) (River Mouth)	168/314 WS	7.5/5.8 WS	—	6.1/4.3 WN	19.4/21.5 WN	0.01/0.23 WS	0.043/0.145 WS	0.01/0.15 WS
E08/F10-(7/03–3/04) (Center 1.5 km)	56.5/75.8 WS	7.9/7.9 WN	—	6.1/10.5 WS	19.9/21.3 WN	0.015/0.023 WN	0.031/0.031 WN	0.005/0.011 WN
E09/F16-(7/03–3/04) (Center 5 km)	28.7/19.4 WN	7.8/7.7 WN	—	7.8/6.2 WN	20.2/20.9 WN	0.03/0.05 WN	0.032/0.052 WN	0.005/0.005 WN

Note: Data were averaged over 9- to 12-month periods from 1992 to 2004 and analyzed with *t*-tests and Wilcoxon Means tests for differences (see Appendix II).
WN, not sig. (p = 0.05).
WS, sig. (p = 0.05).

Table 4.5 Long-Term Comparisons of Chlorophyll *a* Averaged over 12-Month Periods and Analyzed with *t*-Tests and Wilcoxon Means Tests for Differences (see Appendix II)

Cognate (Station Pair)	Chlorophyll *a* April 1992–March 1993 (μg L^{-1})	Chlorophyll *a* April 1998–March 1999 (μg L^{-1})	Chlorophyll *a* April 1999–March 2000 (μg L^{-1})
E06/F06 (River Mouth)	**5.58/2.81** **WS (P < 0.01)**	3.0/4.3 WN	3.0/3.9 WN
E07/F09 (River Mouth)	**8.3/4.04** **WS (P < 0.05)**	3.6/6.5 WN	3.0/5.0 WN
E08/F10 (Center 1.5 km)	**4.1/8.0** **WS (P < 0.01)**	**3.8/7.4** **WS (P < 0.02)**	**1.0/4.7** **WS (P < 0.02)**
E09/F16 (Center 5 km)	3.4/3.2 WN	4.3/3.3 WN	0.7/1.3 WN
E10/F11 (W. Shore)	3.7/5.7 WN	4.1/5.4 WN	**1.0/5.7** **WS (p<0.02)**
E12/F12 (W. 3 km)	8.1/10.4 WN	2.7/5.7 WN	0.9/1.6 WN
E11/F14 (E. Shore)	2.6/6.1 WN	5.1/9.3 WN	4.3/6.4 WN
E13/F15 (E. 1.5 km)	5.5/7.3 WN	3.7/6.4 WN	**1.0/4.9** **WS (p<0.02)**
E06/F06 (River Mouth)	**4.4/7.4** **WS (P < 0.01)**	**2.2/4.7** **WS (P < 0.02)**	1.2/2.7 WN	2.3/3.0 WN

Table 4.5 (continued) Long-Term Comparisons of Chlorophyll *a* Averaged over 12-Month Periods and Analyzed with t-Tests and Wilcoxon Means Tests for Differences (see Appendix II)

Cognate (Station Pair)	Chlorophyll *a* April 2000–March 2001 (μg L)	Chlorophyll *a* April 2001–March 2002 (μg L)	Chlorophyll *a* April 2002–March 2003 (μg L)	Chlorophyll *a* April 2003–March 2004 (μg L)
E07/F09 (River Mouth)	3.3/8.8 WN	**1.6/9.1** WS (P < 0.02)	6.9/6.8 WN	**8.2/4.3** WS (P < 0.05)
E08/F10 (Center 1.5 km)	**1.1/6.2** WS (P < 0.02)	**0.9/3.9** WS (P < 0.02)	**0.8/4.2** WS (P < 0.02)	1.0/16.6 WS (P < 0.05)
E09/F16 (Center 5 km)	0.5/1.4 WN	0.4/0.9 WN	0.9/1.0 WN	10.2/5.3 WS (P < 0.05)
E10/F11 (W. Shore)	**0.7/3.6** WS (P < 0.05)	1.6/1.9 WN	1.6/1.9 WN	**11.4/17.5** WS (P < 0.05)
E12/F12 (W. 3 km)	**0.5/2.0** WS (P < 0.02)	**0.5/2.0** WS (P < 0.02)	1.4/2.7 WN	8.0/8.7 WN
E11/F14 (E. Shore)	**3.6/9.9** WS (P < 0.05)	2.8/6.1 WN	**1.9/7.8** WS (P < 0.02)	12/14.1 WN
E13/F15 (E. 1.5 km)	0.9/5.8 WN	0.9/4.0 WN	0.9/3.4 WN	**11.5/22.5** WS (P < 0.05)

Note:
WN, not sig. (p = 0.05).
WS, sig. (p = 0.05).
Boldface = bloom species

Table 4.6 Summary of Phytoplankton Data Taken in the Econfina and Fenholloway Offshore Systems from 1999 to 2004 (distribution of the top 35 species are by numbers by station. Also shown is the distribution of *Karenia brevis* (red tide species)

Species	E06	F06	E07	F09	E08	F10	E09	F16
Johannesbaptistia pellucida	3,471	4,995	27,031	26,593	36,591	31,251	57,816	25,162
Leptocylindrus minimus	**0**	**222**	**499**	**101,977**	**1,300**	**104,483**	**17,879**	**51**
Leptocylindrus danicus	**0**	**10,157**	**1,217**	**88,599**	**461**	**178,367**	**456**	**3,682**
Cyclotella choctawhatcheeana	1,140	1,510	16,067	2,996	6,557	2,310	8,121	409
Cyclotella cf. *atomus*	5,086	38,397	6,584	30,866	5,250	15,178	4,471	1,345
Proboscia alata	211	6,826	948	17,356	7,973	24,149	29,223	2,567
Dactyliosolen fragilissimus	2,209	1,109	8,906	1,602	3,504	14,628	4,542	2,524
Cerataulina pelagica	18	422	885	4,623	431	11,892	5,057	4,566
Thalassiosira cedarkeyensis	5,259	24,454	10,080	13,961	8,970	8,871	1,866	2,940
Pseudonitzschia pseudodelicatissima	**0**	**0**	**0**	**0**	**0**	**0**	**5,308**	**0**
Thalassiosira minima proschikinae complex	1,332	15,251	2,549	23,730	2,190	7,552	968	6,315
Asterionellopsis glacialis	88	543	1,940	4,393	2,228	3,237	48,649	2,714
Rhizosolenia setigera	**99**	**3,054**	**4,045**	**5,989**	**1,334**	**28,896**	**551**	**5,088**
Cylindrotheca closterium	4,540	8,913	5,504	8,009	1,302	4,158	566	2,200
Chattonella sp.	**0**	**44**	**0**	**0**	**0**	**147**	**0**	**743**
Bacteriastrum hyalinum	140	44	3,136	0	608	100	2,569	57
Chaetoceros laciniosus	1,036	178	24,461	95	7,055	1,040	284	7,100
Cerataulina pelagica (veg.)	0	0	0	0	0	0	0	0
Thalassiosira proschkinae	88	11,518	346	19,484	0	2,711	19	0
Chaetoceros compressus	105	0	7,977	650	3,614	7,458	463	500
Helicotheca tamensis	105	1,187	448	1,237	243	6,380	12	1,902
Navicula sp. 54 small (F09 8/01)	18	2,952	13	13,471	0	4,509	0	6
Cocconeis scutellum	1,176	121	2,993	215	4,176	1,154	1,263	502
Skeletonema sp. 1 (2 celled)	193	12,918	410	4,348	13	2,957	0	0
Skeletonema costatum	245	932	633	1,927	57	1,618	252	595
Peridinium quinquecorne	5,225	200	268	762	19	46	43	44

Table 4.6 (continued) Summary of Phytoplankton Data Taken in the Econfina and Fenholloway Offshore Systems from 1999 to 2004 (distribution of the top 35 species are by numbers by station. Also shown is the distribution of *Karenia brevis* (red tide species)

Species	E06	F06	E07	F09	E08	F10	E09	F16
Pyramimonas-like sp.	1,875	0	16,906	87	0	0	0	0
Rhizosolenia imbricata	18	0	218	278	58	207	555	3,987
Cyclotella atomus var. *nov.*	105	266	77	0	179	453	25	0
Hemiselmis sp. A	0	0	0	0	0	0	0	0
Chaetoceros tortissimus	0	0	884	0	365	2,438	1,024	0
Thalassionema nitzschioides	308	0	527	349	486	3,864	380	604
Chaetoceros subtilis f. abnormis	0	11	13	2,253	167	287	0	0
Merismopedia tenuissima	**0**	622	**0**	**1,205**	**0**	**0**	**0**	**0**
Karenia brevis	**0**	**9,146**	**0**	**0**	**0**	**0**	**0**	**205**
TOTAL	52,214	164,683	167,264	385,571	104,121	486,745	247,899	85,902
Cumulative Species Richness	97	88	123	102	130	135	127	141
Shannon Diversity	2.98	2.64	3.00	2.43	2.65	2.33	2.24	2.93
Shannon Evenness	0.65	0.59	0.62	0.53	0.54	0.47	0.46	0.59

Species	E10	F11	E11	F14	E12	F12	E13	F15	TOTAL
Johannesbaptistia pellucida	25,879	111,853	11,675	13,284	24,310	19,689	44,077	16,333	480,010
Leptocylindrus minimus	**391**	**14,716**	**611**	**12,103**	**135**	**6,161**	**142**	**61,899**	**322,569**
Leptocylindrus danicus	**557**	**5,258**	**3,398**	**1,065**	**3,425**	**1,066**	**357**	**730**	**298,795**
Cyclotella choctawhatcheeana	49,565	3,691	130,485	3,489	341	758	429	412	228,280
Cyclotella cf. *atomus*	6,362	40,395	17,305	31,920	1,673	2,039	2,154	388	209,413
Proboscia alata	9,408	15,082	2,244	9,524	12,231	17,732	18,862	22,319	196,655
Dactyliosolen fragilissimus	608	24,169	55,741	10,336	571	2,831	27,005	7,945	168,230
Cerataulina pelagica	187	19,659	863	86,345	191	18,907	64	12,805	166,915
Thalassiosira cedarkeyensis	8,601	10,216	16,977	18,482	3,671	5,470	10,294	8,840	158,952
Pseudonitzschia pseudodelicatissima	**0**	**26,766**	**1,481**	**0**	**3,868**	**0**	**5,767**	**43,925**	**134,888**
Thalassiosira minima proschkinae complex	538	4,048	1,079	36,611	745	4,612	1,720	18,989	128,229

Asterionellopsis glacialis	1,781	4,546	572	22,461	1,588	1,275	556	15,961	112,532
Rhizosolenia setigera	**1,549**	**5,227**	**1,131**	**4,307**	**13,281**	**5,028**	**599**	**16,035**	**96,213**
Cylindrotheca closterium	664	4,062	4,291	11,910	1,734	1,881	808	4,680	65,222
Chattonella sp.	**0**	**19,156**	**25**	**0**	**0**	**0**	**0**	**42,813**	**62,928**
Bacteriastrum hyalinum	0	51	6	13	41,154	0	106	16	48,000
Chaetoceros laciniosus	45	90	111	173	0	0	0	0	41,668
Cerataulina pelagica (veg.)	0	36,912	0	0	0	0	0	0	36,912
Thalassiosira proschkinae	0	391	148	267	0	0	0	0	34,972
Chaetoceros compressus	1,185	3,414	1,672	1,426	1,349	125	207	3,902	34,047
Helicotheca tamensis	589	3,464	604	4,036	32	3,463	817	5,867	30,386
Navicula sp. 54 small (F09 8/01)	0	1,140	0	3,050	0	541	0	1,087	26,787
Cocconeis scutellum	2,031	1,365	4,732	593	1,944	1,200	1,624	906	25,995
Skeletonema sp. 1 (2 celled)	64	814	197	2,945	0	0	0	0	24,859
Skeletonema costatum	250	1,363	717	10,204	215	660	238	1775	21,681
Peridinium quinquecorne	92	480	11,895	471	0	66	12	56	19,679
Pyramimonas-like sp.	0	0	0	0	0	0	0	0	18,868
Rhizosolenia imbricata	180	2,293	266	518	1,301	5,320	16	880	16,095
Cyclotella atomus var. nov.	0	789	8,304	466	0	0	2,389	1,173	14,226
Hemiselmis sp. A	0	0	0	13,560	0	0	0	0	13,560
Chaetoceros tortissimus	2,241	2,190	648	80	317	0	0	2,046	12,233
Thalassionema nitzschioides	413	525	240	651	1,120	583	207	1,532	11,800
Chaetoceros subtilis f. *abnormis*	0	0	7,461	0	0	0	48	0	10,851
Merismopedia tenuissima	**0**	**0**	**0**	**0**	**0**	**0**	**0**	**0**	**10,351**
Karenia brevis	**0**	**0**	**0**	**0**	**0**	**8**	**0**	**0**	**213**
TOTAL	**120,103**	**376,194**	**297,853**	**309,925**	**123,097**	**111,411**	**123,471**	**306,169**	**3,462,622**
Cumulative Species Richness	118	136	137	112	122	114	110	131	339
Shannon Diversity	2.13	2.66	2.15	2.64	2.37	2.83	2.07	2.71	3.28
Shannon Evenness	0.45	0.54	0.44	0.56	0.49	0.60	0.44	0.56	0.56

Boldface = bloom species.

Table 4.7 Long-Term Comparisons of Phytoplankton Community Indices Averaged over 12-Month Periods and Analyzed with *t*-Tests and Wilcoxon Means Tests for Differences (see Appendix II)

Factor (Station pair)	June–November 1992		June–November 1999		June–November 2000	
	Phytoplankton (# cells L^{-1}) × 1000	#Taxa	Phytoplankton (# cells L^{-1}) × 1000	#Taxa	Phytoplankton (# cells L^{-1}) × 1000	#Taxa
E06/F06 (River Mouth)	**265/18** **WS (P < 0.01)**	**28.5/14.5** **WS (P < 0.05)**	736/802 WN	22.7/28.8 WN	nd	nd
E07/F09 (River Mouth)	341/173 WN	31.2/20.5 WS (P < 0.05)	**798/1,001** **WS (P < 0.01)**	37.6/28.8 WS (P < 0.05)	**340/816** **WS (P < 0.05)**	**34.0/30.1** **WS (P < 0.05)**
E08/F10 (Center 1.5 km)	465/443 WN	35.0/29.3 WN	171/775 WS (P < 0.05)	37.6/28.0 WS (P < 0.05)	171/775 WS (P < 0.05)	38.0/33.5 WS (P < 0.05)
E09/F16 (Center 5 km)	206/344 WN	30.7/32.0 WN	770/294 WS (P < 0.05)	32.6/34.2 WN	137/121 WN	35.8/32.8 WN
E10/F11 (W. Shore)	**513/344** **WS (P < 0.05)**	**37.1/31.1** **WS (P < 0.05)**	511/1,583 WS (P < 0.05)	26.8/34.2 WN	161/338 WS (P < 0.05)	32.8/33.7 WN
E12/F12 (W. 3 km)	nd	nd	nd	nd	181/242 WN	34.5/33.7 WN
E11/F14 (E. Shore)	233/353 WN	27/26 WN	1,441/1,082 WN	34.6/33.5 WN	323/507 WS (P < 0.05)	37.2/33.0 WS (P < 0.02)
E13/F15 (E. 1.5 km)	nd	nd	nd	nd	208/581 WS (P < 0.05)	34.5/35.0 WN

Cognate (Station pair)	Phytoplankton (# cells L⁻¹) × 1000	June–November 2001 #Taxa	Phytoplankton (# cells L⁻¹) × 1000	June–November 2002 #Taxa	Phytoplankton (# cells L⁻¹) × 1000	August–November 2003 #Taxa
E06/F06 (River Mouth)	**31/180 WS (P < 0.02)**	**12.2/9.3 WS (P < 0.05)**	nd	nd		
E07/F09 (River Mouth)	35/150 WS (P < 0.05)	13.7/10.2 WS (P < 0.05)	nd	nd		
E08/F10 (Center 1.5 km)	17/447 WS (P < 0.05)	11.7/15 WS (P < 0.02)	5/101 WS (P < 0.05)	15.4/16.4 WN		
E09/F16 (Center. 5 km.)	11./27 WS (P < 0.05)	9.8/13.2 WS (P < 0.02)	6./16 WN	13.0/19.6 WN		
E10/F11 (W. Shore)	**8./41 WS (P < 0.05)**	11.8/10.7 WS (P < 0.02)	**4./167 WS (P < 0.05)**	12.8/12.8 WN		
E12/F12 (W. 3 km.)	8./30 WN	11.3/16.5 WS (P < 0.02)	9/118 WS (P < 0.05)	16.8/19.2 WS (P < 0.05)		
E11/F14 (E. Shore)	252/736 WS (P < 0.05)	14.7/10.0 WS (P < 0.05)	5/167 WS (P < 0.05)	14.8/12.8 WS (P < 0.01)		
E13/F15 (E. 1.5 km.)	**149/69 WS (P < 0.05)**	12.3/14.7 WN	**4./98 WS (P < 0.05)**	13.8/19.4 WS (P < 0.05)		

Note: nd = no data.
WN, not sig. (p = 0.05).
WS, sig. (p = 0.05).
Boldface = bloom species.

Table 4.8 Summary of SAV Data Taken in the Econfina and Fenholloway Offshore Systems from 1999-2002. (Distribution of the top 35 species is by dry weight (gm²) by station.)

Species	E07	F09	E08	F10	E09	F16	E10	F11	E11	F14	E12	F12	E13	F15	FT29	Grand Total
Syringodium filiforme	0.5	0.1	1,276.3	0.3	655.0	521.8	1,023.8	32.1	0.4	0.3	258.5	1,454.1	1,135.4	8.5	0.2	6,367
Halimeda incrassata	0.1	0.1	39.3	0.0	20.4	18.8	1,562.2	1.7	0.0	0.0	232.3	319.9	229.6	0.1	0.0	2,424
Thalassia testudinum	0.3	0.1	42.1	0.1	285.4	5.0	126.7	0.0	0.0	0.0	1,125.3	159.7	105.6	0.0	0.1	1,850
Digenia simplex	31.2	0.9	347.8	0.5	2.9	1.1	389.8	0.2	33.4	0.2	16.0	1.1	723.0	0.5	0.3	1,548
Spyridia filamentosa	71.3	0.5	326.4	3.9	408.0	31.4	213.1	4.7	31.8	6.8	106.0	127.1	171.0	1.5	0.3	1,503
Laurencia poitei	19.2	16.0	191.1	9.7	185.9	83.9	240.6	9.6	37.1	24.1	1.5	9.4	277.0	35.0	14.2	1,154
Laurencia intricata	77.8	4.2	209.8	6.3	32.3	40.9	102.9	8.1	40.1	2.3	14.7	36.6	235.1	6.0	5.2	822
Caulerpa prolifera	231.9	0.6	45.6	1.0	7.5	4.7	128.6	1.2	236.1	0.3	15.4	13.8	12.1	19.0	2.3	720
Halophila englemannii	13.9	0.7	18.8	24.9	15.3	68.6	14.6	32.0	48.4	4.8	8.2	3.5	34.5	141.3	89.4	519
Halodule wrightii	3.3	0.2	22.4	2.0	50.7	193.6	28.9	5.3	49.4	0.6	21.7	29.3	38.1	16.0	4.1	465
Gracilaria debilis	0.2	8.1	11.1	6.4	0.0	0.3	8.9	2.5	5.0	24.0	0.0	8.1	278.4	22.4	11.2	386
Gracilaria cylindrica	5.0	6.6	24.6	20.1	0.1	0.2	12.6	5.5	4.3	19.9	0.1	6.0	139.1	44.9	11.0	300
Jania sp.	0.0	2.6	6.6	5.3	0.8	0.1	8.9	91.1	0.0	0.3	0.0	0.2	6.3	16.1	29.1	167
Ruppia maritima	26.3	0.2	11.3	1.1	0.0	0.0	0.1	0.7	110.8	0.7	0.0	0.0	1.1	4.8	5.3	162
Gracilaria compressa	0.3	20.5	0.7	7.4	0.0	0.0	3.3	5.2	0.3	18.8	0.1	0.0	14.7	22.2	19.6	113
Chondria spp.	8.4	1.7	5.9	1.1	13.6	4.6	9.6	2.8	7.4	1.0	0.8	17.0	26.7	3.0	1.4	105
Penicillus capitatus	0.2	0.0	13.8	0.0	4.4	2.1	37.2	0.0	0.0	0.0	27.0	0.1	7.6	0.1	0.0	92
Ulva lactuca	0.0	11.9	0.0	2.2	0.0	0.0	0.0	1.1	0.0	34.6	0.0	0.3	0.0	6.6	1.8	58

Anadyomene stellata	0.0	0.1	0.0	0.0	0.0	0.3	7.2	0.0	0.0	0.0	0.0	3.3	7.3	0.0	0.0	46
Gracilaria foliifera	0.3	3.3	0.0	6.0	0.0	0.1	0.1	1.1	0.1	7.3	0.0	0.0	2.8	7.2	5.8	34
Polysiphonia spp.	0.8	0.3	2.0	0.7	8.9	1.3	0.2	1.3	0.5	0.3	0.2	0.1	12.8	1.6	1.2	32
Codium isthmocladum	0.0	1.9	0.0	18.9	0.3	0.7	0.0	0.6	0.0	1.8	0.0	1.0	0.2	1.9	0.2	27
Gracilaria cervicornis	6.5	1.1	0.9	0.4	0.0	5.7	0.3	1.3	0.0	6.3	0.1	0.2	4.0	0.3	0.4	27
Syringodium filiforme (w/flowers)	0.0	0.0	7.4	0.0	8.3	2.1	1.1	0.0	0.0	0.0	0.2	0.6	5.3	0.0	0.0	25
Cladosiphon occidentalis	0.5	0.0	0.3	0.3	0.0	0.9	0.0	0.3	0.2	0.0	0.0	17.3	0.7	0.9	0.1	21
Unknown green filamentous algae	0.0	0.0	2.2	0.0	0.0	0.0	0.0	0.0	0.0	0.0	13.4	0.0	1.1	0.0	0.0	17
Derbesia sp.	0.0	0.0	0.6	0.0	0.2	0.0	0.2	0.1	0.1	0.0	13.4	0.6	0.7	0.9	0.0	17
Caulerpa cupressoides	0.0	0.0	0.0	0.0	15.6	0.0	0.0	0.0	0.0	0.0	0.0	0.0	0.0	0.0	0.0	16
Sargassum pteropleuron	1.2	0.0	0.6	0.8	0.8	0.7	3.0	0.2	0.0	0.3	2.1	1.0	0.4	0.7	0.3	12
Hypnea spinella	0.0	2.5	0.0	1.8	0.0	0.1	0.0	0.4	0.0	4.6	0.0	0.0	0.2	0.7	1.1	11
Halophila englemannii (w/flowers)	1.1	0.0	0.2	1.5	0.0	0.0	0.0	0.7	0.1	0.0	0.0	0.0	0.0	4.3	3.3	11
Champia parvula	0.7	0.2	0.4	0.8	0.0	0.0	0.1	1.1	1.6	0.5	0.0	0.1	0.6	1.8	1.2	9
Solieria tenera	0.1	0.0	2.7	0.1	0.0	0.0	0.0	3.8	0.0	0.1	0.0	0.0	1.0	0.0	0.1	8
Polysiphonia hapalacantha	0.3	0.0	3.7	0.0	0.0	1.4	0.1	0.0	0.8	0.2	0.8	0.0	0.4	0.0	0.0	7
TOTAL	505	85	2,654	125	1,717	1,000	3,927	218	611	169	1,859	2,212	3,480	374	210	19,148

Table 4.9 Long-Term Comparisons of SAV Indices Averaged over 12-Month Periods and Analyzed with *t*-Tests and Wilcoxon Means Tests for Differences (see Appendix II)

Factor/Station Pair	SAV Comparison: April 1972–March 1973		SAV Comparison: April 1974–March 1975		SAV Comparison: April 1992–March 1993	
	Biomass (gm⁻²)	#Taxa	Biomass (gm⁻²)	#Taxa	Biomass (gm⁻²)	#Taxa
E07/F09 (River Mouth)	9.0/0.25 WS (P < 0.01)	4.8/1.3 WS (P < 0.01)	188.5/2.4 WS (P < 0.01)	6.0/2.8 WS (P < 0.01)	2.8/3.9 WN	3.3/0.6 WN
E08/F10 (Center 1.5 km)	1196/20 WS (P < 0.01)	8.7/5.3 WS (P < 0.02)	895/30.1 WS (P < 0.01)	10/5.7 WS (P < 0.01)	26.0/4.0 WS (P < 0.02)	3.4/0.9 WS (P < 0.01)
E09/F16 (Center. 5 km)	101/47.2 WN	8.6/9.3 WN	412/ WN	12.0/ WN	20.3/14.8 WN	6.2/4.6 WN
E10/F11 (W. Shore)	627/.17 WS (P < 0.01)	9.5/2.7 WS (P < 0.01)	641/5.8 WS (P < 0.01)	10.2/4.0 WS (P < 0.01)	7.1/0.41 WS (P < 0.01)	5.7/0.6 WS (P < 0.01)
E12/F12 (W. 3 km)	1007/94.8 WS (P < 0.01)	10.8/12.3 WS (P < 0.02)	1007/227 WS (P < 0.01)	12.0/10.5 WN	57.6/14.0 WS (P < 0.01)	8.0/6.3 WS (P < 0.02)
E11/F14 (E. Shore)	523/1.2 WS (P < 0.01)	8.8/2.8 WS (P < 0.01)	521/5.3 WS (P < 0.01)	12.8/3.0 WS (P < 0.01)	1.4/0.1 WS (P < 0.01)	1.3/0.3 WS (P < 0.02)
E13/F15 (E. 1.5 km)	513/23.8 WS (P < 0.01)	10.1/6.2 WS (P < 0.01)	438.3/279 WS (P < 0.01)	11.9/10.4 WN	8.7/0.4 WS (P < 0.01)	4.5/0.6 WS (P < 0.01)

Cognate (Station Pair)	SAV Comparison: April 1999–March 2000		SAV Comparison: April 2000–March 2001		SAV Comparison: April 2001–March 2002	
	Biomass (gm⁻²)	#Taxa	Biomass (gm⁻²)	#Taxa	Biomass (gm⁻²)	#Taxa
E07/F09 (River Mouth)	**29.0/1.1** **WS (P < 0.01)**	5.0/2.6 WN	**20.8/5.3** **WS (P < 0.02)**	6.1/5.9 WN	**35.5/7.5** **WS (P < 0.01)**	**7.8/5.2** **WS (P < 0.01)**
E08/F10 (Center 1.5 km)	**15.6/7.2** **WS (P < 0.01)**	**11.3/5.1** **WS (P < 0.01)**	**144/5.5** **WS (P < 0.01)**	**12/7.8** **WS (P < 0.01)**	**151/7.9** **WS (P < 0.01)**	**13.5/9.9** **WS (P < 0.01)**
E09/F16 (Center 5 km)	**100.3/41.7** **WS (P < 0.01)**	8.3/7.6 WN	**77.8/38.4** **WS (P < 0.01)**	9.5/8.3 WN	107/85.9 WN	**10.8/8.8** **WS (P < 0.01)**
E10/F11 (W. Shore)	**216/1.6** **WS (P < 0.01)**	**11.1/3.3** **WS (P < 0.01)**	**233/7.8** **WS (P < 0.01)**	**11.8/6.1** **WS (P < 0.01)**	**216/25.4** **WS (P < 0.01)**	12.4/10.8 WN
E12/F12 (W. 3 km)	123/125 WN	10.5/8.8 WN	93/103 WN	6.9/7.9 WN	**95.7/138** **WS (P < 0.01)**	7.6/8.3 WN
E11/F14 (E. Shore)	**35.1/7.6** **WS (P < 0.01)**	**9.1/4.2** **WS (P < 0.01)**	**35.6/10.5** **WS (P < 0.01)**	**8.6/7** **WS (P < 0.01)**	**23.9/7.6** **WS (P < 0.01)**	**8.8/6.3** **WS (P < 0.02)**
E13/F15 (E. 1.5 km)	**212/19.5** **WS (P < 0.01)**	**12.5/8.1** **WS (P < 0.01)**	**201/18.8** **WS (P < 0.01)**	13.6/13.4 WN	**178/23.2** **WS (P < 0.01)**	13.6/13.4 WN

WN, not sig. (p = 0.05).
WS, sig. (p = 0.05).
Boldface = bloom species.

Table 4.10 Summary of Invertebrate Data Taken in the Econfina and Fenholloway Offshore Systems from 1999 to 2002. (Distribution of the top 35 species are by numbers by station.)

Species	E07	F09	E08	F10	E09	F16	E10	F11	E11	F14	E12	F12	E13	F15	TOTAL
Hippolyte zostericola	2,641	31	10,394	2,145	1,004	4,089	4,578	1,115	8,980	4,974	1,101	6,522	13,142	5,624	66,340
Tozeuma carolinense	919	0	3,822	154	3,976	4,230	3,600	204	1,809	11	8,094	12,464	5,782	3,402	48,467
Pagurus sp.	428	136	8,131	962	2,027	2,150	7,375	1,321	769	932	3,122	3,776	9,609	3,756	44,494
Anachis avara	2,186	1,274	4,379	939	134	1,432	5,139	582	5,428	3,862	361	1,418	4,914	3,637	35,685
Periclimenes longicaudatus	2,159	40	2,195	2,461	911	2,647	1,671	1,355	3,073	2,090	655	2,211	2,755	4,776	28,999
Thor dobkini	386	42	7,194	208	1,330	885	2,358	603	2,255	740	833	1,344	4,510	913	23,601
Palaemon floridanus	1,688	925	1,441	941	106	509	803	1,105	3,793	2,354	69	1,414	2,058	2,856	20,062
Astyris lunata	658	242	1,435	553	2	68	171	502	2,990	8,798	32	305	377	1,638	17,771
Columbella rusticoides	496	59	3,033	32	254	1,994	1,543	154	642	140	282	1,467	2,526	266	12,888
Neopanope packardii	540	77	1,973	943	365	480	1,006	628	741	394	415	605	1,851	1,554	11,572
Palaemonetes intermedius	2,924	897	5	103	4	2	9	56	4,548	1,369	1	3	5	328	10,254
Neopanope texana	1,696	192	377	894	26	39	324	427	3,444	842	51	141	252	948	9,653
Periclimenes americanus	1,004	166	1,326	151	379	443	1,043	561	1,547	640	141	581	644	222	8,848
Crepidula maculosa	564	48	2,470	55	35	70	1,543	107	881	132	248	554	1,242	245	8,194
Aequipecten gibbus	71	1	728	27	115	53	598	29	284	85	53	83	1,251	94	3,472
Brachidontes exustus	65	127	1,098	63	4	38	103	50	309	483	12	62	732	209	3,355
Aequipecten irradians	43	0	761	24	165	205	341	21	119	6	359	135	538	38	2,755
Ophiothrix angulata	44	54	62	165	63	635	19	256	91	152	90	620	69	127	2,447

Metaporhaphis calcarata	12	1	46	294	353	116	32	266	55	33	273	80	47	198	1,806
Alpheus normanni	67	44	33	43	26	240	11	176	226	210	2	223	19	99	1,419
Turbo castanea	3	6	228	21	88	277	246	12	5	26	141	134	167	31	1,385
Lysmata wurdemanni	62	2	319	16	28	45	331	31	84	15	73	109	192	64	1,371
Echinaster sp.	60	2	196	3	122	34	158	5	67	1	255	116	206	13	1,238
Penaeus duorarum	151	103	21	68	20	195	29	14	239	174	6	44	40	67	1,171
Callinectes sapidus	281	231	6	43	0	3	8	37	213	257	1	15	9	29	1,133
Latreutes fucorum	1	0	12	8	196	479	4	16	0	3	62	143	43	6	973
Libinia dubia	105	4	60	81	32	0	94	56	140	43	36	11	61	69	792
Penaeus sp.	147	88	45	46	0	2	53	21	173	73	0	15	21	27	711
Urosalpinx tampaensis	50	2	139	0	3	0	251	4	133	30	1	2	47	4	666
Nassarius vibex	59	12	54	22	5	2	36	184	81	53	18	35	38	53	652
Crassostrea virginica	58	89	115	7	0	15	6	5	84	91	0	13	85	42	610
Neopanope sp.	105	17	18	143	2	4	4	40	95	65	2	2	6	62	565
Crepidula convexa	93	1	100	9	2	7	59	6	97	12	9	28	94	39	556
Penaeus aztecus	53	2	37	4	12	25	88	10	142	21	3	23	47	40	507
Total # Individuals	**20,025**	**5,118**	**52,750**	**11,847**	**12,290**	**21,982**	**34,047**	**10,202**	**43,868**	**29,530**	**17,183**	**35,238**	**54,217**	**32,003**	**380,300**
Shannon Diversity	**2.75**	**2.46**	**2.60**	**2.59**	**2.40**	**2.58**	**2.55**	**2.86**	**2.66**	**2.35**	**2.02**	**2.26**	**2.48**	**2.55**	**2.79**
Shannon Evenness	**0.68**	**0.62**	**0.61**	**0.62**	**0.58**	**0.59**	**0.62**	**0.67**	**0.66**	**0.57**	**0.49**	**0.53**	**0.59**	**0.60**	**0.58**

Table 4.11 Long-Term Comparisons of Invertebrate Community Indices Averaged over 12-Month Periods and Analyzed with *t*-Tests and Wilcoxon Means Tests for Differences (see Appendix II)

Factor/ Station Pair	Invert Comparison: April 1972–March 1973		Invert Comparison: April 1992–March 1993		Invert Comparison: April 1999–March 2000	
	Numbers	#Taxa	Numbers	#Taxa	Numbers	#Taxa
E07/F09 (Mouth)	286.5/101 WS (P < 0.01)	13.4/9.8 WS (P < 0.01)	161/60 WS (P < 0.05)	6.1/5.5 WN	975/185 WS (P < 0.01)	23.3/15.2 WS (P < 0.01)
E08/F10 (Center 1.5 km)	418.3/99.8 WS (P < 0.01)	15.3/12.3 WS (P < 0.01)	424/154 WS (P < 0.01)	6.7/6.1 WN	2438/384 WS (P < 0.01)	27.5/19.7 WS (P < 0.01)
E09/F16 (Center .5 km)	238.4/77.8 WS (P < 0.01)	25.6/15.5 WS (P < 0.02)	633/363 WN	7.16/6.44 WN	453/425 WN	18.6/22.7 WS (P < 0.02)
E10/F11 (W. Shore)	424.3/94.2 WS (P < 0.01)	15.2/10.3 WS (P < 0.01)	456/80 WS (P < 0.01)	18.0/14.7 WS (P < 0.05)	1424/411 WS (P < 0.01)	24.7/19.9 WS (P < 0.05)
E12/F12 (W. 3 km)	82.5/219 WS (P < 0.01)	11.5/16.8 WS (P < 0.05)	652/1332 WS (P < 0.01)	14.8/25.6 WS (P < 0.02)	624/1426 WS (P < 0.05)	19.9/25.9 WS (P < 0.02)
E11/F14 (E. Shore)	203/60.0 WS (P < 0.01)	11.8/7.0 WS (P < 0.05)	180/65 WS (P < 0.01)	14.5/10.6 WS (P < 0.02)	2126/970 WS (P < 0.05)	25.1/21.1 WN
E13/F15 (E. 1.5 km)	333.6/360.8 WN	16.5/18.3 WN	670/356 WS (P < 0.01)	20.5/16.8 WN	1989/1042 WS (P < 0.02)	29.0/23.5 WS (P < 0.05)

Cognate/ Station Pair	Invert Comparison: April 2000–March 2001		Invert Comparison: April 2001–March 2002		Invert Comparison: April 2003–March 2004	
	Numbers	#Taxa	Numbers	#Taxa	Numbers	#Taxa
E07/F09 (Mouth)	**349/119** WS (P < 0.05)	**20.5/9.8** WS (P < 0.01)	**796/253** WS (P < 0.01)	**21.1/9.1** WS (P < 0.01)	nd	nd
E08/F10 (Center 1.5 km)	**919/321** WS (P < 0.01)	**25.5/19.4** WS (P < 0.05)	**1436/459** WS (P < 0.05)	26.4/21.8 WN	nd	nd
E09/F16 (Center .5 km)	**232/471** WS (P < 0.01)	16.7/19.4 WN	**496/1406** WS (P < 0.05)	19/22.5 WN	nd	nd
E10/F11 (W. Shore)	**525/256** WS (P < 0.05)	20.3/20.5 WN	**1338/280** WS (P < 0.05)	**23.4/20.3** WS (P < 0.05)	nd	nd
E12/F12 (W. 3 km)	697/1043 WN	17.2/18.7 WN	296/786 WN	**14.5/25.4** WS (P < 0.05)	nd	nd
E11/F14 (E. Shore)	**1007/502** WS (P < 0.02)	**27.3/16.9** WS (P < 0.01)	**1378/280** WS (P < 0.05)	23.4/21.4 WN	nd	nd
E13/F15 (E. 1.5 km)	**1432/845** WS (P < 0.02)	**27.9/24.9** WS (P < 0.05)	**1578/1205** WS (P < 0.02)	25.9/27.8 WN	nd	nd

Note: nd = no data.
WN, not sig. (p = 0.05).
WS, sig. (p = 0.05).
Boldface = bloom species.

relative to the Econfina cognate stations. Numbers and species richness were significantly higher at Stations F12 and F16, which was considered a result of the changes associated with dilution of the effluent as it moved offshore. During 1992–1993, these trends were again obvious, with some minor differences such as significantly reduced numbers at Station F15. In 1999–2000, again with the exception of seagrass areas at F12 and F16, there were significant reductions in invertebrate numbers and species richness in the Fenholloway offshore system relative to the reference stations. Species diversity trends tended to follow the species richness differences. There were increased numbers of invertebrates in both systems in 1999–2000 relative to previous years, and there was a trend toward reduced invertebrate numbers and species richness in areas of the Fenholloway system that were characterized by debilitated seagrass beds. Species richness and diversity increased at Station F15, possibly as a response to increased SAV species richness in this area. However, overall, this analysis indicates that the continued absence of SAV biomass and species richness was accompanied by reduced numbers of invertebrates and reduced invertebrate species richness in the Fenholloway system relative to reference areas.

4.10.6 Fishes

A list of the numerically dominant fishes by station is given in Table 4.12. The primary grass bed species (i.e., *Lagodon rhomboides, Diplodus holbrooki, Orthopristis chrysoptera, Bairdiella chrysoura*) were largely absent in the Fenholloway system where overall numbers of fishes were substantially less than those at cognate Econfina stations. The only part of the Fenholloway system where equivalence of fish numbers occurred between cognate stations was in offshore areas (E09, F16). Cumulative species richness, on the other hand, was generally comparable between cognate stations. Species diversity indices were actually higher at most Fenholloway stations. This counterintuitive finding was due to the fact that fish associations in the natural inshore seagrass beds in Apalachee Bay are highly dominant. This dominance is reduced with the altered trophic organization from seagrass order to a plankton-dominated hierarchy (Livingston, 1975a).

A comparative statistical review of fish community indices in the Econfina and Fenholloway systems from 1972 to 2004 is given in Table 4.13. During 1972–1973, with the exception of Stations F12 and F16, fish numbers and species richness were significantly lower in the Fenholloway system compared to reference areas. In the 1992–1993 sampling period, fish numbers were down in both systems, and there were relatively few significant differences of the fish indices between the Econfina and Fenholloway systems. During 1999–2000, offshore Fenholloway areas again were noted by comparable fish faunas in the Econfina and Fenholloway systems. During the next 2 years, there were actually significantly higher species richness levels at Station F16. However, by 2003–2004, fish species richness was down at most stations, including F12 and F16. This pattern was reversed in eastern sections of the Fenholloway system (F13, F15) where there was equivalence of fish species richness with that in reference areas. Fish abundance was still significantly lower in such Fenholloway areas. These data are compatible with the SAV analyses in these areas where SAV biomass was down and SAV species richness was up at Station F15.

This analysis indicated that by the end of the survey, the fish fauna in the Fenholloway offshore system was in a deteriorated state, with relatively low numbers and species richness in most areas relative to the reference Econfina system. This would indicate that there has been no substantial recovery of fishes in the Fenholloway system over the past 5 years.

Table 4.12 Averaged Numbers of the Top 35 Fish Species in the Econfina and Fenholloway Offshore Areas Taken from 1999 to 2004; Also Shown Are Cumulative Numbers and Species Richness and Overall Shannon Diversity and Evenness Indices

Species	E07	F09	E08	F10	E09	F16	E10	F11	E11	F14	E12	F12	E13	F15	TOTAL
Lagodon rhomboides	7,533	706	4,851	584	1,201	1,338	5,806	530	4,400	600	6,569	5,214	4,695	2,076	46,103
Diplodus holbrooki	1	0	1,218	1	1,322	1,449	1,760	8	13	0	6,777	2,967	1,461	126	17,103
Anarchopterus criniger	486	64	896	216	259	328	640	230	2,304	222	134	338	1,203	504	7,824
Orthopristis chrysoptera	598	15	268	243	323	792	300	149	316	54	1,812	560	179	541	6,150
Eucinostomus argenteus	914	597	240	292	222	92	224	180	422	142	262	201	174	208	4,170
Bairdiella chrysoura	368	113	344	69	163	270	335	47	249	150	248	388	402	520	3,666
Leiostomus xanthurus	474	1,342	25	397	120	22	12	71	92	398	17	51	9	294	3,324
Paraclinus fasciatus	101	21	561	82	216	185	389	93	87	44	142	221	417	135	2,694
Anchoa mitchilli	2	1,243	0	340	1	45	0	58	1	744	0	110	1	59	2,604
Sciaenidae sp.	412	217	436	43	65	18	426	95	101	54	49	83	102	270	2,371
Syngnathus floridae	61	6	126	68	128	154	206	56	93	25	429	332	104	194	1,982
Opsanus beta	122	36	212	78	39	57	192	42	450	180	35	50	163	136	1,792
Gobiosoma robustum	143	35	185	25	17	29	172	71	378	96	33	130	214	54	1,582
Monacanthus ciliatus	3	1	48	4	247	282	82	4	4	0	359	315	56	24	1,429
Centropristis striata	11	1	165	10	158	152	110	9	14	1	180	306	134	46	1,297
Syngnathus scovelli	275	94	23	129	4	2	23	94	233	177	8	18	11	114	1,205
Haemulon plumieri	4	0	148	8	122	172	127	1	1	1	165	180	93	21	1,043
Chasmodes saburrae	228	23	27	11	0	0	29	4	279	113	0	1	19	31	765
Chilomycterus schoepfi	18	13	63	28	57	63	48	33	22	9	140	134	51	52	731
Brevoortia patronus	90	124	36	2	0	0	3	1	14	25	0	1	11	0	307

Table 4.12 (continued) Averaged Numbers of the Top 35 Fish Species in the Econfina and Fenholloway Offshore Areas Taken from 1999 to 2004; Also Shown Are Cumulative Numbers and Species Richness and Overall Shannon Diversity and Evenness Indices

Species	E07	F09	E08	F10	E09	F16	E10	F11	E11	F14	E12	F12	E13	F15	TOTAL
Cynoscion nebulosus	63	31	14	10	6	15	16	15	33	13	9	35	17	30	307
Lactophrys quadricornis	19	1	13	20	13	20	8	18	13	1	33	32	14	10	215
Calamus arctifrons	0	0	6	3	70	58	5	3	0	0	3	23	10	7	188
Paralichthys albigutta	19	17	5	12	5	7	9	10	18	10	8	3	12	6	141
Synodus foetens	21	1	15	10	6	14	9	14	7	3	6	13	6	10	135
Sparidae sp.	0	0	0	0	120	0	0	0	0	0	0	0	0	0	120
Paraclinus marmoratus	0	1	3	3	11	73	0	1	2	1	2	7	1	1	106
Urophycis floridana	3	7	3	6	3	18	4	5	2	12	4	17	1	11	96
Chloroscombrus chrysurus	36	17	0	2	3	1	1	6	2	5	2	1	3	16	95
Blenniidae	12	0	7	13	0	0	6	1	28	11	3	0	7	1	89
Sphoeroides parvus	28	4	6	7	2	2	1	9	9	3	1	2	5	9	88
Sphoeroides nephelus	16	8	9	4	2	2	7	6	6	4	3	5	6	9	87
Syngnathus louisianae	23	11	1	9	2	2	0	13	6	6	1	2	1	4	81
Stephanolepis hispidus	0	0	5	0	33	3	0	0	0	0	16	9	13	0	79
Total # individuals	12,226	4,886	10,026	2,831	5,008	5,800	11,010	1,949	9,716	3,171	17,530	11,885	9,652	5,612	111,302
Cumulative # taxa	64	62	53	61	47	49	47	54	57	54	46	55	48	56	118
Shannon Diversity	1.70	2.15	2.00	2.67	2.45	2.41	1.81	2.72	1.90	2.50	1.59	1.95	1.87	2.40	2.29
Shannon Evenness	0.41	0.52	0.50	0.65	0.64	0.62	0.47	0.68	0.47	0.63	0.42	0.49	0.48	0.60	0.48

Table 4.13 Long-Term Comparisons of Fish Community Indices Averaged over 12-Month Periods and Analyzed with *t*-Tests and Wilcoxon Means Tests for Differences (see Appendix II)

Factor/ Station Pair	Fish Comparison: April 1972–March 1973		Fish Comparison: April 1992–March 1993		Fish Comparison: April 1999–March 2000	
	Numbers	#Taxa	Numbers	#Taxa	Numbers	#Taxa
E07/F09 (Mouth)	**108/45** **WS (P < 0.01)**	**11.2/6.5** **WS (P < 0.01)**	**21/147** **WS (P < 0.02)**	8.3/7.4 WN	**214/123** **WS (P < 0.01)**	12.1/10.3 WN
E08/F10 (Center 1.5 km)	**125/43** **WS (P < 0.01)**	**15.3/4.2** **WS (P < 0.01)**	95/27 WN	5.1/4.4 WN	**207/61** **WS (P < 0.01)**	**13.5/8.1** **WS (P < 0.01)**
E09/F16 (Center .5 km)	**109/80** **WS (P < 0.02)**	**12.2/11.8** WN	147/135 WN	10.3/10.4 WN	134/216 WN	**9.6/13.1** **WS (P < 0.01)**
E10/F11 (W. Shore)	**133/31** **WS (P < 0.01)**	**12.3/4.8** **WS (P < 0.01)**	92/120 WN	**11.6/7.1** **WS (P < 0.02)**	**234/48** **WS (P < 0.01)**	**12.2/7.7** **WS (P < 0.02)**
E12/F12 (W. 3 km)	145/120 WN	11.8/11.1 WN	243/166 WN	13.8/15.0 WN	299/398 WN	11.4/12.8 WN
E11/F14 (E. Shore)	**114/29** **WS (P < 0.01)**	**11.4/6.8** **WS (P < 0.01)**	50.2/27.3 WN	8.5/6.6 WN	**202/87** **WS (P < 0.01)**	12.0/10.9 WN
E13/F15 (E. 1.5 km)	**118/28** **WS (P < 0.01)**	**12.6/7.6** **WS (P < 0.01)**	89.3/48.3 WN	10.1/8.2 WN	**189/110** **WS (P < 0.02)**	**13.3/10.4** **WS (P < 0.02)**

Table 4.14 Long-Term Comparisons of Fish Community Indices Averaged over 12-Month Periods and Analyzed with *t*-Tests and Wilcoxon Means Tests for Differences (see Appendix II)

Cognate/ Station Pair	Fish Comparison: April 2000–March 2001		Fish Comparison: April 2001–March 2002		Fish Comparison: April 2003–March 2004	
	Numbers	#Taxa	Numbers	#Taxa ʹ	Numbers	#Taxa
E07/F09 (Mouth)	**340/157** WS (P < 0.02)	**13.2/7.9** WS (P < 0.01)	**550/70** WS (P < 0.01)	**16.3/7.2** WS (P < 0.01)	81/87 WN	8.2/5.3 WS (P < 0.01)
E08/F10 (Center 1.5 km)	**201/86** WS (P < 0.02)	11.1/9.1 WN	**426/79** WS (P < 0.01)	14.7/10.2 WS (P < 0.02)	110/28 WS (P < 0.01)	8.3/6.9 WS (P < 0.02)
E09/F16 (Center .5 km)	104/140 WN	**9.6/12.5** WS (P < 0.02)	119/150 WN	**8.7/11.8** WS p<0.05	83/9 WS (P < 0.01)	7.6/3.9 WS (P < 0.01)
E10/F11 (W. Shore)	**240/91** WS (P < 0.01)	9.9/8.5 WN	**410/48** WS (P < 0.01)	13.6/10.3 WS (P < 0.02)	140/24 WS (P < 0.01)	9.5/4.8 WS (P < 0.01)
E12/F12 (W. 3 km)	529/237 WN	10.5/11.8 WN	560/411 WN	9.4/14.4 WN	223/47 WS (p< 0.01)	9.8/7.3 WS (P < 0.02)
E11/F14 (E. Shore)	**293/129** WS	**13.2/8.2** WS (P < 0.01)	**365/41** WS (P < 0.01)	**14.1/8.8** WS (P < 0.01)	40/23 WS (P < 0.01)	6.1/4.9 WN
E13/F15 (E. 1.5 km)	133/144 WN	12.4/11.6 WN	**507/202** WS (P < 0.02)	13.7/14.3 WN	96/62 WS (P < 0.02)	7.7/9.5 WN

WN, not sig. (p = 0.05).
WS, sig. (p = 0.05).
Boldface = bloom species.

4.11 Press Coverage, Public Response, and Failure of the Restoration Process

Despite the fact that the environmental problems of the Fenholloway system have been documented in a series of theses, dissertations, unpublished reports and reviewed scientific papers for over 30 years, the regional press has consistently disregarded the scientific data in favor of sensationalistic coverage that pits a small group of "environmentalists" against the pulp mill. This approach emphasizes uninformed controversy instead of scientific facts.

The release of dioxin (2,3,7,8-tetrachlorodibenzo-*p*-dioxin or TCDD) by pulp mills is due to the use of chlorine in the bleaching process. Dioxin is a toxic by-product of papermaking processes. In the aquatic environment, dioxin is harmful to the environment due to its carcinogenic, co-carcinogenic, embryotoxic, teratogenic, and endocrine disruptive effects. This chemical is not manufactured for any direct commercial purpose. There are many dioxin sources, primarily incinerators and power plants that contribute to the widespread presence of dioxins in the environment. The specific dioxin compound (2,3,7,8-TCDD) is usually found within a mixture of similar compounds known as dioxins and furans, which are considered "ubiquitous contaminants that are released into the environment as by-products of incomplete combustion or as chemical impurities" (Lohman et al., 2000). There are approximately 75 different polychlorinated dibenzodioxin (PCDD) isomers and 135 polychlorinated dibenzofuran (PCDF) isomers that have been identified in the environment. Each individual type of dioxin and furan molecule is termed a "congener." Of these various congeners, 17 are toxic; and of these 17, some are more toxic than others (these totals do not include other possible congeners, such as those containing bromine instead of chlorine, or dioxin-like polychlorinated biphenyls [PCBs]). It is these 17 molecules that are generally referred to as "dioxin" or "dioxins." The singular "dioxin" can also be used to refer solely to the most toxic of these 17 compounds (TCDD). Dioxin thus constitutes a real threat to aquatic food webs, which include various forms of human impacts.

The impact of dioxin constitutes one of the primary focal points for "environmentalists" and press coverage of the Fenholloway system.

> DIOXIN TAINTS FENHOLLOWAY FISH: "Anyone fishing on the Fenholloway River near the big paper mill south of Perry ought to throw back the catch state health officials warned Friday… State HRS and DER inspectors found dioxin levels ranging from 11.5 to 19.1 parts per trillion (ppt) in the Fenholloway and from 8.1 to 25.7 ppt in Eleven Mile Creek."

> **—Bill Cotterell,** *Tallahassee Democrat,* **September 22, 1990**

> PLAN OFFERS HOPE FOR POLLUTED FENHOLLOWAY: "…a report by the Florida Department of Environmental Protection indicates that changes in pulp-processing technology, combined with a 17-mile-long pipeline to divert wastes, could make the river usable for recreation. … Fish and shellfish were also tested. Of those examined from the estuary, there were no detectable amounts of dioxin found. But fish from the freshwater samples were found to have 1 to 3 parts of dioxin per trillion in their tissues. … that finding is below the current fish-consumption criteria."

> **—Elsa C. Arnett,** *Tallahassee Democrat,* **September 22, 1994**

EPA TESTS FIND HAZARDOUS COMPOUND AT BUCKEYE PLANT: "Recent tests at the Buckeye Florida plant near Perry have revealed the hazardous chemical compound dioxin. The findings ... could result in more federal scrutiny of Florida's stalled proposal to pipe waste 17 miles from the plant to the Gulf of Mexico. ... The EPA blocked Florida's pipeline proposal in 1998. ... Dioxin was detectable only at the level of 10 parts per quadrillion in the past.... New test methods now make it possible to detect even smaller amounts, such as the 2.4 parts per quadrillion found at the Buckeye plant. ... A representative of a coalition of environmental groups that has fought the pipeline proposal said the findings should negate the need for a pipeline."

—Bruce Ritchie, *Tallahassee Democrat,* **January 13, 2001**

POLLUTION STILL PLAGUES RIVER: "... Dioxin, a group of hazardous chemical compounds that persist in the environment, also has been detected in fish along the river. Chemicals in the pulp-mill waste are suspected of causing some fish in the river to grow both male and female sex organs ... the Florida Department of Health announced in September that it was lifting an advisory against consuming fish from the Fenholloway River. Samples collected from fish along the river showed much lower dioxin levels than in 1990, the department said."

—Bruce Ritchie, *Tallahassee Democrat,* **April 5, 2004**

The fact that dioxin concentrations in organisms in the Fenholloway system have been marginal at best, and that the substitution of chlorine dioxide for chlorine in the paper-making process has brought regional paper mills into compliance with federal regulations, have not deterred regional press coverage to continue to cover the dioxin problem to the exclusion of the primary environmental problems associated with pulp mill discharges to the Fenholloway River. This continuous drumbeat of stories concerning dioxin trivializes the real problems that dioxin poses to the environment, and is a good example of an emphasis on the sensational instead of a factual presentation of the problems in the Fenholloway system.

Likewise, there has been an almost continuous drumbeat of attention given to another well-publicized problem with pulp mill discharges to the Fenholloway River; the masculinization of topminnows. Davis and Bortone (1992) noted that rivers such as Elevenmile Creek (Pensacola) and the Fenholloway River that are associated with pulp mill effluents were associated with changes in the secondary sex characteristics of topminnows such as mosquitofish (*Gambusia affinis*). Female mosquitofish grew a gonopodium, which is the male intromittant organ of this species. This change in secondary sex characteristics was used by "environmentalists" to try and stop a proposal to pipe the effluent to the coast in order to restore the Fenholloway River.

FEMALE FISH SHOWING UP WITH MALE SEX ORGANS: "Female fish in north Florida's Fenholloway River are developing male sex organs and researchers warn it could hurt the food chain, spread to other species and affect people who eat fish. ... 'It's not just our problem here in Florida,' said Linda Young, southeast regional coordinator for the Clean Water Network. 'Our seafood gets shipped all over the country.' Researchers at the University of West Florida and the U.S. Environmental Protection Agency say the mutations have shown up in three fish species so far. ... But what worries EPA and university researchers is a recently announced cleanup plan. The mill plans to pump its treated waste through a 15-mile pipeline to the mouth of the Fenholloway."

—Associated Press, January 10, 1997

MUTATIONS OF THE SEA: "News that female fish in the Fenholloway River are developing male sex organs should alarm everyone, especially because of the repercussions it could have for the entire nation. ... We think it (a pipeline) will just spread the problem to a bigger area and, potentially, to a greater amount of people. ...The waste could affect the oyster communities and other fish on the Gulf Coast. ... Researchers say the long-term effects from eating the fish are unknown, but who wants to find out firsthand after people start developing deformities as well? ... Bye-bye beautiful beaches, hello hermaphroditic fish."

—*Florida Flambeau,* January 14, 1997

SOME SAY RIVER FINDINGS ARE A LITTLE FISHY: "Scientists reported last year that ... they caught 11 tarpon in nets near the plant's wastewater discharge. ... They had to swim through what is considered one of the most polluted rivers in Florida. 'That is why the catches are hard to believe,' said Linda Young, southeast region director for the Clean Water Network. She thinks the report raises questions about the study that led the Department of Health to lift the consumption ban. 'I just don't believe those tarpon found their way up there,' she said. ... The species tested included catfish, gar and warmouth, and had less than 7 parts per trillion of dioxin used as the state's criteria...."

—Bruce Ritchie, *Tallahassee Democrat,* April 5, 2004

REJECT PIPELINE INTO FORGOTTEN FENHOLLOWAY: "...Scientists have found that the few fish remaining in the river are mysteriously changing sex characteristics — the girl fish are becoming partly boy fish. ... If the mill's waste is making fish change sex in the Fenholloway, then what might it do to all the marine species in the Gulf? Tell EPA to write a new permit that doesn't include the pipeline and that contains the protections promised to Florida by the Clean Water Act."

—Linda Young and Joy Towles Ezell, *Tallahassee Democrat,* May 24, 2004

The above articles were just a small sampling of press reports that distorted the scientific findings (Davis and Bortone, 1992) that actually found the following:

1. The phenomenon of the purported changes in secondary sex characteristics involves masculinization (arrhenoidy) of female poeciliids (topminnows). The initial discovery was made with mosquitofish (*Gambusia affinis*) in Elevenmile Creek (Perdido Bay system) below a pulp mill, and involved modification of the anal fin into an organ resembling the male gonopodium, an intromittent organ common to live-bearing topminnows.
2. Two species (*Gambusia affinis* and *Heterandia formosai*) were experimentally exposed to treated phytosterol. No chemical identifications of specific androgenic steroidal compounds were made concerning kraft mill effluents (KME). Field collections

were made concerning KME-exposed fishes (*Gambusia affinis, G. holbrooki, Heterandia formosa*) in the Fenholloway River. Very few incidences of hermaphroditism or sex reversal were noted in the field collections of such fishes. Induction of KME-exposed fishes to hermaphroditism was not found, and there was no indication that the masculinization was due to genetic alterations. There was no evidence of so-called mutant fishes, as the changes noted were not proven to be inherited.

3. Masculinized females reproduced in laboratory conditions and produced non-arrhenoiid offspring. There was no evidence that environmentally induced masculinization was passed on to offspring.

4. Field analyses corroborated that KME were associated with masculinization of females. However, the fate of these fishes in the field was not determined, and there was no evidence of an adverse ecological impact due to the masculinization of the females.

5. The authors noted the potential importance of *Gambusia* as an indicator of potential effects of hormonal disruption due to hormone-mimicking compounds in anthropogenous discharges. However, such effects were not found in fishes higher in the freshwater food webs, and no such effects were found in marine species.

6. Although KME-induced masculinization of certain poeciliid species has been experimentally induced in the laboratory with phytosteroids, the specific substances in KME that are operational remain unidentified. Researchers do not consider dioxin a likely candidate for such effects.

The distortions and exaggerations of the masculization of mosquitofish in Elevenmile Creek and the Fenholloway River by "environmentalists" and the regional press were in keeping with the worst form of sensationalism that caters to anything sexual for the American public. This evidently applies to the sex lives of fishes. The dire warnings that masculinization endangered estuarine invertebrates such as oysters or marine fish populations in the human food web were gross exaggerations and distortions that were seized upon by journalists to expand the dangers of the pulp mill effluents to humans along the entire Gulf coast. There was no threat to seafood interests. The use of the term "mutants" tended to follow the usual horror that accompanies grade "B" offerings from Hollywood.

The fact that mosquitofish are usually not part of the human food web and the totally unscientific extrapolation to adverse impacts on humans misrepresent the scientific facts, and divert attention from the real impacts of the pulp mill effluents on receiving systems. The well-documented scientific facts regarding the pollution of the Fenholloway system have been almost totally ignored in favor of unsubstantiated ranting by people who are primarily interested in putting the mill out of business. The sensationalist approach of pitting "environmentalists" against the pulp mill sells news, but has also helped to prevent effective restoration of the Fenholloway system. The fact that the public has been misled by such coverage does not apply when news is sold as a commodity rather than a means of education and understanding.

The irony of this sad chapter of environmental journalism is that the one population of fishes in the Fenholloway River that was least affected by the KME was the mosquitofish (see Table 3.1). This was due to the fact that, in the largely hypoxic condition of the Fenholloway River, only those fishes adapted for breathing with lungs such as tarpon or sipping oxygen-rich water at the surface (topminnows) are able to survive in this system. However, low DO is not a very sexy item, so such facts are largely ignored by the press. The recent story in the *Tallahassee Democrat*, where "environmentalists" questioned the presence of tarpon in the Fenholloway River, was thus based on ignorance of basic fish ecology. The Fenholloway River is populated mainly by organisms adapted for life in hypoxic waters. The increased BOD due to high concentrations of DOC discharged into the river by the pulp mill is a primary factor in the deteriorated state of the Fenholloway River. This factor (low DO) lacks the prurient appeal of gonopodial growth in female topminnows or the threat of dioxin. Thus, a primary factor in the ecological demise of the river has been marginalized by press reports. This lack of coverage extends to most of the factors that have led to the damage done to the Fenholloway River and associated bay areas. The failure to take into consideration the scientific determinations of causative factors concerning adverse impacts of the pulp mill on the Fenholloway system has led to the continuing failure to remediate these effects.

The results of a "Use-Attainability" study by Florida state officials in the early 1990s led to the recommendation of a pipeline to take the treated mill wastes to the Fenholloway estuary. The pipeline would have mitigated the adverse impacts of high conductivity, low DO, and high concentrations of nutrients on the Fenholloway River. The impacts of high color and ammonia/orthophosphate loading were not addressed by this plan. These effects were largely ignored by state and federal regulatory agencies, whose attention focused mainly on the masculinized topminnows and the impacts of dioxin on the Fenholloway system. The pipeline was vigorously opposed by the "environmentalists" and the regional news media. A proposal by the Florida Department of Environmental Regulation to issue the pipeline permit in 1997 was opposed by the U.S. Environmental Protection Agency in 1998. In 2002, a review of plant processes was undertaken by a group of scientists and engineers representing various groups (state and federal regulators, "environmentalists," pulp mill interests) to determine how the mill could clean up its effluent without resorting to a pipeline. This group, designated by regional media as the "dream team," failed to come to any decisions concerning acceptable methods to achieve such restoration. Meanwhile, the effort to reduce color and bring back the seagrass beds was not successful, and the various adverse impacts of the paper mill on the Fenholloway River and offshore system remain largely unchanged from those noted in the early 1970s.

EXPERTS JOIN PLAN FOR THE RIVER: "Ten of Florida's top water pollution experts have been tapped for a special Fenholloway River cleanup…"

—Julie Hauserman, *Tallahassee Democrat*, July 20, 1991

BUCKEYE PIPELINE PUT ON THE SHELF FOR NOW: "A controversial waste pipeline for the Buckeye Florida pulp mill in Perry is being placed on hold five years after it was proposed. … Environmental groups have argued that the plant should reduce pollution rather than piping it to the Gulf. The proposal has been stalled since the U.S. Environmental Protection Agency in 1998 objected to a proposed state permit. … Pulp industry experts from the company, state and federal agencies and environmental groups will tour the plant next week to look for possible plant improvements that could reduce pollution. … The review team is expected to issue its recommendations by the summer."

—Bruce Ritchie, *Tallahassee Democrat*, February 5, 2002

DEP DROPS SUPPORT OF MILL'S PERMIT: "…EPA is currently working on establishing TMDLs or total maximum daily loads of individual pollutants for the Fenholloway. … Buckeye may have to weigh between the 'public acceptance' of a pipeline and a river that does not meet 'all Class 3 standards,' but has variances to these standards. … Buckeye officials say seagrass growth has improved 35% in recent years and questioned Livingston's recent statements ['improvements were due to drought conditions and had little to do with process changes at the mill.']."

—*Taco Times*, April 21, 2004

GROUPS SAY EPA RULES ON PIPELINE: "EPA recommended in 1998 that the plant make further improvements to reduce pollution rather than build the pipeline. Federal wildlife officials said moving the wastewater flow would dry up the river, exposing dioxin-laden sediment."

—Bruce Ritchie, *Tallahassee Democrat*, May 27, 2004

BUCKEYE CONTINUES ENVIRONMENTAL IMPROVEMENTS: "In recent years, our manufacturing plant … has spent $64 million to improve the quality of our facility's treated wastewater. These expenditures have reduced waste watercolor by 50 percent; reduced wastewater volume by 22 percent; and reduced sodium discharge by 27 percent. The use of elemental chlorine to purify pulp has been eliminated. As a result of these improvements, seagrasses have begun to return near the mouth of the Fenholloway River and the Florida Department of Health last fall lifted a fish advisory issued in 1990."

—Howard Drew, *Tallahassee Democrat*, May 28, 2004

POLLUTION STILL PLAGUES RIVER: "Young (Clean Water Network) wants the company to clean up its pollution rather than simply move it to the Gulf. She said she thinks the process is back to where it started 10 years ago, except that the DEP and EPA now are opposing the company's push for a pipeline."

—Bruce Ritchie, *Tallahassee Democrat*, April 5, 2004

Both sides of the issue of pollution in the Fenholloway system have misrepresented well-established scientific data on the subject. In the entire long and sad history of press coverage of the Fenholloway situation, the actual scientific facts concerning the impacts of the mill on the river and associated bay areas were largely ignored in favor of sensational and largely unscientific statements. Diversion of attention to environmental non-issues, a lack of understanding of the scientifically demonstrated relationships for restoration of the river and the offshore seagrass beds, and the trivialization of the increasing threat of damaging plankton blooms have contributed to the problems rather than helped to correct them. The distortion of the scientific data to serve the parochial interests of groups involved in the Fenholloway issue have hamstrung a meaningful and effective way to restore the polluted system. The lack of even a basic understanding of the assimilative capacity of the Fenholloway River for a high-volume effluent, the substitution of emotional intrigues for scientific data, and the absence of effective treatment of the effluent for reduction of color and nutrient loading have led to a stalemate that has effectively hamstrung the restoration process.

The use of environmental journalism as a form of entertainment is not new. It is, however, more prevalent as American journalism has become more centralized and directed by business interests. The *Tallahassee Democrat* is somewhat typical of regional news media that are driven by such interests, and has distorted and even covered up important environmental issues ranging from global warming to local lake pollution. One example of the almost humorous aspects of such distortion turned up in a recent Page 1 article in the *Democrat*:

CATFISH LICKING: A NEW HIGH: Tony Bridges, *Tallahassee Democrat*, 29 January 2005.

"…a story's been going around for years about hallucinogenic properties in the slime of a certain kind of saltwater catfish. But whether fact or urban legend is not exactly clear…Most people call them gafftops or sailcats…They're bottom dwellers, comfortable in mud…Not too big, but feisty…And apparently, they're less than tasty…"

Of course, the article is misleading on various points. The gafftopsail catfish (*Bagre marinus*) swims in open water and can get up to 2 feet long. Unlike their cousin, the sea catfish (*Arius felis*), gafftops are good to eat and are considered a food fish. As for the hallucinogenic characteristics of the mucous of gafftopsail catfish, there are none. The myth was actually made up by the editor of a sports fishing magazine. However, it is interesting to contemplate how many citizens started the licking process as word spread from the *Democrat's* front-page epistle on the joys of fish slime in the Gulf of Mexico. Such widespread beliefs are no more fantastic than those engendered by the *Democrat's* coverage of regional environmental problems such as the Fenholloway River system.

chapter 5

Nutrient Loading and the Perdido System

5.1 Phytoplankton Blooms in Coastal Systems

Anthropogenous nutrient loading has resulted in increased incidence and severity of plankton blooms in coastal systems around the world (Hallegraeff et al., 2003). Between 1965 and 1976, the number of confirmed worldwide red tide outbreaks (raphidophytes, dinoflagellates) increased sevenfold concurrent with a twofold increase in nutrient loading (Hallegraeff, 1995; Hallegraeff et al., 1995). Bricker et al. (1999) stated that 44 estuaries in the conterminous United States suffer "high expressions of eutrophic conditions," with an additional 40 estuaries having "moderate [eutrophic] conditions." In the United States, adverse effects due to plankton blooms are most pronounced along the coasts of the Gulf of Mexico and the Middle Atlantic states. Bricker et al. (1999) projected that eutrophic conditions would worsen in 86 estuaries by the year 2020.

Of the approximate number of marine phytoplankton species (5000), some 300 species are considered to occur at numbers high enough to discolor seawater (Sournia et al., 1991). Many algal specialists consider a bloom to be defined as a population density in excess of 1×10^6 cells L^{-1} although such numbers are not necessarily required for adverse impacts on other species (Hallegraeff et al., 2003). We used this criterion to identify ten bloom species in the Perdido system (Figure 5.1) over the 16-year study period (Livingston, 2000, 2002). About 40 or 50 of the known bloom species produce toxins that can affect both natural marine populations of plants and animals as well as human beings (Hallegraeff et al., 1995). Anderson (1996) included in the concept of blooms the concentration(s) of one or more species that cause harm to other species or that cause accumulations of toxins in such a way as to cause harm to those who might eat the toxic species. These so-called harmful algal blooms (HABs) produce and release substances having direct and/or indirect effects on associated plant and animal populations. We have made quantitative assessments of the bloom species to identify those associated with adverse effects on animal populations and the food web structure in the Perdido Bay system (Livingston, 2000, 2002).

Specific bloom species are responsible for severe damage to estuarine resources. *Heterosigma akashiwo* frequently causes heavy and extensive red tide events (Hara and Chihara, 1987) and has been associated with fish kills in New Zealand, Chile, and British Columbia (Chang et al., 1990). Imai et al. (1997) described the life-cycle and bloom dynamics of *Chattonella*, a genus with two known fish-killing species (*C. antiqua* and *C. marina*). The dinoflagellate *Prorocentrum minimum* is a toxic bloom species associated with postulated shellfish poisoning and fish kills. Nakazima (1965) indicated poisonous effects on shellfish

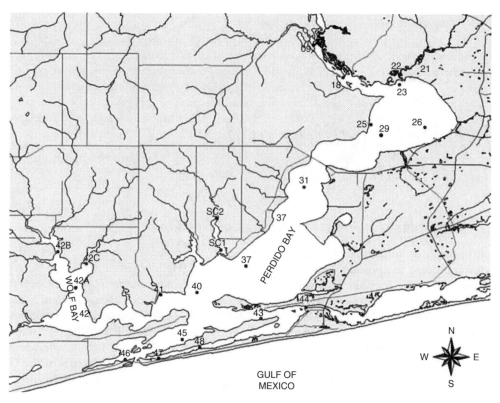

Figure 5.1 Perdido drainage system, contributing rivers, and near-shore parts of the Gulf of Mexico with distributions of permanent sampling stations used in the long-term studies of the area. The Florida Geographic Data Library (FGDL) provided geographic data.

feeding on *Prorocentrum* sp. Lassus and there are also reports that *P. minimus* caused mortalities in old oysters. Woelke (1961) found that this species caused oyster (*Ostrea iurida*) mortalities and cessation of feeding at high densities. The above bloom species have been found in concentrations greater than 1×10^6 cells L^{-1} in Perdido Bay.

5.2 Research in the Perdido River–Bay System

A long-term (16 years of sampling by September 2004), interdisciplinary study has been carried out to determine the response of the Perdido drainage system (northeast Gulf of Mexico; Figure 5.1) to effluent loading from a pulp mill and other sources of nutrients that include a sewage treatment plant (STP) and agricultural/urban runoff (Livingston, 2000, 2002). The research effort is based on written, peer-reviewed protocols for all field and laboratory operations, which are given in Appendix I. Water-quality methods and analyses and specific biological methods have been certified through the Quality Assurance Section of the Florida Department of Environmental Protection (Comprehensive QAP #940128 and QAP #920101). These methods have also been published in peer-reviewed journals (Flemer et al., 1997; Livingston, 1975a, 1976a, 1980a, 1982a, 1984a,b, 1985a, 1987c, 1988b,c, 1992b, 1997a–f, 2004a,b; Livingston et al., 1974, 1976a,b, 1997, 1998a,b).

All sampling for water quality, nutrient loading, phytoplankton, infaunal and epibenthic macroinvertebrates and fishes in the Perdido River and Bay system (Figure 5.1) has been carried out monthly to quarterly on a synoptic basis (i.e., all samples taken within one tidal cycle). River stations used for the determination of nutrient loading are shown

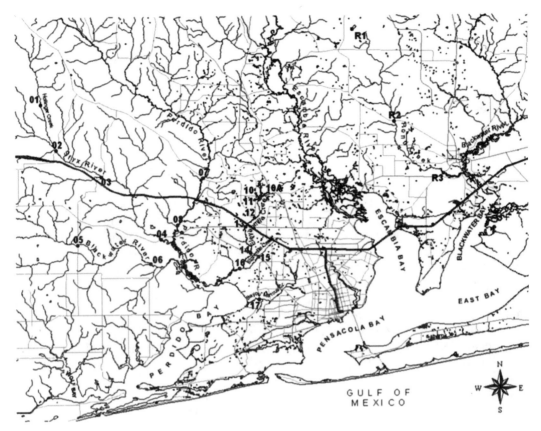

Figure 5.2 River stations used for sampling of chemical and biological factors and nutrient loading. The Florida Geographic Data Library (FGDL) provided geographic data.

in Figure 5.2. Loading to the upper bay from various sources included the following parameters:

1. Nitrogen nitrate and nitrite
2. Ammonia nitrogen
3. Organic nitrogen and organic phosphate
   a. Total organic nitrogen and total organic phosphate
   b. Dissolved organic nitrogen and dissolved organic phosphate
   c. Particulate organic nitrogen and particulate organic phosphate
4. Total nitrogen
5. Orthophosphate
6. Total dissolved phosphorus
7. Total phosphorus
8. Dissolved reactive silicate
9. Inorganic carbon
10. Organic carbon
    a. Dissolved organic carbon
    b. Particulate organic carbon
    c. Total organic carbon
11. Total carbon

Detailed protocols for the various field and laboratory operations are given in Appendix I.

5.3 History of Results

Using a database that includes long-term, species-specific phytoplankton analyses, nutrient loading, water/sediment quality data, biological (infauna, invertebrate, fish) collections, and food web determinations, we have been able to quantify the origin, succession, and impact of phytoplankton blooms on the Perdido Bay system (see Figures 5.1 and 5.2). Analyses of the data have been published by Livingston (2000, 2002).

5.3.1 River Flow Trends

River flow trends in the Perdido system over the study period are shown in Figure 5.3. There were three droughts during the study period 1988–2004: 1988, 1993–1994, and 1999–2002. The most recent dry period was considered the drought of the century, comparable only to the drought of the mid-1950s (Livingston et al., 2003). Livingston (2002) noted that there were two basic components of Perdido River flow: (1) winter–early spring highs and (2) late summer–fall lows. Two-way ANOVAs, run by year and season using monthly averages within each season as replicates, indicated significant (P = 0.05) differences between winter–early spring and late summer–fall flows. Aperiodic freshwater influxes to the bay due to storm activity accounted for occasional peak flows during most months of the year over the 16-year study period. Droughts were defined by continuous low summer flows and relatively flat river curves during winter–early spring flood periods. Relatively heavy river flow events occurred during winter–spring 1990, over a prolonged period from spring 1995 through winter 1996, during winter 1998, and during five

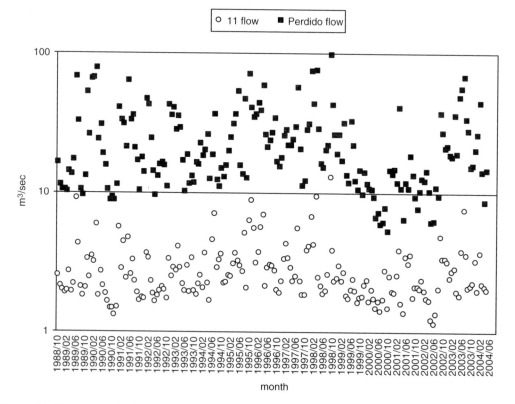

Figure 5.3 Flow rates of Elevenmile Creek and the Perdido River system at monthly intervals from October 1988 to June 2004.

storm periods. A trend analysis of monthly river flows indicated that storm-related increases in river flows were most noticeable during 1995 and 1997–1998. River flows during drought periods were significantly (P = 0.05) different from those during the peak years of 1990 and 1995–1996. With the exception of the 1994 storm, most storm events occurred during summer low flows.

5.3.2   Nutrient Loading

In Upper Perdido Bay during 1988–1991, there were no plankton blooms and areas associated with Elevenmile Creek had relatively high secondary production (Livingston, 1992b). Nutrients (orthophosphate, ammonia) released from a pulp mill on Elevenmile Creek were not associated with phytoplankton blooms. Phytoplankton productivity actually stimulated secondary production: there were peaks of estuarine fish and invertebrate populations in areas surrounding the mouth of Elevenmile Creek. However, beginning in 1993–1994, increased nutrient discharges from the mill into upper Perdido Bay (Figure 5.4) were associated with a series of plankton blooms. These blooms followed patterns of increased orthophosphate and ammonia loading by the pulp mill from 1993 to 1999.

The two primary sources of nutrient loading to upper Perdido Bay included the Perdido River system and Elevenmile Creek (Livingston, 2000, 2002, 2003). Loading of TN, TP, TON, nitrite/nitrate (NO$_2$+NO$_3$), total carbon (TC), and silica (SiO$_2$) was dominated by the Perdido River (Livingston, 2002, 2003). This loading usually peaked in

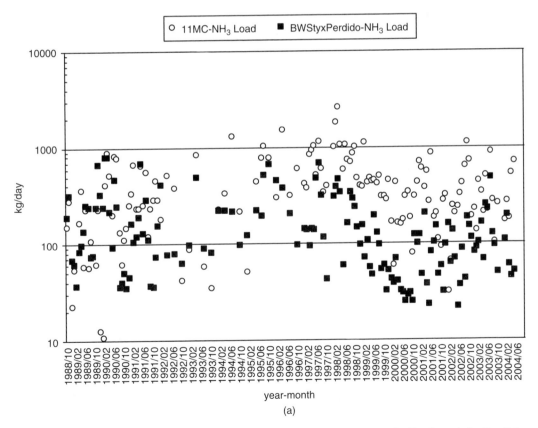

Figure 5.4 Nutrient loading (ammonia and orthophosphate) in Elevenmile Creek and the Perdido River system, monthly from October 1988 to June 2004.

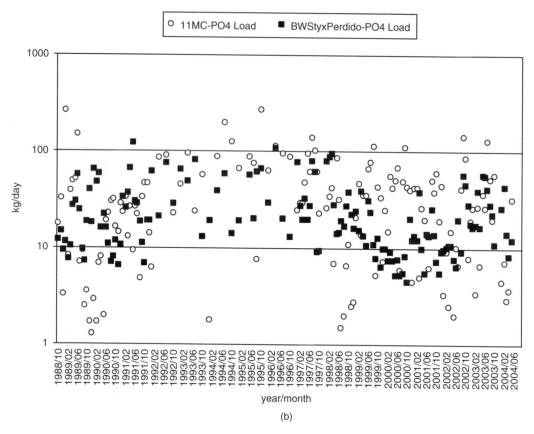

FIGURE 5.4 (continued)

winter–spring periods. Ammonia and orthophosphate loading was highest in Elevenmile Creek as a function of the contribution from the pulp mill (Figure 5.4). Ammonia loading from the creek peaked from February to April and July. Ammonia loading from Elevenmile Creek was relatively low during the early years of analysis. Ammonia loading from Elevenmile Creek to Perdido Bay was significantly higher during the period 1995–1999 than that of the preceding periods (Livingston, 2000, 2002). Ammonia loading rates less than these levels were comparable to those of Elevenmile Creek and the Perdido River system during the period of relatively low ammonia and orthophosphate loading (1989–1991). There was a progressive reduction of ammonia loading from 1999 to 2004, with occasional peaks in 2001, fall 2002, and summer 2003.

Orthophosphate loading tended to be highest from the creek during October and July. Orthophosphate loading was generally low during the first few years of sampling, with occasional peaks in 1989 (Figure 5.4). There was an increase of such loading from 1992 to 1993, with relatively high orthophosphate loading from 1994 to 1997. These loadings were significantly higher than orthophosphate loading during the early years of sampling (Livingston, 2000, 2002). From fall 1997 to 1998, treatment of orthophosphate with alum by the mill reduced such loading to levels noted during the early sampling period, after which there was a general increase in 1999. With the exception of peaks during 2000, fall 2002, and summer 2003, there was a general decrease in orthophosphate loading from the mill from 1999 to 2004 to levels that approximated those during the first 3 years of sampling (1989–1991).

5.3.3 Nutrient Concentrations and Ratios

Orthophosphate concentrations were usually highest at Station P23 at the mouth of Eleven-mile Creek during the early years of the study. From 1993 through the third quarter of 1997, there was a consistent pattern of high orthophosphate concentrations at the mouth of Elevenmile Creek. There was a precipitous decrease in this nutrient throughout the bay during late 1997–1998; this decrease was associated with reductions of mill orthophosphate discharges from August 1997 through March 1999 (Livingston, 2000, 2002). Orthophosphate levels during late 1997–late 1998 resembled those during the period 1989–1990. This trend was followed by increased orthophosphate concentrations during the spring and summer of 1999, at which time the mill resumed relatively high loading of orthophosphate to Elevenmile Creek. By September 1999, the mill again reduced orthophosphate loading to the bay, and upper bay concentrations of this nutrient reflected this change. This trend continued through the 2004 sampling period.

Ammonia concentrations in bay waters were relatively high during the 1989–1991 period (Livingston, 2000, 2002). There were periodic increases in ammonia during 1990 and 1991. By 1993, mean ammonia concentrations in the bay were somewhat lower; this pattern of reduced ammonia continued, with further decreases from late 1997 through early 1998. Increased ammonia concentrations were noted during the summer–fall months of 1998, with reduced ammonia during 1999. Low ammonia was associated with bloom activity in the bay (Livingston, 2000, 2002). Thus, higher ammonia loading from Eleven-mile Creek from 1995 to 1999 was accompanied by periodic reductions in ammonia concentrations in the upper bay due to bloom activity. Variables other than loading were involved in the trends of ammonia concentrations in the bay.

The ratios of surface ammonia and orthophosphate entering the bay from Elevenmile Creek (Station 22/23) followed the same general pattern as the loading (Figure 5.5) (Livingston, 2000, 2002). One difference was the influence of droughts (especially the drought of 1999–2002) when the ammonia and orthophosphate ratios were proportionately higher than the absolute loading rates. The general declines from 1999 to 2003 were evident; concentration ratios were relatively low during the resumption of increased river flows from 2002 to 2003 and the reduced loading from the paper mill. These gradients, together with high nutrient loading rates and drought–flood cycles, tended to define the nature and extent of seasonal and interannual bloom successions. The drought from 1998 to 2002 had an important influence on these relationships (Livingston, 2002).

By averaging differenced data for (orthophosphate + ammonia) loadings and concentration ratios, the data were standardized so that the combined effects of these two nutrients could be evaluated. These transformations were expressed as percent (%) ratio and loading differences from the mean. The long-term changes of such indices are shown in Figure 5.6. Ratio differences of loading were highest from 1993 to 1999, with a general decrease in this index from 1999 to 2003. Ratio differences were highest during 1995–1996 and fall 2002, and the differences from the loading trends reflected the effects of drought on such ratios. The decrease in such ratios over the past few years corresponded to reductions of mill loading of ammonia and orthophosphate.

5.3.4 Phytoplankton Trends: Bloom Distribution

Livingston (2002) noted that river-dominated estuaries in the northeast Gulf of Mexico are highly productive due to factors such as nutrient enrichment from land runoff, the shallow nature of the receiving system, and energy supplements from wind, tidal currents, and thermohaline circulation. Processes that determine primary productivity (based largely on phytoplankton activity) and associated food webs in coastal areas vary widely

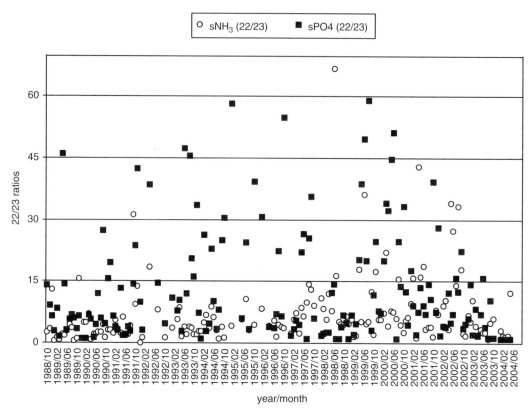

Figure 5.5 Ammonia and orthophosphate ratios at the mouth of Elevenmile Creek as it enters the bay system, taken monthly from October 1988 through June 2004.

due to differences in nutrient loading, the physiography of the receiving area, and habitat features such as temperature, salinity, stratification characteristics, currents, light transmission, and sediment quality. Nutrient loading is fundamental to the growth of coastal phytoplankton (Livingston, 2000). Light availability is an important determinant of estuarine and coastal phytoplankton communities (Philips et al., 2000). Biological processes (competition, predation) also influence phytoplankton production. Different combinations of the set variables thus determine the highly individual rates of primary production and food web responses that differentiate one system from another with resulting differences in population dynamics, community structure, and overall secondary production (Livingston, 2000, 2002).

The response of phytoplankton to nutrient loading in coastal systems has been well studied (Anderson and Garrison, 1997). Specific effects of nutrient loading on phytoplankton assemblages can be related to currents and salinity distribution (Squires and Sinnu, 1982), the physiography of contributing systems (Marshall, 1982a,b, 1984, 1988), and effects of human activities. Industrial wastes such as pulp mill effluents are known to affect phytoplankton (Reddy and Venkateswarlu, 1986). Resulting changes included increased domination by Cyanophycean types, along with the reduction of green algae. Sewage wastes have been associated with *Oscillatoria* spp., *Rhopalodia gibberula,* and *Nitzschia pale.* Blue-green algae (*Oscillatoria nigroviridis*) are often indicators of waters affected by sewage (Premula and Rao, 1977), although blue-green algae are also abundant in marine areas under natural conditions (Potts, 1980). A *Prorocentrum micans* bloom in a New Zealand estuary was coincident with increased nitrogen from upwelling (Chang, 1988).

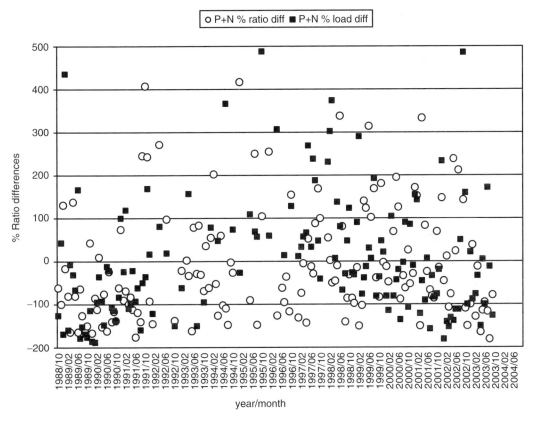

Figure 5.6 P+N percentage ratio differences (and 3-month moving average) in Elevenmile Creek, taken monthly-to-quarterly from October 1988 through September 2003.

Flemer et al. (1997) described results of nutrient limitation experiments conducted with water taken from Perdido Bay during 1991. Six experimental treatments were established in triplicate for each of three bay stations (P23, P31, and P40). Treatments included three control tanks, three P-enriched tanks at 10 μM PO_4-P above ambient, three N-enriched tanks at 50 μM NH_3-N above ambient, and three combined NH_3+PO_4 (referred to as N+P) above ambient, as described for single additions. Primary P limitation occurred mostly during cooler months at upper (tidal brackish) and mid-bay (lower mesohaline) stations. Primary N limitation occurred mostly during warmer months (late summer–fall) in mid-bay areas and infrequently at upper and lower bay stations (upper mesohaline). Apparent N+P co-limitation occurred throughout the year, with peaks during spring and fall in the upper bay. Winter and summer–fall peaks were noted in mid-bay areas, with summer peaks in the lower bay. Primary orthophosphate limitation was associated with high dissolved inorganic nitrogen (DIN); DIN/dissolved inorganic phosphorus (DIP) ratios ranged from 20 to 200. Conversely, primary N and N+P co-limitation were associated with decreasing DIN/DIP ratios.

Phytoplankton assemblages were not strongly nutrient limited, but, given a nutrient increase, these groups responded differentially to nitrogen and phosphorus, both seasonally and along the longitudinal salinity gradient. The combination of phosphorus and nitrogen was usually more stimulatory to phytoplankton growth in Perdido Bay than either of these nutrients alone. Overall, nutrient limitation in Perdido Bay was seasonal, with phosphorus limitation during cold months and nitrogen and/or (nitrogen + phosphorus) limitation during warm months.

During early years of analysis (1988–1991), orthophosphate and ammonia loading from the pulp mill enhanced secondary production in the immediate receiving area of the upper estuary (Livingston, 2000, 2002). Plankton blooms were not present during this period. The chrysophytes and, to a lesser degree, the chlorophytes tended to be abundant and were associated with a balanced food web and relatively high secondary production in the upper bay. During the winter–spring drought in 1993–1994, the pulp mill increased orthophosphate loading to the bay. The combination of drought conditions and increased orthophosphate loading was associated with phytoplankton blooms dominated by diatom species. There was an orderly seasonal succession of these bloom species in the bay. Continued high orthophosphate loading over the next 3 years led to increases in phytoplankton bloom frequency and intensity throughout the bay. From 1996 through 1998, there was increased ammonia loading to the upper bay by the pulp mill. Plankton response included a pattern of the individual bloom species that was attributed to seasonal differences in nutrient requirements of the bloom species. Interannual qualitative and quantitative phytoplankton dominance shifts occurred from 1993 through 1999, whereby larger-celled raphidophyte and dinoflagellate species replaced diatom species that had been predominant during the initial blooms. Increased dominance of bloom species was usually accompanied by reductions in phytoplankton species richness. There was a long-term increase in plankton numbers and biomass during the years of increased nutrient loading.

From late summer 1997 through spring 1999, the pulp mill reduced its orthophosphate loading and there were concurrent reductions of the high relative dominance of bloom species, especially during winter–spring periods (Livingston, 2000, 2002). There was a partial recovery of the phytoplankton associations during the period of low orthophosphate loading to upper Perdido Bay. However, by spring 1999, the mill again resumed high orthophosphate loading to the bay. This loading was accompanied by bay-wide spring and early summer blooms, increased dominance of bloom species, and associated reductions in phytoplankton species richness. The postulated effects of increased orthophosphate and ammonia loading were usually correlated with general sequences of bloom species and associated changes in the phytoplankton community structure, which were consistent with observed natural history characteristics of the diatoms, raphidophytes, and dinoflagellates that comprised the bloom types. There were distinct concentration gradients of the orthophosphate and ammonia as water from Elevenmile Creek entered the bay; such areas were noted as primary sites of bloom origin. The concentration gradients appeared to provide the spark that ignited at least some of the plankton blooms.

The spatial–temporal distribution of the plankton blooms in the Perdido system is shown in Table 5.1. There was a distinct species-specific pattern to the seasonal occurrence of the bloom species in Perdido Bay. In the upper bay, *Prorocentrum minimum* and *Leptocylindrus danicus* were restricted to winter peaks, whereas the raphidophyte blooms (*Heterosigma akashiwo* and *Chattonella subsalsa*) occurred during warm months of the year. The diatoms *Cyclotella choctawhatcheeana* and *Miraltia throndsenii* bloomed mainly during spring months, whereas *Synedropsis* sp. bloomed during the summer. The raphidophyte and blue-green algae blooms occurred later in the interannual bloom succession, and it is possible that *H. akashiwo* displaced *C. choctawhatcheeana* when it reached bloom numbers.

Seasonal changes of monthly averages of nutrient loading in the Perdido system indicate that the highest ammonia loading in Elevenmile Creek occurred during early summer months (June–July) (Livingston, 2002). Ammonia loading from the Perdido River followed the same seasonal pattern, peaking from May through July. Orthophosphate loading in the creek was highest during summer months. In the Perdido River, such loading again followed a similar pattern, peaking in May. Relative abundance (% total numbers) of diatoms occurred during late winter–spring months. There was no statistical

Table 5.1 Occurrences of Blooms ($>1 \times 10^6$ cells L^{-1}) in the Perdido Bay System from October 1988 through June 2002 (data are shown by month and year in various areas of the bay)

Month	Bloom Area	Bloom Species	Bloom Year	Bloom Area	Bloom Species	Bloom Year
December						
January	Baywide	*C. choctawhatcheeana*	2000, 2001			
February						
March				Upper	*M. throndsenii*	1994, 1997
April	Baywide	*C. choctawhatcheeana*	1994, 1998, 1999, 2001	Upper	*M. throndsenii*	1999, 2001
May	Baywide	*C. choctawhatcheeana*	1998, 1999			
June						
July						
August						
September						
October						
November						
December						
January				Upper	*P. minimum*	1999
February						
March	Upper	*H. akashiwo*	1997			
April	Upper	*H. akashiwo*	2001			
May	Upper	*H. akashiwo*	1997, 1998, 2000			
June	Upper	*H. akashiwo*	1998, 1999			
July	Upper	*H. akashiwo*	1996			
August	Upper	*H. akashiwo*	1998	Lower	*P. minimum*	2000
September	Upper	*H. akashiwo*	1997, 1999			
October						
November						
December						
January						
February	Upper	*L. danicus*	1994			
March						
April						
May						
June				Upper	*Synedropsis* sp.	1996, 1998
July				Upper	*Synedropsis* sp.	1996, 1997, 1998
August				Upper	*Synedropsis* sp.	1996
September						
October						
November						
December	11-Mile Ck	*M. tenuissima*	1998, 1999			
January	11-Mile Ck	*M. tenuissima*	1999, 2000, 2001			
February	11-Mile Ck	*M. tenuissima*	2000			
March	11-Mile Ck	*M. tenuissima*	2000			
April	11-Mile Ck	*M. tenuissima*	2000, 2001			
May						
June						
July	11-Mile Ck	*M. tenuissima*	2000			
August	11-Mile Ck	*M. tenuissima*	1999, 2000			

Table 5.1 (continued) Occurrences of Blooms (>1 × 10⁶ cells L⁻¹) in the Perdido Bay System from October 1988 through June 2002 (data are shown by month and year in various areas of the bay)

Month	Bloom Area	Bloom Species	Bloom Year	Bloom Area	Bloom Species	Bloom Year
September	11-Mile Ck	*M. tenuissima*	1999, 2000			
October	11-Mile Ck	*M. tenuissima*	1999, 2000	Upper	*U. eriensis*	1998
November	11-Mile Ck	*M. tenuissima*	1997			

Note: C. choctawhatcheeana = Cyclotella choctawhatcheeana; H. akashiwo = Heterosigma akashiwo; L. danicus = Leptocylindrus danicus; M. tenuissima = Merismopedia tenuissima; M. throndsenii = Miraltia throndsenii; P. minimum = Prorocentrum minimum; U. eriensis = Urosolema.

relationship between diatom abundance and silica loading. Raphidophyte relative abundance peaked in June and July. The raphidophytes were virtually absent from the bay during cold months of the year. Dinoflagellate relative abundance peaked in January. The lowest dinoflagellate relative abundance occurred during April when diatoms were at peak levels of relative abundance. There was little overlap in the relative abundances of the numerically dominant phytoplankton groups. Total phytoplankton biomass peaked in June, reflecting the greater size of the raphidophytes, whereas phytoplankton numbers peaked during April, a reflection of the generally smaller diatoms (i.e., *C. choctawhatcheeana*). Phytoplankton species richness was highest during March, with the least numbers of phytoplankton species noted from August through October.

A summary of the top 30 phytoplankton species organized by average biomass per station over the 16-year sampling period is shown in Table 5.2. Fully eight of the top ten species in such order were bloom types. The blue-green alga *Merismopedia tenuissima* was taken mainly at Stations P22 and P23, and was primarily associated with Elevenmile Creek. The raphidophytes *Heterosigma akashiwo* and *Chattonella* cf. *subsalsa* were found mainly at Station P23 near the mouth of Elevenmile Creek; a substantial showing of *Heterosigma Akashiwo* was also noted at Station P26 off Bayou Marcus Creek and the sewage treatment plant in upper Perdido Bay. The dinoflagellate *Prorocentrum minimum* occurred mainly in the upper and middle parts of the bay, with highest biomass off Bayou Marcus Creek. The diatom *Cyclotella choctawhatcheeana* showed a similar distribution to that of *Prorocentrum*, but occurred bay-wide. *Miraltia throndsenii*, along with other species such as *Cerataulina pelagica*, was found mainly in the lower bay. Elevenmile Creek had the highest numbers per liter, whereas species richness was highest at Stations P23 and P40. Species diversity and evenness, on the other hand, were highest in areas of the upper bay grass beds (Station P25). These data reflect the importance of the bloom species in the Perdido system over time.

Seasonal and interannual successions of plankton blooms could have been affected by interspecific competition and predation within the plankton community. High nutrient loading tended to favor cryptophytes, cyanophytes, dinoflagellates, and raphidophytes. During periods of high winter orthophosphate loading, the dinoflagellates were numerically dominant, whereas summer ammonia loading was associated with increased dominance by the raphidophytes. The chrysophytes and, to a lesser degree, the chlorophytes tended to be more abundant during periods of low nutrient loading. Diatom blooms were noted during winter–spring periods and could have been affected by the predominant raphidophytes during periods of high raphidophyte biomass.

5.3.5 Response to Nutrient Restoration Program

Nutrient control that addresses specific phytoplankton requirements should take into consideration both the seasonal aspects of nutrient loading and long-term temporal

Table 5.2 Top 30 Species of Whole Water Phytoplankton Ordered by Total (Averaged) Biomass (μg) at Stations in Perdido Bay over the 16-Year Study Period; Also Shown Are the Total Numbers L^{-1}, Cumulative Numbers of Species, and Average Shannon-Wiener Species Diversity and Evenness Indices by Station

Species	18	22	23	25	26	29	31	37	40	TOTAL
Heterosigma akashiwo	36.33	3.65	90.48	45.52	78.11	16.12	11.60	4.23	2.55	288.59
Cyclotella choctawhatcheeana	4.00	1.68	18.16	26.58	37.92	30.98	24.87	17.28	11.97	173.44
Chattonella cf. *subsalsa*	0.78	0.23	63.44	16.62	16.47	25.61	5.28	0.09	0.00	128.52
Merismopedia tenuissima	0.01	92.70	9.36	0.46	0.60	0.78	0.20	0.03	0.04	104.17
Prorocentrum minimum	2.10	0.10	13.62	16.97	24.33	19.09	12.44	3.06	2.53	94.25
Gymnodinium cf. *splendens*	2.83	0.07	9.16	10.49	13.46	9.33	3.53	1.33	0.65	50.86
Miraltia throndsenii	0.34	0.02	1.41	4.24	5.60	6.84	3.36	15.97	4.60	42.37
Cerataulina pelagica	0.71	0.29	1.07	4.22	5.49	3.74	4.22	7.81	7.28	34.82
Leptocylindrus danicus	0.02	0.01	0.82	0.06	9.84	7.72	0.30	1.33	2.14	22.23
Dactyliosolen fragilissimus	0.17	0.04	0.54	1.86	2.15	1.61	2.26	5.69	5.80	20.11
Cryptomonas sp.	0.69	0.04	16.62	0.00	0.00	0.00	0.00	0.00	0.00	17.35
Synedropsis sp. *novum*	0.04	0.01	2.71	1.58	1.13	1.28	0.65	7.35	0.12	14.89
Skeletonema costatum	0.72	0.08	0.51	1.00	0.92	1.09	2.26	3.73	4.08	14.39
Heterocapsa pygmaea	0.44	0.00	1.83	2.07	2.38	2.04	0.93	0.56	0.56	10.81
Chaetoceros simplex	0.07	0.03	0.23	0.95	1.00	0.89	1.32	1.99	1.87	8.34
Ceratium hircus	0.09	0.00	0.86	1.01	1.00	1.03	0.88	1.33	0.92	7.13
Bacteriastrum hyalinum	0.09	0.01	0.29	0.17	0.21	0.37	0.67	2.04	2.21	6.04
Chaetoceros wighami	0.07	0.01	0.87	0.56	0.48	0.27	0.66	1.61	1.50	6.03
Miraltia sp.	0.13	0.01	0.64	0.71	1.66	0.76	0.73	0.65	0.57	5.87
Proboscia alata	0.00	0.00	0.06	0.09	0.20	0.00	0.34	0.61	2.13	3.43
Asterionellopsis glacialis	0.09	0.03	0.06	0.19	0.27	0.22	0.35	1.10	1.07	3.38
Cyclotella meneghiniana	0.07	3.25	0.03	0.00	0.01	0.01	0.00	0.00	0.01	3.38
Chaetoceros diversus	0.01	0.00	0.24	0.30	0.19	0.25	0.37	0.56	0.68	2.62
Chaetoceros affinis	0.02	0.01	0.16	0.19	0.20	0.17	0.23	0.55	0.59	2.11
Chaetoceros subtilis	0.02	0.00	0.11	0.26	0.27	0.23	0.28	0.42	0.41	2.01
Cryptomonas sp. (*Teleaulax*-like)	0.00	0.00	1.62	0.00	0.00	0.00	0.00	0.00	0.00	1.62
Hemiaulus hauckii	0.02	0.01	0.05	0.15	0.08	0.15	0.09	0.43	0.32	1.29
Spermatozopsis exsultans	0.03	0.83	0.14	0.05	0.09	0.05	0.03	0.03	0.01	1.26
Dinobryon divergens	0.44	0.04	0.18	0.22	0.10	0.14	0.07	0.00	0.00	1.17
TOTAL #, L^{-1} × 1000	**396**	**23,473**	**3,676**	**163**	**223**	**1,804**	**1,236**	**1,883**	**848**	**37,184**
Cumulative # Taxa	**136**	**111**	**237**	**116**	**127**	**120**	**160**	**119**	**177**	**404**
Shannon Diversity	**1.98**	**0.10**	**1.40**	**2.13**	**2.02**	**2.02**	**1.99**	**1.80**	**2.19**	**1.27**
Shannon Evenness	**0.40**	**0.02**	**0.26**	**0.45**	**0.42**	**0.42**	**0.39**	**0.38**	**0.42**	**0.21**

changes of interactions among the various plankton species. There is evidence that nutrient removal from some estuaries has resulted in water quality improvements (e.g., phosphorus removal in the upper tidal freshwater Potomac [Jaworski, 1972; 1981]; the upper portion of the Thames River estuary [Gameson et al., 1973]; Kaneohe Bay and Hawaii [Smith, 1981]). In most cases, it was not ascertained if the phytoplankton community and/or associated food webs had actually been either affected by the blooms or restored.

A restoration program of proposed reductions of orthophosphate and ammonia loading by the pulp mill to Elevenmile Creek was undertaken in 1999. This program continues to this day. The restoration program was based on the long-term Perdido Bay database, and, for the first time in Florida, the Florida Department of Environmental Protection (FDEP) designed a permit to the paper mill that used the proposed orthophosphate and ammonia loading rates based on the long-term research in the Perdido system. Orthophosphate and ammonia loading by the pulp mill was targeted for reductions to levels that approximated loadings during periods when the bay was free of blooms. A review was made concerning ammonia loading to upper Perdido Bay and ammonia concentrations in Elevenmile Creek. It was estimated that the level of ammonia loading by the mill should also ensure that ammonia concentrations at the mouth of Elevenmile Creek would remain below concentrations that could be toxic to phytoplankton species (Livingston, unpublished data).

Continuous monitoring of ammonia and orthophosphate gradients at the end of Elevenmile Creek was used to allow a rapid determination of the efficacy of the nutrient loading reductions by the mill. It was estimated that ratios of 8 (ammonia) and 14 (orthophosphate) mark the upper boundary of "safe" concentration gradients for these nutrients. When monthly monitoring indicated exceedances of these ratios, an immediate review was made of the nutrient loading from the mill with the aim of reducing such ratios by applying nutrient treatments such as alum if exceedances of the target ratios were observed. In addition, the mill continued to make changes to the existing treatment system that aided in reducing the loading of ammonia and orthophosphate to Elevenmile Creek and Perdido Bay.

The field data were consistent with the results of nutrient limitation experiments (Flemer et al., 1997): orthophosphate loading was associated with winter bloom species (*Prorocentrum minimum*) and ammonia and orthophosphate loading was associated with warm-water blooms (*Heterosigma akashiwo*) (Livingston, 2000, 2002). The responses of the different groups of plankton to changes in nutrient loading corresponded to the results of nutrient limitation experiments that indicated phosphorus limitation during cool months and phosphorus + nitrogen limitation during warm months. During the period of high ammonia and orthophosphate loading, interannual successions of phytoplankton blooms peaked, especially during warmwater periods (Figure 5.7). Phytoplankton biomass increased from the early sampling (no blooms: 1990–1992) to the period of bloom initiation (1993–1996). With the 1997–1998 cessation of orthophosphate loading by the mill, there was a general decrease in phytoplankton biomass, which was then followed by increased biomass from 1999 to 2001 that coincided with resumed loading of mill nutrients. With the exception of two dates (October 2002 and July 2003), there was a general downward trend in whole water phytoplankton biomass through the summer of 2004. Orthophosphate loading was well controlled but there were periodic problems with ammonia loading during summer months.

There were no blooms in Perdido Bay during the sampling period 1988–1993 (Figure 5.8). The first blooms were noted in the bay during the increase of nutrient loading by the mill in 1993–1994. Peak numbers of plankton blooms occurred from 1996 to 2001, a period of high nutrient loading from the mill. This period included months of high rainfall and a record drought, during which time there was a replacement of the initial diatom blooms with raphidophyte and dinoflagellate blooms. After a period of relatively high numbers of *Prorocentrum minimum* from 1994 to 1997, there was a major decrease in this species during late 1997 to 1998; this coincided with lowered orthophosphate loading from the mill and supported the experimental evidence of orthophosphate limitation of this species (Figure 5.9). From 2001 to 2002, there was a reduction in the numbers of plankton blooms (Figure 5.8). Reduced numbers of the two species that were associated with lowered secondary production (*Prorocentrum minimum* and *Heterosigma akashiwo*) followed the lower levels of nutrient loading from the mill during this period (Figure 5.9). *Prorocentrum* showed a general reduction in numbers, whereas *Heterosigma* showed blooms during October 2002 and April–May 2003. The increased bloom numbers during 2003–2004 were due largely to periodic increases in ammonia loading and to other sources of nutrients, as will be explained below.

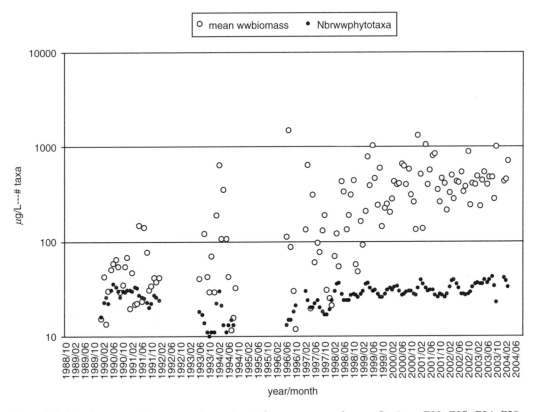

Figure 5.7 Whole water biomass and species richness averaged over Stations P23, P25, P26, P29; P31, P33, P37, and P40 by month during the sampling period October 1988 through June 2004.

5.3.6 Bay Impacts (Fall 2002–Summer 2003)

The Bayou Marcus Water Reclamation Facility, on the eastern shore of upper Perdido Bay (see Figure 5.1), discharges effluents through a marsh and into the upper bay. The permitted discharge to the wetlands is 8.2 million gal day^{-1} (MGD). Flow rates in the year 2000 approximated 3.4 MGD (Florida Department of Environmental Protection, personal communication). Following a lengthy drought from 1998 to 2002, the Perdido Bay area received rainfall from two tropical storms in September 2002. One of the tropical storms dropped approximately 9.5 in. of rain in the area. It was noted that flows went from about 3,500 to over 10,000 MGD after the rainfall. During an October 2002 trip to the bay, field personnel noted (by smell and taste) chlorine in the surface waters at Stations 18, 25, 26, 29, 31, 33, 37, and 40. Following another heavy rain in June 2003, there was another anecdotal observation of chlorine in the water.

Downstream currents from upper Perdido Bay move nutrients and particulate matter into deeper parts of the bay; Station 37 represents an area of maximum deposition (Livingston, 2000). Sediments were analyzed during October 2002 as part of the long-term sediment analysis of the Perdido Bay system (Figure 5.10). During fall 1993, TON sediment concentrations (see Figure 5.10A) were highest in depositional areas of the upper bay (Station 29) and the lower bay (Station 37). By 1998, during a period of heavy ammonia loading from the mill, there was a clear gradient from the mouth of Elevenmile Creek to the depositional areas of the upper and lower bay. However, after reductions of ammonia loading by the mill, the fall 2001 analysis indicated a shift in the location of upper bay

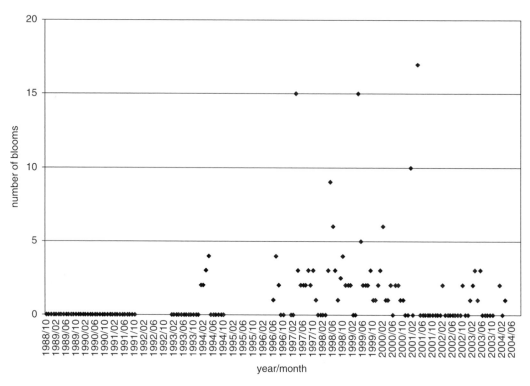

Figure 5.8 Bloom numbers totaled over Stations P23, P25, P26, P29; P31, P33, P37, and P40 by month during the sampling period October 1988 through June 2004.

high sediment TON to Station 26 (i.e., the Bayou Marcus Water Reclamation Facility). There was an increase in the lower bay depositional area (Station 37). By fall 2002, sediment TON concentrations were highest at Station 26 in the upper bay and at Station 37 in the lower bay. These concentrations represent the highest such concentrations ever taken.

Sediment TP concentrations from 1993 to 2002 (Figure 5.10B) indicated that, during 1993, there was a relatively low TP sediment burden that came from Elevenmile Creek. By 1996, after 2 to 3 years of higher orthophosphate loading from the mill, there was a more widespread distribution of sediment orthophosphate, again from Elevenmile Creek. High concentrations also occurred at Station 37 (i.e., the deepest part of the bay). By 1998, after more than a year of reduced orthophosphate loading by the mill, there were general reductions in this nutrient in areas receiving creek loading. However, by 2001, the upper bay source appeared to have shifted to Station 26 (i.e., the station receiving effluents from the sewage treatment facility). By October 2002, the highest sediment concentrations of TP were noted at Station 26 in the upper bay and at Station 37 in the lower bay. Sediment TON followed the same spatial pattern over the sampling period. The sediment TP and TON data are consistent with the hypothesis that temporal changes in sediment nutrients followed loading characteristics in the upper bay.

During the fall 2002 rainfall period, the highest chlorophyll concentrations taken during the entire 16-year sampling period were noted at Station P26, which is located off Bayou Marcus Creek. During October 2002, there was a massive bloom of *Heterosigma akashiwo* (Figure 5.11). The bloom extended from the Bayou Marcus drainage down the entire bay in a pattern similar to that noted in the anecdotal chlorine observations and the quantitative, long-term trends of sediment TON and TP. The October 2002 occurrence

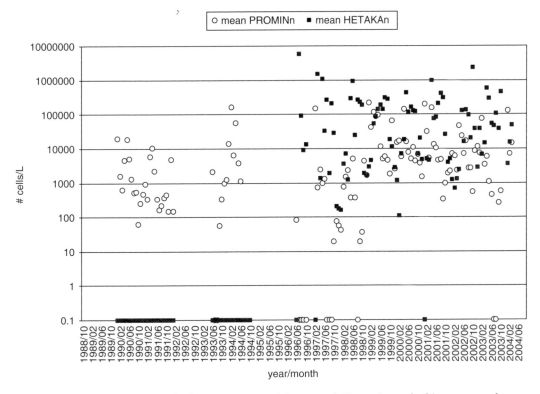

Figure 5.9 Numbers of cells L^{-1} of *Prorocentrum minimum* and *Heterosigma akashiwo* averaged over Stations P23, P25, P26, P29; P31, P33, P37, and P40 by month during the sampling period October 1988 through June 2004.

of this species in upper Perdido Bay represented the most dense such bloom taken over the 16-year survey. The generally decreasing *Heterosigma* numbers at the mouth of Elevenmile Creek indicated that the mill was not implicated in the 2002 and subsequent 2003 increases of *Heterosigma* (Figure 5.11).

Infaunal macroinvertebrates are often used as indicators of water quality. In Perdido Bay, there are various species that are common to various estuarine systems. These species represent an important link to bay food webs, and any reduction in infauna will result in impaired food webs. Bay mollusks can be periodically dominated in terms of biomass by the bivalve mollusk *Rangia cuneata* and the prosobranch *Assiminea succinea*. The polychaetes are represented by dominants such as the spionids *Streblospio benedicti*, *Paraprionospio pinnata*, and the capitellid *Mediomastus ambiseta*. Various subdominants in Perdido Bay include the polychaetes *Parandalia americana*, *Glycinde solitaria*, *Hobsonia florida*; the crustaceans *Grandidierella bonnieroides*, *Edodea* sp.; and the harpacticoid *Scottalana canadensis*. A number of chironomids are also found in the bay, including the *Chironomus decorus* group and the *Chironomus fulvus* group.

Long-term changes in Perdido Bay infaunal species richness are given in Figure 5.12. Seasonally high numbers of this index were taken from 1988 to 1992. The infaunal species richness decreased as plankton blooms occurred in the bay from 1993 to 1999. With decreases in both *Prorocentrum* (winter) and *Heterosigma* (spring–summer) during recent years, there was a general increase in infaunal species richness, which indicated partial recovery of this group of organisms relative to this index taken during periods of no blooms (1988–2002). *Prorocentrum* continued to decrease during the last 2 years of sampling, whereas *Heterosigma* showed marked increases during fall 2002 and summer 2003.

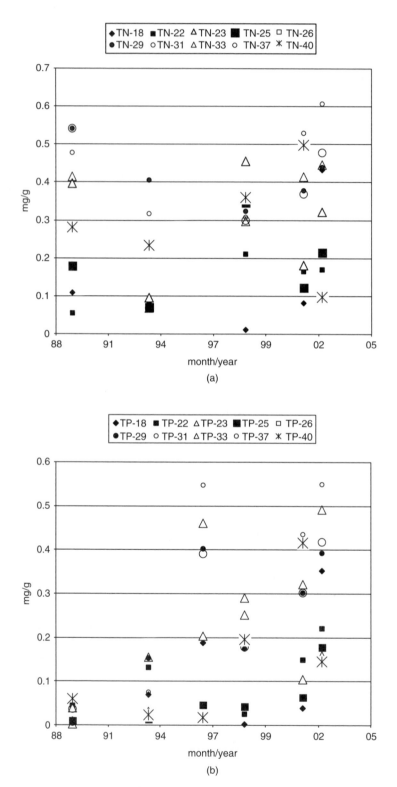

Figure 5.10 Sediment analyses of (A) total organic nitrogen (TON) and (B) total phosphorus (TP) during fall periods from 1993 to 2003.

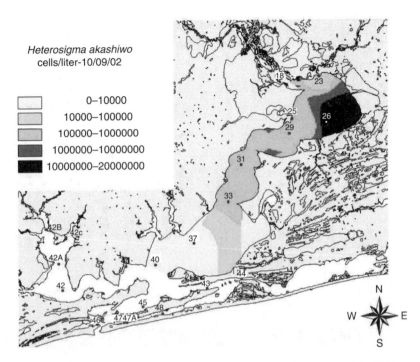

Figure 5.11 Distribution of *Heterosigma akashiwo* in Perdido Bay during October 2002.

It was during this period that infaunal species richness reached the lowest levels of the 16-year survey (Figure 5.12). These decreases coincided with the presence of chlorine in the bay during October 2002 and June 2003. The most marked decreases were noted at Station P26 (Bayou Marcus) (Livingston, 2003). There was recovery noted at Station P23 following the October 2002 incident, whereas the summer 2003 decrease was general throughout the bay at this time. By summer 2003, there was an almost complete elimination of the infauna of the Perdido system, which suggests serial reduction of bay resilience due to repeated incursions of polluted water in the Bayou Marcus region.

The above temporal trends of infaunal species richness are illustrated by the spatial distributions of infaunal macroinvertebrate abundance from 1988 to 2003 (Figure 5.13). During the early years of sampling, the highest numbers of infauna occurred near the mouth of Elevenmile Creek. However, numbers dropped precipitously in such areas during the bloom period from 1995 to 1999. There was a subsequent increase in numbers with a cessation of blooms from 1999 to 2001. From 2002 to 2003, there was an almost complete collapse of infaunal macroinvertebrates in the Bayou Marcus areas and associated parts of the middle and lower bay. A similar pattern in space and time was noted in the distribution of total number of species of infaunal and epibenthic macroinvertebrates and fishes in the Perdido Bay system from 1988 to 2003 (Figure 5.14). The long-term changes of key biological indices in Perdido Bay followed the distribution of nutrient loading, water and sediment quality, and plankton blooms, and these data indicate that the almost total collapse of bay biota by fall 2003 was associated with effluent loading from the STP during periods of heavy rainfall during fall 2002 and spring 2003.

5.3.7 Non-point Nutrient Sources: Agricultural and Urban Runoff

Water quality data for Wolf Bay (Figure 5.1) indicated general increases in watercolor and turbidity during recent years (Livingston, 2002, 2003). Both surface and bottom dissolved

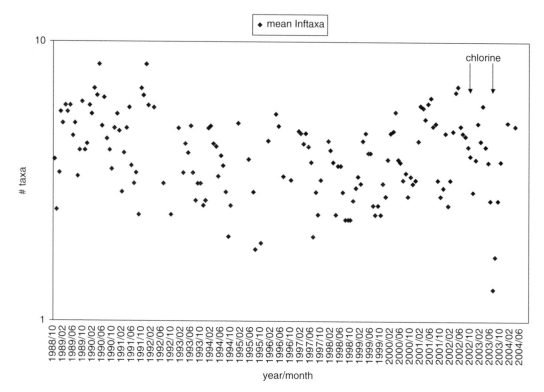

Figure 5.12 Distribution of infaunal species richness averaged over Stations P23, P25, P26, P29; P31, P33, P37, and P40, taken from October 1988 through June 2004.

oxygen also deteriorated from 1991 to 2003, with periodic reductions below 4 mg L^{-1} at depth since 1999. Oxygen anomalies also showed steady deterioration, which was especially pronounced at depth. This indicated a biological source of the reduced oxygen in Wolf Bay. Wolf Bay ammonia concentrations showed recent increases that were most pronounced at depth, indicating sediment involvement (Livingston, 2002, 2003). Both surface and bottom chlorophyll *a* concentrations increased steadily since 1998. The water quality data thus show steady deterioration, which appeared to be related to sediment nutrient concentrations. Based on the experience in the upper bay, there were indications that nutrient loading from upland (agricultural) sources was an important contributing factor to the deterioration of water quality in Wolf Bay.

Phytoplankton collections from Wolf Bay indicated that there were four major bloom species among the numerically dominant populations. These blooms started in August 2000, with major winter and summer blooms occurring periodically to the present. The distribution of the dominant bloom species is shown in Figure 5.15. Bloom numbers of both *Prorocentrum minimum* and *Heterosigma akashiwo* started to occur during the 2000 sampling period, and these blooms continued through winter 2003. There was a reduction in phytoplankton species richness and diversity since the summer of 2000 that coincided with the blooms of these two species The phytoplankton data are consistent with long-term trends of both the sediment quality and water quality in Wolf Bay, and the overall analysis shows that this body of water is deteriorating rapidly due to nutrient loading from upland sources. During the summer of 2003, there were no infaunal macroinvertebrates taken in Wolf Bay.

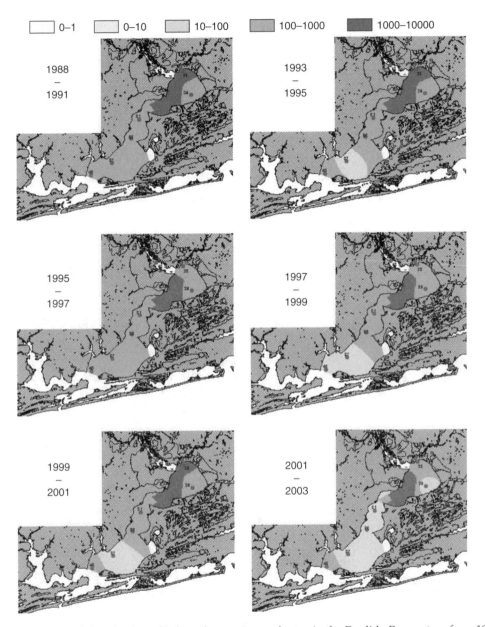

Figure 5.13 Spatial distribution of infaunal macroinvertebrates in the Perdido Bay system from 1988 to 2003. Data were averaged over 2-year periods.

Livingston (1998b) noted salinity stratification and hypoxia in lower Perdido Bay south of Ono Island, an area characterized by generally uncontrolled urban development (see Figure 5.1). There was a general decrease in light penetration over the study period. This appeared to be associated with increased turbidity. Surface and bottom oxygen anomalies showed a downward trend similar to that in Wolf Bay. There were periodic high surface ammonia concentrations with no real temporal trend. Extremely low infaunal numbers and species richness were noted in the lower bay south of Ono Island (Station P48). These data indicate water quality problems in areas receiving runoff from urban areas.

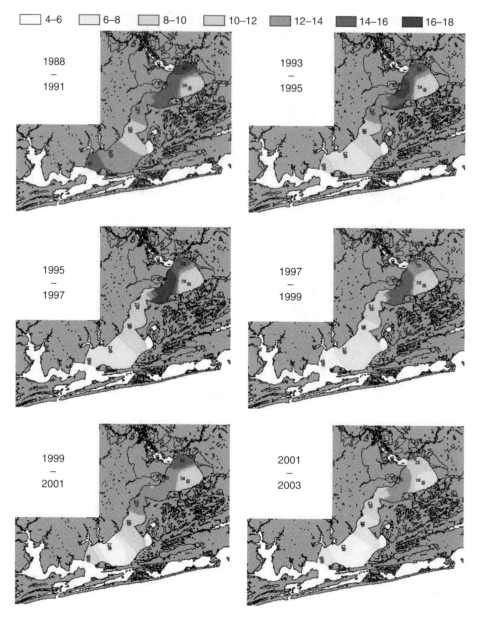

Figure 5.14 Total number of species of infaunal and epibenthic macroinvertebrates and fishes in the Perdido Bay system from 1988 to 2003.

5.3.8 Statistical Analyses of the Long-Term Data

A series of statistical methods were used to analyze the long-term data. Scatter-grams of the long-term field data were examined, and either logarithmic or square root transformations were made, where necessary, to approximate the best fit for a normalized distribution. These transformations were used in all statistical tests of significance. The transformed numerical abundance data for the primary bloom species were reorganized and averaged over bay stations (P23, P31, and P40) and over the months when these species were most abundant (*Prorocentrum minimum* [January–April], *Heterosigma akashiwo* [April–July, October], and *Cyclotella choctawhatcheeana* [April]). All data (250 variables of nutrient loading

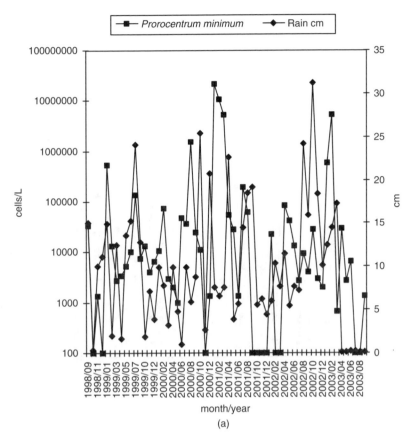

Figure 5.15 Distribution of bloom species in Wolf Bay from August 1998 through February 2003: (A) *Prorocentrum minimum* vs. rainfall and (B) *Heterosigma akashiwo* vs. rainfall.

from all sources, nutrient ratios, river flows and rainfall, water quality, dominant phytoplankton species indicators, whole water phytoplankton community indices [species richness, Shannon diversity and evenness, total numbers and biomass], infaunal, invertebrate, and fish indices [total numbers and biomass, species richness and diversity, total numbers of trophic units, and dominant population data], and summed data for such indices for infauna, invertebrates, and fishes [so-called FII indices]) were averaged over seasonal periods of bloom abundance and by year for 15 years. The various indicators of the dominant bloom species (numbers of cells L^{-1}, dry-weight biomass L^{-1}, % numbers, % biomass) were used as dependent variables for a series of simple regressions and cross-correlation analyses.

Principal Component Analysis (PCA) and associated correlation matrices were determined using the averaged data. Results of the regression analyses were determined to reduce the independent variables into a smaller set of linear combinations that could account for most of the total variation of the original set. For the independent variables in this study, the series were stationary after appropriate transformations; thus, the sample correlation matrix of the variables was a good estimate of the population correlation matrix. In this way, the standard PCA could be carried out based on the sample correlation matrix. The data matrix described above was used to run a series of regression models with the bloom factors as dependent variables.

Results of the PCA regression analyses are given in Figure 5.16. *Prorocentrum* phytoplankton species richness and diversity, and with invertebrate and fish species richness.

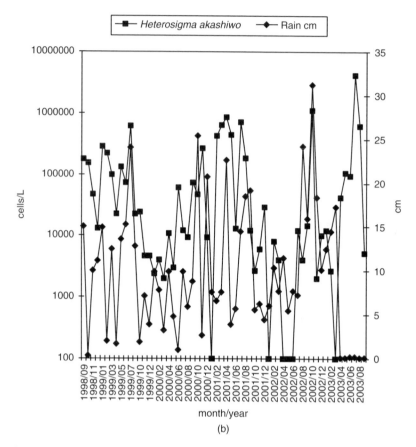

Figure 5.15 (continued)

Heterosigma akashiwo was positively associated with ammonia and orthophosphate loading and the P+N loading difference indicator. It was also positively associated with bottom chlorophyll *a* concentrations, possibly as an indicator of the propensity for this species to occupy benthic habitats. *H. akashiwo* was negatively associated with various phytoplankton indicators and fish species richness. *Cyclotella choctawhatcheeana* was positively associated with ammonia ratios and the P+N ratio difference indicator. It was negatively associated with invertebrae species richness.

Results of the PCA regression analysis are given in Figure 5.16. *Prorocentrum minimum* (dinoflagellate) blooms peaked during January, with population abundance high from January through April. Orthophosphate loading and orthophosphate ratios are rate-limiting factors, although P+N loading and ratio differences can come into play. Results of the PCA regression analyses (Figure 5.16) indicated adverse effects on whole water phytoplankton species richness, and there were positive associations with bottom chlorophyll *a*. There were also significant associations of *Prorocentrum minimum* numbers with reduced invertebrate trophic units and species richness, as well as fish numbers. *Prorocentrum minimum* was thus considered to have a major adverse impact at various levels of biological organization.

Heterosigma akashiwo (raphidophyte) had bloom peaks in May and September, with high abundance from April through July. Results of the PCA regression analyses

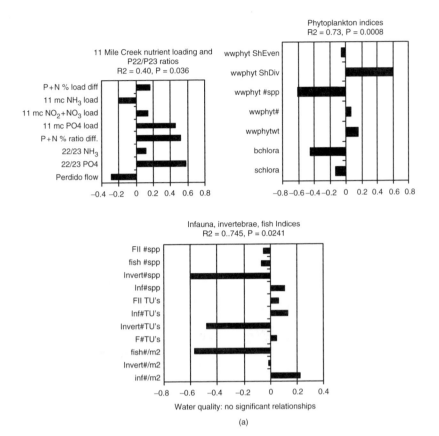

Figure 5.16 PCA regression analysis using averaged data from Stations P23, P31, and P40 concerning the relationship of (a) *Prorocentrum minimum*, (b) *Heterosigma akashiwo*, and (c) *Cyclotella choctawhatcheeana* (# cells L^{-1}) relative to nutrient loading/nutrient ratios, plankton indices, water quality indices, and biological (infauna, invertebrate, fish) indices taken in the Perdido Bay system from October 1988 through September 2003.

(Figure 5.16) indicated that this species was significantly associated with PO$_4$ and NH$_3$ loading and with P+N% loading differences. *H. akashiwo* was also associated with high concentrations of various nutrients, with the exception of silica. *H. akashiwo* adversely affected phytoplankton species richness, diversity and evenness indices, while being significantly associated with reduced herbivore and FII biomass, FII species richness, invertebrate trophic units, and fish/invertebrate species richness. This raphidophyte was also associated positively with numbers of phytoplankton blooms and phytoplankton numbers and biomass. *H. akashiwo* was thus considered to have a major adverse impact at various levels of biological organization.

    Cyclotella choctawhatcheeana is a diatom species with peaks of abundance during April. This species was positively associated with ammonia loading and P22/P23 ratios, P+N % loading differences, and P+N % ratio differences (Figure 5.16). Although *Cyclotella* was positively associated with the number of blooms and phytoplankton numbers and biomass, it was negatively associated with phytoplankton Shannon diversity and evenness indices. There were no significant relationships with any of the other biological factors, and *C. choctawhatcheeana* was not considered to have a major adverse impact at various levels of biological organization.

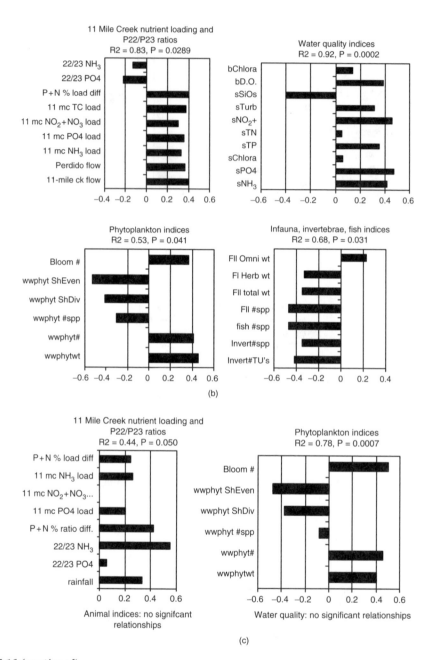

Figure 5.16 (continued)

Overall, two plankton species (*Prorocentrum minimum* and *Heterosigma akashiwo*) were noted to have adverse effects on various parts of the food webs and infauna, invertebrate, and fish indices in the Perdido Bay system.

5.4 The Press and the Perdido System

The 3-year (1988–1992) intensive study of the Perdido system was carried out by more than a dozen scientists and engineers from universities and state/federal agencies throughout the country in response to a court order to get to the bottom of the long-standing controversies concerning the impact of a pulp mill on the Perdido system (Appendix I; Livingston, 2000, 2002). A comprehensive report of the data (Livingston, 1993b) that was based on the research (Appendix I) was met with controversy when the results did not corroborate the beliefs of the "environmentalists." The report did not say that the pulp mill had no effect on the bay. As shown above, the mill had multiple serious adverse effects on Elevenmile Creek and parts of the upper bay. The problems with hypoxia due to high DOC loading, combined with high conductivity and high concentrations of ammonia and orthophosphate, were outlined in detail, along with the potential problems of algal blooms associated with mill nutrient loading (Livingston, 1993b, 2000, 2002). Various lines of evidence indicated that the mill was not primarily responsible for low DO at depth in deeper parts of Perdido Bay (Livingston, 1993b). Nutrient and organic carbon loads from the mill were not high enough to eliminate the entire DO regime of Perdido Bay. High nutrient loadings to the upper bay, associated with mill discharges, presented a potential problem as such loadings were associated with phytoplankton outbursts that could eventually be detrimental to the bay. This projected problem eventually was shown to be a reality, with increased nutrient loading by the mill during the 1990s (Livingston, 2000, 2002). However, such boring facts were largely ignored by the regional press interests that continued to play up the "environmentalist" claims that the pulp mill was the only source of pollution to the bay.

One element of the publicity concerning Perdido Bay was the discharge of toxic agents such as dioxin by the pulp mill. The fact that none of the federal or state agencies found high concentrations of toxic agents in Perdido Bay was not reported. The dioxin testing program in the Perdido system was carried out under the auspices of the U.S. EPA. Dioxin was considered a public health problem that was not part of the routine ecological studies of the bay. However, a field-screening program for extractable organically bound chlorine (EOCl) was undertaken (Livingston, 1993). The EOCl component represents the highly lipophilic, potentially bioaccumulable compounds that are often found in pulp mill effluents. These compounds tend to bioconcentrate in aquatic food webs and are therefore a good measure of potential harm to such systems when chlorinated by-products are discharged. In general, no EOCl residues were noted in sediments of Elevenmile Creek, and there were relatively low levels of EOCl in the sediments of the Perdido Bay system. Moderately low concentrations of EOCl were noted in the muscle tissue of various freshwater fishes and were attributed to the Pensacola mill. However, concentrations of EOCl in organisms taken from Perdido Bay were not substantially different from EOCl in organisms taken from Apalachicola Bay where there is no paper mill. When compared to concentrations of EOCl in other aquatic areas near pulp mills, the concentrations in animals of Perdido Bay and Apalachicola Bay were low (Livingston, 1993b).

5.4.1 The Dioxin Issue

As in other controversial issues before the public (i.e., the Fenholloway River; see above), changes in the secondary sex characteristics of fishes and the dioxin issue were emphasized in press reports of the status of the Perdido system. There was no real review of the scientific data, and in many instances, the publicized "reviews" were made by people who had not even read the report.

POLLUTION FINDINGS BLASTED: "Environmentalists in Alabama and Florida angrily denounced the findings of a study absolving a paper mill of blame for Perdido Bay pollution. ... 'He is absolutely out of his mind,' said Joy Morrill.

... 'Any kind of development does have an impact, but the lower part of this bay is in far better shape than the upper part of the bay.' Jackie Lane...also criticized Livingston's conclusions. Like Morrill, she had not seen a copy of his report."

—Associated Press, August 8, 1992

HEALTH OF PERDIDO BAY STILL AT ISSUE: "Dr. Ishphording (University of South Alabama) concluded from his mish-mash of studies that Champion was having no impact...he had to ignore the 6,000 pounds of sludge that come down Eleven-Mile Creek every day. And what about the dioxin?"

—Letter to the Editor, "environmentalist,"
The Islander Gulf Shores, **Alabama, April 21, 1992**

TEST SIGNIFICANCE, METHODS AT ISSUE: "Champion's effluent has 'hormone-like' effects on many animals and causes reduced growth and reproduction. This hormone effect may very well be due to dioxin."

—Letter to the Editor, "environmentalist,"
The Islander Gulf Shores, **Alabama, April 29, 1992**

FISH FROM ELEVEN MILE CREEK SAID SAFE TO EAT: "Fish from Eleven Mile Creek are safe to eat, the Florida Department of Health and Rehabilitative Services announced last week. ... Recent tests of fish caught in the creek have shown that dioxin has dropped to below the detectable level of one part per trillion, said Dr. Charles S. Mahan, HRS's deputy health secretary. 'Consuming fish from the creek no longer poses a health threat,' he said."

—Michael Hardy, *Mobile Register*, September 5, 1992

PERDIDO BAY ENVIRONMENTAL GROUP SKEPTICAL: "That waste water is composed of substances such as dioxins, ammonia and organic carbon, materials that consume oxygen said Dr. Jackie Lane ... According to (Ms.) Young, the study did not directly address the problem of dioxin, which causes cancer and birth defects."

—Tammy Leytham, *Onlooker*, September 19, 1992

"The Coalition cited a report from the U.S. Fish and Wildlife Service which shows turtles and sediment in Perdido Bay and Elevenmile Creek in Escambia County are heavily polluted with dioxin. 'Environmentalist': "'My face. I've got this adult acne, I'm sure it is due to exposure to, you know, the dioxin... the Coalition complains the bills are weak. One way to make them stronger, they say, would be to supersede the current federal dioxin standard of 7 parts per trillion down to zero.'"

—Florida Public Radio Report, February 11, 1994, WFSU-TV

STUDY DOCUMENTS ENVIROMENTAL DISASTER ON PERDIDO BAY: "In late 1994, the paper mill converted to 100% chlorine dioxide bleaching. In

November 1995, we measured Eleven Mile Creek to see if we could find evidence of chlorine dioxide. It was there. ...it is possible that chlorine dioxide, which is very close to carbon dioxide in chemical structure, could interfere with the photosynthetic process in plants.... It is possible that chlorine dioxide is continuously released along the creek? ...Chlorate is the product from which chlorine dioxide is generated at the paper mill. We found it as well... With all the possibility of damaging effects from both chlorine dioxide and chlorate, no wonder Livingston found an environmental disaster in Perdido Bay. But of course, he blames this on too much phosphate. Why? Because phosphate can easily be controlled. ... Livingston has chosen the 'politically correct' answer and solution to Perdido Bay's problems."

> —*Tidings* (newsletter of the Friends of Perdido Bay),
> 13: No. 4, August 2000

NOTHING NEW: "...since the paper mill in Cantonment installed chlorine dioxide bleaching, much of the vegetation in Perdido Bay has disappeared — grass beds, aquatic vegetation, and even the scum algae."

> —Letter to the Editor, "environmentalist,"
> *Pensacola News Journal,* June 11, 2004

The dioxin question in the Perdido system was emphasized in media reports. The fact that the "environmentalist" claims did not match the scientific data was not questioned by anyone. For example, the U.S. Fish and Wildlife Service report (Brim, 1993) that supposedly reported high dioxin concentrations in the Perdido system actually had the following to say about such pollution:

"A review of the most important chemical data bases for the Perdido Bay ecosystem reveals that Perdido Bay is generally free of toxic compounds. However, sediments in some discrete areas are contaminated. Contamination has been identified in Elevenmile Creek, Saufley Field ditch, and Bayou Marcus. Detectable chemical contaminants have also been identified at one marina and in some locations within Perdido Bay. Contaminants of concern at some sediment sites include: mercury, silver and dioxin compounds. ...Based on the dioxin residue in the turtles, long-lived resident species using Elevenmile Creek may be at some, as yet unquantified, risk to either individual animals or local populations. Most fish species within Elevenmile Creek do not currently appear to be accumulating dioxin above EPA or FDA concern levels."

Actual scientific data concerning dioxin in the Perdido system obviously were not as exciting to various press interests as the unsubstantiated claims of various "environmentalists."

5.4.2   Cumulative Impacts of Development on Perdido Bay

Following the prolonged drought of 1999–2002, there was a series of major rain events in the Perdido drainage basin. These events in the fall of 2002 and summer of 2003 were associated with major changes in the ecological processes of the Perdido system. These changes were reviewed within the context of long-term (16-year) changes in the system. Reports (Livingston, 2003, 2004b) were issued concerning the findings that included a response of the bay to nutrient loading associated with the rainfall events of 2002–2003.

1. Orthophosphate and ammonia loading by a pulp mill into Elevenmile Creek and nutrient concentration ratios in bay receiving areas were associated with a series of phytoplankton blooms from 1994 to 2001. Initial diatom blooms in the bay in 1993–1994 were replaced by raphidophyte (*Heterosigma akashiwo*) and dinoflagellate (*Prorocentrum minimum*) blooms. These species were statistically associated with deterioration of phytoplankton assemblages, reduced invertebrate and fish populations, and disruptions of bay food webs.

2. Blue-green algae blooms (*Merismopedia tenuissima*) in Elevenmile Creek were associated with the recent drought although mill nutrients constituted the probable cause of the blooms. There were statistical associations of reduced biological activity with the *Merismopedia* blooms from 1999 to 2004.

3. Reduced nutrient (PO_4, NH_3) loading by the pulp mill in 2001–2002 was associated with a reduction in bloom activity in the bay. There were marked reductions in *Prorocentrum minimum* and *Heterosigma akashiwo* during this period, and these trends were followed by recovery trends of infaunal macroinvertebrates. This recovery was interrupted by rainfall events in fall 2002 and summer 2003.

4. The opening of Perdido Pass caused increased salinity that moves along the bottom to the upper bay. Salinity stratification was associated with hypoxia in deeper parts of the bay.

5. Deeper (depositional) areas in lower Perdido Bay (Stations P33, P37, and P40) have higher silt and nutrient concentrations than shallow, sand-dominated parts of the upper bay. Sediment deterioration caused by phytoplankton blooms in the upper bay was eventually transmitted to sediments in the deeper, depositional mid- to lower bay areas. Downstream currents from upper Perdido Bay move nutrients and particulate matter into deeper parts of the bay; Station 37 represents an area of maximum deposition.

6. The Bayou Marcus Water Reclamation Facility (Escambia County Utilities Authority ECUA) is located on the eastern shore of upper Perdido Bay. Flow rates in the year 2000 approximated 3.4 MGD. The permitted discharge to the wetlands is 8.2 MGD. During the fall 2002 rainfall event, discharge rates exceeded 10.5 MGD. During this period, anecdotal data indicated chlorine in the upper, mid, and lower parts of Perdido Bay. Following another rainfall event in June 2003, chlorine was again noted in the bay. These events were associated with major declines in infaunal macroinvertebrates for affected areas of Perdido Bay. The origin of the chlorine remained undetermined.

7. A review was made concerning various lines of scientific evidence with respect to the possible effects of sewage releases by the Bayou Marcus Water Reclamation Facility into upper Perdido Bay:

   a. Long-term (1993–2002) changes of TP and TON indicated that sediment nutrient distribution during fall periods followed closely the annual nutrient loading trends in upper Perdido Bay. The distribution of sediment TP and TON in space and time indicated that the Bayou Marcus Facility was the source of nutrients in recent years (2001–2002).

   b. From 1998 to the present, there was a steady decrease in bottom ammonia at Station 23 (mouth of Elevenmile Creek), whereas bottom ammonia has increased at Stations 26 (mouth of Bayou Marcus Creek) and 31 during fall 2002. At this time, the highest chlorophyll *a* concentrations taken during the entire 16-year sampling period were noted at Station 26. There was increased chlorophyll *a* activity at depth in areas receiving flows from the Bayou Marcus Water

Reclamation Facility. These observations are consistent with reduced nutrient loading from Elevenmile Creek and increased nutrient loading from the Bayou Marcus area during the period 2001–2003.

c. The plankton species (*Prorocentrum minimum, Heterosigma akashiwo, Merismopedia tenuissima*) were closely associated with both loading and concentration ratios of ammonia and orthophosphate on a seasonal basis, and were the main effectors of the reduction of phytoplankton species richness and diversity indices in Elevenmile Creek and Perdido Bay. These species were also the main effectors of the loss of secondary production in the Perdido Bay system, with major adverse effects on estuarine food webs.

d. During October 2002, there was a massive bloom of *Heterosigma akashiwo* that extended down the entire bay in a pattern similar to that noted in the long-term trends of water quality and sediment TON and TP. The October 2002 bloom was the most dense concentration of *H. akashiwo* over the 16-year survey. Such numbers went down at Station 23 while they increased at Station 26. The high numbers of *Heterosigma* and *Prorocentrum* in recent years off the Bayou Marcus drainage area indicate a reversal in what had been a recovery of the bay due to reduced nutrient loading from the pulp mill.

e. Annual trends of the numbers of organisms in the Perdido Bay system were significantly lower during the 2002–2003 period than levels taken over the previous 13 years of sampling. Infauna also showed the lowest numbers ever taken during the 2002–2003 period. There was partial recovery of infauna following fall 2002, which was followed by a major reduction in infauna during the summer 2003 incident. There was an almost total collapse of the mean numbers of trophic units (TUs) and total number of fish, invertebrate, and infaunal taxa. The data indicated that there was an almost complete collapse of secondary production in the Perdido Bay system during the 2002–2003 period.

8. Wolf Bay was subjected to unregulated agricultural runoff for years. Sediment TP and TON concentrations were relatively low in Wolf Bay during fall 1993: there were increases in these nutrients in the bay during 1996–1998, and by 2001–2002, sediment TP and TON were high throughout most of Wolf Bay. Light penetration showed a steady decline over the study period, due in large part to general increases in watercolor and turbidity. Both surface and bottom DO and oxygen anomalies deteriorated during 1991–2003 with periodic reductions of DO below 4 mg L^{-1} at depth since 1999. Ammonia concentrations showed recent increases that were most pronounced at depth, indicating sediment involvement. Both surface and bottom chlorophyll *a* concentrations increased steadily since 1998. The water and sediment quality data thus showed steady deterioration.

9. Based on the experience in the upper bay, there were indications that nutrient loading from upland (agricultural) sources was an important contributing factor to the deterioration of sediment and water quality in Wolf Bay. Long-term phytoplankton collections from Wolf Bay indicated that there were major blooms of *Prorocentrum minimum* and *Heterosigma akashiwo* that started in August 2000, with major winter and summer blooms occurring periodically up to the present. These blooms were associated with reductions in phytoplankton species richness and diversity. No infaunal macroinvertebrates were noted in Wolf Bay during summer 2003.

10. Previous studies indicated stratification and hypoxia in lower Perdido Bay south of Ono Island, an area characterized by generally uncontrolled urban development.

Extremely low infaunal numbers and species richness were noted in lower portions of Perdido Bay. As noted in Wolf Bay, there was a general decrease in light penetration over the study period that was associated with increased turbidity. Surface and bottom oxygen anomalies showed a downward trend, and there were periodic high surface ammonia concentrations. Both surface and bottom chlorophyll *a* showed increasing concentrations with time. These data indicate water quality problems in areas receiving runoff from urban areas.

11. Agricultural (and urban) runoff (non-point sources of nutrients) differ from point sources (i.e., the pulp mill) in that mill effluents are continuous whereas non-point sources follow rainfall patterns. The data indicate a basic difference between the mechanisms of bloom stimulation by continuous point sources and discontinuous non-point sources. The Bayou Marcus Sewage Treatment Facility may represent a hybrid of these forms of runoff, with some nutrient loading during drought periods and major nutrient loading during periods of heavy rainfall.

12. There is an almost complete lack of regulation of nutrient loading from sewage facilities and non-point sources (agricultural and urban runoff). In recent years, there has been a reduction in nutrient loading from the pulp mill in the upper bay with associated improvements in bloom frequency. However, Elevenmile Creek continues to be damaged by blooms associated with mill nutrient loading during drought periods. The combined effects of the pulp mill and the Bayou Marcus Facility in the upper bay, and agricultural and urban runoff in the lower bay, have led to an overall deterioration of environmental conditions in the Perdido Bay system.

A summary of the status of Perdido Bay during fall 2003 is given in Table 5.3. In the upper bay, associated with the pulp mill effluents and discharges from an STP, recovery associated with reduction of mill loading of nutrients was temporarily offset by STP discharges. Data comparing the distribution of *Prorocentrum minimum* and *Heterosigma akashiwo* numbers L^{-1} with regional rainfall patterns indicated that there was a relationship between long-term trends of bloom species appearances in Wolf Bay and rainfall. The trends noted in Wolf Bay and upper Perdido Bay were comparable. Similar patterns were noted in the bay's response to urban runoff in the lower bay. Agricultural (and urban) runoff (non-point sources of nutrients) differ from point sources (i.e., the pulp mill) in that the pulp mill effluents are more or less continuous, whereas non-point sources follow rainfall patterns. Overall, the pulp mill was one of the various sources of nutrient loading from human activities that was adversely affecting the Perdido Bay system.

With the exception of the pulp mill, nutrient loading in the bay has not been regulated by state and federal agencies. The noted impacts from sources other than the pulp mill have been totally ignored by "environmentalists" and the regional news media. The impact on the bay of the sewage treatment plant has not been reported by the *Pensacola News-Journal*.

5.4.3 The News Media and Perdido Bay

During early 2003, the Perdido report (Livingston, 2003) outlining problems associated with the sewage treatment plant on upper Perdido Bay was given to personnel in charge of the STP. The facility is run by the Emerald Coast Utilities Authority (ECUA). ECUA officials responded with a memo that accused the author (Livingston) of misrepresenting the facts, among other things. Around this time, a reporter for the *Pensacola News-Journal* called and asked about our analyses concerning sewage spills in Perdido Bay. I informed him of our analyses, but emphasized that I would not have anything public to say until: (1) we reviewed the data, (2) we issued a written report (to be completed by summer

Table 5.3 Summary of the Status of the Perdido Bay System by Fall 2003

Factor	Nutrient Loading	Water Quality	Sediment Quality	Bloom Occurrence	Status Plankton	Status Infauna	Status Invertebrates	Status Fishes
Bay Area Elevenmile Creek	Reduced $PO_4 + NH_3$ loading from pulp mill	High nutrient concentrations	Reduced TP TON*	Blue-green algae blooms *Merismopedia tenuissima*	Damaged	Bloom associated reductions	Bloom associated reductions	Bloom associated reductions
Upper Bay	Undetermined loading from sewage plant	Chlorine smell Oct-02 Increased bottom NH_3*	Increased TP TON* in Bayou Marcus*	Raphidophyte blooms *Heterosigma akashiwo*	Damaged	Bloom associated reductions	Lowest # in 15 years*	Bloom associated reductions
Mid-Bay	nd	Chlorine smell Oct-02	Increased TP and TON sewage smell*	Raphidophyte blooms *Heterosigma akashiwo*	Damaged	Bloom associated reductions	Lowest # in 15 years*	Bloom associated reductions
Lower Bay	nd	Chlorine smell Oct-02 Jun-03	Increased TP and TON sewage smell*	Raphidophyte blooms *Heterosigma akashiwo*	Damaged	Bloom associated reductions	Lowest # in 15 years*	Bloom associated reductions
Wolf Bay	Suspected agricultural loading	Low DO deteriorating conditions*	Increased TP and TON*	Blooms of varioius HAB species	Damaged	Bloom associated reductions	nd	nd
Ono Island Area	Suspected urban loading	Low DO deteriorating conditions*	Increased TP and TON*	nd	nd	Bloom associated reductions	nd	nd

* 2002–2003.

nd = no data.

2003), and (3) we discussed our findings with personnel from the ECUA. The reporter informed me that he was anxious to do a story on the sewage issue. Under the Freedom of Information Act, he obtained the unpublished ECUA memo. In reaction to repeated calls from the reporter, I told him that I did not want to blind-side the ECUA with our information, and intended to release our full report in July 2003 when our analyses of the problems in the upper bay during 2002–2003 were completed. My intention was to work with the ECUA to research the problems that were indicated by our research findings. In July 2003, I sent a copy of our full report to the reporter with the proviso that he would not publish the findings until I had met with ECUA officials. He then told me he was going to write some stories on the issue before our meeting, and that he deliberately did not look at my report so that he would not be "biased" by the facts.

Using the documents obtained from the ECUA, the reporter then wrote three articles a few days before I met with the ECUA. One story was a review of all the false accusations that various "environmentalists" had issued over the years that attacked our Perdido Bay research. Another story outlined how well the sewage treatment plant was designed, and how it was impossible for such a facility to pollute the bay. A third outlined the unflattering response of the ECUA to my original report concerning the sewage spill. These press reports indicated that I was absolutely wrong because I had associated the plant with chlorine discharges. I had actually qualified our observations on chlorine as anecdotal in my report, all of which was deliberately ignored by news accounts. In other words, by delaying the actual factual accounts of the impacts on the bay by the STP, I was villified by press accounts that completely ignored the scientific basis for impact analyses.

After the meeting with ECUA officials, the reporter, who was present at the meeting, then wrote another article that quoted me out of context in such a way as to discredit what we had found. His report largely ignored the detailed scientific data that indicated a sewage spill into the bay. The press accused me of back-tracking on my original (unpublished) report. This story was then followed by an editorial in the *Pensacola News Herald*, which again concentrated on the chlorine issue with a series of completely erroneous statements regarding our work. Another opinion writer then further ridiculed our results. Of course, this was followed by a letter to the editor by a prominent "environmentalist" that further attacked my character and research. Of course, none of the people reporting on our findings had read my report. Associated Press articles based on the *Pensacola News Herald* reports then told the world of my perfidy, causing me further embarrassment. To this day, there has been no investigation of how the STP is operating.

The *Pensacola News Herald* choreographed its reportage to protect ECUA officials from any responsibility in the pollution of Perdido Bay. By concentrating on a preliminary report and the aggressive response by the ECUA, the actual facts were largely ignored in favor of creating a confrontation between the ECUA and the paper mill that was the actual intent of the series of newspaper articles. Although mill personnel had nothing to do with our findings, the *Pensacola News Herald* made it seem as if the data concerning the ECUA discharges were a trumped-up diversion to the problems associated with pulp mill efflu-ents on Perdido Bay. Effects of the mill on the bay have been well described in reports, peer-reviewed scientific papers, and books. The fact that the mill had reduced its nutrient loading to the bay was completely ignored in the effort to stir up controversy between the ECUA and the pulp mill. Meanwhile, our reports of chlorine discharges and the subsequent problems of the bay were ignored by state and federal regulatory agencies. All of this was complicated by the fact that reporting of orthophosphate concentrations and nutrient loading by sewage treatment plants was not required by the Florida Depart-ment of Environmental Protection.

The press coverage of Perdido Bay pollution problems was stereotypically sensationalist, with the usual bow to controversy instead of fact. The problems of omission, exaggeration, and outright lying regarding media coverage of the Perdido system prevailed. The results of scientific studies were trivialized and misrepresented by so-called environmental groups whose primary purpose was to get rid of the pulp mill. The press again sensationalized false issues by pumping up the conflict between a group of hysterical "environmentalists" and apologists for the mill. The dull, old scientific facts came in third in the hierarchy of the press reports. However, all of this was simply a replay of what occurred in the Pensacola Bay system (see Chapter 6).

chapter 6

The Pensacola Bay System

6.1 Background

Reviews of studies in the Pensacola Bay system (Figure 6.1) are given by Livingston (1999c, 2000, 2002). An outline of these analyses is given in Appendix I. A comprehensive review of the Pensacola Bay System has been provided by Thorpe et al. (1997). There are three major rivers associated with the Pensacola Bay system: the Escambia, Blackwater, and Yellow Rivers. The Pensacola Bay system includes five interconnected bay components: Escambia Bay, Pensacola Bay, Blackwater Bay, East Bay, and Santa Rosa Sound (Figure 6.1). The Escambia River system extends northward about 386 km (240 mi) from the north end of Escambia Bay. It runs through Alabama as the Conecuh River. The drainage area of the Escambia River basin includes about 10,880 km^2 (4200 mi.^{2}), with approximately 90% of the basin in Alabama. The Blackwater River basin includes about 2230 km^2 (860 mi.^{2}), of which about 81% is located in Florida. The Yellow River basin extends 177 km (110 mi.) from Blackwater Bay.

Escambia Bay is located east of the City of Pensacola, with the Garcon Point peninsula to the east and the Escambia River delta to the northwest (Figure 6.1). The primary source of fresh water to the bay is the Escambia River; other sources include the Pace Mill Creek and Mulatto Bayou drainage basins in the upper bay, and the Bayou Chico and Bayou Texar basins in the lower bay. Blackwater Bay receives freshwater inflow from the Blackwater River, with East Bay immediately downstream from the Blackwater Bay system. East Bay receives freshwater flows from the Blackwater River–Bay system, the Yellow River, and the East Bay River. East Bay is bordered to the south by the Gulf Breeze peninsula. Pensacola Bay receives flows from Escambia Bay, East Bay, and Bayou Grande, and is bordered to the north by the City of Pensacola and to the south by Santa Rosa Island. Pensacola Bay empties into the Gulf of Mexico through a pass at its southwestern terminus. Santa Rosa Sound is a lagoon between the mainland and Santa Rosa Island. This sound connects Pensacola Bay in the west with Choctawhatchee Bay to the east.

The Pensacola Bay system is characterized by a relatively shallow shelf, a sloped area, and deeper plain. The upper parts of Escambia Bay and Blackwater Bay are relatively shallow. Depth in the central parts of both systems tends to increase moving southward, with the deepest parts of the system in Pensacola Bay.

6.2 Purpose of Study

A field and modeling study of the Pensacola system was made from May 1997 through October 1998 as part of a comprehensive analysis of the Escambia River and Black-water/Yellow River systems and associated estuaries. Field experimental and descriptive

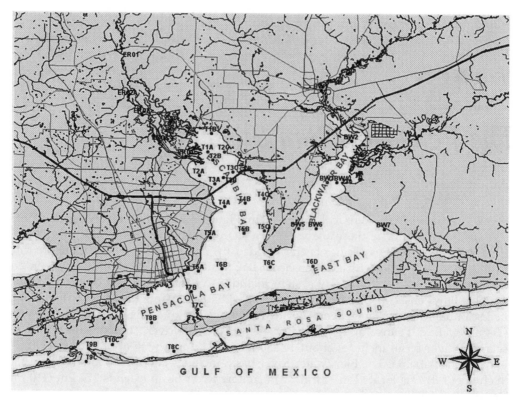

Figure 6.1 The Pensacola Bay system, showing sampling stations for the basin study. Geographic data was provided by the Florida Geographic Data Library (FGDL).

data were obtained to increase our understanding of the Pensacola Bay system and evaluate the assimilative capacity of this system for nutrient loading from potential Cantonment mill discharge. The study thus was designed to determine whether a pipeline from the Cantonment paper mill to the Escambia River would be feasible based on established adverse impacts on the Perdido system due to the discharge of mill effluents into Elevenmile Creek.

6.3   Summary of Results

The Pensacola Bay system is represented by impacts from various human activities. The following is a brief summary of the results of the analyses carried out during 1997 and 1998 (Livingston, 1999c):

1. The Pensacola Bay system is composed of a series of diverse habitats that are commonly found in river-dominated estuaries of the northern Gulf of Mexico.
2. The Escambia River is an alluvial stream that is highly colored and turbid with low light penetration relative to the Blackwater River. The lowest dissolved oxygen in the Escambia and Blackwater Rivers occurred at depth in areas having the highest salinity stratification indices during warm periods. The numerical abundance and species richness of infaunal macroinvertebrates in these rivers are mainly associated with sediment characteristics: high species richness correlates well with sandy conditions, high percent organics, low silt/clay fractions, and large particle size. Low infaunal numbers and low species richness periodically occur

in the lower Escambia River. These conditions are correlated with seasonal changes in river flow, salinity stratification, associated benthic hypoxia, and sediment quality. The Yellow River had the highest species richness of the three rivers, which could be due, in part, to the fact that the sampling area in the study was not influenced by salinity.

3. Salinity variation appears to be an important stress factor for biota in the lower parts of the Escambia and Blackwater Rivers. The Escambia River infaunal assemblages were characterized by generally low numbers of individuals and species. Trends in the infaunal indices reflected the highly stressful habitat conditions in tidal portions of an alluvial river, including salinity variation and hypoxic conditions at depth that accompany the movement of the salinity wedge upstream.

4. River flow, wind and tides, the depth of the receiving basin, and the penetration of salinity from the Gulf caused periodic water column stratification within the Pensacola Bay system. Stratification was maximal in deeper parts of middle and lower Escambia Bay as well as in deeper parts of the Blackwater–East Bay system. Bottom DO was controlled by salinity stratification in the upper and mid-parts of Escambia Bay and the Blackwater–East Bay system, whereas these relationships were not evident in Pensacola Bay.

5. River-associated nutrient loading was a prime determinant of overall primary and secondary production in Escambia Bay and Blackwater Bay. Orthophosphate concentrations, the probable limiting factor for primary production in both estuaries, were highest in upper Escambia Bay, with the Escambia River and possible point source releases in eastern sections of the upper bay as the chief sources of this nutrient. Peak orthophosphate concentrations also occurred in mid- and lower Escambia Bay during periods of high rainfall. Ammonia concentrations were high in upper parts of both Escambia Bay and Blackwater Bay. Relatively high ammonia concentrations were also observed in mid- and lower Escambia Bay and Pensacola Bay. The highest mean concentrations of chlorophyll *a* occurred in upper Escambia Bay.

6. Seasonally averaged light and dark bottle data indicated that the highest net production occurred in upper Escambia Bay, with positive net productivity down to a depth of 1.5 m. Light, temperature, and nutrient (phosphorus) limitation appear to control primary production in upper Escambia Bay. The overall chlorophyll *a* concentrations in the entire system were relatively low, and there were no signs of hypereutrophication in the Pensacola Bay system over the 12-month sampling period.

7. Qualitative and quantitative whole water phytoplankton data indicate that Escambia Bay is currently in a modestly eutrophic state. Sub-dominant phytoplankton populations known to be noxious bloom species in Perdido Bay were found mainly in mid- and lower Escambia Bay. Urban storm water runoff may be a factor in the maintenance of these populations. Based on long-term studies of nutrient loading and phytoplankton occurrence in Perdido Bay, it is likely that excessive nutrient loading in upper Escambia Bay could cause future problems if these species reach bloom status. It is also possible that the past history of excessive nutrient loading, hypereutrophication, and damage to Escambia Bay food webs has sensitized this system to these bloom species.

8. Sediment quality is an important component of the distribution of secondary production in the Pensacola Bay system. Herbivorous infaunal populations depend on sediment quality; this component constitutes the base of the primary estuarine food webs. The overall composition of sediments in the Pensacola system is related

to a combination of factors: depth, proximity to the Escambia and Blackwater/Yellow River mouths, salinity stratification, dredging activities, and pollution sources.

9. The main concentrations of infauna (polychaete worms), epibenthic macroinvertebrates (brown shrimp, blue crabs), and fishes (Sciaenids) were found in the relatively shallow parts of upper Escambia Bay and Blackwater Bay. Upper Escambia Bay was the center of primary and secondary production in the entire Pensacola Bay system, a probable result of nutrient loading from the Escambia River. The trophic organization emanating from the higher primary production of the upper bay was the main determinant of the distribution of dominant, commercially important populations in the Pensacola Bay system. Habitat factors, such as sediment type, DO, and salinity distribution, while important in determining the general distribution of organisms in the bay system as a whole, had effects that could only be understood within the context of the distribution of primary (phytoplankton) production. Predation and competition also could have contributed to population distribution in this system.

10. Overall, there was a relatively low biomass of infauna, epibenthic invertebrates, and fishes along eastern sections of upper Escambia Bay and in lower Escambia Bay and Pensacola Bay. Indicator species for pollution were found in these areas of the bay. According to results of other studies, these areas are stressed by loading of nutrients and toxic wastes due to anthropogenous runoff from point/non-point sources.

11. Compared to other alluvial river-dominated Gulf estuaries such as the Apalachicola, Choctawhatchee, and Perdido systems, the Pensacola Bay system appears to have relatively low overall secondary productivity. In fact, lower Escambia Bay–Pensacola Bay areas were among the most faunally depauperate of all systems sampled. Currently, non-point pollution from the City of Pensacola and other urbanized areas of the Pensacola Bay basin, together with major pollution of the bayous (Texar, Chico, and Grande) in this region, present a serious problem in terms of continuing, and possibly increasing, multi-source urban contamination.

12. The Pensacola Bay system is one of the most polluted bays in the state in terms of concentrations of toxic agents in sediments. High levels of metal contamination have been found in Bayou Grande (Cd, Pb, and Zn) and Bayou Chico (Cr, Zn). Bayou Chico is also contaminated with polynucleated aromatic hydrocarbons (PAHs) and polychlorinated biphenyls (PCBs). In Pensacola Bay, PAHs and PCBs have been found in sediments close to shore and in central parts of the bay. Phenolic compounds have been found at one site near Pensacola Harbor. The most important generalization to date from all of the studies in the Pensacola Bay system is how difficult it will be to restore aquatic systems that are adversely affected by urban runoff.

13. Current levels of nutrient loading in the Escambia system are comparable to loading in other (largely unpolluted) alluvial systems along the northeast Gulf coast. Nutrient loading represents the chief concern of the potential discharge of effluents from the paper mill into the lower Escambia River. Analyses of simulated nutrient loading in the Escambia River have indicated that the overall increases in orthophosphate loading in Elevenmile Creek of the Perdido Bay system during the 1994–1995 period, when added to current Escambia River loadings, would lead to relatively high summer–fall loadings and river nutrient concentrations.

14. Because the Escambia Bay system has already been adversely affected by nutrient loading in the past, and because this bay appears to be in a recovery phase,

increased nutrient loading, such as that of Elevenmile Creek during 1994–1995, was considered problematic. There were indications of the presence of bloom species in Escambia Bay as sub-dominant populations that could cause problems if stimulated by increased nutrient loading. Increased Escambia River nutrient loading could adversely affect areas of the bay that are currently the most productive in terms of populations of important sports and commercial fisheries. The presence of a commercial oyster fishery in Escambia Bay also creates a situation where nutrient-stimulated plankton blooms could cause public health problems.

15. The Escambia Bay system is most vulnerable to hypereutrophication from increased nutrient loading during periods of drought and low river flow, either as a function of seasonal and/or interannual trends of river flow. Thus, both the timing and extent of nutrient loading and resultant nutrient concentrations entering upper Escambia Bay should be taken into account in any review of increased nutrient loading to the Escambia River. Modeling efforts to date have not allowed the determination of effects of nutrient loading on projected changes in phytoplankton species composition.

16. The main residential areas in the Pensacola drainage basin have the highest normalized non-point source pollution loadings of the entire system. The primary threat to the Pensacola Bay system remains non-point sources and sub-surface seepage from toxic waste sites.

6.4 Contamination of the Pensacola System

Escambia Bay receives freshwater flow from the Escambia River. Other water sources in upper Escambia Bay include the Pace Mill Creek and Mulatto Bayou drainage basins, among others (Thorpe et al., 1997). Sources of water in lower Escambia Bay include the river via the upper bay and the Indian Bayou, Trout Bayou, and Bayou Texar basins. Tidal flushing in Escambia Bay is considered poor by various investigators (Thorpe et al., 1997). Circulation is strongly influenced by wind and tidal action, as well as inflow from the Escambia River. There is a net southward flow of river water along the western shore, with water of higher salinity intruding along the eastern shore. This tends to produce a generally counterclockwise circulation pattern. Railroad and highway bridges may inhibit flushing and exchange between the upper and lower bay.

Historically, Escambia Bay has had high levels of toxic agent contamination and nutrient loading. It has received substantial industrial and domestic wastewater discharges, and is still affected by various point and non-point sources. Non-point source pollution is received from the City of Pensacola, unincorporated areas, and the river basin. Pensacola Bay receives direct and indirect runoff from the City of Pensacola and unincorporated areas. On an outgoing tide, surface waters tend to move toward the pass from the more northerly and western portions of the bay. Olinger et al. (1975) found that circulation within the bay could be strongly influenced by surface winds, with effects not necessarily limited to the upper layers. Point source discharges to Pensacola Bay include the Main Street and NAS Pensacola wastewater treatment plants. Component bayous, formerly centers of productivity in the system, are now among the most polluted in the entire Pensacola system. In Bayou Texar, an area noted for fish kills related to nutrient overenrichment and hypoxia, nitrate has been reported to control primary productivity to a greater degree than phosphorus (Moshiri et al., 1987). Bayou Texar is also contaminated by toxic agents, with two U.S. EPA-designated Superfund sites. These bayous act as sinks for sustained urban runoff and other non-point source pollution. Bayou Chico has also received substantial historic point source discharges. Incoming waters from these bayous tend to move along the bottom into the bay and then eastward along the southern part of the bay.

Santa Rosa Sound is a lagoon between the mainland and Santa Rosa Island. Most waters within the sound are designated as Class II, and waters within the National Seashore are designated Outstanding Florida Water (OFW). The Intracoastal Waterway (ICW) transects the sound and supports moderate commercial barge traffic. The Navarre Bridge Causeway divides the sound into eastern and western regions, leading to a bi-directional tidal flow with relatively little freshwater inflow (Hand et al., 1996). This area has some of the most diverse and stable seagrass beds in the region. Human impacts on the lagoon's environment include non-point source pollution and habitat loss resulting from increasing development on Santa Rosa Island and along the U.S. Highway 98 corridor. The Navarre Beach and Pensacola Beach wastewater treatment plants (WWTPs) discharge to the sound (Hand et al., 1996) along with runoff from several golf courses.

The Pensacola Bay system has been subjected to continuous anthropogenous stress for decades. Submerged aquatic vegetation, formerly abundant in this system, has been largely absent since the mid-1970s (Collard, 1991a,b). Formerly productive oyster bars are no longer commercially viable, although a small oyster fishery still exists. The most extensive fish kills in the scientifically documented literature occurred in Escambia Bay during the late 1960s and early 1970s (U.S. Environmental Protection Agency, 1971).

Livingston (1997c), in a detailed study of the Mulat–Mulatto Bayou, indicated the processes involved in the observed fish kills. One classic response of an estuary to anthropogenous nutrient loading is extreme hypoxia with resultant disruption of the biological relationships of the receiving system through changes in the natural food webs. During the late 1960s and early 1970s, Escambia Bay was subject to nutrient loading from various point and non-point sources, resulting in extreme dissolved oxygen fluctuations (U.S. Environmental Protection Agency, 1971; Livingston, 1973, 1997c). The Mulat–Mulatto Bayou was a focal point for the extensive fish kills. This area had been subjected to physiographic changes due to dredging associated with the construction of a nearby highway. In 1965, the Florida Department of Transportation removed approximately 1,028,933 cu. yd. of sediment from Mulatto Bayou. During this period, nutrients and toxic agents were released into Escambia Bay by a broad array of sources (Olinger et al., 1975), severely altering the aquatic habitat. Dye studies in the Mulat–Mulatto Bayou (R.J. Livingston, unpublished data) indicated that the dredging created cul-de-sacs (i.e., the finger-fill canals) that were isolated from direct tidal current exchanges. This resulted in exacerbation of the effects of nutrient loading of the bay by changing turnover rates and the general exchange patterns of this part of the bay.

Massive fish kills occurred in Escambia Bay during the late 1960s and early 1970s. The timing and location of the fish kills indicated that the combination of hypereutrophication due to nutrient loading and habitat alterations due to dredging were responsible for the fish kills. Dredging activities enhanced the effects of cultural eutrophication by inhibiting tidal and wind-produced exchanges of bay water. The fish kills were the final consequence of a distorted DO regime, which, in turn, was influenced by the sequential interaction of temperature, precipitation, altered water circulation patterns, nutrient-enhanced primary production, and a disjunct food web organization.

6.4.1 Upper Escambia Bay

Eastern sections of upper Escambia Bay are characterized by water quality changes and some weakness in biological indicators. Hudson and Wiggins (1996) identified two main industrial sites in eastern sections of upper Escambia Bay: Air Products and Chemicals,

Inc., and Sterling Fibers, Incorporated (formerly Cytec Industries [American Cyanamid]). Air Products discharged at mean flow rates of 1.18 MGD, with discharges of chemical oxygen demand (COD), total suspended solids (TSS), phosphorus, nitrogen, and toxic agents (cyanide, chromium, silver, mercury, polychlorinated biphenyls) (Florida Department of Environmental Protection, 1996). Analysis was also made of the Sterling Fibers effluent site (Florida Department of Environmental Protection, 1997). Sterling Fibers, Incorporated, had a design flow of 1.53 MGD and discharges BOD, TSS, nitrogen, and orthophosphorus. There are also discharges of toxic agents such as cyanide, silver, and copper (Florida Department of Environmental Protection, 1997). According to Hudson and Wiggins (1996), the Pace wastewater treatment plant discharges at the head of the Escambia estuary (Figure 6.1). Tide, wind, and river effects add the Pace effluent to the observed changes in eastern parts of upper Escambia Bay attributed to the two industrial dischargers. Because of the postulated poor tidal flushing and the highly organic sediments, together with a long history of pollution from industrial and domestic wastewater discharge, upper Escambia Bay has long been considered the most stressed of all the various parts of the Pensacola system.

6.4.2 Lower Escambia Bay

Relative to the upper bay, lower Escambia Bay and Pensacola Bay have improved circulation and turnover rates due to the physiographic characteristics of this part of the Pensacola Bay system and the proximity to open Gulf waters. However, the non-point source pollution from the City of Pensacola and the Naval Air Station, together with major pollution of the bayous (Texar, Chico, and Grande) in this region, present a serious problem in terms of continuing, multisource contamination that is difficult to evaluate and even more difficult to mitigate. The most significant point source discharges in this region are the Main Street and Naval Air Station STPs (Thorpe et al., 1997). Storm water outfalls are particularly numerous from Devil's Point to Magazine Point (City of Pensacola) and on the inshore side of the Gulf Breeze Peninsula. In addition, septic tank numbers are extremely high in the vicinity of lower Escambia Bay. The bayous in this area collect urban runoff and have a long history of poor water and sediment quality and fish kills. The Northwest Florida Water Management District (1994a,b) conducted a study of non-point source loading in the Pensacola Bay system. The findings included the fact that Mulatto Bayou and the main residential areas of the Pensacola Bay system have the highest normalized loadings of the entire system. The district suggested that undeveloped areas currently zoned for residential development be targeted for storm water management plans (including specifications or designs for alternative land use practices and storm water treatment facilities) if water quality is to be improved in the Pensacola Bay system.

Bayou Texar has been well studied (Hannah et al., 1973; Moshiri, 1976, 1978, 1981; Moshiri and Crumpton, 1978; Moshiri et al., 1978, 1979, 1980). According to Thorpe et al. (1997), various studies (Hood and Moshiri, 1978; Raney, 1980; Northwest Florida Water Management District, 1978; Moshiri, 1981) were associated with recommendations designed to improve water quality in the Bayou Texar and the Carpenters Creek System. However, water quality remains generally poor (Thorpe et al., 1997). Recent analyses by the U.S. EPA (Pensacola Bay System Technical Symposium, 1997) indicate that, due to increasing residential and commercial development in the Carpenter Creek basin, there has been an accumulation of sediments, forming a delta at the confluence of Carpenter Creek and Bayou Texar. This has required increased dredging by the City of Pensacola. These sediments have elevated concentrations of Al, Cd, Co, Cr, Cu, Fe, Mn, Pb, and Zn as well as increased concentrations of total phosphorus (TP) and total Kjeldahl nitrogen (TKN). Continued maintenance dredging may result in the release of these pollutants from the sediment into the water. Bayou Chico receives runoff from various industrial and urban sources. Storm water entering the bayou failed state water quality standards for turbidity, suspended solids, DO, BOD, nutrients, total and fecal coliforms, copper, lead, mercury, zinc, oils and greases, and phenols. New studies under the state of Florida's SWIM program are continuing, and the U.S. EPA is working on an Ecosystem Criteria Research Program. However, no written reports on this endeavor were available for this review.

Seal et al. (1994) conducted statewide sediment tests for metals and organic contaminants. They found that urban storm water runoff was the major cause of contamination of sediments in Florida's coastal areas. The highest concentrations of toxic agents were found in coastal sediments at sites near Tampa, Pensacola, Miami, and Jacksonville. The Pensacola Bay system was one of the most heavily impacted bays in the state in terms of contamination from highly toxic and long-lasting substances. High levels of metal contamination were found in Bayou Grande (Cd, Pb, Zn) and Bayou Chico (Cr, Zn). Bayou Chico was also contaminated with PAHs and PCBs. In Pensacola Bay, PAHs and PCBs were found in sediments close to shore and in central parts of the bay. Phenolic compounds were found at one site near Pensacola Harbor. Long et al. (1997) carried out a study of the "magnitude and extent of sediment toxicity in four bays of the Florida Panhandle." The sediments of the Pensacola Bay system were compared to those of the Choctawhatchee, St. Andrews, and Apalachicola Bays. The greatest toxicity of all bays analyzed occurred in Bayou Chico, with the other developed bayous in the Pensacola area showing "relatively severe toxicity." Toxicity of the Pensacola Bay sediments was attributed to high-molecular-weight PAHs, zinc, dichlorodiphenyl-dichloroethane/dichlorophenyl-trichloroethane (DDD/DDT) isomers, total DDT, and the highly toxic organochlorine pesticide known as dieldrin. There were elevated PAHs at the National Oceanic and Atmospheric Administration (NOAA) and National Status & Trends (NS&T) Program stations in Pensacola Bay and Indian Bayou. In addition to the bayous and portions of Pensacola Bay, toxicity was also detected in upper and mid-portions of Escambia Bay. Toxicity was also noted in Blackwater Bay and East Bay. The NOAA study confirmed high concentrations of lead, mercury, and PCBs in the sediments of the urbanized bayous and in Pensacola Bay.

More recent studies by various government agencies are being conducted in the Pensacola Bay system, including evaluations of Bayou Texar, Bayou Chico, and Bayou Grande. Analyses by the U.S. EPA (Gulf Breeze, Florida) using annual samples of infaunal macroinvertebrates (Benthic Index) indicate good indices in upper and mid-parts of Escambia Bay and in the Blackwater–East Bay areas (Pensacola Bay System Technical Symposium, 1997). Poor indices were noted in lower Escambia Bay and Pensacola Bay (Pensacola Bay System Technical Symposium, 1997). Metals, PAHs, and pesticides were implicated in the deteriorated conditions in lower Escambia Bay and Pensacola Bay (associated with the City of Pensacola). Metals (As, Ni, Cr) and pesticides were also found in areas associated with the Gulf Breeze Peninsula. According to recent data (Pensacola Bay System Technical Symposium, 1997), the bayous of Pensacola are the most contaminated areas in both the Pensacola Bay system and the state of Florida. It is hypothesized that the three bayous act as contaminant sinks that reduce the impact of toxic agents on the bay system. Other studies by the U.S. EPA (Pensacola Bay System Technical Symposium, 1997) indicate that benthic conditions in Escambia Bay degraded between 1992 and 1996. According to recent FDEP studies (D. Heil, Pensacola Bay System Technical Symposium, 1997), Pensacola Bay continues to have increasing oyster closings due to urban pollution with moderate to high fecal coliform levels. Every 5 to 10 years, there is a major oyster die-off; reasons for this phenomenon remain unknown. Non-point sources of pollution continue to be the most problematic because most point sources in the Pensacola Bay system have been controlled.

PAHs are among the most widely distributed toxic agents in sediments of aquatic systems in Florida (Livingston and McGlynn, 1994; Seal et al., 1994). These agents represent a family of planar organic compounds composed of benzene units. PAHs are derived from petroleum products such as coal, oil, asphalt, gasoline, diesel, rubber, and plastics. The major routes of entry into the aquatic environment are biosynthesis, spillage and seepage of fossil fuels (a major source of bi- and tricyclic PAHs), discharge of domestic and industrial wastes, dry and wet deposition from the atmosphere, and urban runoff. Freshwater and marine biota rapidly accumulate PAHs upon exposure to low concentrations. PAHs have been associated with a wide variety of biological effects, including immunosuppression in fishes and assorted acute and chronic toxic effects. Many PAH compounds are on the EPA's priority pollutant list and some are considered carcinogenic.

Lower Escambia Bay and Pensacola Bay have the lowest biological values (i.e., low biomass, species richness) of any of the systems that have been part of the long-term studies by the Center for Aquatic Research and Resource Management (Florida State University) (Livingston, 2002). The current condition of the Pensacola Bay system is an example of a lack of attention to urban development in coastal areas. Overall, non-point sources of pollution in the Pensacola Bay system will have to be cleaned up for any measurable improvement of water quality and natural productivity. Evaluation of the point sources should also be made, especially because sewage spills constitute a continuing source of pollution in some parts of the bay system. Mitigation of the effects of future non-point sources of storm water loading to the bay is essential for recovery.

6.5 The Press and the Pensacola Bay System

The proposed move of the pulp mill effluent to the Escambia River galvanized the *Pensacola News Journal* as a leader in opposition to such a move. The Perdido Bay "environmentalists," on the other hand, were enthusiastically in favor of the pipeline for obvious reasons. "Environmentalist" letters to the editor were now full of praise for the researchers of the Perdido system. However, those opposed to the move, egged on by the news media, attacked the researchers with the usual accusations of selling out to the pulp mill. However, the data showed that the mill effluent should not be placed into the Escambia River. The assimilation capacity of Escambia Bay had been reduced to such an extent by anthropogenous nutrient loading and toxic waste dumping that any added nutrient loading would only cause further harm. These problems were due to various point and non-point sources that included industrial waste sites, sewage treatment plants, and hundreds of storm water pipes. Of course, when the decision was made for the mill effluent to remain in the Perdido system, the results of the Pensacola Bay Study were not publicized. In fact, the news media rarely if ever reported on the actual perpetrators of the pollution of the Pensacola system as many of these interests represented clients of the newspaper. Rare indeed was the reportage of impacts on the rivers and bay system due to construction activities and the urban runoff created by such activities. The *Pensacola News Herald* rarely saw a development it did not like. As for the toxic waste problem, everyone in the area was strangely silent on the issue.

Based on the publicity and adverse press concerning the mill's effect on the Perdido system, one would think that some kind of public discussion based on the science would have been appropriate. However, even after the decision was made, such a discussion never came. And the tactics of the *Pensacola News Journal* in its campaign against the move are illustrative of the ways in which the news media take advantage of their political power in such decisions. And, as in the past, what better way to oppose the move than attack the research associated with it. Consequently, before the study even ended, there was a full-scale attack on the research.

CRITICS QUESTION OBJECTIVITY OF CHAMPION-FUNDED RESEARCH: "'They are trying to clearly show what they want,' said the chief of the EPA's Coastal Ecology Branch at Gulf Breeze. 'It's a question of what you model,' said a Pace resident. 'You can go out and hire anybody to support your point of view. You can prove anything you want to.' ... The EPA official questions some of Livingston's and HydroQual's methodology. He said some of the data measures only water conditions, but does not say what the effect on algae growth and fish would be."

—**Scott Streater,** *Pensacola News Journal,* **June 16, 1998**

An outline of the proposal for the Pensacola system research is given in Appendix I. The Study Plan was distributed to the state and federal regulatory agencies for public review. The plan included an 18-month analysis of river and bay components in the entire Pensacola system (Figure 6.1). Water and sediment quality were analyzed, and extensive data sets were taken to model salinity–temperature relationships and the distribution of anoxia and hypoxia in the bay. Quantitative, species-specific collections were made concerning seasonal changes in phytoplankton, infauna, epibenthic macroinvertebrates, and fishes. The biological data were transformed into detailed food web distributions in space and time, with specific application to present and future impacts on algal growth and fishes/invertebrates. The entire study was aimed at the potential impact of the mill effluent on the entire Pensacola Bay system. This included processes such as wind/rain events, nutrient loading from the rivers, sediment oxygen demand, light transmission, and the various phases of primary and secondary productivity based on both water and sediment interactions and biological processes. Detailed timetables were provided, along with protocols for all sampling and analytical methodology.

In response to the blatant misrepresentation of the study by the *Pensacola News Journal*, the project leader wrote a letter to the reporter explaining that he wanted to issue a response that would correct open falsehoods in the article. This included corrections concerning the number of measurements and stations in the bay, the frequency of sampling, and the overall objectives and scope of the work. A conversation with the EPA official indicated that the official had not even seen the study outline. The response included an objection to the insinuations that the study team would come to conclusions that were based on pressure from the pulp mill rather than the facts. The response from the *Pensacola News Journal* was not unexpected.

"...in checking with...the chief of the coastal ecology branch of the Environmental Protection Agency's National Health and Environmental Effects Laboratory in Gulf Breeze, (he) tells us he believes he was accurately quoted and does not take issue with the facts or tone of the article you are questioning. Since your letter deals with statements attributed to (the chief), I trust that you understand that we believe the article to which you refer was factually accurate. Thus, we stand by the story."

—**Letter to Project Leader, Bob Bryan, Deputy Managing Editor,**
Pensacola News Journal, **July 6, 1998**

The refusal of the newspaper to explain both the purpose and the scope of the Pensacola Study was a blatant attempt to impeach the research before it even began. Misrepresentation and omission were used to make sure that the public was not informed on the issue. This little exercise in sophistry was followed by full articles with titles such as "Beware of science that is for sale," and editorials headed by titles such as "Politics, 'facts,' collide at Champion." There were even cartoons that indicated that the scientists were taking their time in the final analyses so as to bias the data in favor of the pulp mill.

When the decision was made, based on the scientific evidence of the poor shape of the Escambia Bay system, not to run a pipe to the Escambia River, there was a general silence in the Pensacola news media except for the resurrection of news quotes and letters to the editor from the Perdido "environmentalists" with renewed attacks on the scientific efforts in the Perdido system. All of a sudden, the Perdido crowd was disappointed with the decision to move the mill pipeline, and the Perdido research was, once again, unreliable. The polluted condition of the Pensacola system was ignored by the *Pensacola News Journal* and the regional news media. Things were back to normal.

In the afterglow of the great triumph of the *Pensacola News Journal*'s effort to protect the Escambia County Utilities Authority (ECUA) from unwarranted attacks by ecologists in Perdido Bay (see Chapter 5), virtually nothing was done to evaluate the findings of that study. The origin of the chlorine spills was not investigated. The urban development of Perdido Key was ignored, along with the agricultural wastes coming out of Alabama into Wolf Bay. Likewise, the *Pensacola News Journal* ignored the findings in the Pensacola Report concerning the loading of toxic agents into the bay from surface and groundwaters that drained waste sites around the Pensacola area.

A Grand Jury was formed to review the situation concerning drinking water conditions in the Pensacola area. The following are excerpts from the May 2004 Grand Jury findings:

> "Groundwater contamination is widespread in Escambia County, with more than half the county's wells contaminated. Sources of the contamination include six Superfund sites, dozens of dry-cleaning sites, hundreds of petroleum storage sites and numerous abandoned landfills."

> "The U.S. Environmental Protection Agency, the Florida Department of Environmental Protection, and the Escambia County Utilities Authority failed to do all they could to prevent or remediate groundwater contamination."

> "The EPA and the DEP have not been sufficiently concerned with the health, safety, and welfare responsibilities they bear, or the consequences of their decisions."

> "ECUA staff knew for several years that drinking water from some wells in the ECUA system was contaminated with radium and other harmful substances but did not tell the ECUA board or the public until 1998."

> "The ECUA publicly stated that federal radium standards were likely to be loosened and that the heightened level of radium in some public wells created no health risk. The ECUA knew that neither statement was accurate."

> "In 1999, the ECUA did not inform customers using water from radium-tainted wells to seek an alternative source of supply, as directed by the DEP. The Department of Health compounded the problem when it misinterpreted scientific data to conclude that the radium level at two wells did not constitute a health risk."

"At heavily contaminated sites... the costs of contamination have been shifted to the public because the polluting companies closed their businesses and abandoned their properties."

What is surprising is not that the drinking water of the region has been contaminated for years by a long list of toxic substances or that information of such contamination was kept from the public, but rather that such a finding should be a "surprise" to the news media and the public in this region. As noted above, various sources of toxic wastes have been identified in the Pensacola region with associated contamination of the Pensacola Bay system. Scientific reports concerning the toxic substances going into the waters of the region were ignored by the news media in favor of their own self-serving actions. The fact that water contaminated with high levels of radium from 1996 to 2000 with the knowledge by ECUA authorities represents a major failure of trust.

Earlier reports of increased cancer rates also indicated problems:

EXPERTS LINK CANCER TO POLLUTION: "A three-month study by the *Pensacola News Journal* found that death rates from... two counties (Escambia, Santa Rosa) far exceed national rates and incidences of prostate, brain and lung cancers there consistently rank higher than the national level. ... Escambia and Santa Rosa also far exceed state rates for some major birth defects associated with exposure to toxins. The cancer rates have caused some health insurance companies to stop covering residents in the area."

—Associated Press, February 19, 2001

The most obvious problem with this process is that the scientific data are often ignored in favor of pet projects by various NIMBY (Not in My Back Yard) groups. The hype and controversy engendered by the unscientific statements of such groups mislead the public; however, such statements stir the controversy that sells the news. Unfortunately, the ensuing public dialogue does not address the real problems involved with human impacts on aquatic systems. The overall restoration process is compromised by this pattern of misrepresentation of the scientific facts by the news media. The result is continued ignorance on the part of the public concerning important environmental issues in the region.

chapter 7

Sulfite Pulp Mill Restoration

7.1 Introduction

Sulfite paper mills use ammonia in various processes. A pulp mill on the Amelia estuary (north Florida) was responsible for extremely high ammonia loading (Livingston, 1996b; Livingston et al., 2002). Ammonia toxicity to marine phytoplankton has not been well established in the scientific literature. The U.S. Environmental Protection Agency (1976, 1989) proposed a limit of 0.02 mg L^{-1} as un-ionized ammonia for protection of freshwater aquatic life. Admiraal (1977) showed that toxicity to phytoplankton is due to ammonia (NH_3) rather than ammonium (NH_4^+), and that concentrations of 0.247 mg L^{-1} ammonia retarded the growth of seven species of benthic diatoms. Concentrations of 0.039 mg L^{-1} ammonia reduced reproduction of a red macroalga, *Champia parvula* (Admiraal, 1977). These concentrations were within the range of ammonia found in polluted parts of the Amelia system (Florida Department of Environmental Regulation, 1991). Ammonia is also an important nutrient for coastal phytoplankton, with studies that indicate preferential uptake by individual plankton species that sometimes leads to blooms (Admiraal and Peltier, 1980; Flemer et al., 1997; Livingston, 2000; U.S. Environmental Protection Agency, 1989). Ammonia has been shown to be a selective factor in the species composition of benthic diatoms due to species-specific variation of the ammonia toxicity (Van Raalte et al., 1976; Sullivan, 1978; Admiraal and Peletier, 1980). Thus, the potential effects of ammonia discharges on coastal phytoplankton can be both stimulatory and inhibitory with species-specific responses to ranges of ammonia concentrations.

Ammonium toxicity in water can be due to the effects of both the ionized (NH_4^+) and un-ionized (NH_3) forms with the relative concentration of each dependent on ambient pH and temperature (Kórner et al., 2001). Un-ionized ammonia toxicity increases with increased pH and temperature (U. S. Environmental Protection Agency, 1989). Whitfield (1974) defined the relationships of the ionized and un-ionized forms under different conditions of temperature, atmospheric pressure, and pH. Downing and Merkens (1955) found that the un-ionized form is the most toxic as it is uncharged and therefore traverses the cell membrane more readily. Some authors (Clement and Merlin, 1995) attributed toxicity to NH_3 only. Other studies (Monselise and Kost, 1993) attributed toxicity to both forms. With respect to the effects of ammonium on duckweed (*Lemna gibba*), Kórner et al. (2001) did not have a firm conclusion regarding the relative toxicity of the ionized (NH_4^+) and un-ionized (NH_3) forms. In this study, we determined the ammonium concentrations as un-ionized ammonia with the pH of the study areas being relatively constant (mean, 7.64; S. D., 0.32; Livingston, 1996b).

7.1.1   Study Area

The Amelia and Nassau River estuaries (Figure 7.1) are located in coastal northeast Florida, and are characterized by extensive marsh development and relatively high salinities. Tidal ranges approximate 2 to 3 m. The study area is a maze of channels and bayous with direct connections to the Atlantic Ocean. Major parts of the Nassau system are within a state park, and preliminary water quality analyses indicated relatively high water quality (Livingston, 1996b). The climate along this part of the coast is mild. Annual rainfall averages around 120 cm, with peaks during summer months.

A sulfite pulp mill discharges effluents (approx. 114 million gallons day^{-1} [mgd]) into the Amelia River-estuary (Figure 7.1). Mill effluents currently are discharged into a 125,000 m^2 mixing zone on outgoing tides. Dennison et al. (1977) found that effluent-receiving areas of the Amelia system were characterized by low dissolved oxygen (DO) and pH,

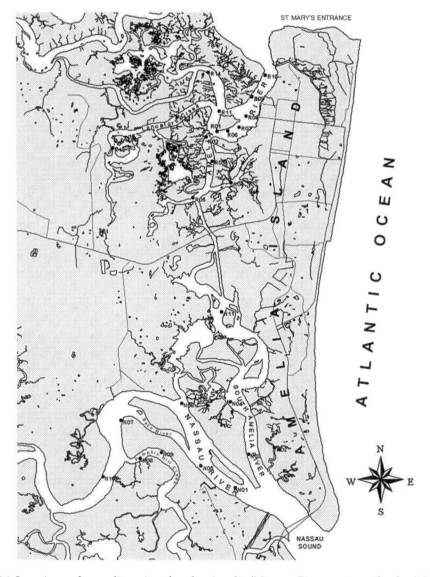

Figure 7.1 Locations of sampling sites for the Amelia/Nassau River estuary Study (1994–1995; 1997–1998; 2000–2001). Geographic data provided by the Florida Geographic Data Library (FGDL).

high watercolor, and low primary production relative to reference sites. Secchi depths were relatively low and total organic carbon (TOC) levels were relatively high in receiving areas of the Amelia estuary. Generally, phytoplankton and zooplankton numbers and diversity in the Amelia system were comparable to those in reference areas elsewhere (Dennison et al., 1977). The Florida Department of Environmental Regulation (1991) reported high (approx. 1.7 mg L^{-1}) concentrations of free ammonia in areas affected by the pulp mill; recent analyses (Livingston, 1996b) corroborated these findings. There were also indications of low phytoplankton species richness in the discharge areas (Florida Department of Environmental Regulation, 1991).

Based on a 12-month field analysis (1994–1995), Livingston (1996b) found that the Nassau system was an adequate (i.e., unpolluted, with comparable habitat distribution) reference area for studies of the Amelia system. Determinations of water quality and phytoplankton/zooplankton distributions in the Amelia and Nassau River estuaries indicated that ammonia was present in significantly high concentrations in the Amelia system. A combined field descriptive and field/laboratory experimental program (1997–1998) was then established to determine the effects of ammonia on phytoplankton assemblages. Specific research questions for this study were (1) whether pulp mill effluents were associated with observed reductions of phytoplankton assemblages in the Amelia system and (2) whether ammonia and/or light transmission were responsible for such effects.

7.2 Methods and Materials

Detailed protocols for this work are given in Appendix I. Detailed descriptions of methods for collection of physicochemical field data are given by Flemer et al. (1997); Livingston (1979, 1982a, 2000): and Livingston et al. (1997, 1998a, 2000). Stations were determined that defined the distribution of mill effluents in the receiving area. Matching stations, chosen for comparability of habitat characteristics (temperature, salinity), were established in the Nassau River estuary as reference sites for comparative analyses (Figure 7.1). Data were taken monthly over three 12-month sampling periods (1994–1995, 1997–1998, and 2000–2001). Microcosm and mesocosm experiments with pulp mill effluents and ammonia were carried out during the 1997–1998 sampling period under conditions approximating those observed in the field.

Net phytoplankton samples were taken with two 25-μm nets (bongo configuration) in duplicate runs for periods of 1 to 2 minutes. Repetitive (3) 1-L whole water phytoplankton samples were taken at the surface. Zooplankton were taken with two 202-μm nets (bongo configuration) in duplicate runs. Methods used for the comparison of monthly data (water quality, biological factors) were developed to determine significant differences between matching Amelia and Nassau sites (polluted and unpolluted) over the 12-month study periods (Livingston et al., 1998; Livingston, 2000) (Appendix I). Field data were analyzed using a Principle Components Analysis (PCA) as a preliminary review of the water quality variables (Appendix II; Livingston et al., 1998b). The PCA was used to reduce the physicochemical variables into a smaller set of linear combinations that could account for most of the total variation of the original set. Significant principal components were then applied to regression models, with phytoplankton and zooplankton abundance and species richness as dependent variables.

A combination of background field monitoring, controlled laboratory experiments using microcosms of *Skeletonema costatum* (Grev.) Cleve, and field mesocosm experiments (multispecies) was used to evaluate the effects of pulp mill effluents and ammonia on plankton assemblages in the Amelia River estuary. Measured solutions of ammonia were used to evaluate the effects of ammonia by itself relative to the effects of ammonia as part of the whole mill effluent. Target concentrations for the ammonia experiments were based on known field concentrations in polluted areas of the Amelia system. We carried out one microcosm test (June 29–July 4, 1998) using lab-cultured *Skeletonema* with measured injections of ammonia, and two tests (July 7–July 22, 1998 and August 28–September 1, 1998) with pulp mill effluents, with ammonia concentrations approximating those in the field. We performed six larger-volume mesocosm tests in the field, with natural phytoplankton assemblages taken from the reference Nassau area. Two tests (August 19–August 21, 1997 and October 27–October 29, 1997) were run with measured injections of ammonia, and four tests (May 20–May 22, 1998; June 24–June 26, 1998; August 4–August 6, 1998; and September 23–September 25, 1998) were carried out with pulp mill effluents that were added to basal mixtures to approximate ammonia concentrations determined in the field. For all tests, ammonia dosages were tested daily and ammonia was added where necessary to maintain target concentrations.

During 2000–2001, the mill reduced ammonia loading to near-natural conditions, and another field survey of water quality, phytoplankton, and zooplankton was carried out in the manner described above and in Appendix I. A detailed account of the results of the first two sampling periods is given by Livingston et al. (2002).

7.3   Results

7.3.1   Water Quality Data

There were no consistently significant differences in surface temperature, salinity, Secchi depths, BOD, DOC, TSS, silica, TP, POC, or sulfide between cognate station pairs during the survey periods 1994–1995 and 1997–1998 (Livingston et al., 2002). Surface watercolor was significantly ($P < 0.05$) higher at stations R03, R04, R10, N06, N09, and N11 than their paired matches during 1994–1995 and at stations R01, R08, and R11 during 1997–1998. Color was highest in the upper parts of both estuaries during winter months of increased rainfall. There were no significant differences in mean orthophosphate concentrations between cognate stations during both sampling periods although the upper Nassau system (Stations N07, N08, N09, and N10) had uniformly higher concentrations of orthophosphate than the paired stations in the Amelia system (Table 7.1). Total phosphorus (TP) was significantly higher at Stations R01, R04, R05, and R11 during 1994–1995 and was higher at Stations R02, R03, R09, N08, and N10 during 1997–1998.

During 1994–1995, surface ammonia concentrations were significantly ($P < 0.05$) higher at all stations in the Amelia system with the exception of R04 and R12 (Table 7.1). The relatively high ammonia concentrations near the mill outfall and gradients of surrounding stations indicated the pulp mill as the source. Mean annual surface ammonia concentrations ranged from 0.19 to 0.43 mg L^{-1} in the Amelia estuary during 1997–1998 (Table 7.2). No such gradient was noted in the Nassau estuary with annual means ranging from 0.09 to 0.11 mg L^{-1}. The highest ammonia concentrations in the Amelia system appeared during spring/summer months during both sampling periods (Livingston et al., 2002). Mean nitrite/nitrate concentrations near the outfall (Stations R01, R03, and R06) followed this trend, although differences were not statistically significant in the upper parts of the respective study areas. Surface total nitrogen was generally higher throughout the Amelia system during both sampling periods with significant ($P < 0.05$) differences at Stations R01, R02, R03, R06, R09, and R11. Mean surface chlorophyll *a* concentrations were generally lower in the lower Amelia River estuary than matched stations in the Nassau system during both sampling periods, and were significantly reduced ($P < 0.05$) at Stations R01, R08, R11, and R12 during 1997–1998. Spatial and temporal chlorophyll *a* trends followed (inversely) those of ammonia.

Table 7.1 Statistical Comparison (means tests: paired *t*-test, Wilcoxon signed rank test) of Physicochemical Data Collected at Stations in the Nassau and Amelia River Estuaries Monthly for 15 Months from November 1994 through October 1995

Cognate Station Pair	NH_3 (mgN L^{-1})	NO_3 (mgN L^{-1})	TN	PO_4 (mgP L^{-1})	TP	SiO_2 (mg L^{-1})	Chlor. *a*	Sulfide
				Surface				
R01/N04	0.274/0.083 WS (P < 0.005)	0.016/0.006 WS (P < 0.05)	1.211/0.640 WS (P < .05)	0.016/0.018 WN	0.086/0.052 WS (P < 0.05)	0.364/0.358 WN	5.28/5.59 WN	0.328/0.141 WS (P < 0.005)
R02/N04	0.138/0.083 WS (P < 0.025)	0.009/0.006 WN	0.946/0.640 WS (P < 0.005)	0.018/0.018 WN	0.087/0.052 WN	0.347/0.358 WN	5.38/5.59 WN	0.262/0.141 WS (P < 0.025)
R03/N04	0.141/0.083 WS (P < 0.05)	0.013/0.006 WS (P < 0.005)	1.133/0.640 WS (P < 0.005)	0.020/0.018 WN	0.083/0.052 WN	0.379/0.358 WN	6.50/5.59 WN	0.226/0.141 WS (P < 0.005)
R04/N02	0.150/0.118 WN	0.016/0.013 WN	0.978/0.680 WN	0.026/0.022 WN	0.087/0.066 WS (P < 0.025)	0.569/0.397 WN	7.88/6.34 WN	0.232/0.128 WS (P < 0.025)
R05/N11	0.157/0.104 WS (P < 0.02)	0.016/0.014 WN	1.202/0.815 WN	0.022/0.022 WN	0.150/0.073 WS (P < 0.01)	0.421/0.390 WN	6.13/7.24 WN	0.274/0.216 WN
R06/N03	0.234/0.104 WS (P < 0.025)	0.017/0.011 WS (P < 0.025)	1.092/0.619 WS (P < 0.005)	0.017/0.022 WN	0.103/0.068 WN	0.352/0.297 WN	4.68/5.53 WN	0.264/0.195 WS (P < 0.025)
R07/N03	0.179/0.104 WS (P < 0.025)	0.012/0.011 WN	0.860/0.619 WN	0.017/0.022 WN	0.112/0.068 WN	0.341/0.297 WS (P < 0.025)	5.46/5.53 WN	0.173/0.195 WN
R08/N03	0.156/0.104 WS (P < 0.05)	0.014/0.011 WN	0.864/0.619 WN	0.016/0.022 WN	0.054/0.068 WN	0.289/0.297 WN	5.97/5.53 WN	0.129/0.195 WN
R09/N01	0.147/0.100 WS (P < 0.05)	0.014/0.013 WN	0.798/0.622 WS (P < 0.025)	0.016/0.011 WN	0.092/0.065 WN	0.262/0.241 WN	5.20/6.58 WN	0.214/0.167 WN

Table 7.1 (*continued*) Statistical Comparison (means tests: paired *t*-test, Wilcoxon signed rank test) of Physicochemical Data Collected at Stations in the Nassau and Amelia River Estuaries Monthly for 15 Months from November 1994 through October 1995

Cognate Station Pair	NH_3 (mgN L^{-1})	NO_3 (mgN L^{-1})	TN	PO_4 (mgP L^{-1})	TP	SiO_2 (mg L^{-1})	Chlor. a	Sulfide
R10/N01	**0.147/0.100** **WS (P < 0.05)**	0.017/0.013 WN	0.683/0.622 WN	0.017/0.011 WN	0.055/0.065 WN	0.323/0.241 WN	5.68/6.58 WN	0.181/0.167 WN
R11/N05	**0.141/0.082** **WS (P < 0.01)**	0.010/0.008 WN	**0.782/0.582** **WS (P < 0.05)**	0.017/0.014 WN	**0.170/0.052** **WS (P < 0.025)**	0.265/0.263 WN	5.48/6.81 WN	0.212/0.146 WN
R14/N06	**0.146/0.086** **WS (P < 0.005)**	0.010/0.011 WN	0.765/0.610 **WS (P < 0.025)**	**0.015/0.022** **WS (P < 0.01)**	0.102/0.065 WN	0.473/0.428 WN	5.57/5.41 WN	0.199/0.234 WN
R12/N09	**0.162/0.081** **WS (P < 0.01)**	0.010/0.013 WN	0.685/0.664 WN	0.016/0.027 WN	0.088/0.071 WN	0.348/0.523 WN	6.06/6.81 WN	**0.161/0.233** **WS (P < 0.05)**
R12/N07	0.162/0.085 WN	0.010/0.013 WN	0.685/0.655 WN	0.016/0.020 WN	0.088/0.059 WN	0.348/0.389 WN	6.06/6.00 WN	0.161/0.167 WN
R13/N10	**0.181/0.091** **WS (P < 0.01)**	0.016/0.015 WN	0.863/0.689 WN	0.020/0.027 WN	0.063/0.076 WN	0.501/0.493 WN	6.00/6.68 WN	0.209/0.201 WN
R13/N08	**0.181/0.083** **WS (P < 0.01)**	0.016/0.010 WN	0.863/0.689 WN	0.020/0.024 WN	0.063/0.077 WN	0.501/0.490 WN	6.00/6.43 WN	0.209/0.197 WN

WN, *not sig.*

WS, sig.

Boldface = bloom species.

Table 7.2 Statistical Comparison (means tests: paired *t*-test, Wilcoxon signed rank test) of Physicochemical Data Collected at Stations in the Nassau and Amelia River Estuaries Monthly for 15 Months from July 1997 through September 1998

Cognate Station Pair	NH_3 (mg L⁻¹)	NO_2+NO_3 (mgN L⁻¹)	PO_4 (mgP L⁻¹)	TP (mgP L⁻¹)	SiO_2 (mg L⁻¹)	POC (mg L⁻¹)	Chlor. *a* (μγ L⁻¹)	Sulfide (mg L⁻¹)
Surface								
N04/R01	**0.11/0.43** WS (P < 0.01)	**0.07/0.02** WS (P < 0.01)	0.03/0.03 WN	0.14/0.09 WN	1.27/1.55 WN	1.68/1.47 WN	**7.65/5.95** WS (P < 0.05)	0.19/0.24 WN
N03/R08	**0.10/0.37** WS (P < 0.01)	**0.06/0.08** WS (P < 0.05)	0.03/0.03 WN	0.08/0.07 WN	1.12/1.29 WN	1.17/1.76 WN	**7.84/5.09** WS (P < 0.01)	0.18/0.16 WN
N05/R11	**0.09/0.29** WS (P < 0.01)	**0.05/0.09** WS (P < 0.01)	0.02/0.03 WN	0.09/0.08 WN	**0.84/1.41** WS (P < 0.01)	1.76/1.97 WN	**7.31/5.60** WS (P < 0.05)	0.17/0.21 WN
N06/R12	**0.11/0.27** WS (P < 0.01)	0.08/0.10 WN	0.28/0.30 WN	0.08/0.07 WN	1.31/1.61 WN	1.94/1.34 WN	**7.31/4.98** WS (P < 0.05)	0.20/0.27 WN
N08/R13	**0.10/0.19** WS (P < 0.01)	0.08/0.10 WN	0.03/0.03 WN	**0.10/0.08** WS (P < 0.05)	1.61/2.03 WN	2.45/1.97 WN	8.35/8.79 WN	0.20/0.21 WN
Bottom								
N04-R01	**0.11/0.30** WS (P < 0.01)	0.07/0.09 WN	0.03/0.03 WN	0.09/0.09 WN	1.51 WN	**1.51/2.05** WS (P < 0.05)	**8.13/5.99** WS (P < 0.02)	0.17/0.19 WN
N03/R08	**0.11/0.29** WS (P < 0.01)	**0.06/0.09** WS (P < 0.01)	0.02/0.03 WN	0.09/0.08 WN	1.06/1.21 WN	1.9/2.4 WN	**8.57/4.95** WS (P < 0.01)	0.19/0.21 WN
N05/R11	**0.11/0.27** WS (P < 0.01)	**0.04/0.09** WS (P < 0.01)	0.03/0.03 WN	**0.11/0.08** WS (P < 0.05)	**0.73/1.40** WS (P < 0.01)	3.52/1.82 WN	**9.3/5.8** WS (P < 0.01)	0.17/0.22 WN
N06/R12	**0.12/0.26** WS (P < 0.01)	**0.07/0.11** WS (P < 0.01)	0.03/0.03 WN	0.09/0.08 WN	**1.14/1.66** WS (P < 0.01)	2.53/1.85 WN	**7.56/5.6** WS (P < 0.05)	0.21/0.18 WN
N08/R13	0.26.0.23 WN	0.07/0.11 WN	0.03/0.03 WN	0.13/0.10 WN	**1.50/2.05** WS (P < 0.05)	3.32/2.24 WN	9.17/7.97 WN	0.20/0.23 WN

WN, not sig.
WS, sig.
Boldface = bloom species.

7.3.2   Light Transmission

Light data indicated that during 1994–1995, there were no major differences in light penetration between the Amelia and Nassau systems (Table 7.3). Although euphotic depths at Station R01 were lower than at Station N04 during 1997–1998, this was not consistent throughout the entire sampling period. In three of the eight noted readings, the differences were negligible. When viewed as differences in euphotic depths at different wavelengths, there were no significant differences between paired Stations N04 and R01. The lowest euphotic depths (and highest extinction coefficients) in both systems were noted during February 1998, a period of low chlorophyll *a* concentrations. With the exception of Station R11 at the 430-nm level, light extinction coefficients in the Amelia system were not significantly higher than those in the Nassau system. There were no significant reductions in euphotic depths in the Amelia system.

In both systems, there was evidence of a "gelbstoff shift" (Livingston et al., 1998b), whereby humic substances absorb light at lower wavelengths. Extinction coefficients were significantly higher and euphotic depths were significantly lower in the upper Nassau system where the highest levels of color were noted. The highest light extinction coefficients were noted at Station N08. Although there was thus no evidence of a significant mill effect on light transmission in the Amelia River estuary relative to the reference system, the upper parts of the Nassau River estuary were subject to the effects of runoff that affected both color and light transmission.

7.3.3   Phytoplankton and Zooplankton

Nearly 250 species of whole water phytoplankton were identified in the two study areas during the 1994–1995 survey. Numerical abundance of phytoplankton was reduced in the Amelia system relative to the reference area, and totaled only about 57% of the phytoplankton numbers found in the Nassau system. *Skeletonema costatum* was dominant in both study areas. Other dominants included *Cylindrotheca closterium*, *Thalassionema nitzschioides*, and *Asterionellopsis glacialis*. Major reductions were noted for *S. costatum*, *A. glacialis*, *T. nitzschioides*, and *Pseudonitzschia* sp. in the Amelia system relative to the reference area. The 1997–1998 results were similar to those of 1994–1995. Of the ten top dominant species, representing over 75% of the numbers of phytoplankton taken during the 1997–1998 survey, seven such species had considerably higher numbers in the Nassau system than in the Amelia system. The top dominant in the Nassau system was *S. costatum*, whereas phytoplankton assemblages in the Amelia system were dominated by *Chaetoceros socialis*. Cryptophytes and nannoflagellates were somewhat higher in the Amelia system than in the reference system. Nannococcoids were noted primarily at the outfall station. Blooms of *Navicula* sp. were found at Station R13 during July 1998. In addition to *S. costatum*, several species were notably higher in the Nassau system; these included *Thalassiosira proschkinae* and *T. decipiens*, *Asterionellopsis japonica*, *C. closterium*, *T. nitzschoides*, *Chaetoceros curvisetus*, and *Chaetoceros laciniosus*.

With the exception of Station R13, densities of diatoms (Class Bacillariophyceae) were lower in the Amelia system than in the Nassau system during both sampling periods. The silicoflagellates were often more abundant in the Nassau system. The cryptophytes (Division Cryptophyta), green algae (Division Chlorophyta), dinoflagellates (Division Dinophyta), and blue-green algae (Division Cyanophyta) were generally found in higher concentrations in the Amelia system.

Phytoplankton numbers and species richness of the net (25 μm) and whole water phytoplankton were higher in the Nassau system during 1994–1995 (Table 7.4); such differences were statistically significant ($P < 0.05$) in the whole water phytoplankton but

Table 7.3 Statistical Comparison (independence tests, Chi-square normality tests, variance tests, means tests, paired *t*-test, Wilcoxon signed rank test, and distribution shape tests) of Light Transmission Data (Kd, Euphotic Depths) Collected at Stations in the Nassau and Amelia River Estuaries Monthly for 12 Months from August 1997 through August 1998 (no data taken, July 1998). Euphotic Depth and Kd Given for the Overall Spectrum and at Specific Wavelengths (430, 550, 665 nm)

Cognate Station Pair	Kd	Kd-430	Kd-550	Kd-665	Eup. Dpth. (m)	430-Eup. Dpth. (m)	550-Eup. Dpth. (m)	665-Eup. Dpth. (m)
N04/R01	2.91/2.37 WN	5.65/4.31 WN	3.13/2.85 WN	2.56/2.22 WN	1.87/2.11 WN	1.01/1.28 WN	1.77/1.87 WN	2.05/2.33 WN
N03/R08	2.16/2.36 WN	4.43/5.17 WN	2.6/2.89 WN	2.01/2.27 WN	2.38/2.15 WN	1.22/0.97 WN	2.20/1.89 WN	2.48/2.38 WN
N05/R11	2.59/2.81 WN	**4.37/6.82 WS (P < 0.05)**	3.04/3.60 WN	2.28/2.33 WN	2.26/1.75 WN	**1.22/0.81 WS (P < 0.05)**	2.16/1.55 WN	2.38/2.11 WN
N06/R12	2.94/2.21 WN	4.34/4.26 WN	2.83/2.83 WN	2.56/1.90 WN	2.08/2.25 WN	1.25/1.33 WN	1.97/1.89 WN	2.24/2.57 WN
N08/R13	**3.74/2.61 WS (P < 0.05)**	4.59/5.91 WN	3.67/2.81 WN	**3.11/2.35 WS (P < 0.05)**	1.53/2.04 WN	1.68/1.34 WN	1.48/1.73 WN	**1.63/2.31 WS (P < 0.05)**

WN, not sig.
WS, sig.
Boldface = bloom species.

Table 7.4 Statistical Comparison (independence tests, Chi-square normality tests, variance tests), means tests (paired *t*-test, Wilcoxon signed rank test, and distribution shape tests) of Whole Water and 25-μm Phytoplankton and Zooplankton Data Collected at Stations in the Nassau and Amelia River Estuaries Monthly for 12 Months in 1994–1995 and 1997–1998 (data given for numbers of cells L^{-1}, species richness, and Shannon–Wiener diversity)

A. 1994–1995

Cognate Station Pair	WW Phytoplankton (Numbers L^{-1})	WW Species Richness	WW Shannon Diversity	25 μm Phytoplankton (Numbers L^{-1})	25 Species Richness	25 Shannon Diversity
R01/N04	**137707.5/308265.8 WS (P < 0.01)**	**14.2/17.8 WS (P < 0.02)**	1.550/1.508 WN	34340.8/93061.6 CV	32.5/38.0 WN	1.837/1.821 WN
R03/N04	**126539.2/308265.8 WS (P < 0.02)**	**15.3/17.8 WS (P < 0.05)**	1.437/1.508 WN	37181.7/93061.6 WN	38.5/38.0 WN	2.023/1.821 WN
R04/N02	**115747.5/238463.3 WS (P < 0.01)**	**13.6/16.8 WS (P < 0.02)**	1.494/1.493 WN	36003.3/40900.4 WN	35.1/33.3 WN	2.071/1.901 WN
R08/N03	**169075.8/285895.8 WS (P < 0.05)**	**14.3/19.7 WS (P < 0.05)**	1.523/1.509 WN	**42883.8/82276.3 WS (P < 0.05)**	36.2/42.1 WS (P < 0.05)	1.923/1.796 WS (P < 0.05)
R10/N01	**134003.3/334136.7 WS (p < 0.01)**	15.4/18.9 WN	**1.620/1.312 WS (p < 0.05)**	50691.7/96422.1 WN	37.7/42.7 WN	1.953/1.683 WN
R11/N05	154910.8/340943.3 WN	15.2/17.7 WN	1.527/1.344 WN	44840.83/92260.0 WN	39.1/40.8 WN	2.009/1.762 WN
R12/N06	**109955.8/222757.5 WS (P < 0.05)**	14.0/17.1 WN	1.633/1.572 WN	35049.6/77239.2 WN	36.0/37.3 WN	2.081/1.978 WN
R12/N07	109955.8/353772.5 CV	14.0/17.4 WN	1.633/1.491 WN	35049.6/74410.0 WN	36.0/34.3 WN	2.081/1.815 WN
R13/N08	**107782.5/282419.2 WS (P < 0.05)**	**11.9/15.4 WS (P < 0.05)**	1.580/1.434 WN	**32555.8/76985.4 WS (P < 0.05)**	33.1/35.8 WN	2.121/1.778 WN

Cognate Station Pair	Zooplankton (Numbers L^{-1})	ZPL Species Richness	ZPL Shannon Diversity
R01/N04	728.8/1760.5 WS (P < 0.01)	11.6/10.8 WN	1.083/0.963 WN
R03/N04	701.3/1760.5 WS (P < 0.05)	10.6/10.8 WN	1.320/0.963 WS (P < 0.02)
R04/N02	407.4/919.9 WS (P < 0.01)	10.3/11.0 WN	1.1667/1.009 CV
R08/N03	1213.3/1650.0 WN	11.9/11.8 WN	1.258/0.992 WN
R10/N01	601.7/1860.3 WS (P < 0.01)	12.0/11.2 WN	1.500/1.169 WN
R11/N05	1295.7/987.6 WN	11.9/11.4 WN	1.073/1.202 WN
R12/N06	1280.9/1584.6 WN	9.6/10.0 WN	0.993/1.138 WN
R12/N07	1280.9/1997.4 WN	9.6/10.0 WN	0.993/1.179 WN
R13/N08	546.8/1489.2 WS (P < 0.05)	8.917/9.833 WN	0.983/0.884 WN

WN, not sig.

WS, sig.

CV = RHO > Critical Value.

Boldface = bloom species.

Table 7.4 (continued) Statistical Comparison (independence tests, Chi-square normality tests, variance tests), means tests (paired *t*-test, Wilcoxon signed rank test, and distribution shape tests) of Whole Water and 25-μm Phytoplankton and Zooplankton Data Collected at Stations in the Nassau and Amelia River Estuaries Monthly for 12 Months in 1994–1995 and 1997–1998 (data given for numbers of cells L^{-1}, species richness, and Shannon–Wiener diversity)

B. 1997–1998

Cognate Station Pair	WW PHYT (Cells L^{-1})	WW Species Richness	WW Shannon Diversity	ZPL (Numbers m^{-3})
R01/N04	237,111/285,219 WN	27.3/34.8 WS (p < 0.05)	1.798/2.055 WN	728.8/1760.5 WS (p < 0.01)
R08/N03	249,286/342,517 WN	31.3/35.6 WN	2.068/2.060 WN	701.3/1760.5 WS (P < 0.05)
R11/N05	185,109/272,787 WS (p < 0.01)	28.9/37.1 WS (p < 0.02)	1.953/2.158 WN	407.4/919.9 WS (p < 0.01)
R12/N06	139,010/319,956 WS (p < 0.05)	25.2/32.2 WS (P < 0.01)	1.898/1.865 WN	601.7/1860.3 WS (p < 0.01)
R13/N08	1,361,296/223,795 WS (p < 0.01)	23.7/32.0 WS (p < 0.02)	1.728/1.985 WN	546.8/1489.2 WS (p < 0.05)

nd = no data.

not in the net phytoplankton. The most pronounced differences in phytoplankton numbers and species richness were noted during warm months (March–July 1995). Shannon diversity tended to be similar among the various station combinations, with no significant differences except at Station R10. Zooplankton numbers were significantly lower (Stations R01, R03, R04, and R10) in the Amelia system during 1994–1995 (Table 7.4). The most pronounced zooplankton differences between the two study areas occurred during March and April 1995. Zooplankton species richness was not significantly ($P < 0.05$) different between the two systems (Table 7.4).

During 1997–1998, phytoplankton numbers and species richness were generally lower at Amelia stations (Table 7.4); such differences were usually statistically significant ($P < 0.05$). Numerical abundance data showed that the primary reductions at Station R01 occurred during November and December 1997 and April, July, and August 1998. Higher numbers were noted at Station R01 during June 1998; this increase occurred during the period of relatively lower ammonia concentrations. Differences in phytoplankton numbers and species richness between Stations R11 and R12 and their matching stations were significant. Significantly higher phytoplankton numbers were noted at Station R13 than at its Nassau equivalent, a result of the July 1998 *Navicula* bloom. Species richness, however, was significantly lower at Station R13 than at Station N08.

7.3.4 Multivariate Statistical Analyses

Detailed descriptive and statistical analyses were carried out with field data taken during spring–summer 1994–1995 using factors that were significantly different between the two study areas. Particular attention was given to warm-water periods when phytoplankton differences between the study areas were greatest. Watercolor and Secchi depths were comparable between the two systems. Chlorophyll *a* concentrations were somewhat lower in the Amelia system during April, May, and June 1995, although the reductions in this factor were not as pronounced as in phytoplankton. Ammonia concentrations were higher in the Amelia system during April, May, June, and July, with somewhat higher concentrations in the Nassau system during August although overall averages were generally higher in the Amelia system. The occurrence of high ammonia concentrations tended to be the primary factor associated with reduced phytoplankton numbers in the Amelia system. This was generally true of phytoplankton species richness indices. Zooplankton numerical abundance followed the phytoplankton trends (Table 7.4). With the exception of August 1995, concentrations of 0.1 mg L^{-1} ammonia appeared to be the dividing line between the two study areas.

A PCA/regression analysis was run with the 1994–1995 data (Table 7.5). This analysis was run two ways: (1) for all stations and all dates over the 12-month sampling period, and (2) for data taken during warm months of the year. The analysis run over the entire sampling period indicated that whole water phytoplankton numbers were negatively associated with color and positively associated with salinity and chlorophyll *a*. During summer months, whole water phytoplankton numbers were negatively associated with ammonia and positively associated with temperature and chlorophyll *a*. During the 12-month period, net phytoplankton numbers varied negatively with color and BOD, and positively with salinity, chlorophyll *a*, and DOC. Net phytoplankton numbers were negatively associated with ammonia and positively associated with temperature and chlorophyll *a* during summer months. Whole water phytoplankton species richness was negatively associated with color and BOD during the 12-month period and negatively associated with TN during the summer months. Net phytoplankton species richness was negatively associated with color and BOD during the 12-month period, and was negatively associated with ammonia

Table 7.5 Results of Principal Components/Regression Analyses of Field Data for Whole Water Phytoplankton, Net Phytoplankton, and Zooplankton (numbers L^{-1}, number of taxa) Taken in the Amelia and Nassau River-Estuaries over a 12-Month Period (1994–1995) and during Summer Months (1994–1995) (independent variables listed in order of predominance)

A. 12-Month Period (1994–1995)

Dependent Variable	Independent Variable	R^2	Significance	Sign
WW Phytoplankton				
# Cells L^{-1}	Salinity	0.46	0.0001	+
	Chlorophyll *a*	0.46	0.0001	+
	Color	0.46	0.0001	–
# Taxa	Color	0.54	0.0001	–
	Salinity	0.54	0.0001	+
	Chlorophyll *a*	0.54	0.0001	+
	BOD	0.54	0.0035	–
	DOC	0.54	0.0035	+
Net Phytoplankton				
# Cells L^{-1}	Color	0.59	0.0001	–
	Salinity	0.59	0.0001	+
	Chlorophyll *a*	0.59	0.0001	+
	BOD	0.59	0.0030	–
	DOC	0.59	0.0030	+
# Taxa	Color	0.59	0.0001	–
	Salinity	0.59	0.0001	+
	Chlorophyll *a*	0.59	0.0001	+
	BOD	0.59	0.0035	–
	DOC	0.59	0.0035	+
Zooplankton				
# L^{-1}	Ammonia	0.28	0.0059	–
	Secchi	0.28	0.0012	+
	TN	0.28	0.0059	–
# Taxa	Secchi	0.37	0.0012	+
	Turbidity	0.37	0.0012	–
	TSS	0.37	0.0009	–

B. Summer Period (1994–1995)

Dependent Variable	Independent Variable	R^2	Significance	Sign
WW Phytoplankton				
# Cells L^{-1}	Ammonia	0.50	0.0001	–
	Temperature	0.50	0.0001	+
	Chlorophyll *a*	0.50	0.0001	+

Table 7.5 (continued) Results of Principal Components/Regression Analyses of Field Data for Whole Water Phytoplankton, Net Phytoplankton, and Zooplankton (numbers L^{-1}, number of taxa) Taken in the Amelia and Nassau River-Estuaries over a 12-Month Period (1994–1995) and during Summer Months (1994–1995) (independent variables listed in order of predominance)

Dependent Variable	Independent Variable	R^2	Significance	Sign
# Taxa	Temperature	0.19	0.0393	+
	Secchi	0.19	0.0184	+
	Chlorophyll *a*	0.19	0.0393	+
	TN	0.19	0.0184	−
Net Phytoplankton				
# Cells L^{-1}	Ammonia	0.58	0.0001	−
	Temperature	0.58	0.0001	+
	Chlorophyll *a*	0.58	0.0001	+
# Taxa	Ammonia	0.19	0.0434	−
# L^{-1}	Temperature	0.58	0.0148	+
	DOC	0.34	0.0035	−
	Chlorophyll *a*	0.34	0.0035	+
	Ammonia	0.34	0.0035	−
	Sulfide	0.58	0.0148	−
# Taxa	Temperature	0.55	0.0001	+
	Color	0.55	0.0087	−
	Sulfide	0.55	0.0087	−

during summer months. Thus, there appeared to be a negative response of phytoplankton numbers and species richness to ammonia during warmer periods.

Zooplankton numbers varied negatively with ammonia and total nitrogen and positively with Secchi depths during the 12-month period, and were negatively associated with ammonia and sulfides and positively associated with temperature and chlorophyll *a* during summer months. Zooplankton species richness during the 12-month period was positively associated with high Secchi depths and negatively associated with turbidity and TSS. During summer months, zooplankton species richness was positively associated with temperature and negatively associated with color and sulfide.

7.3.4.1 Laboratory Microcosms

The microcosm experiments were designed to evaluate the effects of ammonia on the growth of *Skeletonema costatum*, and to determine the potential influence of mill effluents (with ambient field ammonia concentrations) on such effects (Livingston et al., 2002). In the first experiment with ammonia additions (no mill effluents), Tukey's HSD test, Bonferroni's test, and Scheffe's *S* test indicated that addition of low concentrations of ammonia increased microalgal production; the final chlorophyll *a* concentrations were highest at mean ammonia concentrations around 0.06 mg L^{-1}. Chlorophyll concentrations were significantly lower at ammonia concentrations from 0.11 to 0.24 mg L^{-1} and much lower at concentrations greater than 0.46 mg L^{-1}. A second experiment (ammonia as part of mill effluents) showed ammonia stimulation of *S. costatum* at relatively high concentrations of ammonia (0.06 to 0.62 mg L^{-1}). By day 5, there were significant reductions in chlorophyll at mean ammonia concentrations of 0.71 mg L^{-1}. Compared to experiment 1 (ammonia only), these results indicated a difference in the action of ammonia on the growth of *S. costatum* in the presence of mill effluents.

The third experiment (with pulp mill effluents) showed that, by day 3, there was significant stimulation of chlorophyll production at mean ammonia concentrations of 0.07 mg L^{-1}. These results showed adverse effects on *Skeletonema costatum* at mean ammonia concentrations of 0.27 mg L^{-1}. These differences were significant at mean ammonia concentrations of 0.46 mg L^{-1}. The results of Experiment 3 resembled those of Experiment 1. although the shape of the chlorophyll curves during day 5 was somewhat different. Based on the differences in the results of the ammonia and pulp mill effluent tests, it is likely that factors other than ammonia in the mill effluents affected *S. costatum* growth.

7.3.4.2   Field Mesocosms

Results of field mesocosm tests with mill effluents and ammonia are given in detail by Livingston et al. (2002). Compared to untreated controls, there was usually an increase in chlorophyll *a* at the lowest ammonia concentrations, indicating the stimulatory effects of ammonia. With stabilization of experimental conditions, chlorophyll *a* reductions were seen as indicators of ammonia inhibition. Results of Experiment 1 (mill effluent, mean temperature of 30°C) showed chlorophyll *a* peaks at mean ammonia concentrations of 0.20 mg L^{-1}, with inhibition commencing at mean ammonia concentrations of 0.31 mg L^{-1}. Results of Experiment 5 (mill effluent, mean temperature of 30.3°C) indicated increased chlorophyll *a* (nutrient enhancement) at mean ammonia concentrations of 0.10 mg L^{-1}, with inhibition noted at 0.43 mg L^{-1} ammonia. Experiment 6 (mill effluent, mean temperature of 28.7°C) indicated ammonia stimulation at 0.36 mg L^{-1} and ammonia inhibition at 0.69 mg L^{-1}. The results of the last two experiments showed somewhat higher toxicity thresholds for ammonia than the first pulp mill experiment. The second experiment (ammonia; mean temperature of 21.7°C) gave results that resembled those of Experiment 1 (mill effluents). Chlorophyll *a* peaked at 0.14 mg L^{-1} ammonia, with inhibition at 0.20 mg L^{-1} ammonia. Experiment 3, run with ammonia concentrations at a mean temperature of 28.4°C, gave similar results to those of the first two experiments with stimulation at 0.11 mg L^{-1} ammonia and inhibition at 0.20 mg L^{-1} mean ammonia concentrations. Water temperature did not appear to have a major effect on the results of the mesocosm experiments.

7.4   Discussion and Conclusions

The basic questions of the effects of pulp mill effluents and associated high ammonia concentrations on plankton assemblages in a high-salinity estuary should be answered within the context of major loading of ammonia by a paper mill into a physically variable coastal environment. Phytoplankton abundance and species richness in the Amelia and Nassau systems were seasonally variable, with peaks usually occurring during summer months although there was evidence of winter increases in phytoplankton abundance. Zooplankton abundance peaked during spring months, which coincided with declines in phytoplankton abundance. Summer increases of phytoplankton numbers were correlated with relatively low zooplankton numbers. Nutrient limitation experiments (Livingston et al., 2002) indicated that nitrogen was the chief limiting nutrient to phytoplankton in the Nassau and Amelia River-estuaries during all seasons.

The diatom *Skeletonema costatum* is ubiquitous in coastal waters worldwide, and is frequently dominant in inshore phytoplankton blooms (Bonin et. al., 1986; Young and

Barber, 1973; Stockner and Costella, 1976; Hulburt and Rodman, 1963; Hulburt and Corwin, 1970). In these studies, *S. costatum* was often dominant due to its relatively high growth rate under a wide range of temperature/light conditions. There is also evidence that *S. costatum* grows well under high nutrient conditions, and it can use organic phosphorus as a source of growth. It can also assimilate organic molecules such as urea (Round, 1981). It grows in New York Bight waters where massive amounts of chemicals were dumped (Young and Barber, 1973). This species was a major constituent of the phytoplankton community in the Nassau and Amelia systems during different times of the year. However, *S. costatum* abundance was severely reduced in the Amelia system compared to the reference Nassau system.

Field analyses showed that watercolor, temperature, and salinity play important roles in seasonal changes of phytoplankton associations in the Amelia and Nassau systems. However, during warm periods, ammonia was a leading factor associated with reductions in phytoplankton numbers and species richness. Field results indicated that phytoplankton abundance and species richness were significantly lower in the Amelia system relative to the Nassau system. Microcosm results indicated that ammonia had a stimulatory effect on *Skeletonema costatum* at mean concentrations of 0.06 mg L^{-1}, with negative effects of ammonia occurring within a range of 0.1 to 0.24 mg L^{-1} and major impacts at concentrations greater than 0.46 mg L^{-1}. Mesocosm experiments with ammonia indicated stimulatory effects from 0.11 to 0.14 mg L^{-1} and inhibition of phytoplankton growth beyond 0.20 mg L^{-1}. Results of microcosm and mesocosm experiments with ammonia were generally consistent with field estimates of ammonia effects on phytoplankton. Light penetration and temperature could have obfuscated ammonia effects on phytoplankton in the field. Increased ammonia concentrations during winter (high color, low temperature) did not have as pronounced adverse effects as those during summer months. Multivariate statistical analyses of the field data confirmed that multiple factors determined phytoplankton distribution and there were seasonal differences in plankton response to ammonia. Reduced zooplankton numbers in the Amelia system could have been related to changes in phytoplankton assemblages rather than direct effects of ammonia because zooplankton species richness did not appear to be affected by high ammonia concentrations.

The presence of blooms in upper parts of the Amelia estuary complicated direct evaluations of the influence of ammonia because urban storm water effects were indicated both in terms of water quality and phytoplankton response. Because the lower experimental inhibitory concentrations of the ammonia tests were comparable to the field results, concentrations of 0.20 mg L^{-1} can be taken as conservative estimates of ammonia toxicity. The field results also presented a more representative analysis of the effects of long-term exposure to ammonia than the short-term experimental tests. Based on projections of ammonia inhibition of phytoplankton at 0.20 mg L^{-1}, a restoration effort initiated by the pulp mill was consistent with comparable levels of ammonia concentrations in the Amelia system (range of average ammonia concentrations: 0.19 to 0.43 mg L^{-1}) relative to the reference Nassau River estuary (range of average ammonia concentrations: 0.09 to 0.11 mg L^{-1}). Recommended long-term average ammonia concentrations at Station R01 (Amelia River estuary) were set at 0.11 mg L^{-1} ammonia, with short-term average increases not exceeding 0.20 mg L^{-1}. The pulp mill undertook a restoration program based on these estimates.

7.5 Restoration Program

During 2000–2001, the mill reduced ammonia loading to near-natural conditions, and another field survey (April 2000–March 2001) of water quality, phytoplankton, and zooplankton was carried out. Sampling design/methods and statistical analyses remained the same as those described above. Detailed analyses of the data are given by Livingston (2001a).

An analysis of rainfall trends (M. Franklin, personal communication; United States Geological Survey, Tallahassee, Florida) over the period from April 1992 through April 2001 indicated that, during 1994–1995 and 1997–1998, rainfall was relatively abundant, especially during summer months. However, the 2000–2001 study period occurred during a significant drought, with a relatively low annual mean and extremes of monthly minima. Peak precipitation occurred during late summer. Temperatures were similar during the three study periods. Long-term rainfall trends of drought and flood are important to trends in phytoplankton occurrence and bloom frequency (Livingston, 2000, 2002).

7.5.1   Physicochemical Conditions

Surface salinity was comparable between the Nassau and Amelia systems (Livingston, 2001a). During the first two studies (1994–1995, 1997–1998), there were latitudinal gradients, with the lowest salinities in upper parts of each estuary. The uniformly higher annual mean salinity during the 2000–2001 study period reflected drought conditions. During all three study periods, the highest watercolor conditions were found in upper parts of the respective drainages. The generally lower watercolor during the drought of 2000–2001 was evident. Watercolor peaks tended to occur during fall and winter months. Ratio turbidity in the study areas was not related to the mill outfall. Annual average Secchi depths tended to be lower in the upper parts of the respective drainages during the three sampling periods. There was a trend toward deeper Secchi depths during the 2000–2001 sampling period. There was no evidence that Secchi depth distribution was related to the mill outfall, but reflected changes associated with rainfall trends in the study area.

During the final sampling period (2000–2001), there were significantly increased oxygen anomalies (i.e., lower DO) at Stations R01, R03, R04, and R08 (Table 7.6A). Lower oxygen readings occurred from surface to bottom during spring/summer months at various stations in both systems. However, only during July 2000 were stations in the direct line of the mill outfall (i.e., R01 and R03) associated with hypoxic conditions. Most hypoxic readings were taken in Jackson Creek (R04) during summer 2000. Some of the low DO concentrations occurred in the upper Amelia basin during summer months. A few such readings also occurred in the cul-de-sac station (N02) in the Nassau system. Therefore, only during the July 2000 period were hypoxic conditions noted in areas associated with the mill outfall.

Statistical comparisons of nutrient/chlorophyll a data taken at cognate stations in the Amelia and Nassau systems are given in Table 7.6B. During 2000–2001, ammonia was generally lower in the Amelia system than that noted in previous surveys, although it was significantly higher at Stations R01 and R03 than cognate Nassau stations, average ammonia concentrations remained less than 0.1 mg L^{-1}. Ammonia concentrations were significantly higher in the upper Amelia basin; these effects were associated with rainfall events and urban runoff. Orthophosphate concentrations were somewhat higher in the Amelia system, but not uniformly so. The highest such concentrations were noted in upper parts of both systems. Chlorophyll a was significantly lower at Stations R01, R08, R10, and R12, whereas the highest such concentrations were noted in upper parts of both systems. With the exception of the upper basin, where lower light transmission coincided with higher chlorophyll, there was a generally direct relationship between this factor and the depth of the euphotic zone. These conditions again were associated with rainfall patterns and observed trends of urban runoff.

Table 7.6 Statistical Comparison (means tests: paired *t*-test, Wilcoxon signed rank test) of Surface Physicochemical Data Collected at Stations in the Nassau and Amelia River Estuaries Monthly for 12 Months from April 2000 through March 2001

A. Nassau and Amelia Systems

Cognate Station Pair	Salinity (ppt)	Oxygen Anom. (mg L^{-1})(−)	Secchi (m)	Color (Pt-Co Units)	DOC (mg L^{-1})	BOD (mg L^{-1})	TSS (mg L^{-1})	Turbidity (NTU)
R01/N04	34.3/33.7 WN	**1.79/0.79** WS (P < 0.02)	0.91/0.84 WN	**26.1/19.4** WS (P < 0.05)	20.9/19.6 WN	**1.59/1.14** WS (P < 0.05)	40.3/47.29 WN	7.36/7.83 WN
R03/N04	34.3/33.6 WN	**1.78/0.79** WS (P < 0.02)	0.84/0.84 WN	**24.1/19.4** WS (P < 0.02)	20.9/19.6 WN	**1.45/1.14** WS (P < 0.02)	51.48/47.29 WN	8.41/7.83 WN
R04/N02	34.1/33.2 WN	**2.55/1.75** WS (P < 0.05)	0.76/0.71 WN	**28.8/21.6** WS (P < 0.05)	**22.9/19.4** WS (P < 0.02)	1.44/1.43 WN	49.79/62.04 WN	8.7/10.3 WN
R08/N03	34.6/34.3 WN	**1.16/0.42** WS (P < 0.02)	**1.12/0.82** WS (P < 0.05)	18.4/16.3 WN	18.6/19.2 WN	1.35/1.38 WN	38.61/48.02 WN	6.08/7.4 WN
R10/N01	34.8/34.5 WN	1.08/0.52 WN	0.86/1.06 WN	**18.2/12.7** WS (P < 0.01)	20.0/18.4 WN	1.75/1.43 WN	52.0/54.11 WN	8.48/7.41 WN
R11/N05	34.7/34.1 WN	1.22/0.87 WN	0.92/0.95 WN	**19.3/14.5** WS (P < 0.05)	20.9/19.6 WN	1.24/1.31 WN	40.8/54.9 WN	7.21/9.43 WN
R11/N06	33.6/34.1 WN	1.22/1.16 WN	0.92/0.83 WN	**19.3/27.3** WS (P < 0.05)	20.9/20.5 WN	1.24/1.38 WN	40.8/49.98 WN	7.70/8.58 WN
R12/N07	**32.3/34.1** WS (P < 0.05)	1.56/1.16 WN	**0.95/0.86** WS (P < 0.02)	22.5/26.7 WN	19.9/21.3 WN	1.46/1.25 WN	**47.76/66.03** WS (P < 0.02)	**7.7/11.5** WS (P < 0.05)
R13/N08	33.3/33.9 WN	**2.09/1.21** WS (P < 0.02)	0.78/0.75 WN	26.9/38.1 WN	21.5/21.6 WN	1.43/1.29 WN	49.23/53.25 WN	8.01/8.94 WN

Table 7.6 (continued) Statistical Comparison (means tests: paired *t*-test, Wilcoxon signed rank test) of Surface Physicochemical Data Collected at Stations in the Nassau and Amelia River Estuaries Monthly for 12 Months from April 2000 through March 2001

B. Nassau and Amelia Systems

Cognate Station Pair	NH$_3$ (mgN L^{-1})	NO$_3$ (mgN L^{-1})	TN (mg L^{-1})	PO$_4$ (mgP L^{-1})	TP (mg L^{-1})	SiO$_2$ (mg L^{-1})	Chlor. *a* (μγ L^{-1})	Sulfide (mg L^{-1})
R01/N04	**0.097/0.057** WS (P < 0.01)	0.021/0.017 WN	0.25/0.32 WN	**0.018/0.015** WS (p > 0.05)	0.064/0.054 WN	1.21/1.17 WN	4.60/6.68 WS (P < 0.05)	0.07/0.08 WN
R03/N04	**0.084/0.057** WS (P < 0.01)	0.024/0.017 WN	**0.21/0.31** WS (P < 0.02)	**0.020/0.015** WS (P < 0.02)	0.071/0.054 WN	1.21/1.17 WN	5.3/6.68 WN	0.03/0.08 WN
R04/N02	0.096/0.076 WN	**0.022/0.013** WS (P < 0.02)	0.18/0.20 WN	0.026/0.032 WN	0.084/0.097 WN	1.62/1.80 WN	5.89/7.15 WN	0.05/0.11 WN
R08/N03	0.076/0.066 WN	0.019/0.013 WN	0.19/0.16 WN	0.014/0.012 WN	0.050/0.046 WN	0.96/0.83 WN	**4.25/5.51** WS (p > 0.05)	0.08/0.05 WN
R10/N01	0.063/0.058 WN	**0.018/0.012** WS (P < 0.01)	0.16/0.16 WN	**0.015/0.011** WS (P < 0.01)	0.051/0.052 WN	**0.93/0.63** WS (P < 0.01)	**5.17/7.18** WS (P < 0.01)	0.05/0.05 WN
R11/N05	0.075/0.061 WN	0.022/0.017 WN	0.16/0.17 WN	0.017/0.014 WN	0.055/0.047 WN	0.95/0.80 WN	5.11/5.09 WN	0.07/0.04 WN
R11/N06	0.080/0.075 WN	0.021/0.018 WN	0.16/0.16 WN	**0.017/0.020** WS (P < 0.02)	0.055/0.058 WN	0.95/1.14 WN	5.11/5.73 WN	0.07/0.10 WN
R12/N07	0.080/0.070 WN	0.018/0.015 WN	0.19/0.14 WN	0.017/0.019 WN	0.051/0.075 WN	1.17/1.08 WN	**4.65/6.67** WS (P < 0.02)	0.06/0.04 WN
R13/N08	**0.094/0.060** WS (P < 0.02)	0.020/0.018 WN	0.17/0.17 WN	0.022/0.021 WN	0.069/0.071 WN	**1.69/1.18** WS (P < 0.02)	6.13/7.38 WN	0.05/0.07 WN

WN, not sig.
WS, sig.
Boldface = bloom species.

7.5.2   Light Trends

Statistical comparisons of spectroradiometric data taken at cognate stations in the Amelia and Nassau systems are given in Table 7.7. During 2000–2001, there were significant increases in extinction coefficients and reductions in euphotic depths (400–700, 430, 550, 655 nm) at Stations R01, R03, R04, and R10. Extinction coefficients were highest, however, in the upper parts of both systems where euphotic depths were consistently lower than those taken in the lower parts of the respective drainages. This complex pattern indicated major reductions in light penetration in the upper parts of the Amelia and Nassau systems that were subject to urban runoff with somewhat lower such reductions in areas affected by the mill outfall.

7.5.3   Phytoplankton Analyses

During the 2000–2001 sampling period, phytoplankton assemblages were marked by winter blooms of *Skeletonema costatum*. For the first time, a bloom species (*Leptocylindrus danicus*) was found in the numerically top ten species. In contrast to the previous two sampling periods, there was no pattern of generally increased phytoplankton abundance in the Nassau system compared to the Amelia system. Phytoplankton blooms in the upper Amelia system (*S. costatum*) were similar in dominance to the *Navicula* blooms taken in this area during the 1997–1998 sampling period.

During the 1997–1998 and 2000–2001 study periods, the blooms in the upper Amelia system became obvious. There were general increases in annual average phytoplankton abundance during the last sampling period. The discrepancy of phytoplankton species richness during 1994–1995 was enhanced during the 1997–1998 period. These differences were not obvious during the 2000–2001 period. Relatively low species richness in the upper Amelia system corresponded to the relatively high numbers of phytoplankton during the last two sampling periods. Seasonal changes in phytoplankton abundance during the three study periods indicated the disparities of this factor between the Amelia and Nassau systems during the first two sampling periods. These differences were not obvious in the last sampling period; the winter–early spring blooms in both systems were dominant. These blooms were in the upper parts of both systems. The same pattern prevailed for the phytoplankton species richness distributions, with less disparity between the two systems shown during the 2000–2001 period relative to the previous sampling periods. The species richness was relatively low in areas affected by the blooms.

During 2000–2001, phytoplankton numbers were significantly higher at Stations R08 and R10 than at their Nassau reference sites (Table 7.8); no other significant differences in phytoplankton abundance were noted. Species richness was significantly higher at Stations R08 and R13 when compared to cognate Stations N03 and N08. Both species richness and species diversity were higher in the upper Amelia system than in the upper Nassau system. Comparison of phytoplankton indices at stations common to all three sampling periods indicated that phytoplankton assemblages in the Amelia system were diminished relative to the reference Nassau system during the first two sampling periods, but not during the last sampling period. These data indicate a recovery of phytoplankton assemblages that occurred during a period of reduced ammonia loading from the pulp mill and significant reductions of ammonia concentrations in the Amelia system.

7.5.4   Zooplankton Analyses

During the most recent study (2000–2001), zooplankton species numbers and richness were not significantly different between the two systems in areas affected by the pulp mill

Table 7.7 Statistical Comparison (independence tests, chi-square normality tests, variance tests, means tests, paired *t*-test, Wilcoxon signed rank test, and distribution shape tests) of Light Transmission Data (Kd, Euphotic Depths) Collected at Stations in the Nassau and Amelia River Estuaries Monthly for 12 Months from April 2000 through March 2001. Euphotic Depth and Kd Are Given for the Overall Spectrum and at Specific Wavelengths (430, 550, 665 nm)

Cognate Station Pair	Kd	Kd-430	Kd-550	Kd-665	PFD-Eup. Dpth. (m)	430-Eup. Dpth. (m)	550-Eup. Dpth. (m)	665-Eup. Dpth. (m)
R01/N04	**2.092/1.634** WS (P < 0.01)	**5.430/4.106** WS (P < 0.02)	**2.274/1.626** WS (P < 0.02)	**1.935/1.534** WS (P < 0.02)	**2.349/2.898** WS (P < 0.02)	**0.916/1.25** WS (P < 0.02)	**2.260/2.957** WS (P < 0.02)	**2.543/3.069** WS (P < 0.02)
R03/N04	**2.306/1.767** WS (P < 0.05)	**5.27/4.106** WS (P < 0.02)	**2.350/1.626** WS (P < 0.01)	**2.088/1.534** WS (P < 0.02)	**2.235/2.898** WS (P < 0.02)	**0.967/1.245** WS (P < 0.05)	**2.115/2.957** WS (P < 0.01)	**2.430/3.069** WS (P < 0.05)
R04/N02	**2.332/1.767** WS (P < 0.01)	**6.188/4.615** WS (P < 0.01)	**2.426/1.770** WS (P < 0.01)	**2.094/1.655** WS (P < 0.02)	**2.078/2.699** WS (P < 0.01)	**0.798/1.074** WS (P < 0.01)	**2.028/2.741** WS (P < 0.01)	**2.313/2.891** WS (P < 0.01)
R08/N03	1.729/1.699 WN	3.486/4.051 WN	1.753/2.085 WN	1.570/1.829 WN	3.073/2.747 WN	1.476/1.251 WN	3.367/2.926 WN	3.471/2.931 WN
R10/N01	**2.042/1.605** WS (P < 0.05)	**3.940/2.514** WS (P < 0.01)	1.855/1.473 WN	1.757/1.564 WN	2.458/3.077 WN	1.295/1.550 WN	2.705/3.451 WN	2.780/3.115 WN
R11/N05	1.888/1.718 WN	4.503/3.518 WN	1.834/1.709 WN	1.747/1.610 WN	2.797/3.101 WN	1.237/1.585 WN	3.055/3.311 WN	2.956/2.235 WN
R11/N06	1.888/1.805 WN	4.503/4.554 WN	1.834/1.846 WN	1.747/1.682 WN	2.707/2.744 WN	1.237/1.158 WN	3.055/2.940 WN	2.956/2.897 WN
R12/N07	2.047/2.348 WN	5.118/4.908 WN	2.096/2.514 WN	2.086/2.222 WN	2.395/2.119 WN	0.955/10028 WN	2.369/2.120 WN	2.689/2.235 WN
R13/N08	2.514/2.079 WN	5.263/4.597 WN	2.695/2.283 WN	2.396/1.933 WN	2.063/2.341 WN	**0.893/1.074** WS (P < 0.02)	1.938/2.263 WN	2.225/2.515 WN

WN, not sig.
WS, sig.
CV = RHO > Critical Value.
Boldface = bloom species.

Table 7.8 Statistical Comparison (means tests: paired *t*-test, Wilcoxon signed rank test) of Whole Water Phytoplankton Data (numbers L⁻¹, species richness, Shannon diversity and Shannon evenness) Collected at Stations in the Nassau and Amelia River Estuaries Monthly during the 1994–1995, 1997–1998, and 2000–2001 Sampling Periods

Cognate Station Pair	WW PHYT (Numbers L⁻¹)	WW Species Richness	WW Shannon Diversity	WW Shannon Evenness
	WW Phytoplankton Comparison: 4/2000–3/2001			
R01/N04	957,917/697,023 WN	20.7/18.0 WN	1.78/1.78 WN	0.60/0.63 WN
R03/N04	857,917/1,020,923 WN	20.6/20.3 WN	1.79/1.87 WN	0.60/0.65 WN
R04/N02	601,588/741,674 WN	23.2/17.0 WN	1.80/1.73 WN	0.60/0.62 WN
R08/N03	**921,885/447,310 WS (P < 0.01)**	**27.1/21.8 WS (P < 0.02)**	2.04/1.94 WN	0.62/0.64 WN
R10/N01	**681,610/361,193 WS (P < 0.01)**	26.4/23.8 WN	2.06/2.05 WN	0.63/0.67 WN
R11/N05	694,165/781,122 WN	23.3/20.4 WN	1.91/2.03 WN	0.61/0.68 WN
R11/N06	974,712/781,122 WN	21.0/20.4 WN	1.9/2.03 WN	0.63/0.68 WN
R12/N07	353,308/447,595 WN	20.5/21.8 WN	2.02/2.03 WN	0.67/0.68 WN
R13/N08	561,542/1,226,242 WN	**22.5/16.4 WS (P < 0.01)**	**2.07/1.87 WS (P < 0.05)**	0.68/0.68 WN

WN, not sig.
WS, sig.
Boldface = bloom species.

(Table 7.9). The only significant decrease of zooplankton species richness in the Amelia system was noted at Station R04 (Jackson Creek) that is generally not associated with the mill outfall. There was a significant increase in zooplankton numbers in the upper Amelia system (Station R13).

7.5.5 Multivariate Statistical Analyses

A PCA/regression analysis was run with the database taken during 2000–2001 (Livingston, 2001a). Analyses of phytoplankton and zooplankton data taken during the entire year indicated no significant associations with ammonia. Rather, phytoplankton numbers and species richness were negatively associated with orthophosphate, TP, silica, color, Secchi depths, and temperature. There were positive associations of these factors with salinity and oxygen anomaly. These results are consistent with the reduction of the importance of ammonia as a determinant of phytoplankton numerical abundance and species richness.

Table 7.9 Statistical Comparison (means tests: paired *t*-test, Wilcoxon signed rank test) of Zooplankton Data (numbers L^{-1}, species richness, Shannon diversity and Shannon evenness) Collected at Stations in the Nassau and Amelia River Estuaries Monthly during the 2000–2001 Sampling Period

	Zooplankton Comparison: 2000–2001			
Cognate Station Pair	ZPL (Numbers m^{-3})	ZPL Species Richness	ZPL Shannon Diversity	ZPL Shannon Evenness
R01/N04	1374/2598 nd	17.9/19.8 nd	1.34/1.42 nd	0.47/0.48 nd
R03/N04	1108/2598 nd	18.8/19.8 nd	1.51/1.42 nd	0.52/0.48 nd
R04/N02	481/944 nd	13.4/19.7 WS	0.59/0.52 nd	0.56/0.51 nd
R08/N03	1418/1494 nd	17.2/20.5 nd	1.30/1.54 nd	0.45/0.52 nd
R10/N01	1611/2741 nd	17.3/20.2 nd	1.33/1.51 nd	0.46/0.51 nd
R11/N05	1341/2837 nd	19.9/18.8 nd	1.47/1.37 nd	0.50/0.48 nd
R11/N06	1341/2456 nd	18.9/20.3 nd	1.46/1.33 nd	0.50/0.44 nd
R12/N07	1138/2109 nd	18.2/18.2 nd	1.54/1.30 nd	0.49/0.45 nd
R13/N08	972/2986 WS	15.6/17.6 nd	1.21/1.24 nd	0.45/0.43 nd

Note: nd = no data.
WS = sig.

It was also an indication of the adverse impact of upland urban runoff factors and watercolor.

Zooplankton species richness was unaffected by ammonia, negatively associated with DO anomalies, and positively associated with salinity and temperature. The negative association of zooplankton species richness with orthophosphate and silica is further indication of the negative effects of urban storm water. When the warm-water data were analyzed using the PCA/regression analysis, whole water phytoplankton numbers were inversely associated with color and the nitrogen compounds; there were positive associations with Secchi depths, salinity, temperature, turbidity, and TSS. Warm weather, whole water phytoplankton species richness was negatively associated with orthophosphate, TP, and silica; positive associations included turbidity, TSS, and oxygen anomaly. Warm weather zooplankton numbers and species richness were negatively associated with watercolor, ammonia, nitrate, and silica, and were positively associated with salinity, temperature, and Secchi depths. Thus, with the exception of some of the summer results, ammonia was generally less important as a determinant of phytoplankton and zooplankton community characteristics. Watercolor and various factors (including the possible associations of ammonia) were generally negatively associated with phytoplankton and zooplankton numbers and species richness, an indication of the growing importance of urban storm water on the Nassau and Amelia systems.

7.6 Discussion of the Amelia Program

The use of microalgae as indicators of water quality is well developed in the scientific literature and includes both qualitative and quantitative lines of research (Patrick 1967, 1971; Patrick and Reimer 1966; Maestrini et al., 1984a,b). There are numerous references to the use of microalgae as indicators of nutrient input and toxic compounds (Livingston, 2000, 2002). Previous information indicates that clean water supports speciose assemblages of microalgae, whereas polluted water tends to reduce species richness with increased relative dominance of a few resistant forms. Some microalgae flourish in highly eutrophic conditions; others are sensitive to different types of toxic effects. However, despite the efficacy of using phytoplankton as indicators, there is usually limited information concerning quantitative, species-specific distributions of phytoplankton (Livingston, 2000, 2002) in most coastal systems. Chlorophyll *a* is usually used as an indicator of phytoplankton activity.

Marshall (1982a,b, 1984, 1988) and Marshall and Ranasinghe (1989) have described phytoplankton distributions along the eastern coast of the United States, with numerical dominance by pico-nanoplankton, centric diatoms, and *Skeletonema costatum*. However, the responses of coastal microalgae to primary habitat variables are complex and not well understood (Malone, 1977; Malone et al., 1988). Water circulation, temperature, light, and nutrients, along with competition and predation, appear to be the dominant controlling factors in the phytoplankton associations of a given area. Extreme spatial/temporal variation often confounds simple explanations of the interrelationships of controlling factors. Natural variation complicates generalizations concerning specific anthropogenous effects on coastal phytoplankton.

Existing scientific literature indicates that state variables that control estuarine phytoplankton assemblages vary widely in space and time. Boynton et al. (1982), in a review of data from a series of estuaries, found that phytoplankton production was highest during warm periods. There were seasonal changes in the phytoplankton distributions in the Amelia and Nassau systems, with peaks usually occurring during summer months although there was also evidence of winter increases of phytoplankton abundance. Phytoplankton species richness tended to be higher during warm months. Zooplankton abundance peaks occurred during spring months, coinciding with declines of phytoplankton abundance; summer increases in phytoplankton numbers were correlated with relatively low zooplankton numbers. Nutrient limitation experiments (Livingston, unpublished data) indicated that nitrogen was the chief limiting nutrient to phytoplankton in the Nassau and Amelia River estuaries.

The fact that *Skeletonema costatum* was severely reduced in the Amelia system compared to the reference Nassau system during periods of high ammonia loading (1994–1995, 1997–1998) was qualified by winter blooms of this species in both systems. The association of bloom activity with high orthophosphate concentrations in the upper Amelia system indicted possible effects due to nutrient loading due to urban storm water runoff. These effects were indicated during previous surveys (Livingston, 1998b). Land use surveys in the region during 1995 indicated increased residential development in upper parts of Lanceford Creek and on areas of the Nassau system associated with Nassauville. There has also been extensive urban development on Amelia Island (Figure 7.1) that is undoubtedly affecting lower parts of the Nassau/Amelia system. The appearance of massive *Skeletonema* blooms in both systems during winter 2000–2001, together with dominance of known bloom-forming diatoms (*Leptocylindrus danicus*), indicated that urban storm water nutrient loading has entered a new phase that could be extremely damaging to the Amelia and Nassau systems. Livingston (2000, 2002) has shown how blooms associated with nutrient loading progresses in terms of interannual responses to cumulative nutrient effects that end in blooms by noxious and toxic plankton species that eventually destroy secondary production (i.e., invertebrates and fishes). Unless such non-point sources of nutrients are effectively reduced by known methods of treatment, the cumulative effects will eventually destroy what is still an extremely productive ecosystem.

The weight of field evidence indicated that high ammonia levels (range: 0.19 to 0.43 mg L^{-1}) in areas affected by mill discharges in the Amelia system were significantly associated with observed impairment of key indices of phytoplankton assemblages especially during summer months. The experimental data indicated that ammonia acts as a nutrient as well as a toxic substance with regard to phytoplankton productivity, and that ammonia stimulation and inhibition occur along a continuum with possible changes of inflection points under varying ecological conditions. This finding was consistent with the field data. The relatively broad range of inhibitory ammonia concentrations in the experimental data was consistent with the high variation of phytoplankton response to ammonia in the field. The complex reactions of phytoplankton assemblages to both the stimulatory and inhibitory effects of ammonia could account for the difficulty in locating the precise ammonia concentration levels that were inhibitory. The range of results of the single-species microcosm tests indicated intraspecific changes in response to ammonia. Light penetration was an obvious source of obfuscation of the effects of ammonia on the phytoplankton in the field. There were also indications of seasonal differences (i.e., water temperature) in the effects of ammonia, with increased ammonia concentrations during winter periods not having as pronounced adverse effects as those during warmer months. The results of the multivariate analyses of the field data confirmed that multiple factors determined phytoplankton distribution with seasonal differences in the weighting of such responses.

Overall, many factors influenced the phytoplankton associations in the Nassau and Amelia estuaries. The timing and intensity of ammonia loading to the Amelia system appeared to be associated with observed changes in phytoplankton production and community structure, especially during warm months of the year. However, the influence of periodic reductions in light penetration due to effluent loading could not be ruled out as a contributing factor, especially during winter periods of increased concentrations of color. Reduced zooplankton numbers in the Amelia system were probably related to changes in phytoplankton rather than direct effects of ammonia since zooplankton species richness did not appear to be affected. The presence of blooms in upper parts of the Amelia estuary complicated direct evaluations of the influence of ammonia as urban storm water effects were indicated both in terms of water quality and phytoplankton response. The combined field descriptive and laboratory approach facilitated a more precise definition of the limits concerning ammonia loading to the Amelia system, but such results should be evaluated within the context of the continuously changing habitat structure in the areas of study.

The experimental evidence indicated that other constituents in the mill effluent did not significantly influence demonstrated ammonia effects. Within the context of the noted ranges of impacts in both the field and laboratory results, the data suggested that long-term, average concentrations in the Amelia River estuary at the outfall station should not exceed 0.11 mg L^{-1} ammonia, with short-term increases not exceeding 0.20 mg L^{-1}. These target concentrations could be considered conservative with regard to the propagation of natural levels of phytoplankton in the Amelia River-estuary relative to those observed in the reference area. Based on these projections, a restoration effort was initiated that was consistent with noted levels of ammonia and phytoplankton assemblages in the reference Nassau River estuary.

Based on this evidence, the Fernandina Beach mill reduced ammonia loading to the Amelia system to levels that resulted in ammonia concentrations that remained largely within the proposed guidelines, and these concentrations did not differ significantly from ammonia concentrations in the reference system. Another year of field studies indicated that there were no significant differences in phytoplankton/zooplankton numbers and species richness. However, these results were complicated by several factors. Although not statistically significant, there were some reductions in zooplankton numbers in the Amelia system. In addition, during winter 2001, a series of *Skeletonema costatum* blooms in both systems occurred that appeared to be related to nutrient loading (probably ortho-phosphate) in the upper parts of both systems. These blooms qualified the results in that the reference site could no longer be considered a natural system unaffected by human activities. The blooms in the upper Amelia system also qualified the results in that system. In addition, there were enhanced reductions in light penetration in the Amelia areas most affected by mill effluents. Although such effects were qualified by the general increase in light penetration in both systems due to prolonged drought conditions, direct or indirect effects of reduced light in the lower Amelia system could not be entirely discounted as a factor in the plankton distributions in the Amelia system during the 2000–2001 period. When the phytoplankton species richness factor was taken into consideration, the overall weight of the scientific evidence indicated recovery of the Amelia phytoplankton assemblages as a response to the reduction in ammonia loading by the Fernandina Beach mill.

Several factors combined during the 2000–2001 survey to complicate an interpretation of the phytoplankton data. This study was undertaken during a severe drought that created conditions (higher salinity, increased clarity of the water and increased euphotic depths, and reduced overall nutrient loading to the system). The reduction of ammonia by the mill to near-natural conditions should thus be evaluated within the context of the drought-induced changes in habitat conditions. Phytoplankton species richness during 2000–2001 was generally lower in both systems than in the previous survey, but there were no significant differences in species numbers between the Amelia and Nassau systems. It is likely that this finding was related to the reduced toxicity of ammonia in the Amelia River estuary due to reduced mill loading during 2000–2001. However, there were general increases in phytoplankton numbers in both systems compared to previous surveys. These increases were related to the appearance of the massive *Skeletonema costatum* blooms in both systems during winter–early spring 2001 in areas characterized by increased orthophosphate concentrations, the result of urban storm water runoff in newly urbanized areas of both basins. These blooms should be placed within the context of what is known concerning the origin, seasonal and interannual succession, and impact of such blooms in other coastal systems.

## chapter 8

Research, News Reports, and Restoration Success

8.1 Restoration Processes and Public Opinion

It should be stated from the beginning that there is a long and distinguished history of environmental reporting at the regional, state, and national levels. The list of excellent environmental reporters is too long to review here. Even small newspapers have, in the past, supported environmental reporters of high caliber. That is not to say that the American news media have always covered themselves in glory concerning environmental issues. Newspaper articles that followed a pattern of distortion and omission not dissimilar to that of today's reporting contributed to the genocide of Native Americans in past years. Even today, the difficulties of this oppressed minority are still largely ignored, and even distorted by media reporters. However, the present situation concerning environmental reporting is a product of complex, new trends both at the economic and cultural levels of American society. The takeover of public radio and television airwaves by large corporations has had an important influence on what is reported to the American public. Big business in the form of Disney (ABC), Viacom (CBS), and General Electric (NBC) has literally taken over television reporting. The right-wing empire of Rupert Murdoch (Fox) continues to distort everything environmental. The cable news channels (CNN, MSNBC) are often primary outlets for right-wing, anti-environmental propaganda. The overwhelmingly negative publicity concerning environmental issues generated by a broad range of right-wing radio stations is a disturbing fact of life in today's America. There has been an increasing trend of business interests taking control of newspapers throughout the country. It is not, therefore, very surprising that the regional news media in north Florida continue to use omission, distortion, and outright lying to protect their business and political interests.

There is no simple way to define the relationships among the various societal forces that control public opinion and affect the complex process of ecosystem restoration. These forces include political and economic interests, the news media, environmental organizations (from NIMBY groups to organized environmental societies), educational and research institutions, and legal groups. There has been considerable debate concerning the interaction between the news media and the public. There is little doubt that the media dominate public debate concerning environmental issues with advocates on both sides of any given issue. To be sure, there is a synergistic relationship concerning the interactions of news reporting and public opinion. Both media and public opinion are inextricably linked with continuous feedback that complicates simplistic generalizations concerning control. However, in the final analysis, the news media have a direct impact on public opinion, and this impact drives related factors such as political decisions and the generation of laws and regulations that control environmental matters. There are cultural aspects to the public's perception of environmental issues that color the media's approach to environmental reporting. A mere recitation of these interactions does not, however, explain the inexorable deterioration of public knowledge and interest in things environmental since the peak of such interest in the 1970s.

8.2   Florida as a Microcosm for Restoration Activities

During the past 30 to 40 years, the record of destruction of Florida's aquatic resources has been clear regarding the debilitation or even loss of many of the primary waterways and river–estuarine systems. A list of damaged and destroyed systems includes most of the state's major lakes, rivers, and bays: the Kissimmee–Okeechobee–Florida Everglades– Florida Bay–Florida Keys/coral reef system, the Miami River–Biscayne Bay area, the Indian River system, the St. Johns drainage system, Naples Bay, Sarasota Bay, the Hillsborough River–Tampa Bay area, the Fenholloway River, Choctawhatchee Bay, the Pensacola Bay system, and the Perdido Bay system. A comparison of the scientific data concerning such areas with news coverage and the public's perception of environmental matters is not encouraging. The generally rosy reporting of restoration activities is hardly consistent with what is actually happening. The gap between public knowledge of the key environmental issues and what is scientifically demonstrated continues to widen. Public ignorance and indifference to environmental matters contribute to the scandalous level of posturing by a political establishment that is anything but environmentally oriented.

The situation in north Florida is probably representative of what is going on at the state and national levels. There are instances of the appropriate use of scientific studies in the initiation of successful restoration programs. The problem of ammonia loading in the Amelia Estuary was solved with the direct application of the results of an impact analysis and the cooperative efforts of the pulp mill and the FDEP. This success was tempered by the discovery of urban storm water impacts on both the Amelia and Nassau estuaries, which have little chance of successful remediation. Pollution caused by urbanization is not usually covered by the news media, and is not regulated by local, state, or federal agencies. After all, the development forces that control the politics and regional media are not about to pay for such "progress." The north Florida Lake situation represents the deliberate obfuscation of scientific data by regional news outlets that are literally run by dominant political and economic interests. The situation in Leon County, Florida, is a clear example of the deliberate omission and misrepresentation of scientific facts when confronted with a propaganda machine that is led by the regional Chamber of Commerce approach to environmental reporting. The public remains protected from the unpleasant reality of resource destruction in a totally managed news situation. This control includes the firing and/or personal attacks of anyone who gets in the way.

In Perdido Bay, there was partial success of the restoration of a polluted system with the reduction of nutrient loading from the pulp mill. However, nutrients remain an issue with various sources that include a pulp mill, a sewage treatment plant, agricultural loading, and urban runoff. The need for an ecosystem approach to restoration activities is exemplified in the Perdido and Pensacola drainage systems. The association of a regional news media that is in an unhealthy synergism with NIMBY operatives who are indifferent to and/or ignorant of scientific research results has culminated in the public's preoccupation with false issues while their children drink radioactive water. The deterioration of the Perdido and Pensacola Bay systems is a classic example of problems associated with the influence of local interests on the regional press, and the almost complete indifference of the public to cumulative impacts of the various forms of urbanization and industrialization. Throughout the north Florida area, from Apalachee Bay to the Perdido system, the public remains in a state of almost complete ignorance of what is happening.

The role of the press in environmental matters cannot be underestimated — both as a positive force that brings public attention to the need for restoration and as a negative force that often leads to public ignorance and confusion concerning environmental problems. As noted above, this can occur in various ways: deliberate omission of scientific information, transmission of misinformation and outright lies, and the unspoken protection of economic interests that dominate the political atmosphere. In some cases, coverage of complex environmental issues by people who are not trained in the scientific facts that are crucial to restoration attempts plays an important part in the imperfect process of information transfer. The rise of the NIMBY phenomenon in recent years that has replaced the more responsible environmental interests that were once dominant has contributed to the gap between the perception of environmental progress and the actual success of the growing restoration movement. Parochial concerns rarely address system-wide environmental problems, and the constant erroneous chatter from people who eschew scientific research constitutes a major distraction, especially when the press uses irrelevant proclamations by NIMBY groups to sell news through the creation of controversy. When the actual scientific data are lined up, what is reported by the press — the general failure of the restoration process — is not surprising. The proliferation of simplistic solutions to inherently complex environmental processes has contributed to the current and continuing situation of lost natural resources.

There is another problem associated with the replacement of responsible action with the false enhancement of public image. The restoration movement has, in some ways, replaced the practice of effective preventive action through resource planning and attention to the infrastructure costs associated with the protection of natural resources. According to J.B. Battle and M.I. Lipeles (*Water Pollution*, 1998, Anderson Publishing Co.),

> "America has made essentially no progress in addressing polluted runoff over the past quarter-century, ... there is compelling data indicating that non-point pollution has gotten worse rather than better over the past 25 years. ...polluted runoff is the largest reason that roughly 40% of America's surveyed waters are too polluted for basic uses..."

The complex interactions that have contributed to this situation are often related to the political control of environmental research at all levels of government. The real authors of the lack of enforcement of the water quality laws are often elected officials who keep a tight grip on the budgets of state and federal regulatory agencies. Their actions are largely kept quiet by collusion of many elements of the news media that contribute to the proliferation of ignorance and confusion regarding ecological processes in aquatic systems. Consequently, the American public remains ignorant of the most damaging environmental problems in favor of more sensational but less important issues made popular by elements of the news media.

8.3 Comparison of Research Results, Media Coverage, and Public Response

A review of the comparison of what is known scientifically vs. what is reported and assimilated by the public follows:

1. Research concerning the impacts of urbanization on north Florida lakes indicated that periodic increases in polluted runoff led to increased nutrient loading, which resulted in phytoplankton blooms. The long-term changes associated with such blooms included habitat deterioration, adverse impacts on associated biota, and the debilitation of valuable species such as largemouth bass due to food web response to the blooms. The data indicated that most of the lakes in the region were adversely affected by urban runoff, and that so-called restoration attempts that relied mainly on holding ponds (mainly holes in the ground) rather than combinations of retention and detention ponds were not adequate to protect natural lakes.

    a. There was a change in the reporting of lake issues as the regional newspaper was taken over by a conservative publisher who was closely aligned with development interests. The news accounts went from accurate reporting of the above impacts to omission of any scientific accounts showing adverse impacts.

    b. News media of the region have resorted to propaganda that distorted and even promoted lies about what was happening to the lakes in the region. The media blackout was extended to even the publication of letters to the editor so that any objective scientific results concerning water quality issues were kept from the public.

    c. Recent analyses indicated that the lakes continue to deteriorate, and that anyone who gets in the way of development interests will be fired or otherwise discredited.

    d. Research funds for objective scientific efforts have been withdrawn if results showing adverse impacts are made public.

    e. The control of information and the proliferation of open propaganda by self-serving political interests concerning sensitive environmental issues have replaced objective media reports, and the so-called open process of information dissemination remains compromised.

    f. There has been virtually no regulation of urban runoff by local, state, or federal agencies.

2. Long-term research concerning a pulp mill on the Fenholloway River in the Apalachee Bay basin indicated that the river has been severely affected by low dissolved oxygen (DO), high levels of nutrients such as ammonia, increased watercolor, and high levels of specific conductance. Comparisons of long-term efforts to remediate such impacts indicate that the situation has not changed substantially since the early 1970s, and the river remains polluted. Dioxin levels in fishes meet federal requirements, and changes in secondary sex characteristics of certain top-minnow species have not been associated with adverse impacts in the field. The primary impacts on associated offshore areas include the loss of seagrass beds due to high watercolor levels, and the periodic increase in plankton blooms due to nutrient loading (ammonia, orthophosphate) from the pulp mill. There is no evidence of the impact of toxic substances (metals, organochlorine compounds) on offshore sediments.

    a. Media coverage concerning the problems associated with the Fenholloway River has concentrated largely on the dioxin issue and changes in the secondary sex characteristics of mosquitofish. The emphasis has been on sensationalized accounts of the impacts of the mill that are largely without scientific confirmation.

    b. The real causes of the environmental impacts of the mill on the Fenholloway River and associated parts of Apalachee Bay have been largely ignored by regional news media.

c. Recent efforts to reduce watercolor have not been effective in the restoration of offshore seagrass beds and the associated invertebrate and fish biota that have been adversely affected by the loss of the benthic plants. Both sides of the issue have distorted media accounts of such impacts, and the pulp mill, "environmentalists," and state and federal regulatory agencies have largely ignored the scientific data.

d. Regulatory actions by state and federal agencies remain inadequate, and attempts to save the river using a pipeline have been blocked by "environmentalists" and the U.S. Environmental Protection Agency (EPA) in an effort that precludes the use of scientific information to resolve the problem.

e. Media tactics that include omission, misinformation, and sensationalism have resulted in a lack of public knowledge concerning the pollution of the Fenholloway system, and the resulting misdirection of attention to the actual impacts of the paper mill have contributed to the lack of effective restoration of the damaged system.

3. A long-term scientific database concerning the response of the Perdido drainage system to impacts due to pulp mill effluents indicated that the immediate receiving system (Elevenmile Creek) has been adversely affected by such discharges. Although there are similarities with impacts on the Fenholloway River such as the general lack of assimilative capacity for high-volume effluents, there were differences. The successful treatment of color impacts in Elevenmile Creek have led to adverse impacts associated with blue-green algae blooms during drought periods. The U.S. Army Corps of Engineers stratified Perdido Bay due to the opening of the bay to the Gulf. Resulting salinity stratification exacerbated hypoxia at depth in major parts of the bay. Nutrient loading by the pulp mill was associated with plankton blooms, which led to adverse impacts on bay food webs and the decline of various fish and invertebrate species. Rainfall events were associated with nutrient loading from a sewage treatment plant that seriously damaged the bay. Similar impacts were noted in the lower bay due to nutrient loading from agricultural interests in Alabama and urban runoff from development. By 2003, the Perdido Bay system was largely bereft of fishes and invertebrates due to the combined impacts of the various activities that loaded nutrients to the bay.

a. Like the Fenholloway River, the emphasis of media reports has been on sensationalized accounts of the impacts of dioxin and hormone changes causing changes in the secondary sex characteristics of topminnows. These claims, although discredited in reports by state and federal agencies that include findings of no dioxin levels in fishes that exceed federal standards, continue to dominate media coverage of the Perdido system.

b. "Environmentalists" who claim the real impacts are due to chlorinated compounds such as dioxin have denied impacts on Perdido Bay due to plankton blooms associated with mill nutrient loading.

c. The *Pensacola News Herald*, which ran a whitewash of such impacts on the bay, deliberately distorted findings concerning the impacts of a sewage treatment plant on Perdido Bay run by the Escambia County Utilities Authority (ECUA). The Associated Press broadcast this false information even though the impacts were documented in a public research report of the findings. There was no regulatory response to the report.

d. Media tactics that include omission, misinformation, and sensationalism have resulted in a lack of public knowledge concerning the pollution of the Perdido system, and the resulting misdirection of attention to the actual impacts of various pollution sources has contributed to the lack of effective restoration of the damaged Perdido system.

4. A study of the Pensacola Bay system indicated that the Escambia estuary was so polluted by point source pollution (toxic wastes), releases of sewage, and the impacts of polluted urban storm water that the effluents of the pulp mill on Eleven-mile Creek in the Perdido system should not be diverted to the Escambia River.

   a. The *Pensacola News Herald* deliberately misrepresented the scientific effort to analyze the Pensacola Bay system.

   b. The news media in the Pensacola region used misrepresentation and omission to make sure that the public was not informed concerning the state of toxic waste loading into the Pensacola Bay system.

   c. Recently (May 2004), a Grand Jury found that groundwater contamination was widespread in Escambia County, with more than half the county's drinking water wells contaminated. Sources of the contamination include six Superfund sites, dozens of dry-cleaning sites, hundreds of petroleum storage sites and numerous abandoned landfills. The Grand Jury also found that the U.S. EPA, the FDEP, and the ECUA failed to do all they could to prevent or remediate groundwater contamination.

   d. The Grand Jury also found that ECUA staff knew for several years that drinking water from some wells in the ECUA system was contaminated with radium and other harmful substances but did not tell the ECUA board or the public until 1998.

   e. The Grand Jury also found that the ECUA publicly stated that federal radium standards were likely to be loosened and that the heightened level of radium in some public wells created no health risk. The ECUA knew that neither statement was accurate.

   f. The Grand Jury also found that the ECUA did not inform customers using water from radium-tainted wells to seek an alternative source of supply, as directed by the FDEP.

   g. The Grand Jury also found that, at heavily contaminated sites, the costs of contamination have been shifted to the public because the polluting companies closed their businesses and abandoned their properties.

   h. Media tactics that include omission, misinformation, and sensationalism have resulted in a lack of public knowledge concerning the pollution of the Pensacola system, and the resulting misdirection of attention to the actual impacts of various pollution sources have contributed to the lack of effective restoration of the damaged Pensacola system.

8.4 Summary of Recent Trends in North Florida

To summarize, the experience of environmental reporting over the past 35 years has gone from excellent and objective coverage in the 1970s and early 1980s, to a drift toward obfuscation and open camouflage of environmental damage during the 1990s that has not improved to the present day. In areas where there has been extensive and detailed scientific data concerning impacts due to various human activities, the data have been largely ignored by media interests in favor of sensationalism generated by inflated and erroneous (often inflammatory) claims by "environmentalists." Real problems are largely ignored when they involve politically powerful interests such as urban developers and agricultural groups. When the scientific data are reported, the media often treat such information in a way that inflates controversy and obfuscates understanding of the real pollution issues. Unsubstantiated claims by all sides of a given issue are treated as equal by media reports.

There is virtually no effort to substantiate claims by the various sides of a given environmental controversy. The usual tools of investigative reporting are ignored in favor of alternating sensationalism and obfuscation of actual scientific facts. The public remains titillated but largely ignorant of the real issues in terms of impact and remediation.

The problems outlined above are not restricted to north Florida — they are worldwide and afflict some the most prominent aquatic systems in terms of increasing water pollution and the loss of useful productivity.

section III

Major Restoration Programs

In recent years, there have been major efforts to restore ecologically important aquatic areas at the ecosystem level. These efforts have received considerable publicity and are among the most well-funded restoration projects ever attempted. The Chesapeake Bay system is one of the largest water bodies in North America. At one time, the Chesapeake was one of the major producers of seafood in the world. However, with an increase in human population in recent times, there has been a serious reduction in seafood production. Various forms of water pollution have recently become obvious to the public.

Likewise, the Florida Everglades ecosystem, located in south Florida, represents an important part of one of the most extensive and unique ecological resources of the United States. This integrated system, which includes the Kissimmee River, Lake Okeechobee, Florida Bay, and the hermatypic coral reefs along the Florida Keys, has been seriously damaged by various human activities over the past century. A multibillion-dollar restoration program has been established to rehabilitate the Everglades system.

Both projects have been supported by extensive press coverage as examples of our progressive concern for the environment. Politicians who avidly opposed environmental concerns in the past have been lionized by the press as $billions have been projected for "restoration." In the process, the science has been muffled as the political aspects have been emphasized. The blatant ignorance and greed that led to the destruction of these resources have been covered up so that those who were responsible for the damage escape notice. The questions remain: What is that nature of the problems in these formerly productive systems, and is there reason for hope of their recovery?

chapter 9

The Chesapeake Bay System

9.1 A Declining Resource

The Chesapeake Bay system (over 41 million acres and includes 18 trillion gallons of water) is one of the largest water bodies in North America. It is composed of a series of major rivers that flow into a major estuary. From 1950 to 2000, the human population in the Chesapeake Basin increased from around 8 million to almost 16 million (Ernst, 2003). Historically, the Chesapeake system was the premier producer of oysters, blue crabs, and finfishes in the northern hemisphere. The Chesapeake Bay also represents one of the most studied estuaries in the world with long-term, high-level funding for an extensive series of research projects.

In recent decades, however, there have been unprecedented declines in fisheries production in what was this once-rich estuarine system. The history of this decline has been well documented. The Chesapeake was one of the primary sources of oysters (*Crassostrea virginica*) in the world. The loss of oyster production in Chesapeake Bay in recent years has been attributed mainly to disease, which has also "devastated" oyster production areas of Delaware Bay. In addition to the loss of the oysters, seagrass beds have been reduced or completely eliminated in major parts of the system. The outlook for near-term SAV recovery in the mesohaline parts of the Patuxent estuary was considered "unlikely" (Stankelis et al., 2003). It has been reported that the high density of phytoplankton in Chesapeake Bay has caused increased hypoxia at depth. Various forms of point and non-point source pollution have been associated with a rapid decline in the famous Chesapeake blue crab industry. Human population growth and development in the Chesapeake drainage basin have also been considered a threat to many fish species such as red drum, bluefish, and tautog. These species have succumbed to both over-fishing and pollution. Various other fish populations continue to fluctuate widely.

Coverage of bay seagrasses in Chesapeake Bay in recent times approximates 10% of what has been estimated to be the historical potential. Expanses of unspoiled seagrasses have been estimated at over 600,000 acres. In recent times, it has been estimated that there were close to 200,000 acres of SAV along the shoreline of Chesapeake Bay (Chesapeake Bay Program, 2003). However, by 1984, this number had dwindled to 38,000 acres. Between 1960–1991 in the Patuxent River Estuary, both water clarity and SAV declined (Boynton, 1997). During 1985-1986, there was a 4.8-fold increase in nitrogen loading and a 19.5-fold increase in phosphorus loading compared to pre-colonial periods (Boynton et al., 1995). D'Elia et al. (2003) noted that Patuxent River nutrient control measures have not yet resulted in enhanced water quality. This decline was attributed (Chesapeake Bay Program, 2003) to reduced light penetration, interference of light by epiphytes, high sedimentation from land runoff, and nutrient-induced algal blooms.

Current trends in Chesapeake Bay fisheries populations are not encouraging. Blue crab (*Callinectes sapidus*) abundance "is approaching a record low and has been declining in recent years" (Chesapeake Bay Program, 2003). There is now evidence that the Chesapeake blue crab population could be decimated by further stress in the form of storms or increased pollution loading to the bay. Oyster (*Crassostrea virginica*) stocks are currently less than 1% of former levels due to a combination of factors that include over-fishing, habitat destruction from different human activities, parasites, predators, water quality deterioration and associated algal blooms, toxic wastes, and siltation from human activities in upland areas. American shad have been virtually extirpated since the 1970s. In short, the loss of fisheries in the Chesapeake Bay system in recent decades has been calamitous.

9.2   Research Results

D'Elia et al. (2003) reviewed the history of the relationship of science and public awareness of the condition of the Chesapeake Bay system. Prior to the 1960s, fish and shellfish harvests were characterized as a "limitless bounty." Trophic organization underwent major changes during the 1970s. The serious declines of water quality noted during the 1980s were associated with the decline in the oyster industry and the near-elimination of seagrass beds. The nitrogen problem was examined amid relatively slow recognition of the problems by federal regulatory agencies. By the 1990s, N+P budgets were created based on a 40-year record of nutrient inputs from both point and non-point sources for the Patuxent River-estuary. There have been detailed analyses of the sources of nutrients in the Patuxent River basin (Jordan et al., 2003). Nutrient loading from urbanized areas thus was noted as a major source of the water quality problems in the Chesapeake system.

There is a direct relationship between the number of people in the Chesapeake basin and the loading of nutrients to the bay. However, the needs of the Chesapeake system have come up against problems associated with the restriction of human migration into the basin. The possibility of the need for more nutrient controls leading to actual restrictions on net immigration to the basin presented a significant problem based on current freedoms associated with development and land ownership. This controversial end game where population limitation for restoration purposes came up against traditional civil rights represents a universal problem in many ecosystem-level restoration programs.

9.2.1   Hypoxia

Smith et al. (1982) found that deep-water hypoxic conditions in Chesapeake Bay resulted from the cumulative effects of biochemical mechanisms that were modified by physical stratification. Nutrient loading from myriad sources enhances phytoplankton productivity, which in turn affects various factors such as dissolved and particulate organic matter. Suspended and sedimentary organic matter in Chesapeake Bay is probably derived from autochthonous sources that include fresh and detrital phytoplankton, zooplankton, and bacteria. The dominant factor that determines temporal variation of the sedimentary pool is phytoplankton productivity. Enrichments of particulate organic carbon (POC), chlorophyll *a*, total fatty acids, total sterols, and various biomarkers specific to phytoplankton have been found at the surface and the bottom of the Chesapeake estuary. Thus, particulate organic matter produced by nutrient loading in the Chesapeake system is the result of various processes that impinge on sediment quality and food supplies for benthic organisms. The dissolved oxygen (DO) regime in the Chesapeake system is directly dependent on the processes involved in the translation of nutrients into the various forms of dissolved and particulate organic matter.

The causes of reduced DO are many, and decisive cause-and-effect relationships can be difficult to ascertain. Hypereutrophication has long been correlated with increased biochemical oxygen demand (BOD), hypoxia, and anoxia in various estuaries. Chronic hypoxia has been a problem in coastal areas such as the Florida Keys, the Pamlico River–estuary, Long Island Sound, and broad areas of the northern Gulf of Mexico associated with runoff from the Mississippi and Atchafalaya Rivers. Hypoxia is also considered a problem in estuaries and bays throughout Europe, the Far East, and Australia (Kennish, 1997). However, long-term reductions in bottom DO in western Long Island Sound have been attributed to changes in vertical temperature stratification rather than changes in point and non-point nutrient loading (O'Shea and Brosnon, 2000). The lower DO concentrations occurred after decades of reduced BOD loading to the system from sewage treatment plants. The DO trends were not associated with changes in Secchi transparency and chlorophyll *a*; the exact causes of the DO trends remained unclear. Low bottom DO was associated with high temperature and salinity stratification in the Perdido Bay system and Choctawhatchee Bay in north Florida (Livingston, 2000).

Cultural eutrophication (nutrient excess leading to overproduction of microalgae and associated trophic imbalances) is common in estuaries near human population centers (Livingston, 1987a, 1997c). This condition is characterized by exaggerated fluctuations of DO from super-saturation during the day to hypoxia and anoxia at night (Livingston, 1997a). The eutrophication process involves complex trophic interactions that often are system specific (Livingston, 1997c). Hypereutrophication can lead to periodic hypoxia and anoxia (Livingston, 1997c). According to Rabalais et al. (1996, 1999), there is a "dead zone" of over 9500 km^2 in the Gulf of Mexico due to hypoxia that has been connected to nutrient loading from the Mississippi River. However, DO levels in coastal areas can be subject to considerable natural variability, largely because of fluctuations in temperature, salinity, basin stratigraphy, productivity, and associated biological conditions. Thus, periodic hypoxia in Gulf estuaries often reflects natural conditions (Turner et al., 1987; Seliger and Boggs, 1988). Episodes of naturally low DO ("jubilees") occur in Gulf coastal areas (Loesch, 1960). There is little doubt, however, that hypoxia in the Chesapeake systems is directly linked to nutrient loading and associated changes in the phytoplankton assemblages (Chesapeake Bay Program, 2003).

DO is considered an important limiting factor in inshore marine systems (Santos and Bloom, 1980; Santos and Simon, 1980a,b). Various field studies have been carried out to evaluate the impact of hypoxia and anoxia on estuarine populations and communities. Santos and Bloom (1980) and Santos and Simon (1980a,b) found that annual defaunation in Hillsborough Bay (Tampa Bay system) was due to hypoxia. A stochastic recolonization response of the soft-bottom macroinvertebrate community was demonstrated following the recurrent defaunation event. Pearson (1980) and Pearson and Rosenberg (1978) published detailed studies on the effects of organic enrichment and low DO on marine systems. Van Es et al. (1980) demonstrated a direct response of meio- and micro fauna in tidal flats along distinct gradients of organic enrichment and oxygen saturation. Lenihan and Peterson (1998) showed that salinity stratification, depth considerations, and periodic hypoxia/anoxia combined to affect oyster mortality and associated fishes and invertebrates. Interactive factors were operational in the specific patterns of the response to different forms of disturbance. Based on the complexity of the interactive controlling factors, the authors indicated the need for more integrative approaches to ecosystem management. Keister et al. (2000) showed how near-bottom hypoxia in the Patuxent River (Chesapeake Bay) affected depth distribution of various organisms with effects on predator–prey relationships and recruitment rates of vulnerable species. Other effects of hypoxia on fish larvae include decreased growth rates and limitation of habitat availability.

Benthic communities below the pycnocline in mesohaline waters were the most degraded in the bay (Holland et al., 1977, 1980; Dauer et al., 1982, 1992). Dauer et al. (2000) used associations of macrobenthic communities as indices of anthropogenous effects in Chesapeake Bay. The authors found that the distribution of low DO was extensive and, based on regression analyses, explained 42% of the benthic index of biotic activity. Sediment contamination (i.e., toxic agents) accounted for 10% of the variation in this index. Residual variation of the condition index was only weakly associated with eutrophication indices (total nitrogen, total phosphorus, chlorophyll *a*) after removal of the effects of DO The benthic condition was negatively associated with urbanization, point source loadings, and total nitrogen loadings. These results were consistent with other studies that associated severely degraded benthic communities with very low oxygen concentrations in the Chesapeake system (Holland et al., 1977; Dauer et al., 1992; Diaz and Rosenberg, 1995). The naturally hypoxic condition due to stratification (Malone et al., 1988) was mentioned as a possible cause in addition to high rates of deposition of particulates to the benthos (Kemp and Boynton, 1992). These effects were attributed to organic matter passed to the stratified mesohaline parts of the bay from oligohaline areas (which were not part of the study).

9.2.2   Phytoplankton

Phytoplankton indicators such as chlorophyll, primary production rates, biomass, and species composition support the hypothesis of a connection of the deterioration of the bay to anthropogenous nutrient loading. With the increase in human population in the Chesapeake basin, nutrient loading from point sources, urban runoff, and agricultural development in the basin has led to damaging algal blooms that have caused a range of impacts, which include reduced habitat for submerged aquatic vegetation (SAV), low DO, and other forms of habitat deterioration. Phosphorus-limited periods have been noted in Chesapeake Bay (Fisher et al., 1988, 1992); these periods occur during winter–spring when temperatures are low. The regulation of nutrient availability by physical flushing rates, mixing, geochemical equilibria reactions, and biological processes are thought to control the temporal successions of nutrient limitation of phytoplankton production (Pennock and Sharp, 1994).

Gilbert et al. (2001), in a description of harmful algal blooms in Chesapeake Bay, noted that nutrient input to the bay in the organic form has been increasing in the past decade, and that the availability of DOC and DOP (dissolved organic phosphorus) may provide a substrate for some bloom species. The authors found that the timing of the nutrient delivery may also be important in the success of some bloom species. According to Newell (1988), algal blooms have resulted, in part, from the loss of the oysters, which, in less polluted times, cropped most of the phytoplankton in a relatively short period of time. Sellner et al. (1995) indicated that *Prorocentrum minimum* did not adversely affect oysters (*Crassostrea virginica*) in Chesapeake Bay; oysters effectively reduced these bloom-forming dinoflagellates. However, Lassus and Berthome (1988) reported that *P. minimum* caused mortalities in old oysters. Woelke (1961) found that this species caused oyster (*Ostrea iurida*) mortalities and cessation of oyster feeding at high densities. Wikfors and Smolowitz (1993) found that increased abundance of *P. minimum* could cause shell losses.

Recent publications concerning Chesapeake Bay have led to a series of hypotheses as explanations for the decline of the habitats and fisheries of the system. Malone et al. (1996), in a review of nutrient limitation of phytoplankton productivity in Chesapeake Bay, found that phytoplankton growth rates were limited by dissolved inorganic phosphorus (DIP) during spring when biomass reaches an annual maximum. Such growth rates were limited by dissolved inorganic nitrogen (DIN) during summer when phytoplankton growth rates are maximal. Riverine DIN inputs are associated with seasonal accumulations of phytoplankton biomass in salt-intruded reaches of the bay, whereas the magnitude of the spring diatom bloom is limited by the supply of dissolved silica.

It has been generally acknowledged that non-point sources are among the leading causes of water quality problems in the Chesapeake system (Dauer et al., 2000). Although agricultural and urban land use was considered responsible for the degraded benthic community condition, the relationship of the phytoplankton communities to these connections was not completely evaluated. At intermediate levels of eutrophication, macrobenthic communities responded with increased abundance and/or biomass, whereas at more advanced stages of the eutrophication processes, there could have been an associated deterioration of the benthos. The observed deterioration of seagrass beds and bivalve mollusks would be consistent with this hypothesis. The negative relationship between nitrogen loadings and benthic condition, and the absence of this relationship with phosphorus loadings. were consistent with the conclusion that nitrogen was more important to eutrophication in the Chesapeake system, at least during summer months. However, phosphorus loading to the upper bay during winter–spring periods could have produced blooms that had either delayed direct effects or indirect effects on the subject (mesohaline) areas of the bay.

Sediment deterioration and food web alterations due to bloom-induced changes in the phytoplankton communities could be a primary reason for the observed changes in the benthic invertebrates in many coastal systems affected by anthropogenic nutrient loading. DO, while seasonally important as a contributing factor to the observed effects, could be another important factor relative to the year-round deterioration of sediment quality. The reduction of SAV in Chesapeake Bay (from 250,000 ha to current areas of 25,000 to 35,000 ha) with regrowth in mesohaline areas defined as "poor" is also consistent with an impact that could be related to phytoplankton responses to nutrient loading. The deterioration of the oyster industry in the Chesapeake system is also consistent with a possible effect of altered phytoplankton communities due to anthropogenous nutrient loading.

9.2.3 Toxic Substances and Over-fishing

Pollution has also led to the accumulation of toxic agents in fishes and other aquatic organisms. Wildlife species that consume these contaminated fish have also been adversely affected by pollutants that include chlordane, polychlorinated biphenyls (PCBs), and mercury. Fish consumption advisories have been issued to preclude human exposure to contaminated fishes and invertebrates.

According to Jackson et al. (2001), over-fishing in the past (sometimes the distant past) led to "simplified coastal food webs." The Chesapeake Bay was given as an example of reduced seagrasses and benthic diatoms due to farming during the 19th century, with subsequent increased phytoplankton populations during the 1930s due to the loss of filter-feeding phytoplankton. Corresponding increases in the flux of organic matter led to widespread anoxia and hypoxia. Reduced oysters due to over-fishing during the 20th century thus led to hypereutrophication. The authors hypothesized that declines in water quality were secondary to over-fishing as a cause of loss of the bivalve populations and subsequent losses of other species in addition to destruction of seagrass beds and other sensitive habitats. If true, restoration activities such as aquaculture of bivalve mollusks would constitute an important part of pollution abatement. Jackson et al. (2001) thus concluded that early over-fishing represented the precondition to the present-day "collapse we are witnessing," and basic changes in our approach to restoration will be required based on the new paradigm if we are to reverse the effects of eutrophication. As is the case with proposed models used as substitutes for long-term scientific data, the case for top-down control of the decline of the Chesapeake system remains hypothetical, and recent events tend to discredit this simplistic explanation. However, the combination of over-fishing and pollution are obviously an important part of the problem, and determinations of the causation of the multiple adverse impacts on the Chesapeake system will be critical to the development of an effective restoration program.

9.3   The Chesapeake Restoration Program

The Chesapeake Bay Program, which was started in the 1970s, has been defined as "America's Premier Watershed Restoration." In terms of financial support, the restoration effort remains one of the most well-funded such projects in history. Recently, Maryland officials announced that the state's share in the Bay Restoration program would be $7 billion (*Washington Post*, September 13, 2003).

> BAY RESTORATION TO COST MD. $7 BILLION: "Maryland officials plan to announce today that the state's share in restoring the Chesapeake Bay will cost $7 billion, more than half of which they say has already been identified."
>
> **—Raymond McCaffrey, *Washington Post*, December 21, 2001**

According to Ernst (2003), officials put the overall cost of restoration at around $20 billion. This was serious money that was proposed to carry out a massive restoration program.

There have been many optimistic news reports regarding the potential return of the once-famous Chesapeake Bay fisheries to their former glory. A complex reporting system that has accompanied the restoration effort has been developed regarding the year-by-year status of the various habitats and fisheries of the system. Reports have been filed concerning recent updates of the research results in a manner that was exemplary in terms of depth of coverage and inclusiveness of the many facets of bay ecology. However, an overwhelming number of reports and news coverage concerning the Chesapeake were glowingly optimistic in terms of the ongoing restoration of the system.

> FOR CHESAPEAKE STATES, CLEANUP HITS HOME: TOUGHER PACT TO HELP BAY: "'There will be big changes in how we farm, where we live, in our personal habits,' says David Carroll, Maryland's Chesapeake Bay coordinator. … The patient in 1972 was admitted into shock-trauma,' Carroll says, 'Now it's in guarded condition and we're hopeful.'"
>
> **—Linda Kanamine, *USA Today*, August 12, 1992**

> "As Don Boesch and Eugene Burreson say, the time for a strong link between science and management has never been better."
>
> **—Chesapeake Bay Foundation, "The State of the Chesapeake Bay: A Report to the Citizens of the Bay Region," July–August 1999**

> "As we approach 2000, striped bass are back in record numbers, underwater grasses have rebounded since the 1980s, and sewage treatment plant upgrades have helped in the ongoing clean-up of rivers. We have made impressive progress toward the ambitious nutrient goal set in 1987. … There's more good

news: in some places, living resources are beginning to respond, especially in areas where management actions have been concentrated."

—Chesapeake Bay Foundation, "The State of the Chesapeake Bay:
A Report to the Citizens of the Bay Region," October 1999

"Chesapeake Bay bald eagle continues resurgence ... Bay Program meets 2000 waterfowl goals for fourteen species... Managers and scientists believe that the creation of oyster sanctuaries is key to their recovery.... Shad populations reach highest levels since 1980s.... The Striped Bass Juvenile Index recorded an increase from 2000–2001 and achieved its restoration goal for the seventh year in a row."

—Chesapeake Bay Foundation, "The State of the Chesapeake Bay:
A Report to the Citizens of the Bay Region,"
Overview of Bay Program, 2001

MASSIVE RESTORATION BEGINS ON RAPPAHANNOCK RIVER TRACT: "The Chesapeake Bay Foundation (CBF), in partnership with the U.S. Fish and Wildlife Service, is restoring 206 acres of wetlands, forested stream banks and forestlands on former farmland in Richmond County. ... This restoration project will make a significant improvement in fisheries, wildlife habitat and water quality. 'It is an honor to be working with our partners on this project.' (Martin MacDonald, Bass Pro Shops)."

—Press release, Chesapeake Bay Foundation, April 29, 2003

In terms of money expended on research, there is little doubt that the Chesapeake effort represents the most ambitious project ever to restore a complex ecosystem. The research was awarded around $282 million from 1984 to 2002 (Ernst, 2003). The joint effort has included the states of Delaware, Maryland, New York, Pennsylvania, Virginia, and West Virginia, together with a group of state and federal agencies led by the U.S. Environmental Protection Agency (EPA). Scientific meetings have been dominated by the comprehensive research results from the Chesapeake system. The high quality of the research effort was never in doubt. However, with few exceptions, the general tone of the research results was optimistic with regard to the potential for the effectiveness of constructive restoration efforts in the Chesapeake drainage basin. Modeling of bay processes indicated a reduction in nutrient loading during a prolonged drought. There was positive support by research foundations and environmental groups devoted to the restoration of the Chesapeake Bay system. The overall support was underlined by a group of outstanding researchers who had worked on the Chesapeake system for most of their distinguished scientific careers. The Chesapeake Bay Program was generally acknowledged as a model for restoration efforts everywhere in the world.

And then it rained.

9.4   Reality Sets In: The Rainfall of 2003

During spring–summer 2003, record rainfall and runoff occurred in the Chesapeake region. This rainfall was associated with dead oysters and massive fish kills (*Bay Journal*, 2003). After 3 years of drought, scientists noted that the increased runoff loaded nitrogen and phosphorus into the Chesapeake system. This loading led to plankton blooms and deteriorated water quality. Excessive nutrient loading was associated with closed beaches and mortality of bay populations. During June 2003, 20,000 fishes died in the Maryland part of the bay, due, reportedly, to low DO. The so-called "dead zone" of hypoxia (low DO) and anoxia (no DO) had expanded to an unprecedented area in the bay that included shallow parts of the system. Oxygen-deprived areas expanded to an extent not seen in nearly 20 years of sampling. Dead fishes and crabs were found in traps throughout the bay. Blue crab abundance had stabilized near historically low levels over the 4 years prior to the 2003 bloom incidents. Female spawning stock was near the historical lows observed in 2000. In one of the many summer incidents, so-called red tide blooms killed 70% of the oysters at the Chesapeake Bay Foundation's Sarah's Creek Oyster Farm in Virginia. Oyster harvests were down to rock bottom in the bay.

Overall water quality continued to decline during 2003 despite the extensive restoration efforts in the Chesapeake system. However, the problems of the bay were now evident to the to public, and did not depend on complex scientific analyses and models that predicted continuing recovery of the Chesapeake system.

LOSS OF CREDIBILITY COULD UNDERMINE CHESAPEAKE CLEANUP EFFORT: "The Chesapeake Bay Program and its partners are perilously close to losing their credibility. By claiming that they have achieved a measure of success toward restoring the Chesapeake and its tributaries in the face of significant evidence to the contrary, they run the risk of being compared to the W.C. Fields character who asks, 'What are you going to believe — me or your own eyes?' And, by pursuing policies that clearly have little chance of success, the Bay Program undermines the general public's faith that it is up to the job with which its has been entrusted. While the Chesapeake Bay is not in as bad a shape as it might have been without a restoration effort, it is still dying. ... And, while computer models purport to show declines in the flow of nutrients to the Bay, the monitoring data from the water column generally show little change in the concentrations of these nutrients. ...the failure of the restoration community to speak publicly against them (public policies) does not inspire confidence... there is little to cheer in the Bay Program's use of science. When these problems are addressed — when research funds better target the problem and its potential solutions, when the Bay Program and its partners speak up about bad resource management, when effort shifts from making the program look good to actually achieving the goals of improved water quality — then there will be a basis for vesting the program with credibility."

—*Bay Journal* (Alliance for the Chesapeake Bay), September 2003

CONTROLS URGED ON NUTRIENTS IN BAY: "William C. Baker, president of the private not-for-profit environmental advocacy group (Chesapeake Bay Foundation) is calling on Virginia, Maryland, the District and Pennsylvania to write nutrient limits into the permits of all wastewater treatment plants... 'it's a chicken and egg argument,' he said. 'There has to be the political will to stand up and say sewage treatment plants must be upgraded... Why allow the bay to continue to die for another three years when the technology is available, when the science is very clear and the government has a responsibility to enforce the law?' Baker said."

—**Anita Huslin,** *Washington Post,* **October 29, 2003**

And what did Chesapeake Bay scientists think about all this?

"Mike Kemp, a longtime bay scientist, spoke a few nights before about how easily sated the appetites of sports people and tourists are, compared with those of commercial watermen."

—**Tom Horton**, *Baltimore Sun*, **June 4, 2004**

"...Walter Boynton, another bay scientist said: 'The bay for so many now has become optional, an hors d'oerves — just like oysters.'"

—**Tom Horton**, *Baltimore Sun*, **June 4, 2004**

"I think we estuarine scientists are very proficient at conducting eco-autopsies on estuaries. I think too many folks prefer 7,000 ft^2 homes to healthy bays. Our eco-religion has failed us."

—**David Flemer, distinguished, long-term researcher on the Chesapeake system, personal communication, 2004**

Press reports following the problems associated with the 2003 rainfall events proved critical of the ongoing efforts to stop polluted water from entering the Chesapeake system.

A LAW DOESN'T STOP SEDIMENT: "It will take a widely shared sense of obligation, stewardship and caring — an environmental ethic. But riding through the developing landscape of Baltimore County on a rainy morning shows that any such ethic continues to elude us."

—**Tom Horton**, *Baltimore Sun*, **May 28, 2004**

COMPROMISING ON POLLUTION LEVELS: "It might turn out what we want is not the sacrifice and expense of a bay restored to the health of half a century ago. Maybe we can maximize happiness with a degraded system which will become more degraded as population grows."

—**Tom Horton**, *Baltimore Sun*, **June 4, 2004**

CHESAPEAKE BAY NEEDS SCIENCE, NOT SLOGANS: "'Progress on reducing the pollution flowing into the Chesapeake bay, north America's largest estuary, has been 'significantly overstated,' the *Washington Post* hyperventilated in a front-page story this week ... It seems that allegedly erroneous estimates of pollution reduction were based on faulty computer modeling, not actually sampling. Politicians from Maryland, Virginia, Pennsylvania, and the District of Columbia were partially to blame, suggested the *Post*, as they were more concerned about saving themselves and their bureaucrat regulator buddies from environmentalist and media criticism than they were about 'saving the bay' — the local mantra." ...What does 'restoration' really mean? ...Are 1900, 1950, or 2000 more 'reasonable' baseline dates for 'restoration'? ... Sound science and reasonable expectations are needed, not sloganeering ('Save the Bay') and unachievable eco-fantasies (e.g.,'restoration'). It was the alleged reductions in phosphorus and nitrogen that were the focus of the *Washington Post* report."

—**Stephen Malloy, "JUNK SCIENCE," Fox News, July 22, 2004**

And the inevitable legal solutions to the Chesapeake Bay problem surfaced with an emphasis on the newly acquired understanding of the causes of pollution to the bay:

CLASS ACTION MAY FOCUS ON FARMING FIRMS, MUNICIPAL PLANTS: "Watermen in Maryland, frustrated by declining seafood harvests and 'dead zones' in the Chesapeake Bay, are preparing a class action lawsuit against polluters that could include municipal sewage plants and farming conglomerates.... These have been cited by scientists as major sources of the pollutants nitrogen and phosphorus, which feed the algae blooms that course 'dead zones' in the bay.... The plans come during a time of high frustration in the Chesapeake region, after a disastrous year last year in which a low-oxygen 'dead zone' encompassed 40 percent of the bay, and scientists saw a sharp drop in the amount of underwater grasses — a crucial part of the bay ecosystem. ... And watermen and recreational anglers have reported a resurgence of the 'dead zone' in the upper bay."

—David A. Fharethold, *Washington Post*, July 28, 2004

Politicians and federal bureaucrats reacted in stereotypical fashion. The Bush Administration responded to the call for action by cutting federal funding for Chesapeake Bay-related activities. Governor Robert L. Ehrlich, Jr., "would not rule out the creation of a regulatory agency but prefers 'a more cooperative approach where all the parties are brought together,' said spokesman Greg Massoni" (Dennis O'Brien, *Baltimore Sun*, August 12, 2003). According to Rebecca Hanmer, head of the EPA's Chesapeake Bay Program, "there has been 'a great deal of progress' in curbing runoff from farms, suburban sprawl, and wastes from outdated sewage plants." (Dennis O'Brien, *Baltimore Sun*, August 12, 2003). The one common denominator of public assessment was denial on the part of politicians and regulatory bureaucrats concerning the failure of state and federal efforts to curb nutrient loading into vulnerable and formerly productive Chesapeake water bodies. The tendency surfaced of press reports that made a controversy out of an obvious dereliction of so-called public servants to protect public resources.

Suddenly, however, there were new assessments of the Chesapeake Bay Program from diverse sources. William C. Baker, President of the Chesapeake Bay Foundation, indicated that the management system for the bay restoration has failed and needs to be replaced (Dennis O'Brien, *SunSpot.net*, August 22, 2003). He said that "the point is that what we're doing now doesn't work. We are struck by the deafening silence of a leadership vacuum.... This summer has been as bad as any I've seen in my 25 years with the program." *Baltimore Sun* columnist Tom Horton wrote a book (*Turning the Tide; Saving the Chesapeake Bay*) that noted evidence of little progress in improvement of the condition of the Chesapeake system over the prolonged restoration period since 1982. Baker emphasized the need for a new "compact" that would bring federal and state regulatory powers to bear on the bay's water quality problems.

Perhaps the most direct and comprehensive evaluation of the Chesapeake restoration effort was presented in a book by Howard R. Ernst (2003), who outlined the reasons behind the acknowledged failure:

1. The greatest danger was the "cozy" political partnership that trumpets "success," based on "few tangible environmental accomplishments."
2. Well-intentioned policymakers credit themselves with reports, agreements, and programs whose funding increases yearly with results.

3. The scientific community continues to monitor without noting significant improvements in water quality and productivity.
4. The Chesapeake restoration program is promoted as the model for worldwide programs while the facts indicate otherwise.
5. Nutrient loading limitation programs have failed, and, although the research has clearly indicated the problems with indications for solutions, the real political problems persist.
6. Economic interests and intergovernmental competition create obstacles for effective restoration efforts. Agricultural pollution problems are not addressed, and enforceable regulations are lacking. Political expediency takes the place of environmental improvements.
7. Fisheries management remains uncoordinated and reactionary as Bay fisheries collapse as a product of a failed political process.

Although the Chesapeake experience of failure to protect aquatic resources is in a more advanced state than that noted in the North Florida lakes, rivers, and estuaries, there is a similar pattern of interactive processes that operate within an umbrella of political and economic forces that control information and the failed restoration process. The political and economic system simply ignores the scientific data that form the basis for evaluating and correcting the problems. The underlying linchpin of such failure resides in a political system that is dominated by economic forces that control most aspects of American society. The press continues to give us images of painted ladies instead of productive aquatic systems. The public remains ignorant of the real problems and continues to be largely indifferent to environmental issues. The underlying lack of an environmentally sensitive culture of ignorance, greed, and indifference lies at the base of the problems related to the restoration of aquatic systems. The pseudo-populist policies of a mainstream American news media that is now run by large corporations feeds this process of environmental destruction as the cumulative impacts of increasing human actions contributes to the death of a thousand cuts, which is the fate of many formerly productive aquatic systems. The issue of preventive medicine in the form of progressive planning and management is not even considered because most restoration programs currently in place are a response to political policies that are self-sustaining prophecies for such failure.

chapter 10

Kissimmee–Okeechobee–Florida Everglades–Florida Bay–Coral Reef System

10.1 The System

The Kissimmee–Okeechobee–Florida Everglades–Florida Bay–coral reef (KOEFR) system, located in central and south Florida, is among the most unique ecological resources in the United States. This system formed over thousands of years as a major wetland represented by the Kissimmee River vegetation, the Florida Everglades, the Big Cypress Swamp, and the coastal mangroves and glades (Mitsch and Gosselink, 1993). The Florida Everglades represent the only such system in the Northern Hemisphere. The KOEFR region, the largest remaining subtropical wilderness in the conterminous United States, is a mosaic of fresh-water and saltwater areas that includes lakes, grassy wetlands, open prairies, pine rock lands, tropical hardwood forests, mangrove forests, a subtropical estuary, a string of keys, and offshore, hermatypic coral reefs. Unlike any other ecosystem in the United States, the KOEFR system supports a diverse mixture of temperate and Caribbean flora and unique fauna (Davis and Ogden, 1994) that includes nesting and over-wintering fishes, reptiles, amphibians, birds, and mammals. Various species of wading birds, such as egrets, herons, spoonbills, and the endangered wood stork, need the specific habitat provided by the Florida Everglades. Grassland birds and the endangered Cape Sable seaside sparrow are also present. Other wildlife includes the Florida panther, alligators, the endangered American crocodile, tropical fish, and crustaceans such as the valuable pink shrimp and spiny lobster.

The effects of changes in the Florida Everglades in recent decades on animal and plant assemblages due to anthropogenous activities have been complex and extensive (Figure 10.1). The losses of major parts of the wetlands of south Florida, from the Kissimmee Valley and Lake Okeechobee to the eastern half of the Everglades, have been the single most important reason for such effects. In the Kissimmee system and the Water Conservation Areas, continuous and stable water levels (as opposed to the natural varied historic levels) have contributed to losses of natural vegetation and invasions of deeper-water and floating vegetation (Kushlan, 1991). Large areas south of Lake Okeechobee are now either farmland or improved pastureland. Plant introductions such as melaleuca (*Melaleuca quinquenervia*), weeping willow (*Salix babylonica* L.), and other swamp species have taken over major parts of the Everglades. Only the southern Everglades still contain extensive acreage of plant associations that resemble the pre-development situation (Kushlan, 1991). The numbers of wading birds, such as egrets, herons, and ibises, have been reduced by 90%, mainly

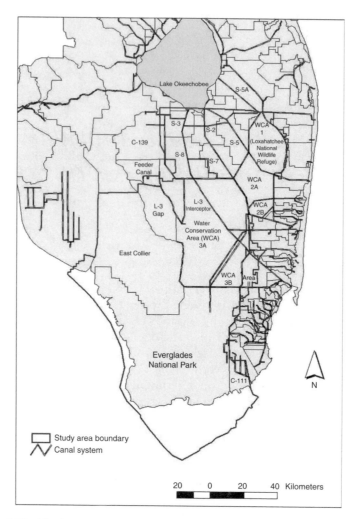

Figure 10.1 South Florida drainage basin, showing boundaries of the study area for the analyses by the South Florida Water Management District (SFWMD) and the canal system built to drain the Florida Everglades. (After South Florida Water Management District, 1999.)

because of habitat loss. By 1982, the wood stork essentially abandoned nesting in the Everglades, dropping by 75% of its 1967 numbers (Kushlan, 1991) due to retardation of the rate of dry seasons' water recession. In addition, the Cape Sable seaside sparrow may be on its way to extinction; its numbers have dropped by nearly half since the 1980s, with the current population somewhere around 3500. The Florida panther, already perhaps beyond recovery, continues to lose important habitat as development encroaches on its natural territory.

The KOEFR system represents an integrated ecosystem where fresh water originally moved from the Kissimmee River to the Florida Keys. Over the past century, the entire system has been threatened by diverse human activities that have had cumulative impacts in conjunction with the pressures of rapid population growth. During this time, the Florida Everglades have shrunk to less than half their original size as a result of the expansion of agricultural and residential development in the region. Accompanying irrigation and flood control demands (Figure 10.1) have compounded problems associated with nutrient loading on what is essentially an oligotrophic system. The rapid proliferation of the sugar industry, other agricultural growth, and the extensive urbanization of Florida's east coast have led to widespread environmental degradation in the region. Channelization and water diversions to satisfy the needs of agricultural and urban development interests have upset the ecological integrity of the entire KOEFR system, resulting in major losses of habitat and biodiversity. The introduction of exotic plant and animal species, and the addition of polluted runoff from agricultural and residential areas, have contributed significantly to the severe degradation of the natural resources of the system.

10.2 Background

10.2.1 Kissimmee River–Lake Okeechobee

The Kissimmee River originally drained about 7000 km^2 of the Osceola and Okeechobee plains, and meandered over 160 km at an average depth of only 1.2 m (Kushlan, 1991). Canalization (1962–1971) by the U.S. Army Corps of Engineers severely altered the functional aspects of the Kissimmee flow rates and water quality. Canalization reduced the river length by 90 km, and increased the width and depth to 60 m and 9 m, respectively (Kushlan, 1991). Approximately 80% of the river wetlands were lost, and the remaining wetlands were severely altered. Dams within the system are currently used to control downstream flows as water is stepped down across five artificial impoundments. Stabilization of the lake/pool water levels has resulted in decreased outflows during dry periods and increased drainage during wet periods (Kushlan, 1991). As a result, the natural functions of the Kissimmee River within the KOEFR system were significantly altered (Livingston, 2000). In an effort to undo some of the harm caused by the river channelization, Congress enacted the 1992 Water Resources Development Act. They authorized a Kissimmee River restoration project, the goal of which is to restore over 40 mi^2 of the Kissimmee River floodplain ecosystem.

Lake Okeechobee is the second largest freshwater lake entirely within the United States. It is shallow, with an average depth of around 3 m and a maximum depth of 5 to 6 m. Originally, Lake Okeechobee was the direct source of water to the Everglades (Harvey and Havens, 1999). After periods of heavy rainfall, water left the lake and entered small tributaries, in addition to draining as broad "sheet flow" at the southeastern lake edge. Water from the lake slowly made its way through the entire southern Everglades system. In the 1880s, the lake was connected to the Caloosahatchee River for increased drainage to the Gulf of Mexico. In 1921, the lake was surrounded by an earthwork levee called the Herbert Hoover Dike, which eliminated surface water connections to surrounding marshes and swamps. The St. Lucie Canal was constructed in the 1920s as an eastern outlet (Figure 10.1). Accordingly, a significant amount of water flow from Lake Okeechobee is now totally controlled by releases through the gated Caloosahatchee and St. Lucie Canal. Water releases are also made through structures on the south rim of the lake.

Alterations of the Kissimmee system along with back-pumping of animal wastes have led to serious eutrophication problems in Lake Okeechobee that, together with major hydrological changes made by the U.S. Army Corps of Engineers, have seriously altered the relationship of the lake and the Florida Everglades. Lake Okeechobee receives significant amounts of nutrients from the Kissimmee River and the back-pumping practices of the Everglades Agricultural Area (EAA) south of the lake. By the mid-1970s, the lake was in an early eutrophic state with periodically severe periods of nutrient enrichment (Kushlan, 1991). Basic changes in the lake have been traced to the decrease of southward discharges and increased nutrient loads from the Kissimmee River. Cattle farming in the region, increased nutrient loads from dairy operations, runoff from suburban areas, and back-pumping of agricultural wastewater into the lake have led to massive algal blooms and encroachment of cattails on the lake. Lake Okeechobee was subject to extensive blue-green algae blooms (Harvey and Havens, 1999), with erratic control of lake water levels and the loading of phosphorus as the primary contributing factors.

By 1999, record levels of phosphorus were loaded into the lake by dairy and citrus farms, cattle ranches, and suburban areas. There was no scientific explanation for such loading. Knowledgeable people could not understand the trends of increased nutrient loading to Lake Okeechobee.

> "I'm desperate. I'm beside myself … this is my single biggest failure…. I haven't been able to convince the powers that be that this is a real emergency."
>
> **—"Environmentalist," Associated Press, May 7, 1999**

The problems associated with eutrophication of the lake are supposed to be addressed by the Everglades Restoration Plan.

10.2.2   Florida Everglades

The hydrology of the 19th-century Florida Everglades was dominated by rainfall, with down-gradients of surface flows having seasonal and interannual changes in depth and hydroperiod. These changes formed the basis for critical aspects of the unique ecology of the system (Kushlan, 1991). The drainage system of the KOEFR system was composed of three primary sub-basins: (1) the Kissimmee River Valley, (2) Lake Okeechobee, and (3) the Florida Everglades (see Figure 10.1). Rainfall in the system peaks from May through October as part of a subtropical pattern that varies within a given water year. Historically, rainfall has been highest in the Everglades due to increased incidence of convective thunderstorms (Kushlan, 1991). Rainfall is least over Lake Okeechobee, and becomes less seasonal northward due to the greater effects of winter storms and relatively dry summers. Limited storage with nearly no carry-over of water from one annual hydrological cycle to another adds another dimension to the naturally high variance of flows through the Everglades system. Evapotranspiration is high as a result of high temperatures, persistent wind effects, slow overland flow rates, high surface-to-volume ratios, and the relatively long residence times of the surface water (Kushlan, 1991).

Flooding in the 1940s led to the formation of a flood control district, the Central and South Florida Flood Control Project ("Project"). The Project became fully operational by 1967. Although local effects on the drainage of the KOEFR system date back more than 100 years, the alteration of the core Everglades is relatively recent (Kushlan, 1991), occurring mainly in the past 50 years. In total, the alterations to the Everglades region have included the construction of 1000 mi. of canals and 720 mi. of levees (see Figure 10.1). Flow through the Florida Everglades is controlled by 16 pump stations and 200 gates, in addition to other water control structures. The result of these water control structures and pumps is that an average of 1.7 billion gallons of water is released to the ocean every day, and flows to the Everglades have been reduced by 70% (Florida Department of Environmental Protection [FDEP], unpublished data). Eastern parts of the Everglades have been drained for farming and urban development. The eastern Everglades, which almost reach the Atlantic coast in some areas, had been pushed back by as much as 32 km by 1991, and currently 65% of the original Everglades marsh, primarily in the east, has been drained. In dry areas in the remaining Everglades, there has been a subsidence of peat by about 3 cm per year.

The remaining Florida Everglades has been divided in two by the Tamiami Trail, a road bounded by a levee and canal. The entire northern Everglades has been enclosed by levees, with the exception of a small portion on the western side. These levees form three shallow reservoirs, the so-called Water Conservation Areas (WCAs). The southern Everglades is bounded by an eastern levee system that effectively holds back Everglades water from the developed East Coast and retains water in the remaining core Everglades. The southern Everglades includes Everglades National Park, as well as remaining marshes to the east that are generally on higher ground with shorter hydroperiods than the Shark River Slough to the west.

Water movement through the Everglades is now controlled by levees and gated structures, and substantially transported through canals. Surface water continues to enter the Everglades from the Big Cypress Swamp. From the north, water moves from the Everglades Agricultural Area into the Water Conservation Areas, with much of this flow discharged south at Tamiami Trail into Everglades National Park and into Taylor Slough via a canal. Some water bypasses the slough into the eastern part of the park and adjacent state-owned lands. This water can be discharged directly into Biscayne Bay (Kushlan, 1991). Thus, the bulk of the water today, prior to entering the Shark River Slough, moves through canals, thus bypassing the marsh. This has substantially increased water levels and hydroperiods over most of the remnant marshes while simultaneously reducing or eliminating standing water on high marshlands, most of which are now developed (Kushlan, 1991). Water flowing into the southern end of the Park has been altered so as to create seasonal and geographical changes in water distribution. These changes have been associated with major alterations of natural animal populations in the Everglades that were adapted to the more natural flow fluctuations (Kushlan, 1991).

Drainage and reclamation of the naturally occurring wetlands in the Everglades has been the most important cause of change. Remaining wetlands have been dried out by drainage practices implemented for the benefit of adjacent developed lands. Nearly 65% of the primitive wetlands of the Everglades had been drained by the 1980s. The second major effect of the hydrological changes to the system was the water flow manipulations that altered such flows to the remaining wetlands due the construction of the Conservation Areas that create deep flooding in southern areas and reduced hydroperiods in upstream northern areas (Kushlan, 1991). This has created discharges to the Everglades that are asynchronous with seasonal rainfall. In this way, major parts of the Florida Everglades have been destroyed by hydrological alterations by the U.S. Army Corps of Engineers. Agricultural interests have benefited from the draining of the swamps and redirection of water flows in one of the most extensive plumbing jobs in history.

10.2.3 Florida Bay

Florida Bay historically represented a unique subtropical estuary characterized by vast seagrass beds and important fisheries that included nurserying pink shrimp and various finfishes. In recent times, Florida Bay has experienced algal blooms with increasing frequency and corresponding losses of seagrass beds. There has been considerable debate concerning the general deterioration of Florida Bay between 1987 and 1991 (Fourqurean and Robblee, 1999). Alteration and inhibition of freshwater flows to many of Florida's coastal areas due to urbanization and agricultural activities, combined with enhanced nutrient loading from these sources, have caused widespread deterioration of aquatic habitats throughout Florida and the United States (McPherson and Hammett, 1991; Estevez et al., 1991). Factors responsible for the observed habitat deterioration of Florida Bay remain "poorly known" (Fourqurean et al., 1999). The deterioration of Florida Bay was, at one time, considered to be associated with the anthropogenous destruction of the natural flow of freshwater from the Everglades into Florida Bay and the rampant urbanization of the Florida Keys. However, the lack of relevant nutrient loading information and water quality data prior to the algal blooms has contributed to confusion and opposing scientific theories regarding the causes and effects of the observed losses of seagrass beds. Boesch et al. (1993), in a review of the Florida Bay research, indicated that algal blooms predated the seagrass die-offs, and that such deterioration may have been initiated by long-term increases in land-based nutrient loading somewhere in the system. This possible explanation differed from the findings of Fourqurean and Robblee (1999), who associated the blooms with preceding deterioration of the seagrass beds. The Boesch panel found that "virtually nothing was published on the Florida Bay algal blooms" and "surprisingly little quantitative historic information exists on the Bay's water quality" (Boesch et al., 1993).

A recent review of some of the Florida Bay studies can be found in the journal *Estuaries* (Fourqurean et al., 1999). This volume consists of the following: three papers on recon-struction of the history of Florida Bay (mollusk shell isotope records, paleoecological analyses, proxy chemical records in coral skeletons), seagrass distribution analyses, recruit-ment records of pink shrimp (*Penaeus duorarum*), a series of descriptive fish studies, and an analysis of the American crocodile (*Crocodylus acutus*) in Florida Bay. Rudnick et al. (1999) evaluated the importance of the Everglades watershed as a source of nitrogen and phosphorus to Florida Bay. It was determined that less than 3% of all phosphorus inputs and less than 23% of all nitrogen inputs were from freshwater runoff from the Everglades. The Gulf of Mexico was viewed as a major source, although nutrient loading from the south was not a focus of the study. Nutrient data used for the loading determinations were derived mainly from reports by federal agencies, such as the U.S. Environmental Protection Agency (1993). No comprehensive nutrient loading analyses were reported in the Florida Bay papers.

Boyer et al. (1999) reviewed water quality in Florida Bay from 1989 to 1997. Phyto-plankton and zooplankton data were not taken; the phytoplankton component was rep-resented by turbidity, total phosphorus, and chlorophyll *a*. Boyer et al. (1999) concluded that "the death and decomposition of large amounts of seagrass biomass can at least partially explain some of the changes in water quality of Florida Bay, but the connections are temporally disjoint and the processes indirect and not well understood." Tomas et al. (1999) found that the blooms were mixed populations of cyanobacteria and diatoms, although no consistent community-level data concerning long-term changes in the phyto-plankton were given. The blooms appeared to be seasonal and varied in different parts of the bay. The authors quoted Fourqurean et al. (1993), who hypothesized that offshore water was a source of phosphorus for Florida Bay. Nutrient ratios were used to estimate nutrient limitation for phytoplankton (Tomas et al., 1999); however, the authors qualified their results by the limitations involved in using static nutrient concentrations to describe a dynamic process. Overall, the source(s) of the nutrients loaded to Florida Bay remained undocumented.

10.2.4   Florida Keys, Coral Reefs

For the past three or four decades, the Florida Keys have undergone massive urbanization with inadequate controls on runoff and discharges. Impacts on water quality have resulted from this development. The hermatypic reefs of south Florida are in a state of decline, with recent reports of extensive algal growths. Macroalgae have overgrown coral reefs with a spreading of coral diseases that has damaged major portions of reef system. Brand (2000) pointed out that these changes represent classic symptoms of nutrification and cultural eutrophication. The potential effects of urban development in the Florida Keys on Florida Bay have not been systematically evaluated. Porter et al. (2002) pointed out that the pattern of measured coral decline in Florida Bay and the Florida Keys is consistent with adverse effects of Florida Bay water.

10.3   Water Quality in the Florida Everglades System

Nutrient loading and water quality in the Florida Everglades should be viewed within the context of the altered hydrology of the system. Core nutrient concentrations in the undisturbed Florida Everglades have been historically low in what has been widely considered a naturally oligotrophic system. Phosphorus is limiting at concentrations below 0.05 mg L^{-1}. Inorganic nitrogen concentrations were found to be less than 0.1 mg L^{-1}. Much of the nitrogen originally came from rainfall and was retained in the plants and sediments (Kushlan, 1991). Due to slow water movement and the absorptive effect of plants, the load-carrying capacity of the Everglades was virtually nonexistent in the natural state (Kushlan, 1991). Nutrients, of limited supply in the undisturbed Everglades, were removed quickly by algae and vascular plants and were sequestered into plant biomass and detritus. During more recent times, water quality in the Florida Everglades has been affected by movement of water in canals through the Water Conservation Areas (WCAs), resulting in increased mineralization due to canal limestones and direct storm water runoff (Kushlan, 1991). Nutrient loading has increased due to the rapid movement of water through the canal system and agricultural and urban discharges.

10.3.1   Mercury

Toxic substances such as mercury have been found in high concentrations in the Everglades system in recent times (South Florida Water Management District, 1994, 1999, 2000, 2001a,b). Methyl mercury is produced from an available supply of inorganic mercury through sulfate-reducing bacteria (SRB) under anoxic conditions. Sulfate stimulates SRB activity although the absence of sulfate is not necessarily associated with inactivity of methylating bacteria. Sulfide is inversely associated with methylation. High phosphate and increased plant production is associated with reduced mercury concentrations in plants and associated food webs (i.e., biodilution). Conversion of inorganic mercury to methyl mercury has occurred with accompanying biological concentration and magnification up food webs in the Everglades system. The South Florida Water Management District (South Florida Water Management District, 2001a) concluded that:

1. It is unlikely that increases in mercury occurred in such a way as to pose an increased risk to wading birds.
2. Farm and urban runoff may be affecting local increases in mercury but the main source of the mercury itself is atmospheric deposition from unknown sources.
3. Something other than sulfate was considered limiting to the concentration of mercury.
4. Peat soil concentrations of mercury appeared to be important in the food web concentration of this element.
5. Biodilution of the mercury effect does not occur along nutrient gradients.
6. There is some evidence that mercury in organisms in the Everglades has decreased over the past decade.
7. Reductions of mercury through manipulation of water quality is unlikely.
8. Production of methyl mercury in the Everglades is greater than that in other areas.

A risk assessment indicated that current water concentrations of mercury are an unlikely source of impact in Lake Okeechobee (South Florida Water Management District, 2001a,b). Recent analyses indicated reductions of mercury in the biota of the Florida Everglades that was attributed to reductions of mercury releases from coal-fired power plants and incinerators.

There is a far-ranging database concerning water quality in the Florida Everglades system. Livingston and Woodsum (2001) examined quality parameters collected by government agencies over the last 5 years in a geographic area that included the southern portion of Lake Okeechobee; the Everglades Agricultural Area (EAA); the Everglades Nutrient Removal (ENR) Project; Water Conservation Areas (WCAs) 1, 2, and 3; the Big Cypress National Preserve; and Everglades National Park (Figure 10.1). According to the Florida Department of Environmental Protection (2000), total mercury concentrations should not exceed 0.012 µg L^{-1} for the protection of human health in Class III freshwaters. Chronic freshwater habitat effects are indicated at concentrations of unfiltered surface water greater than 0.012 µg L^{-1}. Other criteria include 0.2 ng L^{-1} for the protection of fish-eating birds (2000) and 0.4 ng L-1 (2000) for the protection of fish-eating mammals. Current EPA levels of safe exposure for humans are set at 0.1 µg kg (body weight)$^{-1}$ d^{-1} with particular concern expressed concerning exposure for pregnant women and children. The FDA action level for reproducing women and children is 0.0625 µg. (body weight)$^{-1}$ d^{-1}. According to EPA standards, total mercury in fishes should not exceed 0.3 mg kg^{-1} for the protection of fish-eating birds (U.S. Environmental Protection Agency, 1997a). For trophic level three and trophic level four fishes, the numeric criteria are 0.04 mg kg^{-1} and 0.14 mg kg^{-1}, respectively, in the Florida Everglades.

The highest frequency of observations of total mercury and filtered total dissolved mercury concentrations exceeding the criterion were located largely in the upper northeastern section of the study area, near the border of Storm Water Treatment Area 1 and Water Conservation Area 1. The distribution of unfiltered total mercury indicated that the highest numbers of observations of increased concentrations of this form of mercury were along areas of the Everglades closest to urban development east of the study area. Increased methyl mercury concentrations were most often found in the eastern and western areas of the upper parts of the study area, with the highest frequency of such observation in the northeastern part of the study area. The numbers of observations exceeding the numeric criterion for unfiltered methyl mercury were highest in eastern portions of the study area bordering highly urbanized areas. Sediment mercury concentrations showing increased frequency of observations exceeding the criterion (0.49 µg g^{-1}) were located largely in northeastern and southeastern sections of the study area, again bordering the urbanized areas. The highest frequency of high mercury concentrations in shellfish tissue (mg kg^{-1}) was located just south of Lake Okeechobee. In sum, the data indicate that urban areas were associated with the number of observations where dissolved forms of mercury in water exceeded the criterion.

The most recent reviews indicate that restrictions placed on coal-fired power plants and other regional sources of air pollution have succeeded in reducing the amount of mercury in the Florida Everglades.

10.3.2 Nutrients

Dissolved nutrients are rapidly transformed by plant activity with many complex feedback processes that bring major alterations to nutrient concentration gradients. The impacts of nutrients on complex aquatic systems such as the KOEFR system should be viewed as the product of long-term trends of nutrient loading relative to the assimilative capacity of receiving areas. The South Florida Water Management District (SFWMD) (2001a,b) has concluded, based on various studies of the area, that phosphorus is the chief limiting factor for the Florida Everglades. The District proposed the following limitations based on concentration criteria: 10-µg L^{-1} total phosphorus (TP) in the water column and 500-mg kg^{-1} sediment concentration. These are the concentrations at which various biological factors, such as marsh dissolved oxygen (DO), microbiota, periphyton, macrophytes, and benthic invertebrates respond to gradients of total phosphorus (South Florida Water Management District, 1999). Almost the entire research effort of the SFWMD has been directed at the phosphorus question (see Chapter 3, South Florida Water Management District, 1999; South Florida Water Management District, 2001a,b).

Orthophosphorus and total phosphorus data indicated that the highest concentrations were usually in areas receiving runoff from areas to the north of the Florida Everglades. Trend analyses indicated seasonal increases during summer/fall periods. There were no noticeable interannual trends of the orthophosphorus data. Ammonia was highest directly south of Lake Okeechobee and in urbanized areas. Excursions of the state criterion of 0.02 mg L^{-1} for free ammonia occurred most often directly south of Lake Okeechobee and near urban areas. Urban areas at the southern end of the study area had high levels of un-ionized ammonia concentrations. Trend analyses indicated increased concentrations of un-ionized ammonia in various urban areas. Agricultural areas and urban runoff appeared as the main sources of nitrate-nitrogen to the system. There were high concentrations of nitrite + nitrate in urban areas on the southeast coast.

10.3.2.1 Relationships of Nutrient Loading and Water Quality

Although non-acute inputs of nitrogen-based nutrients are currently not considered harmful to the freshwater Everglades, these contaminants are receiving increasing attention with respect to the ecological decline of Florida Bay and the coral reefs of the Florida Keys. Ammonia contaminants are associated with urban as well as agricultural runoff. Until recently, the focus of most nutrient questions in the Everglades has been phosphorus. This is because the system has been viewed as a phosphorus-limited system. Now, data indicates there may be significant sources of nitrogen and ammonia coming from the urban and agricultural areas that contribute to poor water quality in the Everglades.

An important hypothesis of researchers to explain the seagrass deterioration in Florida Bay has been that reduced flow from the Everglades led to hypersaline conditions that, in turn, caused the seagrass die-offs. This hypothesis was used as the rationale for pumping freshwater into Florida Bay. However, Brand (2000) indicated that there was little temporal or spatial correlation between high salinity and the seagrass losses. The die-off occurred in 1987 (a non-drought year), whereas the drought years occurred during 1989–1990. The major drought thus occurred 2 to 3 years after the seagrass die-off and much of the die-off occurred in areas that had near-average salinity. Recent articles have addressed this controversy. Lapointe and Barile (2004) claimed that overstating the hypothesis of hypersalinity as the cause of the loss of seagrasses in Florida Bay undermined the objectives of the Everglades Restoration Plan and Zieman et al. (2004). The authors gave evidence that refuted claims that cultural eutrophication of the water column was not associated with such seagrass losses.

The increased discharges of nutrient-laden water to Florida Bay between 1991 and 1997 as a result of the flawed hypersalinity hypothesis was associated with "irreparable damage not only to the bay, but also to downstream waters of the FKNMS" (Florida Keys National Marine Sanctuary) (Lapointe and Barile, 2004). Due to the "restoration" effort of increased water loading to the bay, bay-wide salinity decreased by 44% while ammonium, chlorophyll *a*, and turbidity increased significantly. This increase was mirrored by significant increases in dissolved inorganic nitrogen (DIN) and chlorophyll *a* at FKNMS and further losses of corals due to infestations of coralline algae and macroalgae. According to this interpretation, discharges of Everglades runoff based on flawed scientific hypotheses led to increased blooms and turbidity, sponge die-offs, and lost macroalgal diversity in Florida Bay and a 38% loss of corals in the FKNMS between 1996 and 1999. If correct, this alternative interpretation would invalidate basic assumptions of the current restoration effort. Zieman et al. (2004) responded with a rationale based on various facets of a group of studies that were carried out in Florida Bay and reported in a compendium published in *Estuaries* (1999). The authors claimed that the Lapointe/Barile hypothesis ignored the published literature.

The deterioration of the bay was, at one time, considered to be associated with the anthropogenous destruction of the natural flow of freshwater from the Everglades into Florida Bay and the rampant urbanization of the Florida Keys. However, the almost complete lack of relevant nutrient loading information and water quality data prior to the algal blooms contributed to confusion regarding the causes and effects of the observed losses of seagrass beds. Boesch et al. (1993), in a review of the Florida Bay research, indicated that algal blooms predated the seagrass die-offs, and that such deterioration may have been initiated by long-term increases in land-based nutrient loading somewhere in the system. This possible explanation was in direct conflict with the findings of Fourqurean and Robblee (1999) that associated the blooms with preceding deterioration of the seagrass beds. A second part of the Florida Bay hypothesis (Zieman et al., 1999) was that organic decomposition of the dead seagrasses released nutrients that led to the algal blooms that persisted in Florida Bay. However, there was no scientific evidence that this happened. In fact, anecdotal information indicated that blooms started well before the seagrass losses. These observations indicated gradual deterioration from 1981 to 1986 with major declines starting in 1987. The algal blooms persisted during the 1990s; long after the effects of the seagrass die-offs would be effective if indeed they contributed to the blooms. In addition, the spatial distribution of the blooms was not consistent with the seagrass die-off hypothesis. Inorganic nitrogen concentrations were highest in eastern sections of Florida Bay from 1991 to 1999, whereas average concentrations of TP occurred farther west (Brand, 2000). These concentrations were not located in the same areas of the seagrass die-offs. The distribution of chlorophyll from 1996 to 2000 indicated that the blooms occurred upstream of the seagrass die-off areas. Based on these data, Brand (2000) proposed an alternative hypothesis whereby P limitation occurred in eastern parts of Florida Bay with N limitation in western sectors. Nutrient bioassays confirmed this distribution of nutrient limitation.

As part of the Brand hypothesis, the most extensive blooms occurred in north-central bay areas characterized by high phosphorus from the west and high nitrogen from the east (Brand, 2000). A possible source of the TP could be phosphorite deposits enhanced by high phosphorus from phosphate mining in areas around the Peace River. This would indicate phosphorus loading toward Florida Bay from the Peace River area. Agricultural areas north of the Everglades are a major source of nitrogen to the system, and increased water flows through the South Dade Conveyance System just north of Florida Bay would be the means of nitrogen loading to the bay. There has been an observed 42% increase in nitrate and a 229% increase in ammonia in Florida Bay from 1989–1990 to 1991–1994. The C111 canal was deliberately altered so that more water would flow to Taylor Slough further to the west and less water would flow to the east, having the effect of injecting N-rich water into the area of high P in western Florida Bay (Brand, 2000). The algal blooms, as indicated by the chlorophyll data, coincided with the nutrient loading patterns described above. Thus, N-rich water from agricultural areas through Taylor Slough was associated with the blooms. There was a temporal correlation of nutrient-rich flows of water from agricultural runoff into the Shark River, Taylor Slough, and Florida Bay, with blooms noted during the early 1980s. Thus, increased runoff appeared to be the cause of the blooms in Florida Bay, which is directly opposite to the hypersalinity hypothesis that is currently popular with the U.S. Army Corps of Engineers and the SFWMD.

Brand (2000) found that the nutrient-rich water from Florida Bay makes its way to the northern bay side of the Florida Keys and the southern ocean side of the keys into Hawk Channel and over the coral reefs, thus associating the Florida Bay situation with the macroalgal overgrowth of the coral reefs. Plumes of turbid, nutrient-rich water from Florida Bay have been observed to move all the way to the reefs, and maps of chlorophyll data confirm this association. In addition, nutrient-rich water in western Florida Bay has been observed to move east over the reefs. Shark River outflow has been shown to move southeast through the passes around or Vaca Key and out into the offshore reefs. This means that nutrient-rich water from agricultural areas is transported into Florida Bay and from the bay into the coral reefs to the south (net flow from northwest to southeast). Proposals by the U.S. Army Corps of Engineers and the SFWMD to open more passages along the Florida Keys between the bay and the reefs would thus compound the problems created by these agencies in the nutrient loading to Florida Bay, thus accelerating the ongoing hypereutrophication of the coral reefs.

10.4 Recent Evaluations of the Everglades Ecosystem

A compendium (Porter and Porter, 2002) outlines the most recent information concerning the Everglades–Florida Bay–coral reef system. The emphasis of the book rightly concerns the connectivity of the freshwater and marine "hydroscapes" of this South Florida system. The book outlines the various hypotheses and models that have been put forward to explain how this complex system works, with an emphasis on the truism that successful restoration attempts are based on good science.

Steinman et al. (2002) gave the historic basis for the current condition of Lake Okeechobee and its relationship to the channelized Kissimmee River. The altered hydrology of the lake has had adverse effects on the two estuaries it now feeds with respect to enhanced nutrient loading and reduced light transmission and salinity. The answer, in part, to such impacts lies in restoration of the destroyed wetlands in the watershed. The original Okeechobee system is gone, and restoration to some degree of its former productivity rests on hydrologic remedies, reduced nutrient loading, source controls on various agricultural practices, and sediment removal from the lake. Sklar et al. (2002) reviewed the impacts of altered hydrology of the Florida Everglades from 1880 to current times. The effects of the altered movement of water through the Everglades was related to biological changes due to landscape fragmentation, reduced overland flows, and habitat loss. The complexity of providing an effective restoration program was shown by the complex ecological response to the anthropogenous changes. McCormick et al. (2002) analyzed the effects of anthropogenous phosphorus loading to the Everglades. The authors reviewed various untested hypotheses related to factors such as differential responses of different habitat types to sensitivity of P enrichment and factors related to habitat diversity and relative to enrichment effects that trend toward spatial homogeneity (e.g., cattail; expansion).

Lee et al. (2002) analyzed the transport processes that link the various coastal ecosystems of South Florida. Detailed analyses were made of the complex wind-driven water circulation trends of the southwest Florida shelf and the Keys Atlantic coastal zone. The interactions of these extensive water bodies and the relationship of this circulation to biological processes in South Florida coastal waters demands that water management policies designed to sustain these ecosystems should be based on analyses of the entire region. This would include upstream parts of the eastern Gulf of Mexico. This ecosystem approach to evaluations of the input of pollutants and nutrients reflects a different approach than has been taken so far in studies of enrichment, eutrophication, and the impact of degraded water on Florida Bay and the coral reef habitats off the Florida Keys. Smith and Pitts (2002) looked at 15 years of studies on both sides of the Florida Keys, and concluded that tidal and wind forcing leads to transport pathways that couple the Gulf and Atlantic sides of the Keys through routes that include Florida Bay and associated tidal channels. This linkage means that water quality in Florida Bay is linked to offshore areas of the Florida Keys, and any region along this path can affect other downstream regions.

Brand (2002) noted hypotheses that linked phosphorite deposits on the western side of Florida with shifts to N limitation in coastal systems in South Florida. This would make such areas more susceptible to anthropogenous N inputs than that which occurs in the N-rich EAA where freshwater runoff meets the P-rich waters of western Florida Bay. This mechanism would then explain the increased turbidity, the location of the algal blooms, and the associated die-offs of seagrasses and sponges. Nutrient-enriched water from Florida Bay and the Shark River could be transported to the middle and lower Florida Keys where it might then adversely affect the naturally oligotrophic coral reefs. Continued input of nutrients from upstream urban and agricultural areas could eventually affect the Ten Thousand Islands area as a result of downstream N-loading. The likelihood of this happening means that pumping freshwater into Florida Bay as a management tool has already aggravated past seagrass die-offs, and will not solve the ecological problems associated with the plankton blooms. If current proposals for additional pumping by programs such as the Everglades Forever Act and the Comprehensive Everglades Restoration Plan (South Florida Water Management District, 1999) are carried out, by ignoring N as an important limiting nutrient in South Florida coastal waters, damaging algal blooms could result in N-limited ecosystems such as central and western Florida Bay.

The Florida Keys and associated coral reef systems represent a unique aquatic system that has complex associations with various water bodies. Porter et al. (2002) documented the significant losses of coral reef species between 1996 and 2000. The authors attributed such losses to multiple stressors: regional changes such as increased nutrients, turbidity and chlorophyll, and global climate changes. The average rate of coral loss was 12.6% between 1996 and 1999, a rate of decline that is unsustainable even in ecological time. Tougas and Porter (2002) noted that there was a link between the lack of juvenile coral recruitment and high death rates of adult corals. Low recruitment rates in the Keys relative to other areas indicates that the Florida corals are in serious trouble.

When compared with water quality data from Boyer and Jones (2002), a review of water quality taken quarterly in the FKNMS from 1995 to 1998 indicated that other water bodies affect various parts of the Keys (Lapointe and Matzie, 2002). This included the Upper Keys by intrusion from the Florida Current, the Back Country by internal nutrient sources, the Middle Keys by southwest Florida Shelf and Florida Bay transport, and the Lower Keys and the Marquesas by the southwest Florida Shelf. Lapointe and Matzie (2002) outlined the hydrological and biogeochemical linkages between the Florida Everglades, Florida Bay, and downstream coral reefs. The authors related the adverse changes noted in seagrass associations and coral reefs to "escalated nutrient loading." Once again, the increased flows and associated nitrogen loads from the Everglades between 1991 and 1995 were associated with phytoplankton blooms and increased turbidity in the central and western parts of Florida Bay that initiated the sponge die-offs and the loss of macroalgal biodiversity in the bay. These events were also connected to the increased nutrients, algal blooms, and coral die-off of downstream areas of the FKNMS. The authors emphasized that oligotrophic seagrass beds and coral reefs are particularly susceptible to even small increases in nutrient concentrations. The omission of the "right science" in the acceptance of the hypersalinity hypothesis by scientists and managers contributed to the water quality deterioration of Florida Bay and the Florida Keys. Lapointe and Matzie (2002) thus claimed that the Everglades Restoration Plan was seriously flawed and was actually responsible for deterioration of major parts of the South Florida aquatic system. Coral reef bleaching was directly connected to nutrient enrichment, according the authors.

Keller and Itkin (2002) noted that inputs of phosphorus and nitrogen to marine waters in the Florida Keys increased phytoplankton concentrations in what may be episodic events. Reich et al. (2002) found that anthropogenous nutrients from injection of sewage, septic tanks, and cesspools increase with increasing population pressure. This polluted water enters groundwater seepages into near-shore areas of the Keys, and it is likely that such pollutants are transported offshore to the reef tract. That is, increased development without proper treatment of the associated pollutants leads to surface and shallow sub-surface injections that end up in near-shore marine waters. Bacchus (2002) noted that the catastrophic declines and die-offs of the hermatypic (reef-building) corals and other reef-dwelling organisms, along with the mass mortalities of the Florida Bay seagrasses, have been subject to controversy concerning the causes of such phenomena. He noted that the FDEP has continued to permit injection of wastes into the groundwater throughout South Florida with claims of no adverse impact. These claims ignore the fact that neither state nor federal regulatory agencies have calculated the loading of nutrients and other pollut-ants that are transported and discharged by the pipelines that carry the wastes into the karstic groundwater systems of the region. Bacchus (2002) hypothesized that injected effluents contribute to the eutrophication of South Florida coastal waters, mass mortalities and disease of coastal organisms, and contributes to hydroperiod disruption of the Florida Everglades. If this hypothesis is true, the restoration of that wetland would be compro-mised. Further, the author suggested that continued deep-well injection of sewage wastes as part of the $213 million Florida Keys Water Quality Improvement Act will exacerbate current environmental problems. The author further suggested that the $7.8 billion Ever-glades Restoration Project and the $52 Harmful Algal Blooms and Hypoxia Research and Control Act of 1998 (Public Law 105-383) are being "circumvented" by the underground discharge of wastes into a famously karstic groundwater system by federal, state, regional, and local government agencies. Bacchus (2002) also disputed the scientific monitoring programs that are not designed to provide data concerning impacts from discharges from the regional karst carbonate aquifer system in South Florida.

10.5   *Management and Restoration*

A federal lawsuit was filed by the U.S. attorney against the SFWMD and the Florida Department of Environmental Regulation. The chief allegation for the plaintiffs was that the state's water quality standards had not been enforced as phosphorus was discharged from agricultural areas into the Florida Everglades. In 1991, the newly elected governor of Florida conceded the case, thus starting discussions for restoration of the Everglades system. The settlement of the case included requirements for specific reductions in phos-phorus discharges by agricultural interests. However, after a series of lawsuits by sugar interests, the federal government handed over the Everglades clean-up to the Florida Legislature, which came up with the "Marjorie Stoneman Douglas Everglades Forever Act of 1994." The namesake for the bill promptly asked that her name be taken off the act because she objected to the lack of a real commitment to the restoration of the Everglades system. Later that year, an amended version of the "Everglades Forever Act" was signed by the Florida governor.

The Act suspended water quality standards until 2003, with empowerment of state officials to determine allowable phosphorus discharge levels. There were also caps on the sugar industry's clean-up costs, with the balance to be provided by public taxes. In 1996, amendments placed on the ballot by Save Our Everglades were passed that placed respon-sibility of restoration on "polluters;" one amendment created a trust fund for restoration based on a sugar tax. In 1999, the Miccosukee tribe won a lawsuit for federal approval of a 10-ppb limit on phosphorus discharged into their reservation. A short time later, the FDEP advocated a limit of 8.5 ppb for such discharges. In 2003, the Florida Environmental Regulation Commission voted to limit phosphorus concentrations to 10 ppb, although this action was criticized on the grounds that the calculations to get to such a limit were based on geometric means. Meanwhile, during the summer of 2003, Governor Jeb Bush signed a bill that rewrote the Everglades Forever Act to the effect that clean-up of agricultural discharges would be delayed for 10 years. The Comprehensive Everglades Restoration Plan quandary has thus deteriorated into a controversy concerning whether or not the restrictions of phosphorus loading from agricultural interests will be effective in the restoration of the Florida Everglades.

Regardless of the political and economic aspects of the Florida Everglades situation, the systematic destruction of the KOEFR system due to cumulative impacts of myriad human activities is indisputable. The emphasis on the Everglades eutrophication problems has led to management efforts that largely ignore the inter-consecutiveness of the KOEFR system. Physical alterations, often funded and even carried out by state and federal agencies at the behest of elected officials, have had major and, in many ways, irreversible impacts on the natural resources of this vast region. Causy (2002) has outlined the path of destruction. This list includes channelization of the Kissimmee River, pollution of Lake Okeechobee with agricultural wastes, physical alteration, water diversion, and nutrient pollution of the Florida Everglades, the introduction of exotic plant and animal species, physical destruction of freshwater and marine wetlands, elimination of important seagrass beds and associated habitat reduction in Florida Bay, elimination of coral reefs, alteration of drought and flood cycles in the region, the unlimited expansion of urbanization on the East Coast, and pollution of surface and groundwater systems by various forms of urbanization and agricultural activities. Nutrification and cultural eutrophication are rampant with associated damage by algal blooms. Davis and Ogden (1994) described the potential for restoration of the Everglades wetlands with descriptions of the complex interaction of hydroperiods, structural habitat features, periodic and aperiodic disturbances, and the complex natural processes that drive this unique ecosystem.

The usual factors — greed, political power, public ignorance and indifference, bureaucratic ineptitude, the lack of adequate and relevant scientific information, poor to nonexistent regulation, misunderstanding and misrepresentation of facts by environmental organizations, the failure of the press to report what was happening, and the overall cultural superstructure that is largely driven by economic rather than environmental factors — have all contributed to this situation. As is usually the case, after the damage is done and the money is made, the loss of resources is "discovered" and the restoration programs begin. And so it is that various local, state, and federal management and restoration programs have been created to address the problems outlined above with regard to the KOEFR system.

Currently, the following programs are in place for the restoration effort (Causy, 2002):

1. Federal South Florida Ecosystem Restoration Task Force (established 1993)
2. Governor's Commission for a Sustainable South Florida (established 1994)
3. Congressional Water Resources Development Act (established 1996)

The Act established a South Florida Ecosystem Restoration Task Force, with funding of projects and state/federal cost-sharing programs administered by the task force. A $7.8 billion program has been designated for restoration of the Florida Everglades. The stated objectives include efforts to improve the quality and quantity of the water with adequate distribution to bring back natural processes in the Everglades system.

Various projects are underway that are currently overseen by governmental and nongovernmental agencies and groups. Catchwords such as "ecosystem approach" and "integrated management" are used by bureaucrats and the news media to describe the effort. However, there are serious questions concerning how this unwieldy and politically charged program is going to be operated. Just how effective the program will be in terms of restoration is still in doubt. Lockwood et al. (2003), in an evaluation of the Everglades Restoration Project, noted that assumptions that conceptualize the restoration process as primarily a water reallocation project ignore basic ecological factors associated with processes, such as fire ecology and biological feedback loops associated with complex habitat interactions. The lack of integration of research and resource monitoring has been characteristic of the Everglades movement from the beginning. The restoration of water flows to the exclusion of other factors that drive the Everglades system place the success of restoration in doubt. The general lack of ecosystem-level considerations related to the interconnectiveness of water flows through the system (Porter and Porter, 2002), together with controversies concerning a limited scientific database, underlie problems with the $7.8 billion restoration effort.

10.6 The News Media and Public Involvement

In the mid-1960s, scientists from the Institute of Marine and Atmospheric Sciences (University of Miami) completed a project in the lower Florida Everglades and Florida Bay and were concerned with what they had found. They felt that ongoing water diversion projects and changes in water quality would result in damage to the Florida Everglades and Florida Bay. They presented their findings to the Florida Legislature and the South Florida news media (i.e., *The Miami Herald*). Their concerns were ignored with disdain although there was ample proof that something was wrong with the system. Similar problems were noted by state environmental personnel concerning the ditching of the Kissimmee River system. Written reports regarding the Kissimmee River were ignored, and some of the more persistent researchers were given the opportunity to seek outside employment. There was no ambiguity in the results of scientific inquiries concerning the physical changes that were being carried out by the U.S. Army Corps of Engineers at the bidding of their superiors in the Florida Legislature and the U.S. Congress. It should be noted that most of the changes in the KOEFR system were being carried out with public funds at the discretion of elected officials. The economic rewards of the destruction of this system were distributed to agricultural interests and urban developers.

The political and economic aspects of the problems that have destroyed the KOEFR system have guaranteed widespread press coverage of the restoration issues. However, complications associated with basic ecological factors have been ignored in the simplistic manner in which the restoration plan has been presented to the public by the press. The usual polarization of opinion that underlined the selling of controversy was a major factor in the press coverage.

> CORAL REEF DAMAGE PUZZLES SCIENTISTS: "Some scientists hypothesize that nitrogen flowing into Florida Bay from the Everglades is quickly killing once-vibrant coral reefs far to the south in the Florida Keys. Others say sewage and nutrients created from humans are flowing from the keys. Damage to corals caused by over fishing and divers, global warming, disease, coral bleaching and coastal development might also be factors in the wrecking of the reef. …there is not much agreement among scientists what is causing the decline of the reefs in the Florida Keys."

> **—Karin Meadow, Associated Press, February 4, 2001**

EVERGLADES IN PERIL: "The most ambitious environmental rescue operation ever tried in this country — $7.8 billion to restore the Everglades — is at risk. The reason is that one of the major players in one enterprise, Florida's politically connected sugar cane industry, wants to postpone into the distant future the deadline for cleaning up the polluted water flowing into the Everglades. ... The main culprit is phosphorus, which flows from the farms and sugar cane fields north of the Everglades, and which 15 years ago topped out at more than 300 parts per billion — 30 times the maximum amount that scientists said the Everglades could handle."

New York Times, editorial, April 21. 2003

A POLITICALLY CONNECTED INDUSTRY DEVASTATES THE EVER-GLADES: "Last spring, in a bravura display of clout, the industry succeeded in ramming a sweetheart deal through the Florida Legislature that gives Big Sugar more time to clean up its act. The measure, supported by Governor Jeb Bush, pushes back a looming 2006 water cleanup deadline to 2013, and gives sugar companies until 2017 to pay a cleanup tax. 'Big Sugar is not only raping the resource; it expects breakfast in the morning,' wrote *Orlando Sentinel* columnist Mike Thomas."

**—Ted Levin, "The Environmental Magazine,"
Environmental News Network, August 2003**

EVERGLADES CLEANUP EXPOSES ENVIRONMENTALISTS: "In knee-jerk fashion, the 'environmentalists' called the 1994 law a sell-out to the sugar industry and claimed it wouldn't work. They favored their standard expensive, command-and-control crackdown on farming and business interests. ... So what if the cleanup is not proceeding according to some arbitrary rigid timeline and standard demanded by activists with dubious goals?"

—Steven Milloy, Fox News, October 3, 2003

As usual, omission and misinformation by press accounts led to the simplistic reviews of the scientific facts, and have contributed to the confusion that pervaded issues regarding the KOEFR system.

FLORIDA BAY ECOSYSTEM COMING BACK TO LIFE: "Dry spell was (the) source of area's decline. ...scientists trying to unravel the mysteries of the die-off believe that trouble started when turtle grass grew too dense during the dry, calm years, then got stressed by high salinities and fell prey to disease."

—Cyril T. Zaneski, *Miami Herald*, November 26, 1999

FROM EVERGLADES TO US: "Municipal water utilities... could feel the ripples of a rock thrown into the pond by the Miccosukee Tribe which wants to slow down the Everglades clean-up. The tribe sued to delay the biggest environmental publics works project in world history saying a federal Clean Water Act permit is needed to move water out of a canal in western Broward County. ... Forty-nine organizations from the National League of Cities, Conference of

State Legislatures and state attorneys are imploring the court to clear up this misinterpretation. It too sweepingly corrects a problem that doesn't exist."

—*Tallahassee Democrat* **editorial, November 16, 2003**

CONFLICT OVER EVERGLADES RESTORATION: "The project... also could help the state and federal taxpayers understand what is considered the most complicated restoration in the world. 'This isn't rocket science. They dug a canal and they are going to fill it back in' (Florida Audubon Society policy director)...state officials have to contend with Miccosukee Indians, who live in the Everglades. ...The tribe refuses to give up a piece of land it owns in the Picayune forest because it is their only land with the habitat needed to make some herbal medicines and native thatched dwellings called 'chickees.' ...the tribe is skeptical of the state's plan because, as many critics have contended, it can sacrifice one area of the Everglades for another.... State officials recently had sent too much water to one area, flooding tree islands needed by deer. ... 'it's clear that everything that's called a restoration effort is not a restoration effort.' (tribe official)."

—**Associated Press, November 27, 2003**

According to the news media, the real culprits in the problems plaguing the Florida Everglades are the Miccosukee Indians who object to the pumping of polluted water from the condos along Florida's East Coast into their lands in the name of restoration. The *Tallahassee Democrat*, whose news coverage has been less than enlightened on most environmental issues, has thus seen fit to continue the Indian wars in the name of restoration. Unfortunately, this kind of cynicism, combined with the usual racial overtones, is not unusual in our part of the world.

Meanwhile, the official effort to silence scientific opinions that run counter to the bureaucratic edicts of the day continue:

RESPECTED SFWMD SCIENTIST DEMOTED: "An awarded and highly respected river scientist has been demoted after he publicly expressed concerns about delays in the Kissimmee River restoration ....(He) has worked for nearly 19 years on the $600 million project to redirect part of the Kissimmee River back to its historical meandering path. ...he said (work on the project) had fallen five years behind schedule and no longer appeared to be the priority it once was."

—**Prakash Gandhi,** *Florida Specifier*, **October 2003**

And the problems with the Everglades Restoration Program have continued to the present time:

EVERGLADES STEWARDSHIP URGED: "Last year, the Florida Legislature moved back deadlines to reduce phosphorus pollution by 10 years.... Jeb Bush's office says that 90 percent of the Everglades already meets the phosphorus pollution requirement of 10 parts per billion... the Sierra Club withdrew its support for the comprehensive Everglades restoration plan, saying it will only advocate projects which have clear environmental benefits, not those that mainly

enhance urban water supply. The group cited the new federal rules for imple-
menting the plan, which has been criticized as vague and lacking in specific
interim goals, and the delay in water-quality-standard deadlines by the state."

—Coralie Carlson, Associated Press, January 25, 2004

EVERGLADES CLEANUP TOP PRIORITY: "'It's an understatement to say that
the Everglades Coalition is disappointed with the way Everglades restoration
is going' (co-chairman of Everglades Coalition). ... The state and federal governments have embarked on a 30-year $8.4 billion project to restore the natural
water flow through the Everglades."

—Coralie Carlson, Associated Press, January 24, 2004

Despite the extensive scientific research concerning the KOEFR system, there has been
no published report that has outlined the overall interrelated factors involved in the past
destruction of the KOEFR system. The results of the scientific effort remain controversial
and there are conflicting accounts of the problems associated with the interactions of the
Florida Everglades and associated systems to the south. The press has played up the
controversial aspects of the restoration effort without presenting a cogent understanding
of the underlying issues. This is due either to a lack of understanding of the scientific
issues by the reporters and/or a reluctance to bother the public with anything more
complicated than a grade-school-level account. Meanwhile, the bureaucrats and politicians
play the usual games of obfuscation and bravado, although the political process is at the
base of the environmental destruction of the KOEFR system. Many of the environmental
groups remain parochial in their interests, and bereft of a basic understanding of the
scientific questions involved in the restoration process.

A few of the outstanding scientific questions that remain unaddressed by the operations currently under way include the following:

1. Is there going to be enough water to sustain the urban interests, the agricultural
   needs, and the proposed reordering of the Everglades system?
2. During drought periods, what will be the order of priority of the distribution of
   water to the competing interests?
3. Will the proposed redistribution of water through the KOEFR system release
   damaging concentrations of toxic substances such as mercury, pesticides, ammonia
   and other nitrogenous compounds, urban wastes, and compounds that currently
   contaminate the various waterways in the system?
4. Will the serious adverse effects of introduced plants and animals on the various
   parts of the system be exacerbated by the reordering of the water flows through
   the system?
5. What scientific markers will be used to determine the effectiveness of the proposed
   restoration of the different parts of the KOEFR system?

Many of those involved in the KOEFR situation have good intentions; however, the
public remains confused and even indifferent to the situation in the KOEFR system, thanks
to the zero-sum controversies that swirl around the main issues and the lack of accurate
and objective reporting by the news media that continue to play up the sensational and
controversial aspects of the current situation without thoughtful accounts that underlie
the scientific questions that remain unanswered and/or disputed. Success of the multibillion-dollar effort of restoration of the various parts of the KOEFR system thus remains
in doubt.

section IV

Restoration of Toxic Waste Sites

There are certain criteria that are commonly used to determine whether or not a given chemical presents a clear threat to the aquatic environmental and/or human health. The chemical attributes of various substances are often indicative of the potential for adverse impacts on biological systems. Compounds with chlorinated benzene rings have been associated with high potential for toxic effects. Toxicity can be organized into two broad categories: (1) acute (short-term, less than 96 hr, effects usually measured as mortality) and (2) chronic (long-term effects, greater than 96 hr, on a range of morphological, physiological, or behavioral attributes of a given organism). Compounds with chlorinated benzene rings are often associated with relatively high persistence through time in aquatic systems. Chemicals that are not readily soluble in water, with relatively high affinities for nonpolar molecules (i.e., high partition coefficients), are often associated with a propensity for enhanced bioconcentration and biomagnification in aquatic systems. Bioconcentration is the ability of a given organism to concentrate a given chemical to levels that are orders of magnitude higher than ambient water concentrations. Biomagnification denotes the increased concentration of a given chemical as it moves up a given aquatic food web. The efficacy of food web concentration depends on three factors: (1) broad distribution of the chemical in the environment, (2) persistence of the chemical in a given system, and (3) exposure of organisms to the compound through their position in the food web.

The ultimate test for risk determinations of toxic effects in the aquatic environment concerns fate-and-effects of the chemical in question. This involves the scientific determination of where a given chemical (and/or its breakdown products) goes in the aquatic environment, how long these compounds last, and what the compounds do to populations in the area of impact. The determination of fate-and-effects, however, often involves well-developed and integrated field and laboratory studies concerning the above criteria, with specific application to how human health concerns are related to the environmental data. Effective studies of this kind are relatively rare.

Background scientific information concerning dioxin (2,3,7,8-TCDD) indicates that this compound fulfills the criteria for environmentally damaging compounds in the aquatic habitat and thus presents a clear threat to environmental integrity and human health when present at relatively low concentrations. The chemical structure (organochlorine, high partition coefficients) of dioxin is consistent with the potential for serious environmental damage through its persistence and capacity to bioconcentrate in aquatic organisms and biomagnify up aquatic food webs. It is extremely toxic and is one of the most dangerous compounds known in terms of its acute and chronic toxicity.

Mercury (Hg) does not degrade once deposited in aquatic habitats, but it can be altered chemically as it is sequestered in sediments. The inorganic form of mercury can be transformed by microbial activity into the extremely toxic compound, methyl mercury. Because of the chemical characteristics of methyl mercury, acute and chronic toxicological aspects of this compound are important in understanding the implications of mercury contamination. Relatively low concentrations of methyl mercury can affect aquatic organisms at various levels of biological organization, with the most severe effects usually at the top of the food web. The relatively extensive scientific literature concerning the toxic effects of methyl mercury on aquatic species has defined the limits of the impacts of this compound on river systems. The methylated form of mercury is environmentally persistent, lasting for decades to centuries in aquatic systems. Methyl mercury can be bioconcentrated in individual aquatic organisms and biomagnified up aquatic food webs, with top predators such as large fishes, birds, and mammals (including humans) having the highest concentrations. The ecological significance of mercury in the aquatic environment is directly related to food web dynamics.

As noted, methyl mercury and dioxin share the primary criteria for compounds that are considered a threat to the environmental integrity of aquatic systems and for the capacity of having adverse impacts on human health. These compounds are at the top of the list of toxic compounds considered as serious problems even at low concentrations in aquatic systems. The key to the threats of both dioxin and methyl mercury lies in the aquatic food webs that eventually lead to human trophic interactions through consumption of aquatic organisms.

chapter 11

Mercury and Dioxin in Aquatic Systems

11.1 Mercury in the Aquatic Environment

Mercury is a metal that occurs naturally and is widely distributed in various forms throughout the aquatic environment. In aquatic systems, the inorganic form of mercury can be transformed by microbial activity, usually in sediments, to the extremely toxic form of methyl mercury. The process involved in this transformation is influenced by environmental factors such as available dissolved oxygen (DO), pH, sulfur, and sediment type. Low DO enhances the methylation process. Methyl mercury is known for its ability to pass through biological membranes, its high chemical stability, slow excretion from aquatic animals, bioconcentration in aquatic animals, biomagnification in aquatic food webs (Regnell and Ewald, 1997), and relatively high acute and chronic toxicity to various aquatic species. Concentration of mercury in human food webs represents an immediate and direct threat to public health.

Mercury in aquatic systems has been extensively studied. Mercury is known to have harmful effects on wildlife and human beings. In recent years, mercury concentrations in the environment have increased in many areas of the United States. Aerial deposition from coal-fired power plants, incinerators, and various forms of industrial wastes (i.e., chlor-alkali plants) are the chief sources of mercury release to the environment. Recent studies (U.S. Environmental Protection Agency, 1997a) indicate that aerial burdens of mercury have increased fivefold during the industrial era. Aerial deposition, sometimes hundreds or even thousands of miles away from the source, has led to severe contamination of various aquatic systems that are distant from the source. Mercury does not degrade, but it can be altered chemically or sequestered in sediments if it is strongly bound to sediment particles.

The inorganic form of mercury can be transformed by microbial activity to the extremely toxic form of methyl mercury. The complex processes whereby mercury is concentrated in aquatic systems have been documented in various studies. Fate-and-effects of mercury within a given freshwater drainage basin and its associated estuary or coastal area are often complicated by specific attributes of the system. These factors include turnover rates, physicochemical microhabitat distributions, sediment/water exchanges, conversion rates between inorganic mercury and methyl mercury, bioavailability of the various forms of mercury, bioconcentration rates of available aquatic populations, and complex food web relationships that can lead to biomagnification of the organic forms of mercury. The considerable scientific literature concerning the environmental impacts of mercury in its various forms clearly establishes the threat of this metal to aquatic systems. Mercury can affect aquatic organisms at various levels of biological organization at relatively low concentrations (Zillioux et al., 1993; Burgess et al., 1998; Wolfe et al., 1998). The human risk is well known, and the Food and Drug Administration (FDA) and the U.S. Environmental Protection Agency (EPA) have continuously reviewed minimal risk levels (i.e., no adverse effects).

Different studies have indicated the modes of mercury transport and concentration in aquatic systems (Locarnini and Presley, 1996; Campbell et al., 1998). The EPA has issued fish consumption advisories in 44 states (U.S. Environmental Protection Agency, 1997a). Advisories have been most common for lakes and rivers in states in the Midwest and New England (Maine, Vermont, and New Hampshire). The release of 22.7 kg mercury a year can be compared to the contamination of 5000 walleye (1 to 1.5-kg class) having average tissue concentrations of 0.5 ppm (a level sufficient to trigger fish consumption advisories). Current EPA levels of safe exposure for humans are set at 0.1 µg kg^{-1} (body weight) d^{-1}, with particular concern for pregnant women and children. A recent review panel of the National Academy of Sciences (NAS) endorsed the strict standards of the EPA (0.1 µg kg^{-1} body weight d^{-1}) based on established low-level effects of methyl mercury on children. This follows a previous NAS study that estimated that 60,000 newborns each year may suffer developmental damage due to fetal mercury exposure, mainly from mothers' consumption of contaminated fishes.

Up to now, the regulatory response of state and federal agencies concerning reductions in the release of mercury has been minimal. It is now estimated that more than one third of the nation's lakes and nearly a quarter of its rivers are contaminated with mercury, PCBs (polychlorinated biphenyls), dioxin, and pesticides. Recent tests show that in a survey of fish samples from 500 lakes and reservoirs across the country, the EPA found mercury in every sample. Mercury levels exceeded "safe limits" for women in 55% of the samples, and 76% of the samples exceeded the "safe limits" for children under the age of 3. Every one of the New England states has issued formal advisories warning people to restrict or avoid consuming certain fish species from lakes due to mercury concentrations. Any statement to the effect that there is not enough information concerning how mercury is transported to aquatic systems, the effects of mercury once it enters such areas, and the human health risks associated with mercury concentrations in food is not reasonable based on ample scientific information. Statements by EPA officials that mercury emissions have decreased in recent decades are misleading and indicate a basic misunderstanding of the cumulative effects of mercury in the aquatic environment. Because the largest source of mercury in the environment is coal-fired power plants, the answers to the publicity associated with mercury pollution are fraught with political overtones, which account for the lack of action regarding regulation of mercury emissions from various sources.

11.2   Penobscot River–Bay System in Maine

The history of the Penobscot River–Bay system in Maine represents the interplay of science, regulation, economics, and politics in the current problems associated with mercury.

11.2.1   Background

The Penobscot River estuary is the largest drainage system in Maine. The combination of high tidal fluxes and high river flows contributes to relatively dynamic processes of particulate transport. There is no organized body of scientific information concerning the Penobscot system because no comprehensive, long-term ecological studies have been undertaken. A chlor-alkali plant on the Penobscot River near Orrington was permitted by state and federal agencies to discharge more than 7 kg mercury per year for more than three decades. Wastes from the plant included brine purification muds, mercury cell process water, mercury-contaminated substances, waste paint and solvents, PCBs, acetone, methyl ethyl ketone, and waste oils (Camp Dresser & McKee, Inc., 1998). Limited sediment mercury monitoring around the plant indicated that extremely high mercury residues (460 mg kg^{-1}) occurred in sediments near the plant. When adjusted for grain size, the distribution of sediment mercury indicated that the chlor-alkali plant was the most significant source of past and ongoing mercury pollution to the Penobscot system (Maine Department of Environmental Protection, 1998; Livingston, 2001b). However, with the exception of Livingston (2001b), virtually no comprehensive analyses had been made concerning mercury concentrations in sediments and aquatic biota of the lower Penobscot River and Bay system. The absence of quantitative mercury loading information, together with the almost complete lack of ecological data concerning potential impacts of the high sediment concentrations of mercury in the Penobscot system, left open questions concerning environmental effects of past and ongoing mercury discharges.

11.2.2   Mercury in the Penobscot River–Bay System

A report by the Maine Department of Environmental Protection (MDEP) on "Mercury in Maine" (Morgan, 1998) found that mercury concentrations in sediments in the Upper Penobscot River at Orrington exceeded any other concentrations found in the state or anywhere in the country by orders of magnitude. One sample from the chlor-alkali plant discharge point exceeded 460 ppm (Morgan, 1998); Morgan noted that "the Upper Penobscot is so atypical that it dwarfed the rest of the data" (in Maine). Previous (limited) sediment analyses in the Penobscot River indicated sediment mercury concentrations that exceeded concentrations in sediments of any of Maine's freshwater, estuarine, or marine areas. According to Mower et al. (1997), many of the sediment samples taken from the Penobscot River system exceeded NOAA-developed effects ranges. All samples exceeded 0.15 ppm and most exceeded 0.71 ppm. These findings led to the conclusion by the MDEP that mercury contamination of the Penobscot River sediments presented "a significant ecological problem." Various results (Sowles, 1997a,b, 1999; Maine Department of Environmental Protection, 1998) confirmed previous analyses that indicated that the chlor-alkali plant was the primary source of mercury in the sediments of the Penobscot River estuary.

Very little scientific information had been generated concerning the loading of mercury to the Penobscot River. Even less was known concerning sources and rates of accumulation of mercury in Penobscot Bay. The report by the Maine Department of Environmental Protection (Morgan, 1998) concentrated on comparisons of mercury concentrations in the Penobscot system with levels of biological impact that had been established in other systems. Sediment concentrations of total mercury below 0.14 ppm (dry weight) were considered natural background. As total mercury concentrations in sediments exceeded 0.71 ppm (dry weight), "the probability of more severe biological impacts becomes more likely." Based on the relative paucity of actual data concerning mercury in the Penobscot system, we remain restricted to reviews of descriptive information concerning the sources and dynamic processes that are involved in the observed high concentrations of mercury in the system. The lack of ecological data in the largest river-bay system in Maine is not atypical of many of the river systems in the United States.

A review of the report by Camp Dresser & McKee (1998) indicated that there had been a history of spills from the chlor-alkali facility into the Penobscot River:

- *August 2, 1995:* 65,000 gal. mercury brine into river (conc. 1.350 ppm)
- *1996:* various rule violations were found; corrective actions 1997
- *January 27, 1997:* seepage noted from HMC facility into Penobscot tributary; included Mercury (pH 9–10), mercury-contaminated waste brine

- *February 19, 1997:* tank leaked mercury brine into river over 13-day period; 270,000 gal. into ground (estimated)
- *May 1, 1997:* 1000 gal. mercury brine spilled into ground; NOV filed by MDEP; following months had further discharges
- *July 19, 1997:* further discharges of mercury to waters of the state
- *August 19, 1997:* 200 gal. mercury-contaminated condensate released to waters of state

"Areas of Concern" (i.e., having hazardous wastes or hazardous constituents) included five landfills, a lined process lagoon, spill areas within the plant, storm water ditches, and sediments in the Penobscot River (Camp Dresser & McKee, 1998). Plant wastewater containing mercury was discharged to the Penobscot River (National Pollutant Discharge Elimination System [NPDES] Permit) through Outfall 001. Sediment had been transported from the plant via the north ditch and southerly stream to the Penobscot River at the southern cove (Camp Dresser & McKee, 1998). Sediment samples taken from the river near the northern drainage channel contained 2.4 to 8.2 ppm total mercury (Camp Dresser & McKee, 1998). It was also reported that mercury runoff from plant soils and surface water were the primary sources of contamination to ditches, which then transported the mercury into the Penobscot River. Discharges via surface water were characterized by mercury in both dissolved and particulate-associated states. However, the combination of no quantitative loading information together with the almost complete absence of ecological data concerning potential impacts of the sediment concentrations of mercury in the Penobscot River and Bay system left an open question concerning the environmental effects of past and ongoing mercury discharges to the Penobscot River. Also, the connection between the river contamination and bay concentrations of mercury in sediments and animals was almost entirely unknown.

A review of the available information concerning mercury in the Penobscot system (Morgan, 1998; Mower et al., 1997; Zeeman, 2001) indicated the following:

1. Total mercury concentrations in sediments of the Penobscot River estuary (range: 0.25–4.69 mg kg^{-1} dry wt.) were generally orders of magnitude higher than concentrations found in sediments from reference areas in Maine (range: non-detect–0.14 mg kg^{-1} dry wt.).

2. Mercury sediment concentrations in the Penobscot River estuary were extremely high when compared to unpolluted sediments in other parts of the country such as Apalachee Bay in the Gulf of Mexico (Livingston, 2001b.; range: non-detect–0.10 mg kg^{-1} dry wt.).

3. The distribution of sediment methyl mercury and sediment total mercury (standardized by % organics) indicated that it was likely that the chlor-alkali plant on the Penobscot River was the primary source of mercury to the Penobscot system.

4. The accumulation of mercury in the tissues of aquatic organisms in the lower Penobscot system was generally higher than such accumulations in the biota of other estuarine systems in Maine.

5. Blue mussel (*Mytilus edulis*) tissue mercury concentrations in the Penobscot system were comparable to Mussel Watch data (NOAA) as reported by the Environmental Working Group (Houlihan and Wiles, 2001). The Penobscot blue mussel mercury concentrations were among the highest in the nation.

6. Body burdens of omnivorous fishes were directly related to sediment mercury concentrations. Omnivores such as lobsters had the highest concentrations of

hepatopancreas mercury of all the estuaries sampled in Maine. Filter feeding mussels in the Penobscot system had the second highest concentrations of mercury (second only to mussels taken from a highly urbanized coastal system).

7. Mercury concentrations in lobster (*Homarus americanus*) hepatopancreas from Penobscot Bay were higher than any mercury concentrations found in lobsters from a series of Maine estuaries in a 1996 analysis (Zeeman, 2001).

8. Average whole-body mercury concentrations in eels (*Anguilla anguilla*) at locations near the chlor-alkali facility averaged 5 mg kg^{-1} dry wt., whereas concentrations in fish from an upstream tributary (Kenduskeag River) averaged 0.3 mg kg^{-1} dry wt. (Zeeman, 2001).

9. Zeeman (2001) reported that cormorants from the Penobscot system had higher blood and feather mercury concentrations than mercury in birds taken from eight other Maine estuaries.

10. The mercury data indicated that a systematic analysis of mercury in the food webs of the Penobscot system should be undertaken not only as a risk assessment for the aquatic organisms in the system, but also as part of a larger study of the risk to humans who consume these organisms on a regular basis.

Data concerning the mercury concentrations at different levels of biological organization indicated that the high sediment burdens of mercury in the Penobscot system were getting into the river estuarine food webs through direct concentration (mussels) and food web biomagnification (cormorants). The data also provided indications of high mercury concentrations in food webs utilized by humans. However, the patch-quilt research effort was inadequate to determine the extent and the potential seriousness of the mercury contamination problem.

Follow-up studies in the Penobscot system (Camp Dresser & McKee, 1998) indicated the following:

1. Topminnows (*Fundulus heteroclitus*) associated with runoff from Penobscot wetlands had higher mercury tissue concentrations that any other part of the system.

2. Sediments taken from areas draining Penobscot wetlands exceeded NOAA ER-M criteria that defined a high likelihood of adverse biological effects.

3. Sediments taken from Penobscot areas subject to runoff from wetlands caused significant reductions in growth of test organisms in 100% of the chronic bioassays.

4. Qualitative field reviews indicated substantial reductions in wildlife in relatively extensive parts of the Penobscot wetlands.

5. One out of two surveys of osprey fledglings in the Penobscot system indicated success rates (numbers of fledglings per active nest) that were below those considered necessary for maintenance of the population.

11.2.3 The "Wetlands Hypothesis"

The limited data concerning mercury impacts on the Penobscot system indicated a potential link with mercury accumulation in associated wetlands over the prolonged period of mercury loading by the chlor-alkali plant. There were indications in the scientific literature that wetlands played a role in mercury contamination in river systems. A recent proposal by a crown-owned utility called Hydro-Quebec to construct hundreds of dams along rivers flowing into the James Bay–Hudson Bay system in Canada led to studies concerning previous dam operations in this area. Habitat changes encountered with flooding of wetlands due to dam construction are well known (Rosenberg et al., 1987). Flooding associated with new impoundments caused reductions in DO, which in turn provided appropriate habitat conditions for methylation of mercury that had been deposited in the previously unflooded wetlands. The newly flooded wetlands were thus associated with increased methyl mercury in the aquatic food webs. Eventually, mercury was concentrated in top predators such as fishes that for centuries had provided a primary source of food for natives in the area (Brouard et al., 1990). It was estimated that it could take up to 50 years for methyl mercury levels in top predators to return to background levels. The contamination of the Florida Everglades with mercury is a wetland phenomenon where mercury has eliminated the consumption of game fishes (South Florida Water Management District, 2000, 2001a,b).

Rada et al. (1986) found that river sediments in Wisconsin were highly contaminated with mercury 20 to 30 years after contamination from human sources had been eliminated. Biological availability was enhanced by rapid methylation of the mercury in surficial sediments, although substantial amounts of the metal were buried. Mean sediment total mercury concentrations — from 0.3 to 0.8 $\mu g\ g^{-1}$ (ppm) dry weight, with ranges of 0.2 to 1.8 $\mu g\ g^{-1}$ — were considered high when compared to reference areas with concentrations of 0.04 to 0.05 $\mu g\ g^{-1}$. Rada et al. (1986) found that fishes had relatively high methyl mercury concentrations, which were attributed to uptake by feeding and respiration. Overall, mercury concentrations in depositional areas were considered mercury sinks. Continual disturbance and resultant exposure of sedimentary mercury to the water column in other areas due to natural river habitat conditions was noted as the source of continuing methyl mercury exposure years or even decades after mercury contamination of the river was eliminated. The increased bioavailability of mercury was thus associated with physical disturbance of the sediments in addition to organic loading and high temperatures. These data indicated that transformation and transport of the mercury in riverine systems is highly dynamic. Once contaminated by loading from an industrial source, the mercury was available to the biota for relatively long periods (years to decades) due to resuspension and subsequent methylation of mercury-contaminated sediments.

Recent evidence indicates that mercury in riverine drainages can last a very long time. The behavior of mercury and the dynamic habitat changes that are associated with the methylation process have been ascribed to specific conditions found in rivers and estuarine systems. Parks et al. (1989), studying mercury contamination of a Canadian river (the Wabigoon River system) by a chlor-alkali plant, found that net production of methyl mercury was related to water column equilibrium conditions of methyl mercury and inorganic mercury. The actual generation of methyl mercury is a complex chemical process that involves both the sediment/water interface and overlying water column equilibria. Continuous methylation in areas distant from the source helped to explain why the methyl mercury concentrations in the river remained high 10 years after cessation of releases of mercury from the plant. These results indicated that the answer to impact evaluations and possible remediation of mercury contamination lies in downstream sediment/water processes and habitat conditions that contribute to production of methyl mercury.

Wiener and Shields (2000) reviewed the transport, fate, and bioavailability of mercury in the Sudbury River in Massachusetts. The Sudbury system was contaminated by an industrial complex that operated with continuous release of mercury for decades up to 1978. High mercury concentrations in sediments and fishes qualified the Sudbury for inclusion in the Super-Fund Program of the U.S. EPA. However, Wiener and Shields (2000) reported that methyl mercury concentrations in aquatic biota did not parallel the concentrations of total mercury in sediments to which they were exposed. Instead, it was found that contaminated wetlands 25 km downstream from the point source of mercury "produced and exported methyl mercury from inorganic mercury that had originated from the site." Mercury was buried in depositional areas along the river, but methylation of inorganic mercury took place in the dynamic and continuous floodplain wetlands of the Sudbury River system. Whereas the relatively high concentrations of mercury in depositional areas such as impoundments were eventually buried, with resultant removal of mercury from biological contamination, natural processes associated with the "lesser contaminated" wetland habitat farther downstream remained available for the methylation process. The eventual and continuous transport and entry of methyl mercury into downstream food webs was thus associated with wetland areas far removed from the original source.

These studies established a viable hypothesis concerning the processes that contribute to long-term mercury contamination of drainage areas. The results showed that mercury transported to wetlands far removed from the origin of the contamination could be more problematic than the mercury levels at the industrial site in terms of long-term, adverse effects on the system. Accordingly, remedial efforts at the contaminated site, without attention to other centers of mercury contamination in wetlands, could be meaningless because the real problems of methylation, transport, bioavailability, bioconcentration, and biomagnification of mercury could be far removed from the site of mercury origin.

11.2.4   Proposed Restoration of the Penobscot System

State and federal regulatory agencies (MDEP and the U.S. EPA) determined a set of criteria for the restoration of the Penobscot system. There was a problem with this effort in that the scientific database for such restoration was incomplete and indeterminate with respect to the identification of the current sources of mercury to the Penobscot system. These shortcomings include the following:

1. There has never been a comprehensive analysis of mercury in sediments and animal tissues taken from the Penobscot system.
2. No mercury bioavailability studies have been performed in the Penobscot system
3. No comprehensive field studies or food web analyses have been undertaken in the Penobscot system. The actual sources of the continuing contamination of the Penobscot River and Bay with mercury have not been identified.
4. Relatively few sediment toxicity tests have been run with samples from the Penobscot system.
5. There have been no comprehensive benthic community analyses in the Penobscot River estuary.
6. There has been no comprehensive wildlife evaluation program undertaken in the Penobscot system.
7. There has been no attempt to determine mercury tissue burdens of people who eat contaminated seafood taken from the Penobscot system.

Despite the almost complete lack of scientific data on the Penobscot River estuary, state and federal regulatory agencies proposed a restoration program whereby sediments exceeding 10.7 mg kg^{-1} dry wt. would be removed from the Penobscot system. This would eliminate the need for remediation everywhere in the Penobscot system, with the possible exception of areas immediately adjacent to the chlor-alkali plant. Aquatic food web considerations and potential human health problems would be simply ignored if such a program was undertaken.

11.2.5   Legal Solution to the Penobscot Mercury Problem

In July 2002, the Maine People's Alliance and the Natural Resources Defense Council, Inc. (NRDC), filed a lawsuit in U.S. District court against the owners of the chlor-alkali plant. The judge ruled that "methyl mercury continually accumulating and biomagnifying in the food web creates a reasonable medical concern for public health and a reasonable scientific concern for the environment downriver of the plant site. The Court will, therefore, order the Defendant Mallinckrodt be responsible for the cost of undertaking a scientific study of mercury contamination downriver of the plant site in the Penobscot River." Currently, a court-appointed panel of experts is overseeing a study to determine if there is a significant mercury problem in the Penobscot drainage system. If there is such a problem, the study will determine if restoration of the system is economically and environmentally feasible.

11.3   Mercury in the South River–South Fork Shenandoah River

11.3.1   Background

A textile manufacturing operation by DuPont de Nemours and Co. in Waynesboro, Virginia, used mercuric sulfate as a catalyst for manufacturing from 1929 to 1950. The mercury (Hg) used by the plant was first found in a private study of the South River–South Fork Shenandoah River system in September 1976. The company delivered the study report concerning the mercury contamination of the river to the U.S. EPA and the State Water Control Board (WCB) in 1977. A report by the Virginia Department of Environmental Quality (DEQ) concerning mercury in fish and sediments from the river system was completed in 1977. The WCB and the EPA reviewed the study results. A ban was then issued on the consumption of fishes along 200 km of the South River-South Fork Shenandoah River system in 1977. The fish consumption ban was reduced to 80 km in 1979. The initial reports stimulated a series of subsequent studies concerning the mercury problem in the receiving South River–South Fork Shenandoah River system.

11.3.2   Mercury in the South River-South Fork System

Bolgiano (1980) found mean mercury concentrations in fishes of 1.4 to 1.8 mg kg^{-1} at stations below the plant. The author found no significant mercury sources other than the DuPont site. Bolgiano (1980) estimated a maximum input to the river from the site of 25 g mercury per day, with an average of 7 g mercury per day. The study indicated contamination of millraces and floodplain soils. Bolgiano (1981) estimated that 37,154 kg mercury were located in the top 0.8 cm of the 100-year floodplain along the 40-km run of the South River below the plant. Todd (1980) found that sediment mercury decreased down-river from the duPont site. Sediment mercury peaked at 116 ppm. Control areas had sediment mercury levels of 0.06 ppm and lower. This study indicated that fish tissue mercury peaks were located 16.8 km from the plant site. Bass mercury averaged 1.94 ppm. By 1982, a Task Force (DuPont, EPA, and state agencies) issued a report showing no temporal trend of mercury concentrations in fishes taken from the South River system. By 1984, the state agencies and DuPont had entered into a 100-year mercury monitoring program that was undertaken by the Applied Marine Research Laboratory of Old Dominion University. This fish tissue mercury program was later taken over by the DEQ. This program was based on the assumption that the mercury in the South River would be reduced with time.

A report by Lawler, Matusky & Skelly Engineers (1982) reported that floodplain mercury was not available to the river. The authors estimated that 44,182 kg mercury were located in the river floodplain. The exact source of the mercury was not found. Over half of the bass mercury tissue burdens from fishes taken along 40 km of the South River downstream from the plant exceeded U.S. Food and Drug Administration (FDA) limits. Around 61% of the fishes exceeded such limits for 112 km of the South Fork Shenandoah River. It was estimated that, with no restoration action, there would be a 74% recovery of the South River in 100 years. The authors predicted that there would be a 38% recovery of the South Fork Shenandoah River over this time period. Partial sediment removal would lead to no immediate improvement of the South River, and it was thought that sediment removal in the South River would have negligible effects on the improvement of South Fork Shenandoah River. Lawler, Matusky & Skelly Engineers (1982) stated that the South Fork Shenandoah River system was too big for sediment removal based on cost estimates. There were not enough data for mitigation, according to the report. No action alternatives were evaluated, and the authors recommended monitoring as the management alternative. These recommendations evidently led to the 100-year monitoring program instead of active restoration initiatives.

A review of mercury concentrations in fishes from the 1981 sampling of the South River–Shenandoah River system indicated no trend in 4 years of sampling, although there were significant increases in mercury fish concentrations with peaks in down-river areas. Predator tissue mercury concentrations exceeded the 1.0-ppm FDA action level. During the 1980s, there were limited ecological studies carried out concerning aquatic insects and caddisflies. However, during this period, no comprehensive ecological studies were made in the South River–South Fork Shenandoah system. A 1989 report by Lawler, Matusky & Skelly Engineers indicated declines in the sediment mercury but increases in fish tissue mercury over previous sampling events.

A series of fish tissue mercury analyses was carried out during the 1990s. In a review of the 1992, 1994, and 1996 fish tissue mercury collections made during the 100-year monitoring program, Messing et al. (1997) found that methyl mercury in fishes from the study area were well above the FDA Action Level (1 ppm), with 95% of such concentrations around 2 to 3 times the Action Level. High mercury concentrations were found at various trophic levels. Fish predator mercury peaks at some stations were between 3 and 3.5 ppm, with an average of 3.116 ppm in 1996; this represented an increase in fish tissue mercury over previous years in the subject river system. The authors pointed out that the FDA Action Level for reproducing women and children was 0.0625 ppm methyl mercury (15 g day^{-1}). Harlow and Alden (1997) noted that that there was an upward fish tissue mercury trend from 1992 to 1994 and from 1994 to 1996. These studies showed an increase in tissue mercury that was contrary to the predictions of the Lawler, Matusky & Skelly Engineers (1982) report. The highest mercury concentrations were in South River predators, followed by foragers and bottom feeders. Harlow and Alden (1997) made conclusions similar to those reported by Messing et al. (1997).

According to a report by the Chesapeake Bay Foundation (1999), the mercury in fish tissues from the South River and South Fork Shenandoah River were considerably worse than Virginia officials had previously reported. The Foundation report quoted results from the Applied Marine Research Laboratory (Old Dominion University) that showed that, from 1992 to 1996, mercury fish tissue concentrations were higher than those found previously in the 1980s. The more recent mercury concentrations also extended farther down the river system, with increasing mercury along these areas. Unsafe fish mercury tissue concentrations were evident 184 km from the DuPont site. The report stated that the methyl mercury concentrations in bass, sunfish, carp, and crappie represented a human health threat. This report pointed to the river–floodplain sink as a continuing source of mercury to system. Contaminated soils washed into the river, and there were no plans for remediation. The basic assumptions of the report by Lawler, Matusky & Skelly Engineers

(1982) were thus disproved by more recent fish tissue findings, and the issue of restoration by proactive methods rather than long-term monitoring (wait-and-see) was discussed due to the postulated threat to humans who consumed fishes from the affected river areas.

To date, there have been no epidemiological studies of the mercury situation in the South River–South Fork Shenandoah River system. A cooperative program (in 2001, 2002) was carried out by DuPont and the Virginia state agencies. Letters were sent to physicians in the region concerning possible symptoms of mercury poisoning, but there was no response to these letters. This program did not represent an actual epidemiological analysis of the mercury threat to human health in the study area.

In 2000, the South River Science Team was formed. This group represented various interests, including DuPont, state and federal agencies, and private groups. The team first met in 2001 for the purpose of information exchanges and the formulation of a technical plan to determine how and why fish tissue mercury concentrations had increased in recent years. Stations had been established for fish mercury tissue samples in 1999. A 2001 report from the team indicated that the samplings from 1981 and 1997–1999 showed significant correlations between sediment and fish tissue. A South River Science Team Report in 2002. indicated that the DuPont site had a karst element, but it remained unclear whether or not groundwater had intercepted and transported mercury off site. It was reported that recent studies indicated that fish mercury concentrations had not declined and in some cases have increased over the past 20 years. Various hypotheses for this observation were offered.

The South River Science Team gave the most recent review of mercury concentrations in South River fishes in an October 2003 report. Data from combined years 1977–1983 were considered baseline for statistical purposes. Mercury concentrations were highest at various downstream areas.

11.3.3   Ongoing Studies

The 100-year mercury sampling program was related to findings by Lawler, Matusky & Skelly Engineers (1982), and was based on several assumptions:

1. Restoration by proactive methods was not necessary, and long-term monitoring (wait-and-see) should be instituted over a 100-year monitoring period.
2. Data from combined years 1977–1983 were considered baseline.
3. With no restoration action, there would be a 74% recovery of the South River in 100 years.
4. Fish and water sampling should be carried out at shorter intervals than sediment and floodplain sampling.
5. Floodplain sources were not important in the transfer of mercury to the river system.
6. Natural reductions of mercury in the South River–South Fork Shenandoah system would occur over the study period.

However, these assumptions were not consistent with what is known concerning the dynamics of mercury in other river systems. As noted above, the predictions of Lawler, Matusky & Skelly Engineers (1982) were not corroborated by recent mercury tissue analyses. The 100-year mercury monitoring program was thus deficient in various ways.

Despite various reports that there were considerable amounts of mercury in the floodplain, relatively few new data have been generated on this topic since Bolgiano (1981) reported that shallow soils in the floodplain of the South River downstream from the DuPont site were contaminated by mercury. The persistence of mercury in sediments and fish tissue over a period of over five decades indicated that the proposed Floodplain Soil Assessment Program was not adequate in scope or duration, and that there was a need to coordinate such a program with experimental work involving the generation and bioavailability of methyl mercury in floodplain areas. It was evident that these analyses should be accompanied by detailed determinations of the seasonal and interannual trends of mercury loading from the floodplain into the river system.

11.3.4 Resolution

In 2003, the Natural Resources Defense Council (NRDC) and the Virginia Chapter of the Sierra Club announced their intention to take DuPont to federal court to force a more complete study of the South River and South Fork Shenandoah River. It was suggested that the research should be integrated with various other ongoing studies that included the long-term sediment mercury work, a bimonthly water mercury program (started in 2000), a preliminary food web analysis, and some experimental clam (*Corbicula*) work. It was suggested that a triad approach be used, whereby chemical determinations of toxic substances would be made in sediments and water, along with an evaluation of the biological communities in the subject river area, and an experimental (bioassay) program with mercury to determine the ecological impact of mercury on key animal populations in the river. It was further suggested that results of this part of the study should be compared with similar results from a reference system. The triad would also be integrated with hydrogeological monitoring and modeling efforts, detailed floodplain analyses of stored mercury, associated mercury loading studies, water and sediment quality determinations, the results of the mercury residue analyses, experiments concerning methylation and bioavailability of mercury in source areas, biological monitoring of key aquatic assemblages, and a descriptive/experimental wildlife program. The feasibility of a restoration program would be based on the results of such a study.

After a series of meetings, DuPont representatives indicated their intent to do a comprehensive study of the mercury situation in the river system and, based on the findings, determine if restoration was economically and environmentally feasible. Currently, this program is being developed by the various parties in what can be viewed as a progressive way to resolve a potentially serious situation.

11.4 Dioxin in the Aquatic Environment

The 2,5,6-T-tetrachlorodibenzo-*p*-dioxins (TCDD) comprise a group of 75 tricylic aromatic compounds that contain various levels of chlorine saturation. One of these compounds, 2,3,7,8-tetrachlorodibenzo-*p*-dioxin (2,3,7,8-TCDD), is known as dioxin. The polychlorinated dibenzo-*p*-dioxins are formed during manufacture of all chlorophenols (Sittig, 1985). The highly toxic dioxin is formed during production of 2,4,5-trichlorophenoxy acetic acid (2,4,5-T).

11.4.1 Background of Dioxin in the Newark Bay Complex

The Newark Bay Complex includes Newark Bay and tidal portions of the Hackensack and Passaic Rivers, the Arthur Kill and the Kill Van Kull, and Newark Bay in northeastern New Jersey. The former Diamond-Alkali plant was located on the lower Passaic River in Newark, New Jersey (Belton et al., 1985), 5 km above Newark Bay. The plant produced an herbicide combination of a 50:50% solution of 2,4-dichlorophenoxy acetic acid (2,4-D) and 2,4,5-tricholorophenoxy acetic acid (2,4,5-T), also known as Agent Orange. The herbicide 2,4,5-T was contaminated by chemical by-products known as dioxins that were also incorporated in the Agent Orange. This herbicide mixture was used extensively by the United States military in Southeast Asia as a defoliant during the Viet Nam war, and there was also extensive use of 2,4-D and 2,4,5-T as herbicides in the United States as applications to various aquatic systems, home sites, forests, and a broad range of crops. Military use of Agent Orange was suspended due to reports of extreme toxicity to humans, and the U.S. EPA halted some domestic usage of this herbicide in 1970 due to extensive reports of adverse toxic effects (Belton et al., 1985).

In 1983, soil samples taken from the former Diamond-Alkali plant were contaminated with dioxins (Belton et al., 1985). Diamond Shamrock produced 15% of the 2,4,5-T in the United States, and was the principal source of 2,3,7,8-TCDD to the Newark Bay Complex and surrounding waters. Together with various publications concerning the environmental persistence and extreme toxicity of dioxin, the data precipitated an investigation of a study of dioxin contamination in the Passaic and Hackensack Rivers and the associated Newark Bay complex in New Jersey. Belton et al. (1985), in an early study of the distribution of dioxin in the Hackensack, Passaic, and Hudson Rivers and the Newark Bay complex, noted that the highest levels of sediment concentrations of dioxin were noted in areas adjacent to the Diamond-Alkali site despite the fact that the plant had not operated for 12 years prior to the survey. Concentrations ranged from nondetectable to 6.9 ppb. Only one site in areas tested showed detectable dioxin concentrations in sediments (and biota) that were not directly related to the Passaic River–Newark Bay system. These data indicated that there was either a continuous release of dioxin from the plant site and/or the dioxin in the Passaic River–Newark Bay system was highly persistent. The 6-mile stretch of the Passaic River that encompasses the plant site was declared an EPA Superfund area; this designation did not include the remaining parts of the Newark Bay complex.

Bopp et al. (1991) identified the Diamond-Alkali site as the dominant source of dioxin to the Newark Bay system; the authors noted that 2,3,7,8-TCDD deposition in Newark Bay sediments since the late 1940s approximated 4 to 9 kg, which represented the largest such releases noted in coastal areas that had been tested for dioxin. This site was also the dominant source of DDT and its derivatives to the Newark Bay complex. The persistence of anaerobic sediment deposits of 2,3,7,8-TCDD was noted on a time scale of several decades. Bopp et al. (1998) considered the lower Passaic River–Newark Bay system to be among the worst in the world with respect to sediment concentrations of 2,3,7,8-TCDD. They found that there was a decrease in 2,3,7,8-TCDD in sediments of the upper bay from around 1100 to 350 ppt in 1985. These numbers were comparable to the concentrations in the lower Passaic River sediments (7600 to 730 ppt).

Wenning et al. (1993) found several different sources of dioxin to Newark Bay. There were considerable ranges of concentrations in bay sediments. Their results did not support the idea of the Passaic River plant being the main source. There were five major fingerprint patterns in surficial sediments in Newark Bay, and none were related to the Passaic River plant. Ehrlich et al. (1994) used a mixing model for the 2,3,7,8-substituted TCDD and dibenzofurans for Newark Bay. The authors noted that recent sediment concentrations of these compounds had multiple sources in the Newark Bay system. Huntley et al. (1993, 1994, 1995) noted that the 2,4,6,8-TCDT concentrations in the Passaic River–Newark Bay system were not related to the Diamond-Alkali site. However, Chaky (2003), using sediment core analyses to identify indicators of sources of the polychlorinated dibenzo-p-dioxins in the lower Hudson River (New York urban environment) region, found that the ratio 2,3,7,8-TCDD:ΣTCDD was indicative of sediments influenced by 2,4,5-T production (i.e., the Passaic River plant). He found that this ratio was particularly high in Western Harbor (Newark Bay) sediments that were affected by the former 2,4,5-T manufacturing site on the Passaic River. Other areas in the lower Hudson drainage were also affected by this source, but to a lesser degree. Chaky (2003) noted that a generalized PCA (statistical) approach was not recommended for determination of sources as a polychlorinated dibenzo-*p*-dioxins and dibenzofurans (PCDD/F) pattern abundance, thus weakening the results of importance of the findings by Wenning et al. (1993) who used statistical methods for source determinations. Historical data indicated that Newark Bay sediments had high levels of 2,3,7,8-TCDD in the 1950s and 1960s, with the Passaic River plant as the source. Similar comparisons of recent samples indicated that the Passaic River plant is still responsible for 50% of the 2,3,7,8-TCDD deposition in Newark Bay (down from the 1960s but still strong). These data thus indicated that the Passaic River plant continues to be a source of 2,3,7,8-TCDD to Newark Bay and other portions of the lower Hudson area.

According to the New Jersey Department of Environmental Protection (NJDEP, 2000), dioxin-contaminated sediments from the Passaic River are "transported and continue to migrate to Newark Bay and other areas of the estuary." Such transport and deposition of dioxins from the site in the Passaic River thus "impact the ecological health and economic viability of the port of NY/NJ." Zongwei et al. (1994a,b), using analyses of tissue samples taken from blue crabs from the Newark-Raritan Bay system, indicated that the dominant source of dioxin was the Diamond-Alkali site. The PCDD/F levels decreased with increasing distance from the source. The authors found that isomers of 2,3,7,8-TCDD concentrations were significantly correlated with a related chlorinated aromatic compound (2,4,6,8-tetrachlorodibenzo-*p*-dioxin), indicating a common source (i.e., the Diamond-Alkali site).

Crawford et al. (1994, 1995) noted high levels of contaminant loading from publicly owned treatment facilities and combined storm water runoff and sewage treatment plants. This included various forms of spills of various contaminants, including petrochemical products. Gillis et al. (1993, 1995) considered mercury pollution a "hazard to aquatic biota" in the Newark Bay system. Gunster et al. (1993) noted the relatively high incidence of spills of petrochemical products into the Newark Bay system. Huntley et al. (1993, 1995) outlined the relatively high levels of polycyclic aromatic hydrocarbon and petroleum hydrocarbon contamination in the Newark Bay estuary. Finley et al. (1997) noted the distribution of the polychlorinated biphenyls in the lower Passaic River; the authors found no such observable PCB contamination in bass or blue crab muscle. Adams et al. (1996, 1998), in studies concerning sediment quality in the Newark Bay Complex, determined that multiple forms of contamination were widespread in this system. The authors determined that "Newark Bay was the most contaminated sub-basin, with 92% of its area exceeding an ERM [Effects Range Median] and 49% of the area showing a toxicological response." Mercury, chlordane, and PCBs were cited as having relatively pronounced effects on the ERM determinations. Dioxin and furan congeners were only analyzed in the Upper Harbor, Jamaica Bay, and the Lower Harbor; this did not allow direct application of the results to the dioxin question in Newark Bay. However, the bay was noted as the most contaminated part of the system, which is consistent with the hypothesis that Newark Bay constitutes a sink for persistent contaminants such as the dioxins.

11.4.2   Dioxin in Fish and Invertebrate Tissues

Belton et al. (1985) found that both resident and migratory fishes and crustaceans in the Passaic River–Newark Bay system had elevated levels of dioxin in edible tissues, and that these concentrations exceeded two "Levels of Concern" of the FDA (>25 parts per trillion [ppt] = no limits on human consumption; >50 ppt = no consumption). Blue crabs had the highest tissue concentrations of dioxin in the system. A risk assessment of the data confirmed that there was an unacceptable risk associated with the use of the blue crab fishery as a human food source. Belton et al. (1985) noted that high levels of boat traffic and dredging in Newark Bay could affect the distribution of dioxin in the bay, and that dioxin-contaminated sediments from the Passaic River would be constantly reworked by tidal displacements and would be transported out of the system by dredging and, secondarily, from disturbance-related sediment plume transport.

The New Jersey Department of Environmental Protection (2002; NJDEP) cited data from Hague et al. (1994) that showed maximum edible blue crab (*Callinectes sapidus*) concentrations of 480 parts per trillion (ppt) and mean edible muscle concentrations of 418 ppt in the area of the Passaic River near the Diamond-Alkali site. The New Jersey Department of Environmental Protection (2002) also cited blue crab data taken in Newark Bay with maximum dioxin levels in hepatopancreas (which is eaten by a certain percentage of people in the area) at 290 ppt and minimum levels at 98 ppt, for an average of 190 ppt. Blue crab muscle concentrations ranged from 17 to 4 ppt, with an average of 8 ppt. The New Jersey Department of Environmental Protection (2002) cited Chemical Land Holdings data 2001 for blue crab hepatopancreas ranges of 195 to 371 ppt (average of 267 ppt) and blue crab muscle concentrations ranging from 22.7 to 10.7 ppt (average of 17.7 ppt) for animals in the Passaic River. It was noted that blue crabs move throughout the Newark Bay Complex. The New Jersey Department of Environmental Protection (2002) conducted a blue crab consumption analysis based on concentrations of 2,3,7,8-TCDD and 2,3,7,8-TCDD equivalents taken from three separate studies in the Newark Bay Complex (see above). The estimated lifetime excess risk from consumption of the blue crabs from the Newark Bay Complex ranged from 5000 per million to 1.0 million per million (i.e., greater than a 100% risk). This was considered to be an "extremely high" risk.

Wolfe et al. (1996), in a review of sediment toxicity of various compounds in the Hudson–Raritan system, found that sediments in Newark Bay were toxic to amphipods. Newark Bay was one of the most toxic of the areas studied; however, the dioxins were not part of the sampling program and therefore could not be included in the analysis. The PAHs, and possibly mercury, were noted as the most likely sources of toxicity. A more recent report (National Oceanic and Atmospheric Administration, 2003) noted that sediment dioxins were toxic (in terms of EPA guidance evaluations) mainly in areas associated with the Newark Bay Complex. Similar toxicity maps were noted for DDT-R, mercury, and the PCBs. All contaminants in Newark Bay showed downward trends from the 1950s to the 1990s.

11.4.3   Newark Bay Ecology: Fate, Effects, and Restoration

Of the materials available for this review, there were virtually no studies that would qualify as fate-and-effects evaluations. The database was typical of most areas where there are numerous limited field and/or laboratory studies in a patch-quilt pattern that precludes useful systemwide conclusions concerning cause-and-effect relationships. Virtually no nutrient studies were noted in addition to a relatively poor ecological literature. This problem of interpretation of the impact of dioxin on Newark Bay is due, in part, to the limited and often fractured (piece-meal) scientific approach used by state and federal regulatory agencies when evaluating the effects of pollution on coastal systems. This problem is further exacerbated by the multiple sources of toxic agents in the study area.

There have been several attempts at ecological evaluations of the Newark Bay Complex. Crawford et al. (1994) reviewed conditions in the Newark Bay estuary over the past century. The authors noted significant reductions in aquatic species since the late 1800s due to increased industrialization and urbanization. Restoration activities over the past two decades indicate some (gradual) improvement in biological conditions over this period. Su et al. (2002) found that removal of submerged barges in the Passaic River resulted in increased concentrations of PCCD/F that exceeded water quality (regulatory) criteria. Models indicated that bioconcentration would not occur due to the instantaneous increases of these chemicals. However, bioconcentration and biomagnification would occur if there were long-term disturbances of contaminated sediments. Wallin et al. (2002), using historical data concerning Passaic River sediments, determined that the contaminated sediments would limit benthic invertebrates through the 20th century even though there has been a general decrease in sediment toxics concentrations over the past 50 to 60 years. None of the available scientific literature determined toxicant-specific alterations of the biota of the Newark Bay system, nor the direct connections concerning relationships of the multiple toxic agents in the system.

Background scientific information concerning dioxin (2,3,7,8-TCDD) in the Newark Bay Complex indicated that this compound fulfilled the criteria for environmentally damaging compounds in the aquatic habitat and thus presented a clear threat to environmental integrity and human health. Human populations at the top of the food web have been endangered due to the ability of dioxin to biomagnify up aquatic food webs and the propensity of this compound for cumulative toxic effects. There is little doubt that the Newark Bay Complex has experienced some of the highest levels of dioxin contamination in sediments and organisms ever recorded. The question of the current status of Newark Bay is not clear, however, due to the relatively poor scientific data in terms of fate and effects of this compound in this estuary. Recent data indicate that the Diamond-Alkali plant continues to be a source of 2,3,7,8-TCDD to Newark Bay and other areas of the lower Hudson area. Thus, it is highly likely that the original source of the contamination continues to load dioxin to the bay via the Passaic River, and that such dioxin continues to be cycled within the Newark Bay Complex. However, the dynamics of such recycling in terms of loading, dredging effects, egress from the systems, and related bioconcentration and biomagnification processes remain largely unknown.

Recent studies indicate that the sediments of Newark Bay are still toxic to various forms of aquatic organisms, and that this toxicity has been noted for DDT-R, mercury, the PCBs, and dioxin in sediments of Newark Bay. Although all these toxic agents show downward trends from the 1950s to the 1990s, Newark Bay is considered biologically impaired due to the presence of these compounds and continuous adverse effects due to discharges from multiple industrial sites and urban storm water effects. Thus, Newark Bay continues to be affected by a wide-ranging mixture of assorted toxic agents and nutrients from multiple sources, which tends to obscure the issue of how dioxin is affecting Newark Bay.

The dioxin problem has not been adequately addressed by state and federal regulators. The lack of epidemiological studies concerning people who eat contaminated food from the Newark Bay Complex further complicates an evaluation of the extent of the problem. The ongoing dredging of contaminated sediments of the bay is also a confounding issue. Because of the existence of toxic compounds in the Newark Bay sediments, a "no action" option (i.e., letting nature take its course in reducing the dioxin concentrations in the sediments and animals of Newark Bay) is not feasible. This underscores the need for a restoration program for the bay in an effort that matches that for the Passaic River. The likelihood of human health impacts due to dioxin contamination of Newark Bay underlies a need for more comprehensive studies of this area leading to a feasibility analysis of the restoration issue. Such studies are not currently being performed by the state and federal agencies. A combined "fate-and-effects" determination of the relative toxicity of dioxin in Newark Bay with a comprehensive epidemiological study of the public health implications of dioxin contamination is thus a logical extension of this review.

11.4.4   Legal Action and Regulatory Response

In 2003, the NRDC filed suit against the company responsible for dioxin contamination of the Newark Bay complex. The suit called for a study of Newark Bay to determine the feasibility of a restoration program for dioxin in this estuary. The legal action was nullified by actions taken by the U.S. EPA, which has now taken over the Newark Bay study.

11.5   Regulatory Requirements and the Restoration Process

The fact that mercury discharges to the Penobscot and Shenandoah River systems were permitted by state and federal regulatory agencies for decades without even a cursory scientific review of the consequences is not reasonable. This same lack of regulatory action regarding dioxin contamination of the Newark Bay Complex denotes a failure of the regulatory process. By avoiding restoration issues in the various systems noted above, the threat to human health has continued for decades in the respective river systems. The general lack of knowledge of fundamental food web issues that are of importance in the evaluation of the loading of toxic agents such as mercury and dioxin compounds the obvious problems related to analysis and initiation of effective restoration programs.

In the examples given, the regulation of important contaminants such as dioxin and mercury has not been taken seriously by state and federal agencies. The lapses in regulatory action illustrated here involve ecological and health issues that have been largely ignored by the mainstream press.

11.6   Press Response to Toxic Substances

The press reports of toxics such as dioxin and mercury are redundant and stereotypic, with an emphasis on human impacts and controversy rather than the causes of such problems and the processes involved in the inevitable threats to the aquatic environment and human health. Typical of such coverage by the press are the following reports:

> MERCURY REGULATIONS UNDER REVIEW: "The Bush administration is working to undo regulations that would force power plants to sharply reduce mercury emissions and other toxic pollutants, according to a government document and interviews with officials....'It looks as if the administration is going totally in the tank with the utility industry, in a flat-out violation of the law,' said David Hawkins of the Natural Resources Defense Council. 'We know this stuff is bad for kids, but they don't care.'...At a time when 41 states have fish-consumption advisories due to mercury poisoning, it is unconscionable that EPA is proposing to postpone and weaken regulatory protection,' (S. William) Becker said."

> —**Eric Pianin,** *Washington Post,* **December 3, 2003**

> FISHY MERCURY WARNING: "The Food and Drug Administration just issued a new warning to pregnant women about mercury in seafood. You can 'protect your baby' from developmental harm by following three rules, claims the FDA. But there's no evidence that the rules will protect anyone and they're only likely to foster undue concern about an important part of our food supply. ... There is no evidence of a general threat to infants and children from typical maternal consumption of fish with typical mercury concentrations.... In fact, 'a surprising

finding in the results of the examination of children at 66 months of age was that several [intelligence] tests scores improved as either pre- or post-natal mercury levels increased ... linear regression analysis reveals significant beneficial correlations.' ...It seems the FDA is warning (scaring?) us about a scenario that has, essentially, never occurred."

—Steven Milloy, Fox News, December 25, 2003

EPA MAY PROLONG MERCURY CLEANUP DEADLINE: "The Bush administration is leaning toward stretching out plans for reducing mercury pollution from power plants until 2018 after concluding that technology for quick cuts isn't available. Some plants would be able to buy their way out of reducing emissions. The Environmental Protection Agency had offered options three months ago for reducing the 48 annual tons of mercury emitted from 1,100 coal-burning power plants, the largest source of the pollution...But studies cosponsored by the Department of Energy and the utility industry have found there was no existing technology to remove mercury equally well from various types and grades of coal. That leaves the second strategy — endorsed by industry — that would establish a nationwide cap of 15 tons on mercury pollution by 2018 by phasing in lower ceilings on each plant's pollution. ...EPA can turn to that approach only because the Bush administration decided in December that mercury should not be regulated as a toxic substance requiring maximum pollution controls, reversing a Clinton administration determination. To meet a court-ordered deadline in a lawsuit brought by the Natural Resources Defense Council 12 years ago, the agency must issue a final decision before the end of 2004."

—Associated Press, March 22, 2004

It would seem, according to Fox News, that children will actually improve their level of intelligence by consuming mercury.

Numerous articles in the news media have appeared through the years concerning mercury pollution. The same theme continues throughout:

1. Documentation of the ecological effects of mercury is rarely mentioned, nor is the nature of the food web, as a key element in the threat to humans, considered.
2. The controversial and political aspects of the mercury problem are emphasized, with little regard for the scientific findings that underlie the reasons for increasing mercury in various aquatic systems.
3. Sensationalism underlies the various discussions of the mercury problem in the mainstream press, and this is met with dubious counterclaims by reactionary press elements (Fox News).
4. The primary issues that have been shown by the vast majority of scientific inquiries into the ecological effects of toxic agents such as mercury and dioxin are trivialized in favor of pumping up the conflict between environmentalist forces and industrial/political/regulatory interests. The lack of regulation of such substances takes a back seat to the sensationalism engendered by such media reports. Amid the controversy, nothing is done to improve the situation, and the current administration continues to dilute the already feeble regulatory process to control the loading of toxic substances into the environment.

In 2003, an article in *Science* (Harris et al., 2003) indicated in experiments with zebrafish that methyl mercury in this species is bound to cysteine, and that this compound is less toxic to such fish than methyl mercury chloride. This finding implied that methyl mercury may be less readily absorbed than previously noted, and consequently would bring into question current health standards that would exaggerate the risks involved in the consumption of fishes contaminated with mercury. The implications of the findings of Harris et al. (2003) were subsequently countered by other scientists. Stern (2004) noted that the complexing of mercury with cysteine was not new to science, and that the current standards and advisories were based on mercury concentrations in blood and hair, with no assumptions concerning the form of methyl mercury in fishes. Thus, Stern noted that the findings of Harris et al. (2003) had no relevance with regard to the public health implications of mercury in fishes.

Hudson and Shade (2004) noted that Harris et al. (2003) ignored the ready bioavailability of the methyl mercury–cysteine complex to mammalian cells, thus compromising the extrapolations of zebrafish to humans made by the authors. Ekino et al. (2004) noted that the epidemic of neurological disorders in the Minamata disease in Kumanoto, Japan, was closely associated with human consumption of fishes contaminated with methyl mercury. Consumption of mercury-contaminated fishes and shellfish caused more than 2000 cases of Minamata disease (degeneration of the human cerebrum) and more than 10,000 suspected cases through 1996. The extreme toxicity of methyl mercury to fetuses and adult humans in the Minamata incident was scientifically demonstrated according to Ekino et al. (2004), thus establishing the toxicity of methyl mercury to humans as well as other mammals. Harris et al.'s (2004) response was simply that it would be "rash to extrapolate our very limited data on zebrafish to humans, and we did not do this. Our point was merely to illustrate that methyl mercury species can have different toxicities."

The significance of the relatively extensive scientific literature regarding the effects of methyl mercury on aquatic systems remains in limbo as far as the overall media response to the political and human health issues regarding this compound. The public remains ignorant of even the most basic concerns regarding the significance of the ever-increasing mercury problem in aquatic systems. Specific issues such as those outlined above regarding mercury contamination in Maine and Virginia have been marginalized by press reports in favor of the usual scare tactics and controversy that is so favored by virtually the entire range of the American press. This pattern has been followed with other toxic agents such as dioxin. It has been exaggerated in areas that are not strongly contaminated (Chapters 4 and 5), whereas areas where people are eating seafood that is highly contaminated, such as the Newark Bay complex (see above), are not only ignored but have remained almost completely unregulated for decades. The results of the avoidance of scientific data by the news media has led to disruption in the remediation of actual ecological impacts, relaxed regulation of compounds such as methyl mercury and dioxin by politicians and their bureaucratic operatives, and confusion of the public concerning important environmental issues. The drive to sell the news through sensationalism, omission of the scientific facts, and promulgation of misinformation prevails.

section V

Alternatives: Planning and Management

There is an alternative to restoration: resource planning and management. This approach is analogous to preventive medicine in that it requires scientific evaluations and effective planning and management programs that are designed to protect important resources from undesignated future impacts. This alternative involves the gathering of scientific information that can be used for management purposes (Livingston, 2002). This information should include inventories of important environmental resources; lists of rare, endangered, and threatened species; a review of key sports and commercial fisheries; other critical habitat information that shows unique and/or economically viable environmental assets; and a definition of the processes that contribute to the productivity of the chosen system. Environmental, cultural, and socioeconomic assets should be inventoried to make an objective case for resource protection.

Based on this information, there should be a cooperative effort at the local, state, and federal levels to form a comprehensive plan to protect the demonstrated environmental assets of the region. The creation of a multidisciplinary task force is necessary in the development of a resource management plan. The economic assets associated with a management program should be part of the overall evaluation. Although few resource management plans can depend solely on the economic aspects of protection of natural productivity, there are often cultural assets associated with the economy of region that supersede the simple worth of the system in terms of dollars alone.

The problem with this approach is that it requires a long-term commitment that depends on objective reasons for the preservation and/or conservation of a given aquatic system. There are systems of preserves, reserves, and other designated "save" areas. While such designations provide some emphasis on conservation, they do not necessarily protect such systems in perpetuity. There has been a long line of successful efforts in the form of national parks and preserved terrestrial areas, but seldom are there aquatic analogs to such parks on a scale that preserves the basic attributes that ultimately protect the natural productivity of the system in question. The lack of regulation regarding both agricultural and urban development in aquatic systems represents a real threat to meaningful resource management planning in most areas, and there usually is a combination of factors that contribute to a successful management program that includes serendipitous factors that are not predictable at the outset. The history of the Apalachicola system bears testament to this generalization.

chapter 12

The Apalachicola System

The Apalachicola estuary (Figure 12.1) approximates 62,879 ha, and is a shallow (mean depth: 2.6 m) lagoon-and-barrier-island complex. The Apalachicola River dominates the bay system as a source of freshwater, nutrients, and organic matter; together with local rainfall, the river is closely associated with the salinity and coastal productivity of the region (Livingston, 1975b, 1976a,b, 1977, 1980b, 1982b, 1983a,b, 1984b,c, 1985c, 1988c, 1990, 1991a,b, 1993c; Livingston and Joyce, 1977; Livingston and Duncan, 1979; Livingston et al., 1974, 1976b, 1978, 1997, 1999, 2000, 2003). The Apalachicola drainage system remains in a relatively natural state with sparse human population, and little industrial and municipal development (Livingston, 1984b). Water movement in the estuary is controlled by wind currents and tides because of the generally shallow depths (Livingston et al., 1999).

12.1 Background

Temperate, river-dominated estuaries are among the most productive and economically valuable natural resources in the world. Loading of nutrients from associated alluvial rivers contributes to such productivity (Howarth, 1988; Howarth and Marino, 1998; Howarth et al., 1995; Baird and Ulanowicz, 1989), and this loading provides the stimulus for autochthonous phytoplankton production. River-driven allochthonous particulate organic matter maintains detritivorous food webs in estuaries (Livingston, 1984a,b). However, the relative importance of various sources of both organic carbon (dissolved and particulate) and inorganic nutrients can vary from estuary to estuary (Peterson and Howarth, 1987). These sources can be related to the specific tidal and hydrological attributes of a given system (Odum et al., 1979, 1982). Human sources of nutrients and organic matter often have the exact opposite effect, leading to cultural eutrophication, phytoplankton blooms, deterioration of the estuarine food webs, and severe loss of secondary production (Livingston, 2000, 2002).

The Apalachicola River–Bay system is part of a major drainage area (the Apalachicola–Chattahoochee–Flint [ACF] basin) of about 48,500 km^2 located in western Georgia, southeastern Alabama, and northern Florida. There are 13 dams on the Chattahoochee River and three dams on the Flint River. The undammed Apalachicola River is 21st in flow magnitude in the conterminous United States, and flows 171 km from the confluence of the Chattahoochee and Flint Rivers (the Jim Woodruff Dam) to its terminus in the Apalachicola estuary. Mean flow rates approximate 690 m^3 sec^{-1} (1958–1980), with annual high flows averaging 3000 m^3 sec^{-1} (Leitman, 2003a,b,c,d; Leitman et al., 1991). The forested floodplain, about 450 km^2, is the largest in Florida (Leitman et al., 1982, 1983), with forestry as the primary land use in the floodplain (Clewell, 1977). Other activities include minor

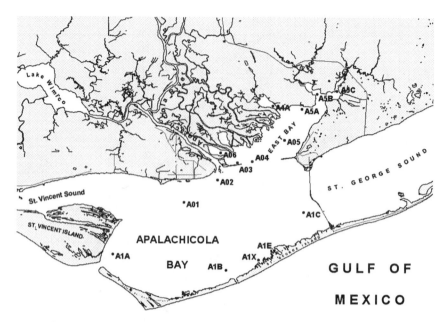

Figure 12.1 The Apalachicola River-Bay system showing long-term sampling stations for studies that were carried out from 1972 to 1991. Geographic data provided by the Florida Geographic Data Library (FGDL).

agricultural and residential use, bee keeping, tupelo honey production, and sports/commercial fishing (Livingston, 1984a,b).

12.2   Apalachicola River Flows

The Apalachicola River is one of the last major free-flowing, unpolluted alluvial systems in the conterminous United States. The importance of freshwater flows to the Apalachicola floodplain has been extensively studied (Cairns, 1981; Elder and Cairns, 1982; Mattraw and Elder, 1982; Light et al., 1998). The Apalachicola River system has the greatest flow rates of all the river drainages along the northeast Gulf. Apalachicola River nutrient loading to the estuary is the highest of the major alluvial river systems along the Gulf coast (Livingston, 2000) and remains relatively high without apparent hypereutrophication in the bay. River flow rates from 1950 to 2003 have been characterized by several major drought events (1954–1955, 1968–1969, 1980–1981, 1987–1988, and 1999–2002). In terms of river flow, the most recent drought was the most extreme, with relatively low minimum and maximum rates of flow.

12.2.1   Apalachicola Floodplain

Based on a long history of management efforts (Livingston, 2002), the unique characteristics of the river–floodplain remain largely intact, a notable exception to the condition of most alluvial waterways in the United States today. The Apalachicola floodplain represents an important source of biological diversity at various levels of organization (Livingston and Joyce, 1977):

1. The Apalachicola River is the only river in Florida to go from the Piedmont to the Gulf of Mexico. The Apalachicola drainage basin receives biotic exchanges from the Piedmont, the Atlantic Coastal Plain, the Gulf Coastal Plain, and peninsular

Florida. This accounts for the high quality of the terrestrial animal biota of the river floodplain (Means, 1977).

2. Floodplain forests include numerous terrestrial plants that are narrowly endemic, endangered, threatened, and rare species (Clewell, 1977).

3. Of all north Florida drainages, the Apalachicola River contains the largest number of freshwater bivalve and gastropod mollusks, with high endemism and a number of rare and endangered species (Heard, 1977).

4. Eighty-six fish species have been noted in the Apalachicola River system, including three endemics, various important anadromous species, and species that form the basis for important sports and commercial fisheries (Yerger, 1977).

5. The Apalachicola River wetlands are a center of endemism for various terrestrial species, to include endangered, threatened, and rare species of amphibians, reptiles, and birds (Means, 1977). Due to the high diversity of wetland and upland habitats, the highest species density of amphibians and reptiles in North America (north of Mexico) occurs in the upper Apalachicola basin.

The importance of the Apalachicola floodplain is also related to various freshwater fisheries, although most of the more important fisheries (e.g., striped bass, *Morone saxatilis*; sturgeon, *Acipenser oxyrhynchus*) have been destroyed or seriously impaired due to habitat destruction by channelization and damming (Livingston and Joyce, 1977; Livingston, 1984a). Dredging activities, mandated by the U.S. Congress and continuing to the present time, have led to serious habitat damage along the river, with a minimum of economic justification for such channelization (Leitman et al., 1991). Nevertheless, the Apalachicola River wetlands system remains largely intact, and is one of the few such systems that is almost completely in public hands.

12.3 Linkage between the Apalachicola River and the Bay

The association between alluvial freshwater input and estuarine productivity has been indirectly established in a number of estuaries (Cross and Williams, 1981). Deegan et al. (1986), using data from 64 estuaries in the Gulf of Mexico, found that freshwater input was highly correlated ($R = 0.98$) with fishery harvest. Armstrong (1982) determined that nutrient budgets in Texas Gulf estuaries were dominated by freshwater inflows, and that shellfish and finfish production was a function of nutrient loading rates and average salinity. Funicelli (1984) found that upland carbon input was in some way associated with estuarine productivity. However, few studies actually evaluated the various facets of linkage of the freshwater river–wetlands and estuarine productivity (Livingston, 1984b). As a response to the projections of anthropogenous freshwater use by the state of Georgia over the next 30 to 50 years (Livingston, 1988c), a long-term analytical program was initiated by our research group, using databases generated during the 1970s and 1980s, to determine how projected reduced flows of the Tri-river system would affect the Apalachicola River–bay system.

Published results of the long-term bay research program included hydrology (Meeter and Livingston, 1988; Meeter et al. 1979), the effects of anthropogenous activities such as agriculture (Livingston et al., 1978) and forestry (Duncan, 1977; Livingston and Duncan, 1979; Livingston et al., 1976b), and the importance of salinity to the community structure of estuarine organisms (Livingston, 1979). The basic distribution of the estuarine populations was analyzed (Edmiston, 1979; Estabrook, 1973; Laughlin and Livingston, 1982; Livingston, 1976a, 1977, 1981, 1983a; Livingston et al., 1974; 1976a,b, 1977; Mahoney, 1982; Mahoney and Livingston, 1982; McLane, 1980; Purcell, 1977; Sheridan, 1978, 1979; Sheridan and Livingston, 1979, 1983). Various studies were also carried out concerning the trophic organization of the estuary (Federle et al., 1983a,b, 1986; Laughlin, 1979; Livingston et al., 1997; Sheridan, 1978; White, 1983; White et al., 1977, 1979a,b). Studies were made concerning the distribution of wetland vegetation in the Apalachicola floodplain (Leitman et al., 1982). It was determined that vegetation type was associated with water depth, duration of inundation and saturation, and water-level fluctuation. Stage range is reduced considerably downstream, indicating a dampening of the river flood stage by the expanding (downstream) wetlands.

Litter fall in the Apalachicola floodplain (800 gm^{-2}) is higher than that noted in many tropical systems and almost all warm temperate systems. The litter fall of these systems is on the order of 386 to 600 gm^{-2} (Elder and Cairns, 1982). The annual deposition of litter fall in the bottomland hardwood forests of the Apalachicola River floodplain approximates 360,000 metric tons (mt). Seasonal flooding provides the mechanism for mobilization, decomposition, and transfer of the nutrients and detritus from the wetlands to associated aquatic areas (Cairns, 1981; Elder and Cairns, 1982) with a postulated, although unknown input from groundwater sources. Studies (Livingston et al., 1974, 1976b) indicated that, in addition to providing particulate organics that fueled the bay system, river input provided ample nutrient loading to the estuary. Of the 214,000 mt of carbon, 21,400 mt of nitrogen, and 1650 mt of phosphorous that is delivered to the estuary over the period of a given year, over half is transferred during the winter–spring flood peaks (Mattraw and Elder, 1982).

The above-mentioned studies noted that the delivery of nutrients and dissolved/particulate organic matter was an important factor in the maintenance of the estuarine primary production (autochthonous and allochthonous). There were distinct links between the estuarine food webs and freshwater discharges (Livingston, 1984b; Livingston and Loucks, 1978). The total particulate organic carbon delivered to the estuary followed seasonal and interannual fluctuations that were closely associated with river flow (Livingston, 1984b; $R^2 = 0.738$). The exact timing and degree of peak river flows relative to seasonal changes in wetland productivity were important determinants of short-term fluctuations and long-term trends of the input of allochthonous detritus to the estuary (Livingston, 1984b). During summer and fall months, there was no direct correlation of river flow and detritus movement into the bay. By winter, there was a significant relationship between micro-detrital loading and river flow peaks.

Up to 50% of the phytoplankton productivity, which is the most important single source in overall magnitude of organic carbon to the bay system, is explained by Apalachicola River flow (Myers, 1977; Myers and Iverson, 1977, 1981). During winter–spring periods of high river flow, there are major transfers of nutrients and organic matter to the estuary. Boynton et al. (1982) reported that the Apalachicola system has high phytoplankton productivity relative to other river-dominated estuaries, embayments, lagoons, and fjords around the world. Wind action in the shallow Apalachicola Bay system is associated with periodic peaks of phytoplankton production as inorganic nutrients, regenerated in the sediments, are mixed through turbulence into the euphotic zone (Livingston et al., 1974; Iverson et al., 1997). Nixon (1988a) showed that the Apalachicola Bay system ranks high in overall primary production compared to other such systems. Iverson et al. (1997) noted that there had been no notable increase in chlorophyll *a* concentrations in Apalachicola Bay during the previous two decades despite increases in nitrogen loading due to increased basin deposition of this nutrient. They found that dissolved silicate did not limit phytoplankton production in the largely mesotrophic Apalachicola Bay.

In the Apalachicola system, orthophosphate availability limited phytoplankton during both low and high salinity winter periods and during the summer at stations with low salinity. Nitrogen, on the other hand, was limiting during summer periods of moderate to high salinity in the Apalachicola estuary (Iverson et al., 1997). Light and temperature limitation was highest during winter–spring periods, thus limiting primary production during this time. High chlorophyll *a* levels during winter periods were attributed to low zooplankton grazing during the cooler months (Iverson et al., 1997). Nitrogen input to primary production was limited by the relatively high flushing rates in the Apalachicola system. Flow rates affected the development of nutrient limitation in the Apalachicola system, with nutrient limitation highest during low-flow summer periods.

Recent studies have documented river influence on nutrient and organic carbon loading to the bay. Chanton and Lewis (1999) found that, although there were inputs of large quantities of terrestrial organic matter, net heterotrophy in the Apalachicola Bay system was not dominant relative to net autotrophy during a 3-year period. Chanton and Lewis (2002), using $\delta^{13}C$ and $\delta^{34}S$ isotope data, noted clear distinctions between benthic and water column feeding types. They found that the estuary depended on river flows to provide floodplain detritus during high-flow periods, and dissolved nutrients for estuarine primary productivity during low flows. Floodplain detritus was significant in the important East Bay nursery area, thus showing that peak flows were important in washing floodplain detritus into the estuary. Peak levels of macrodetrital accumulation occurred during winter–spring periods of high river flow (Livingston, 1984b). These periods were coincident with increased infaunal abundance (McLane, 1980). Four out of the five dominant infaunal species at river-dominated stations were detritus feeders. A mechanism for the direct connection of increased infaunal abundance was described by Livingston (1983a, 1984b), whereby microbial activity at the surface of the detritus (Federle et al., 1983a) led to microbial successions (Morrison et al., 1977) that then provided food for a variety of detritivorous organisms (White et al., 1979a,b; Livingston, 1984b). The transformation of nutrient-rich particulate organic matter from periodic river-based influxes of dissolved and particulate organic matter coincided with abundance peaks of the detritus-based (infaunal) food webs of the Apalachicola system (Livingston and Loucks, 1978) during periods of increased river flooding. Chanton and Lewis (2002) provided analytical support for these observations.

Mortazavi et al. (2000a,b,c) found that phytoplankton productivity in river-dominated parts of the Apalachicola estuary was limited by phosphorus in the winter (during periods of low salinity) and by nitrogen during summer periods of high salinity. The dissolved organic nitrogen (DON) input was balanced by export from the estuary. Mortazavi et al. (2000c) gave detailed accounts of the nitrogen budgets of the bay. However, 36% of the dissolved organic phosphorus (DOP) was retained in the estuary where it was presumably utilized by microbes and primary producers (Mortazavi et al., 2000a). Mortazavi et al. (2000b) determined temporal couplings of nutrient loading with primary production in the estuary. Around 75% of such productivity occurs from May through November, with main control due to grazing. The data indicated that altered river flow, especially during low-flow periods, could adversely affect overall bay productivity. These studies indicated that phytoplankton productivity was an important component of estuarine food webs along the Gulf coast, and that a combination of river-derived organic matter and autochthonous organic carbon provided the resources for consumers in Gulf coast river-dominated estuaries.

Reductions in overall Apalachicola River flow rates due to anthropogenous use of freshwater in the Chattahoochee and Flint Rivers would eventually threaten and destroy the natural biota of this highly productive system (Light et al., 1998). In addition, it would jeopardize millions of dollars of investments by the people of Florida in the various wetlands purchases and management efforts over the past 30 years (see below) as the wetlands would disappear as a result of reduced flooding.

12.4   Freshwater Flows and Bay Productivity

Apalachicola Bay ecology is closely associated with freshwater input from the Apalachicola River and local sources such as drainages in East Bay and St. George Island (Livingston, 1984b). The distribution of epibenthic organisms in the Apalachicola Estuary follows a specific spatial relationship to high river flows. Stations most affected by the river are inhabited by anchovies (*Anchoa mitchilli*), spot (*Leiostomus xanthurus*), Atlantic croaker (*Micropogonias undulatus*), gulf menhaden (*Brevoortia patronus*), white shrimp (*Litopenaeus setiferus*), and blue crabs (*Callinectes sapidus*). The outer bay stations are often dominated by species such as silver perch (*Bairdiella chrysoura*), pigfish (*Orthopristis chrysoptera*), least squid (*Lolliguncula brevis*), pink shrimp (*Farfantepenaeus duorarum*), brown shrimp (*Farfantepenaeus aztecus*), and other shrimp species (e.g., *Trachypenaeus constrictus*). Sikes Cut, an artificial opening to the Gulf maintained by the U.S. Army Corps of Engineers, is characterized by salinities that resemble the open gulf. This area is dominated by species such as squid, anchovies, *Cynoscion arenarius*, *Etropus crossotus*, *Portunus gibbesi*, and *Acetes americanus*.

Cross-correlation analysis of the long-term data indicated that the various dominant Apalachicola Bay system populations followed a broad spectrum of diverse phase interactions with river flow and associated changes in salinity. River flow, as a habitat variable, is thus a controlling factor for biological organization of the Apalachicola estuary (Livingston, 1991c). The long-term (14-year) trends of the distribution of invertebrates such as penaeid shrimp indicate that such numbers are associated in various ways with river flow. Fish populations also follow diverse, species-specific phase angles with river flow trends. Overall fish numbers peak 1 month after river flow peaks (winter periods), whereas invertebrate numbers are inversely related to peak river conditions with increases during the summer months (Livingston, 1991c). These data are understandable in that top fish dominants such as spot are prevalent in winter–spring months of river flooding, whereas peak numbers of penaeid shrimp usually occur in summer and fall months. Other top dominants such as anchovies reach numerical peaks 3 months before the Apalachicola River floods. Fish biomass has a significant positive correlation with river flow at monthly lags 2 and 3, whereas invertebrate biomass showed a significant positive correlation with river flow peaks at monthly lag 4 (Livingston, 1991c). Cross-correlation analyses demonstrated that numbers of species of fishes are positively associated with peak river flows. Fish numbers peak 1 month after river flow peaks, whereas invertebrate numbers are inversely related to peak river flow conditions with major increases during the summer months (Livingston, 1991c). The response of the bay was complex due to species-specific responses to the river-directed habitat changes and responses of the food web to nutrient loading and phytoplankton production.

In terms of frequency of occurrence during the long-term sampling effort (1972–1984), the infaunal macroinvertebrate assemblages in East Bay were dominated by species such as *Mediomastus ambiseta* (below-surface deposit feeder and detritivorous omnivore), *Hobsonia florida* (above-surface deposit feeder and detritivorous omnivore), *Grandidierella bonnieroides* (grazer/scavenger and general omnivore), *Streblospio benedicti* (above-surface deposit feeder and detritivorous omnivore), and *Parandalia americana* (primary carnivore). Larger types of infaunal macroinvertebrates included the plankton-feeding herbivores *Mactra fragilis* and *Rangia cuneata*. Dominant epibenthic macroinvertebrates in East Bay over the period of study included the palaemonetid shrimp (*Palaemonetes* spp.: detritivorous omnivores), xanthid crabs (*Rhithropanopeus harrisi*: primary carnivores), blue crabs (*Callinectes sapidus*: primary carnivores at <30 mm; secondary carnivores at >30 mm), and penaeid shrimp (*Farfantepenaeus setiferus*, *F. duorarum*, and *F. aztecus*: primary carnivores at <25 mm; secondary carnivores at >25 mm). Most of these invertebrate species are browsers, grazers, or seize-and-bite predators.

Dominant fishes in East Bay were the plankton-feeding primary carnivore *Anchoa mitchilli* (bay anchovy) and benthic feeding primary carnivores such as spot (*Leiostomus xanthurus*), hogchokers (*Trinectes maculatus*), young Atlantic croakers (*Micropogonias undulatus*: <70 mm) and silver perch (*Bairdiella chrysoura*: 21–60 mm). Secondary carnivores among the dominant fishes included larger croakers (>70 mm), Gulf flounder (*Paralichthys albigutta*), and sand seatrout (*Cynoscion arenarius*). Tertiary carnivores in East Bay include the larger spotted seatrout (*C. nebulosus*), southern flounder (*P. lethostigma*), largemouth bass (*Micropterus salmoides*), and gars (*Lepisosteus* spp). With the exception of the bay anchovies, all of the above species live near the sediment/water interface, with most of the trophic organization of the bay dependent on interactions among bottom living infaunal and epibenthic macroinvertebrates and fishes.

Factors that determine the currently high production of shrimp, blue crabs, and sciaenid fish populations in the Apalachicola Bay system are related to the river flow effects on habitat variables (salinity), nutrient loading, phytoplankton production, and the response of the estuarine food webs to spatial/temporal trends of primary productivity (Livingston, 2000, 2002). Livingston et al. (1997) found that within limited natural bounds of freshwater flow from the Apalachicola River, there was little change in the trophic organization of the Apalachicola estuary over prolonged periods. The physical instability of the estuary was actually a major component in the continuation of a biologically stable estuarine system. However, when a specific threshold of freshwater reduction was reached during a prolonged natural drought, there was evidence that the clarification of the normally turbid and highly colored river–estuarine system led to rapid changes in the pattern of primary production, which, in turn, were associated with major changes in the trophic structure of the system. Increased light penetration due to the cessation of river flow was postulated as an important factor in the temporal response of bay productivity and herbivore/omnivore abundance.

With trophic organization expressed as total biomass m^{-2} yr^{-1}, there was a clear relationship between the mean annual river flow rates and the overall animal (infauna, macroinvertebrate, fish) biomass in East Bay (Livingston et al., 1997). There were significant ($P < 0.05$) seasonal and interannual differences in biomass; however, during the first 5 years of sampling, river flow and total animal biomass remained within a relatively small level of interannual variance. Significantly ($P < 0.05$) different biomass levels were noted during years 1980, 1981, 1982, and 1983. Peak biomass years (1980–1981) coincided with major reductions in river flow and were due largely to the increases in the herbivore component. The significant decrease in biomass, which began late in the drought, continued throughout the 2-year recovery period (1982–1983). Livingston et al. (1997) noted that there was a dichotomous response of the estuarine trophic organization of the Apalachicola Estuary. Herbivores and omnivores were primarily responsive to river-dominated physicochemical factors, whereas carnivores responded to the trophic organization at lower levels (Livingston et al., 1997). There was a major shift in the overall trophic factors during the drought of 1980–1981. Trophic response time could be measured in months to years from the point of the initiation of low flow conditions. The reduction in nutrient loading during the drought period was postulated as a major cause of the loss of productivity of the river-dominated estuary during and after the drought period. Recovery of such productivity with resumption of increased river flows was likewise a long-term event.

There was considerable interannual variation in river flows, which was reflected by the temporal distribution of the dominant fish species in the bay. Individual estuarine invertebrate and fish species used the estuary as a nursery ground, with species-specific ontogenetic feeding patterns that were defined by the complex productivity patterns of the system. Estuarine food web organization was indirectly responsive to changes in river flow through prey responses to state habitat and productivity variables associated with river flows. This suggests that the fish and invertebrate associations were strongly dependent on interannual patterns of Apalachicola River flow, but that such relationships were primarily caused by biological interactions as defined by specific predator/prey relationships (i.e., food web processes). A prolonged drought during the early 1980s led to reduced fish and invertebrate species richness and trophic diversity (Livingston et al., 1997); such habitat stress was related to enhanced instability of the biological components of the estuary as a function of changes in nutrient cycling. The food web was simplified while overall fish biomass and individual species populations were numerically reduced. Changes in flow rates that exceeded specific natural levels of variance could be followed by identification of the subtle yet important changes in estuarine productivity and related changes of fish representation within the food web.

The individual trophic units of each species represent a series of transitional stages whereby the growth stages, as organized by individual trophic entities, occupy different habitats over a given seasonal period. The general occupation of habitats associated with freshwater runoff by most of the dominant bay species of fishes and invertebrates is qualified by temporal movements and changes in trophic needs, which are identified as species-specific growth patterns. The success of an early trophic unit does not necessarily mean high numbers of successive trophic units. Thus, the varied phase angles of different species to river flow events are further qualified by differential success of the different trophic units over time. The complex shifts of trophic units through time, although generally associated with river-driven primary production in the form of allochthonous and authochthonous food sources, is evidence that the complete range of intra- and interannual changes in river flows is necessary for the long-term productivity and biodiversity of the Apalachicola Bay system. Some species are favored by high flows, some by droughts, and other by intermediate flow rates. Therefore, to maintain bay productivity and biodiversity, river flows should follow historical patterns to which the system has become adapted over thousands of years of co-evolution. Future freshwater needs of the estuary should not be managed by any single species, but should be projected based on the trophic integrity of the river–bay system.

Conditions in the Apalachicola Bay system are highly advantageous for oyster propagation and growth (Menzel, 1955a,b; Menzel and Nichy, 1958; Menzel et al., 1957, 1966; Livingston, 1984b) with reefs covering about 7% (4350 ha) of bay bottom (Livingston, 1984b). Mass spawning takes place at temperatures between 26.5 and 28°C, usually from late March through October (Ingle, 1951). Growth rates of oysters in this region are among the most rapid of those recorded (Ingle and Dawson, 1952, 1953), with harvestable oysters taken in 18 months. Overall, the oysters in the Apalachicola region combine an early sexual development, an extended growing period, and a high growth rate (Hayes and Menzel, 1981); effective spawning is restricted to older oysters, although young-of-the-year are able to spawn.

Livingston et al. (1999) outlined the response of the Apalachicola oyster population to hurricane impacts. A detailed analysis of oyster natural history was provided. Hurricanes are common along the Gulf Coast during the spawning period of the oysters; it appears that *Crassostrea virginica* is well adapted for such natural disturbances, with population recovery dependent to a considerable degree on the nature and timing of the storm relative to specific natural history characteristics of this species. The observed response of the Apalachicola oyster population to successive natural disturbances has significant meaning in terms of the long-term ecological stability of estuarine populations and the evolutionary aspects of such biological response to temporally unstable habitats.

In this case, oyster populations can be viewed as highly resilient under even the most extreme conditions of physical instability.

Livingston et al. (1999, 2000) outlined life history descriptions of the Apalachicola oyster population. Larvae were significantly associated with oyster density, Secchi readings and average bottom salinity. They were inversely related to bottom salinity maxima. In general, larvae and spatfall were usually highest in eastern parts of the bay where oyster densities were highest. Oyster density was highest at the reefs in the eastern parts of the bay. Overall oyster production was concentrated on three eastern bars (Cat Point, East Hole, Platform) and was positively associated with surface watercolor and Secchi readings, and average bottom current velocities. Thus high oyster production in the bay occurs in areas subjected to a convergence of highly colored surface water from East Bay (i.e., influenced by the Apalachicola River/Tate's Hell Swamp drainage) and high-velocity bottom water currents moving westward from St. George Sound. Based on the distribution of oyster density, the primary oyster growing areas were in eastern sections of the bay, with maximum growth during periods of low water temperature and high salinity variation.

Oyster mortality was highest in parts of the bay distant to river influence (i.e., high salinity). These areas are also in closest proximity to the entry of oyster predators from the Gulf through the respective passes. Oyster mortality was generally low at the highly productive reefs in the eastern part of the bay (Cat Point, East Hole). Oyster mortality was significantly (ANOVA; $P < 0.05$) higher in open baskets, which indicates that predation was a major factor in such mortality. Field observations tend to support the experimental findings, with the single most important predator being the gastropod mollusk, *Thais haemastoma*. Statistical analyses indicated that oyster mortality was positively associated with maximum bottom salinity and surface residual current velocity. Mortality was inversely related to oyster density, bottom residual velocity, and bottom salinity.

The scientific data thus showed that the highest levels of primary and secondary productivity of the bay were in areas where there were direct inputs of freshwater, with the river being the most important single form of freshwater input. The entire planning and management program for the bay was thus associated with protection of the primary inputs of freshwater. Wetlands on the river and bay were high-priority areas and there was an emphasis on preventing direct runoff from urban areas into the bay proper.

12.5 Planning and Management of the Apalachicola Bay System

12.5.1 Wetlands Purchases

The results of the overall Apalachicola management effort have been continuously documented (Livingston, 1976b, 1977, 2000, 2002). Early research results, as summarized by Livingston (1984b), linked the river wetlands with the Apalachicola estuary. Major elements of a comprehensive planning and management effort for the Apalachicola River and Bay system have been based on the interactions between river flow and river–bay productivity. The primary objectives of much of the early planning were related to maintenance of natural freshwater flows to receiving areas. Based on these and other data by university scientists, a regional comprehensive plan was developed that included the following:

1. Purchases of environmentally critical lands in the Apalachicola drainage system that now include most of the river and bay wetlands systems
2. Designation of the Apalachicola system as an Area of Critical State Concern, (Florida Environmental Land and Water Act of 1972; Chapter 380, Florida statutes)

3. The creation of cooperative research efforts to determine the potential impact of activities such as ongoing forestry management programs, urban development, and pesticide treatment programs

4. Provisions for aid to local governments in the development of comprehensive land use plans, a function that is vested primarily at the county commission level in Florida

Documentation of facts concerning the Apalachicola Basin (Livingston and Joyce, 1977) provided important information on the overall management approach for the Apalachicola basin. The linkage between the upland freshwater wetlands and the Apalachicola estuary was established by various studies (Cairns, 1981; Elder and Cairns, 1982; Livingston et al., 1974, 1976b; Livingston and Loucks, 1978; Leitman et al., 1982; Mattraw and Elder, 1982; Chanton and Lewis, 2002; see above). The basis for a major effort to protect the enormous intrinsic values of the floodplain forests and river–bay fisheries was provided by basin-wide, scientific documentation of the Apalachicola Resource and the underlying processes that were responsible for the extremely high natural productivity of the system.

Based on information that related the river–wetlands to estuarine production, the Florida Department of Natural Resources, as part of the Environmentally Endangered Land Program (Chapter 259, Florida statutes), purchased 30,000 acres of hardwood wetlands in the lower Apalachicola for $7,615,000 in December 1976 (Pearce, 1977). This was the first of many wetland purchases in the Apalachicola region. Ongoing scientific information provided the basis for further wetlands purchases in the Apalachicola and Choctawhatchee floodplains. Upland and coastal wetlands surrounding East Bay were purchased by state agencies (Table 12.1).

A total of $154,675,315 has been spent by the state of Florida to protect the wetlands system of the Apalachicola River and Bay system. These purchases were based on detailed scientific data connecting the river–estuarine wetlands with river and bay productivity.

Currently, the Apalachicola River wetland system is one of the few alluvial areas in the United States where riverine and coastal wetlands are almost entirely held by public agencies for preservation and management. The state of Florida owns the lower half of the Apalachicola floodplain. The protection of these wetlands provided an important step toward maintaining natural (quantitatively and qualitatively) freshwater flows to the bay. Thanks to the efforts of the same coalition of local and state personnel that instituted land planning in Franklin County in the early 1970s, the scientific database was used to establish the Apalachicola River and Bay Estuarine Sanctuary in 1979. This sanctuary, now called a National Estuarine Research Reserve, included about 78,000 ha and remains the largest such reserve in the country. The original designation included $3.8 million for land purchases in the East Bay wetlands. In an associated effort (1977), the Florida government purchased Little St. George Island. Somewhat later, an area above the East Hole oyster beds was purchased through the efforts of the Trust for Public Lands (Caroline Reusch, personal communication). St. Vincent Island was already a federal preserve administered by the U.S. Department of the Interior. The east end of St. George Island is a state park. After considerable legal proceedings, most of the western section of St. George Island was planned for maximum protection of island freshwater drainages, associated wetlands, and upland vegetation. When combined with a county management program designed to protect the bay from urban runoff, most of the land/water interfaces were thus protected from the effects of human activities.

In this way, over a relatively short period, land purchases on the barrier islands were added to the purchases of the East Bay and Apalachicola River wetlands to complete a

Table 12.1 Purchases Made by Florida State Agencies in the Apalachicola River–Bay

Funding	CARL_TF	P2000	Florida Forever	EEL_TF	LATF	WMD_LOCAL	DON_VALUE
EELTF				58,000			
EELTF				808,100			
EELTF				318,000			
EELTF				1,022,150			
EELTF				196,000			
EELTF				3,500,000			
CARLTF	547,000						
CARLTF	603,500						
CARLTF	348,500						
CARLTF	10,000						
CARLTF	48,500						
CARLTF	182,700						
CARLTF	149,000						
CARLTF	37,000						
CARLTF	60,000						
CARLTF	118,576						
CARLTF	757,980						
CARLTF	748,953						
CARLTF	881,697						
P2000		6,500					
P2000		736,000					
P2000		3,500					
P2000		188,700					
P2000		210,000					
P2000		174,850					
P2000		169,850					
P2000		79,950					
P2000		215,000					
P2000		460,000					
P2000		76,000					
P2000		85,500					
P2000		242,250					
P2000		682,100					
EELTF				1,713,000			
CARLTF	2,923,153						
P2000		10,480					
DONATIONS							12,500
EELTF				2,000,000			
EELTF				6,270,000			
EELTF				568,000			
DONATIONS							270,000
P2000		5,146,111					
P2000		6,401,028					
P2000		5,870,000					
EELTF/LATF				625,000	5,834,200		
P2000		970,500					
P2000		156,000					
P2000		156,000					
DONATIONS							2,000
P2000		7,000,000					
P2000		5,550,000					
P2000/WMD		4,975,000				3,500,000	
P2000		810,000					
CARLTF	1,076,912						
DONATIONS							50,000
FF			7,253,787				

Table 12.1 (continued) Purchases Made by Florida State Agencies in the Apalachicola River–Bay

Funding	CARL_TF	P2000	Florida Forever	EEL_TF	LATF	WMD_LOCAL	DON_VALUE
P2000/WMD		3,500,000				3,500,000	
P2000		790,433					
P2000		19,537,775					
P2000		7,882,000					
P2000		7,651,650					
P2000		105,000					
P2000		715,000					
P2000		24,850,000					
P2000		202,800					
P2000		2,017,630					
P2000		726,000					
SOR							
FF			327,500				
	$8,493,471	$108,353,607	$7,581,287	$17,078,250	$5,834,200	$7,000,000	$334,500

Source: Based on data provided by T. Hoehn, FFWCC, personal communication.

ring of publicly owned lands around the most environmentally sensitive areas of the Apalachicola River–Bay system. These purchases were based almost entirely on scientific reports that prioritized the order of ecological value, in both intrinsic and extrinsic terms, of the various estuarine resources. Such order included the maintenance of natural drainages that delivered freshwater from protected (fringing) wetlands, and, of course, the main stem of the Apalachicola River.

12.5.2   Local, State, and Federal Cooperation

The success of the Apalachicola management program was due primarily to a cooperative effort of local, state, and federal elected officials and environmental agencies. A series of Florida governors, which included Leroy Collins, Bob Graham, and Lawton Chiles, aided in the effort to support the Apalachicola system. The head of the Department of Environmental Protection, Victoria Tschinkel played a vital role in management efforts, along with a series of representatives from the Florida Department of Natural Resources. On another level, there was a direct connection between the results of scientific research in the Apalachicola system and the political processes that direct environmental policies. Experience in the Apalachicola system during the early years indicated that research could have a major impact on the management of an important resource. However, the combined descriptive and experimental approach of ecosystem research is effective only when such a program anticipates resource questions that have not been asked (Livingston 1983a, 2002). This research should be of sufficient scope to address systemwide problems (Livingston 2000, 2002). Some form of popularization of the research results is also needed so that informed laypeople can understand the scientific issues. However, even if the research is correctly carried out and the information is delivered in an appropriate format, there is still no real guarantee of resource protection unless there is a political will to translate the scientific data into an effective management plan and there is an ethical basis for the implementation of such a plan. The cooperation of local, state, and federal interests during the 1970s and early 1980s was the single most important factor in the eventual success of the Apalachicola management program. Livingston (1991c) has outlined the history of the planning and management program.

12.5.2.1   The Beginning: 1972 to 1977

In March 1972, my research group began a field inventory of the Apalachicola Bay system (Livingston, 1984b). The lower river and estuary are located in Franklin County, Florida, a community that has historically depended heavily on the fisheries of the Apalachicola system for its economy. This included the money crops of penaeid shrimp (*Farfantepenaeus* spp.), blue crabs (*Callinectes sapidus*), oysters (*Crassostrea virginica*), and finfishes (primarily sciaenids) that provided the basis for the sports and commercial fisheries in the region. A few months after the initiation of the research, a group of Franklin County commissioners and fishermen met with me to discuss the possibility of an alliance between local residents and university researchers. This agreement led to a 14-year joint program of study that included the direct funding of the Apalachicola research effort by this small fishing community (Robertson, 1982). This association between local users and the scientific community was an important part of the sustained effort to manage the resource. In exchange for local matching funds for federal grants, the concerns of local interests were taken into account and research results were transmitted directly to local political leaders. During this period, the news media were generally supportive of environmental concerns, as opposed to the current atmosphere of obfuscation and outright misrepresentation of scientific facts to the public.

The overall scope and direction of the long-term research effort were developed during this early period (Livingston 1983a, 1984b; Livingston and Loucks, 1978; Livingston et al., 1974). The basic plan included continuous monitoring of the lower river and bay (physical, chemical, biological) with a series of studies that were applied to specific questions such as nutrient limitation and trophic organization. Scientific questions were often associated with the possible impact of various human activities in the region on the Apalachicola resource. The interdisciplinary field research was supplemented by various experimental programs and a series of graduate student efforts that resulted in determinations of the trophic organization of the Apalachicola Bay system (Laughlin, 1979; McLane, 1980; Mahoney, 1982; Laughlin and Livingston, 1982; Sheridan, 1978, 1979; Sheridan and Livingston, 1979, 1983).

The early results (Blanchet, 1979; Edmiston, 1979; Livingston et al., 1974) indicated that the Apalachicola Bay system was extremely productive as a result of a combination of geomorphologic characteristics, salinity distribution, and nutrient relationships. Bay habitats were controlled to a considerable degree by the Apalachicola River. The river was characterized by relatively high levels of color and turbidity. The high estuarine phytoplankton production was dependent on the Apalachicola River system (Estabrook, 1973). The seafood potential of the region was thus identified with Apalachicola River flow at an early stage of the investigation (Livingston et al., 1974). Various threats to this productivity were outlined (Livingston et al., 1974). Local and regional real estate developers viewed St. George Island as a prime area for massive building efforts (Figure 12.1). Physical changes in the river caused by ongoing dredging of the Apalachicola and damming in the Flint and Chattahoochee Rivers in Georgia and Alabama (Leitman et al., 1991) were considered threats to the continued protection of the Apalachicola resource. Dredging of the bay included the opening and maintenance of an artificial connection to the Gulf of Mexico (Sikes Cut) (Figure 12.1). Agricultural activities in the river floodplain caused destruction of wetlands (Livingston et al., 1974). All of these activities were outlined in a series of publications (Livingston, 1984b), and were eventually addressed in a series of planning moves by various agencies.

During the early 1970s, the most important issue in the Apalachicola region was a proposal by the U.S. Army Corps of Engineers (Mobile District: ACE) for the construction of a series of four dams on the Apalachicola River. These dams were supported by a congressional mandate that a channel be created for shipping interests located largely in Alabama and Georgia. It did not matter that there was no demonstrable evidence of an economic need for such damming and channelization; to this day, the maintenance of the Apalachicola-Chattahoochee-Flint (ACF) system remains one of the most expensive such operations in terms of tons of shipping per mile in the United States. Corps scientists determined that nitrogen would not be retained behind the dams as efficiently as phosphorus. It was also assumed that nitrogen was limiting to estuarine phytoplankton productivity. Therefore, they reasoned that dams would have little effect on the trophic organization of the Apalachicola estuary due to major losses of phosphorus. However, studies that found nitrogen to be the chief limiting factor (Ryther and Dunstan, 1971) did not necessarily apply to areas of salinity transition in southeastern estuaries (Howarth 1988).

Unlike many estuaries along the East Coast that have been intensively studied, the Apalachicola Bay system had relatively high nitrogen:phosphorus ratios (Nixon, 1988a,b). Phytoplankton studies (Estabrook, 1973; Livingston et al., 1974) indicated relatively low phosphorus levels in the estuary. Additional experimental work (Myers, 1977; Myers and Iverson, 1977, 1981) showed that phytoplankton productivity in the Apalachicola estuary (and other estuaries along the northeastern Gulf) was phosphorus-limited. Although more recent work (Hecky and Kilham, 1988) indicated that the factors that lead to nutrient limitation in transitional portions of southeastern estuaries are highly complex and somewhat erratic, there was evidence that phosphorus is involved as a limiting factor in these systems (Howarth, 1988). The importance of the shallowness of the bay was indicated (Myers, 1977; Myers and Iverson, 1977) since wind mixing of bottom sediments was correlated with increased nutrients and phytoplankton production in the euphotic zone.

The largely undeveloped Apalachicola drainage system is naturally low in phosphorus. The fact that phosphorus was projected to be lost behind the proposed dams, together with evidence that such phosphorus could be important to bay productivity, indicated that a series of dams along the Apalachicola River would eventually have a direct impact on the nutrient dynamics and primary productivity of the Apalachicola Bay system. These data were used by the author to project the impact of dams on the Apalachicola River. A series of debates between Army Corps representatives and researchers brought out the various aspects of dam impacts on the river. Public opinion was strongly against the construction of these dams. The scientific data were influential in the arguments against the dams. After a long and bitter confrontation, the proposal to dam the Apalachicola River was dropped by the U.S. Army Corps of Engineers.

The basis for the overall management effort for the Apalachicola system was outlined by Livingston (1975b). In the beginning, scientific information was an important part of the decision-making process. The impact of pesticides on the system was analyzed at this time. The Apalachicola system was not seriously contaminated with organochlorine pesticides (Livingston et al., 1978) although locally high concentrations of DDT were associated with long-term biological effects (Koenig et al., 1976). All spray programs (insects, introduced aquatic plants) were evaluated according to specific criteria that included proximity of runoff to commercially important estuarine populations in space and time. Those programs that were considered a risk to the estuary were dropped, an unprecedented move in a state that has little regulation of pesticide use to this day. Despite vigorous opposition from state agencies responsible for coastal spray programs, the Franklin County Commission effectively ended spraying in environmentally sensitive areas.

Research results that linked the river wetlands with the estuary were publicized through radio and television shows, newspaper stories, educational tapes, input to secondary school curricula, and various forms of oral presentations. A regional comprehensive plan was developed.

12.5.2.2   The Middle Years: 1978 to 1982

Documentation of facts concerning the Apalachicola Basin was considered a high priority in the implementation of the proposed management program (Livingston and Joyce, 1977). Various aspects of the Apalachicola system were found to be important in the overall planning approach (see above), and this information was publicized through various media channels. The importance of the Apalachicola River wetlands was connected to various freshwater fisheries, although most of the more important fisheries (e.g., striped bass, *Morone saxatilis;* sturgeon, *Acipenser oxyrhynchus*) had been destroyed or seriously impaired due to postulated habitat destruction by channelization and damming (Livingston and Joyce, 1977; Livingston, 1984b). Such activities, mandated by the U.S. Congress, continue to the present time. Despite the enormous intrinsic values of the floodplain forests and river fisheries, such issues did not provide an adequate economic basis for the expensive wetlands purchases. Thus, the argument for protection of the freshwater wetlands of the Apalachicola system depended on the linkage between these upland areas and the economically important commercial fisheries of the estuary.

The linkage between the upland freshwater wetlands and the rich estuarine biota was the subject of considerable research and public debate (Livingston and Loucks, 1978). The distinctive links between the estuarine food web and freshwater discharges (Livingston and Loucks, 1978) were emphasized. Studies were made concerning the distribution of wetland vegetation in the Apalachicola floodplain (Leitman et al., 1982). It was determined that vegetation type is associated with water depth, duration of inundation and saturation, and water-level fluctuation. Stage range is reduced considerably downstream, which indicated a dampening of the river flood stage by the expanding (downstream) wetlands. Litter fall in the floodplain (800 gm^{-2}) found to be high (Elder and Cairns, 1982). Seasonal flooding provided the mechanism for mobilization, decomposition, and transfer of the nutrients and detritus from this wetland to associated aquatic areas (Cairns, 1981; Elder and Cairns, 1982) with a postulated although unknown input from groundwater sources. In short, various connections were made between the freshwater wetlands and bay productivity.

Based on information that related the river–wetlands to estuarine production, the Florida Department of Natural Resources, as part of the Environmentally Endangered Land Program (Chapter 259, Florida statutes), purchased 30,000 acres of hardwood wetlands in the lower Apalachicola for $7,615,000 in December 1976 (Pearce, 1977). This was to be the first of many wetland purchases in the Apalachicola region, as noted above The shallowness of the bay enhances microbial decomposition of the organic matter during warm periods. Nutrient regeneration and wind-mixed currents are correlated with outbursts of phytoplankton production. Particulate organic matter in the bay undergoes a succession of microbial decomposition (Morrison et al., 1977; Bobbie et al., 1978). Grazing of the microbial elements stimulates microbial growth and alters the composition of the microbial community. The integration of dissolved organic substances into particulate organic matter via microbial action was considered an important process in the overall trophic organization of the bay (Livingston, 1984b).

The published results of the long-term bay research program provided the basis for the main elements of local and regional planning initiatives. The publication of the database at different levels of technical detail was only one part of the effort. The complexity of the successful application of reliable scientific data to management questions is illustrated by the history of land development in Franklin County, a process that centered on St. George Island during the 1970s and early 1980s. This narrow sand barrier, about 50 km long, forms the southern border of Apalachicola Bay (see Figure 12.1). During the early 1970s, politically powerful real estate developers were pitted against the Franklin County Commission and a handful of technical advisors who were opposed to high-density development in close proximity to the richest oyster bars in Florida. A detailed recap of the early history of the oyster wars on St. George Island is given by Toner (1975) and Livingston (1976b). After almost 10 years of acrimonious and costly battles, the county adopted a strict comprehensive plan based on the scientific data accumulated over this period. This plan was designed to permit development in areas that could be serviced by adequate controls such as sewage treatment plants and storm water treatment. The relatively fragile nature of the barrier islands of the Apalachicola system sustained an argument against extensive development of such areas, although relatively high population density along mid-sections of St. George Island was grandfathered into the plan with no real provision for sewage treatment or storm water control in this crucial portion of the system.

By the early 1980s, after an unprecedented effort by local (Franklin County) representatives, outside consultants, university scientists, and state and federal officials, a far-reaching comprehensive land management plan was combined with extensive land purchases by state and federal agencies and the National Estuarine Sanctuary designation to form the most ambitious management effort in the country. The scientific database (Livingston, 1983a, 1984) enabled progressive decisions by local, state, and federal administrators that were designed to protect the Apalachicola resource before it was destroyed, a process that is far more efficient than the usual pattern of destruction and restoration (Livingston, 1991c). This effort occurred at a time when effective environmental action before excessive development in a given drainage basin was not common in Florida. The scientific program provided an objective basis for constructive political action.

12.5.2.3 1983 to the Present

The beginning of the end of the cooperative management effort for the Apalachicola system began with the filing of a civil suit against the Franklin County Commission and its advisors (including the author) by a politically influential developer of St. George Island during the summer of 1982 (U.S. District Court Case Number TCA 82-1033-WS). The litigation, totaling $60 million in claims, alleged violations of civil rights law and state and federal antitrust laws, breach of contract, and taking without compensation. The issue of concern was how much density would be allowed developments on St. George Island. The Franklin County Commission depended on advice from the scientific and planning consultants that noted past adverse impacts of high-density development of barrier islands on coastal resources in Florida and throughout the United States. If the litigation concerning the high-density development of the western section of St. George Island had succeeded in pressuring the county to accept the high-density development of St. George Island, a major part of the bay would have been subjected to a form of urbanization that has proven damaging to coastal resources in other areas. The defendants in the trial included people who were placed under severe political pressure by outside political forces and their apologists in the news media (*Tallahassee Democrat*) who had interests in the development of St. George Island.

The following people should be recognized for the important part they played in the protection of the Apalachicola Resource.

Robert L. Howell, Clerk of the Circuit Court, Franklin County
Cecil Varnes, Head, Franklin County Commission
Ikie Wade, Franklin County Commissioner
William F. Henderson, Franklin County Commissioner
Ed Leuchs, Executive Director, Apalachee Regional Planning Council

At issue was the right of developers from Tallahassee to bring high-density development to St. George Island. Almost 4 years later, after a protracted period of continual strife and intense legal confrontation, a federal judge dismissed the charges based on the opinion that the plaintiff used the suit "to harass and intimidate the defendants." A further order was given for reimbursement of court expenses based on the plaintiff's "intent to use the judicial system to harass those who opposed his development plans." Such reimbursement of legal costs was never made, despite considerable expense to some of the defendants.

Over those years of costly litigation, the original group responsible for the management plans broke up because of death, sickness, and harassment by those who wanted to develop the last unpopulated coast of Florida. Ed Leuchs was fired from his position on the Apalachee Planning Council at the behest of developers. A new group of Franklin County officials, with the support of powerful state officials and developers, moved to consolidate political power in the hands of those who were behind local and regional residential and commercial development. There was a growing adverse reaction locally to the planning process. There are current efforts to dilute the Franklin County Comprehensive Plan. Major new land developments for major parts of the north Florida area have been proposed by a powerful corporate entity. State environmental agencies no longer support local and regional environmental efforts. And the news media, once characterized by a distinguished cadre of environmental writers, has become an integral part of the effort to develop the region at the expense of natural assets that include the Apalachicola system. The question of the use of freshwater in the ACF system has become a central issue in the survival of the Apalachicola resource.

12.6 Water Use in the ACF System

Agricultural and municipal interests along the Tri-river system continue to increase pressure on the freshwater resources of the Tri-river (ACF) basin. There are new proposals to reallocate water in the Lake Lanier storage from hydropower to supply water for the rapidly growing Atlanta, Georgia, metropolitan area. In the past, this impoundment represented about half of the stored water that was used to augment downstream flows (Leitman et al., 1991). Models (Livingston, 1988c) indicated that agricultural use of water in the Tri-river system would eventually lead to serious depletion of freshwater input to the Apalachicola from the Georgia area. These changes were projected to lead to serious problems in the maintenance of Apalachicola Bay productivity (Livingston, 1988c). Recent analyses indicate that total depletions of freshwater in the basin already represent a significant portion of low flow during summer months, and forecasted demands for the ACF basin suggest that such losses will become even greater with time. This indicates more extreme hydrologic drought events during comparable meteorological drought events in future years. Consumptive losses to agricultural irrigation have increased significantly in recent years.

During the most recent drought (1999–2002), there were a series of extremely low-flow periods. Data, provided by the Apalachicola National Estuarine Reserve (Lee Edmiston, personal communication), indicated that eastern sections of the bay had systematically high bottom salinities. The prolonged drought of 1999–2002 was associated with the most consistently high salinities since the initial study period by the Livingston research group (1972–1991). These salinities peaked during 2001. The salinity maxima were consistently high from 1991 to 2001, whereas salinity minima showed pronounced increases during 2001. This trend was consistent with the relatively low standard deviations during 2001, an observation that has significance when oyster trends during this period are taken into account.

During the latest drought, there was a collapse of the oystering in the highly productive Eastern reefs from Cat Point to East Hole. In a 2002 field assessment by the Florida Department of Environmental Protection (G.S. Gunter, personal communication), lowered oyster productivity in eastern bay reefs was accompanied by large numbers of predators that included oyster drills, crown conchs, scallops, and sea urchins. Hard and soft corals were noted on Porter's Bar during these field surveys (G.S. Gunter, personal communication). With a return of higher river flows during 2003, there was an increase in observed oysters at Cat Point and East Hole, with an accompanied reduction in oyster predators (G.S. Gunter, personal communication). These changes in oyster productivity during the recent drought represent field verification of model predictions made by Livingston et al. (2000).

Leitman (2003a,b) found that during average years, the net evapo-precipitation losses from impoundments in the ACF basin were considerable; these losses in the Flint and Chattahoochee areas far exceeded consumptive losses in the year 2000 for municipal and industrial uses for the entire Chattahoochee system (including metro Atlanta and Columbus, Georgia) for all months between May and October except August. If the net evapo-precipitation losses for 1986 were considered, such losses exceeded the consumptive losses for municipal and industrial demands for all months between May and September. The net evapo-precipitation losses for 1999 exceeded the consumptive losses for municipal and industrial demands for April, May, August, and September. The drought of 1999–2001 had lower river flows but higher precipitation levels than those observed during the drought of the 1950s. The differences were related to higher consumptive uses (via the evaporation losses from the impoundments and reservoir management practices) during the 1999–2001 drought. These losses required an adjustment by a factor of 1.30 to account for the additional surface areas of the impoundments.

Livingston et al. (2000) developed a time-averaged model for predator-driven oyster mortality during the summer of 1985 by running a regression analysis with averaged predictors derived from the hydrodynamic model and observed (experimental) mortality rates throughout the estuary. High salinity, relatively low-velocity current patterns, and the proximity of a given oyster bar to entry points of saline Gulf water into the bay were found to be important factors that contributed to increased oyster mortality due to predation. The authors found that oyster production rates in the Apalachicola system depended on a combination of variables that are directly and indirectly associated with freshwater input as modified by wind, tidal factors, and the physiography of the bay. Flow reduction, whether through naturally occurring drought phenomena, through increased upstream anthropogenous (consumptive) water use, or a combination of the two, were considered to have the potential for serious adverse consequences for oyster populations.

Model-projected oyster mortality vs. observed (field experimental) mortality for historical and baseline 2000 flows during the 1985 and 1986 flow periods indicated that mortality was highest in areas of the bay distant to river influence (i.e., with high salinity) and in closest proximity to the entry of oyster predators from the Gulf through the respective passes. Oyster mortality during the moderate flow year (1985) was generally low at the highly productive reefs in the eastern part of the bay (Cat Point, East Hole). Field observations tended to support the experimental findings with the single most important predator being the gastropod mollusk, *Thais haemastoma*. Oyster mortality increased during the drought year 1986 according to the model results. The most important differences in mortality between the historical flows and the baseline 2000 flows were noted during the projected drought (1986) results, with generally increased mortality noted in the baseline 2000 data. The oyster mortality noted during the latest drought was due, in part, to anthropogenous reductions in freshwater flows to the Apalachicola estuary. The implications of such effects include the nursery functions of various key species of fishes and invertebrates that have been associated with areas of the bay that receive freshwater input from the river (Livingston et al., 2003). A lawsuit by the state of Florida against Georgia has recently gone to the U.S. Supreme Court for a resolution of this problem.

12.7 The Apalachicola Model: Management, Not Restoration

The early successes of the Apalachicola management initiatives were due to various, unrelated factors. The close association between local and state officials and university scientists was responsible, in large part, for the coordinated effort to protect the Apalachicola resources. During the 1970s, the seafood industry was strong, and oyster representatives and associated elected officials contributed significantly to the development of a far-reaching management plan. However, the symbolic importance of the Apalachicola oyster industry, in terms of historical, cultural, and economic values, has been both strength and a weakness. Oysters were the rallying point for various actions that eventually led to the preservation of more intrinsic wetlands values. However, the weakening of the oyster industry through natural disasters, poor management, and political manipulation has reduced the effectiveness of the early planning initiatives. There was little organized political opposition to such initiatives during the early stages of the planning process. With time, the various elements of management and research have been systematically eliminated by changes at various political levels to a point where development interests are now in control of the Apalachicola region. There has been an increasing shift from the open generation and public use of objective scientific data to political and bureaucratic control of information for the advantage of narrow economic interests. This change, together with the strictly controlled release of information by local and regional news media, has led to an increasingly uncertain situation with respect to the maintenance of the Apalachicola system.

The recent pattern of official reluctance to address environmental problems associated with ruling local and regional political powers is not uncommon in the rapidly developing state of Florida where natural resources are usually neglected in favor of economic development until a point of no return is reached. By the time such destruction is "discovered," expensive restoration projects are promoted that glorify and sustain the same political and bureaucratic interests that created the problem in the first place. That is, effective and relevant research programs in the Apalachicola region have been actively discouraged so that objective facts can be left out of a planning process that has deteriorated into a political/bureaucratic exercise in public relations. The system of environmental management is thus controlled by influential economic forces that have been successful in minimizing scientific input to the solution of environmental problems.

The basic core of the major planning and management effort in the Apalachicola River-Bay system remains in place. It remains to be seen if the highly successful effort to protect this resource will continue to protect an alluvial system that is one of the last such areas in the United States, and to maintain processes that account for the rich biodiversity and high natural aquatic productivity of this drainage basin. In any case, when compared to other systems such as the Chesapeake and the Florida Everglades, the advantages of progressive management and preventive resource maintenance over resource deterioration and expensive but failed restoration efforts are obvious.

chapter 13

Conclusions

13.1 Introduction

In the United States today, despite the self-image that we are all environmentalists, there is little evidence that our optimistic view of ourselves is realistic. The environmental history of this country is mixed, with periods of extreme disregard for the natural environment punctuated by solid advances in environmental protection.

Recent history of our concern is not encouraging. According to a report by the widely respected Pew Oceans Commission (*America's Living Oceans*, 2003), there are some problems regarding our aquatic systems that are not being addressed:

"Coastal development and associated sprawl destroy and endanger coastal wetlands and estuaries that serve as nurseries for many valuable fishery species. More than 20,000 acres of these sensitive habitats disappear each year. Paved surfaces have created expressways for oil, grease, and toxic pollutants into coastal waters. Every eight months, nearly 11 million gallons of oil run off our streets and driveways into our waters — the equivalent of the *Exxon Valdez* oil spill.

More than 60 percent of our coastal rivers and bays are moderately to severely degraded by nutrient runoff. This runoff creates harmful algal blooms and leads to the degradation or loss of seagrass and kelp beds as well as coral reefs that are important spawning and nursery grounds for fish. Each summer, nutrient pollution creates a dead zone the size of Massachusetts in the Gulf of Mexico. These types of problems occur in almost every coastal state, and the trends are not favorable. If current practices continue, nitrogen inputs to U.S. coastal waters in 2030 may be as much as 30 percent higher than at present and more than twice what they were in 1960.

Many ecologically and commercially crucial fish species, including ground fish and salmon populations along the Atlantic and Pacific Coasts, face over fishing and numerous other threats. Thirty percent of the fish populations that have been assessed are over fished or are being fished unsustainably. An increasing number of these species are being driven toward extinction. Already, depleted sea turtle, marine mammal, seabird, and noncommercial fish populations are endangered by incidental capture in fishing gear. Destructive fishing practices are damaging vital habitat upon which fish and other living resources depend. Combined, these aspects of fishing are changing relationships among species in food webs and altering the functioning of marine ecosystems.

Invasive species are establishing themselves in our coastal waters, often crowding out native species and altering habitat and food webs.... New species are regularly finding a home around our coastlines as hitchhikers in ship ballast water or on ship hulls, escapees from fish farms, and even as discarded home aquarium plants and animals. Of the 374 documented invasive species in U.S. waters, 150 have arrived since 1970.

In addition to these varied threats, climate change over the next century is projected to profoundly impact coastal and marine ecosystems. Sea-level rise will gradually inundate highly productive coastal wetlands, estuaries, and mangrove forests. Coral reefs that harbor exceptional biodiversity will likely experience increased bleaching due to higher water temperatures. Changes in ocean and atmospheric circulation attributable to climate change could adversely affect coastal upwelling and productivity and have significant local, regional, and global implications on the distribution and abundance of living marine resources."

The environmental impacts of urban sprawl alone constitute a major problem for most of the water bodies in the United States, and the many suggestions for planning and management of these aquatic resources remain largely misunderstood and/or ignored by the U.S. public.

Public health is at risk due to a lack of action regarding the release of toxic substances into the air and water. For example, it is no longer safe to eat freshwater fishes in the United States due to the broad distribution of mercury put in the air by emissions from sources such as coal-fired power plants. The recent history of environmental regulation is not reassuring. With regard to the mercury problem, the U.S. Environmental Protection Agency recently bypassed its own professional staff and a federal advisory panel to create a rule favored by the coal-fired power industry that would delay reductions in mercury emissions by decades. A few media outlets have started to have misgivings about government policies with regard to mercury in the aquatic environment.

MERCURY EMISSIONS RULE GEARED TO BENEFIT INDUSTRY, STAFFERS SAY: "EPA veterans say they cannot recall another instance when the agency's technical experts were cut out of developing a major regulatory proposal. The (Bush) administration chose a process 'that would support the conclusion they want to reach,' said John A, Paul, a Republican regulator from Ohio who co-chaired the EPA-appointed advisory panel. ...Russell E. Train, a Republican who headed the EPA during the Nixon and Ford administrations, said: 'I think it is outrageous. The agency has strayed from its mission in the past three years.' ...When the Bush administration took office in 2001, slowing mercury regulation was a priority for the coal and power industries... Since 1999, coal and electricity companies have donated $40 million to Republican candidates and committees, including $1.3 million directly to Bush campaigns according to figures compiled by the Center for Responsive Politics.... The administration's proposed mercury rule, published in the *Federal Register* in December, contains numerous paragraphs of verbatim language supplied by two separate industry advocates... The standard — largely incorporated by the EPA — is enormously beneficial to the industry, according to S. William Becker, executive director of the State and Territorial Air Pollution Program.... Today, coal-fired power plants pump out about 48 tons of mercury annually.... The Bush Clear Skies plan... calls for a national cap of 34 tons in 2010, a level that wouldn't require any extra spending by the industry... officials of the coal-fired utility industry argue that technology would be so expensive that it would lead electric generators to shift from coal to natural gas. 'The result would be increased electricity prices and higher costs'... said Scott Segal, director of a coal utility trade association. ... Meanwhile, longtime EPA employees say the administration exaggerated data on the effectiveness of its proposed rule, which would take effect in December. In announcing the mercury plan, the EPA said it would reduce mercury emissions from power plants by 70% by 2018. However, the EPA's own database shows that emissions would, at best, be reduced by only about half by then. And EPA models suggest that the 70% goal may not be reached until 2025, if ever."

—T.H. Hamburger and A.C. Miller, *Los Angeles Times,* **March 16, 2004**

According to a recent report by the Natural Resources Defense Council (NRDC) (*Testing the Waters 2004: A Guide to Water Quality at Vacation Beaches, 2004*), "our towns and our states do not necessarily protect us from swimming-associated diseases, such as gastroenteritis, hepatitis, salmonellosis, and other infections and viruses." The extent of contamination of beach areas continues to grow, due largely to urban storm water runoff and sewage discharges.

"During 2003, at U.S. ocean, bay, Great Lakes, and some freshwater beaches, there were at least 18,284 days of closings and advisories, 64 extended closings and advisories (7 to 13 consecutive weeks), and 60 permanent closings and advisories (more than 13 consecutive weeks). Including extended days, the total comes to more than 22,201 beach closing and advisory days.

Since 1992, there have been more than 89,296 days of closings and advisories and 333 extended closings and advisories (seven to 13 consecutive weeks).

The number of beach closing and advisory days increased 51 percent in 2003 (6,206 days) from the previous year.... The major factors leading to the increase in 2003 appear to be a greater number of monitored beaches, more frequent monitoring, wider use of BEACH Act required indicator organism and numeric standards, and heavy rainfall in some areas.

The continued high level of closings/advisories is an indication that new and more frequent monitoring continues to reveal serious water pollution at our nation's coastal, bay, Great Lakes, and some inland freshwater beaches.... 16,120 (88 percent) of the 2003 beach closings and advisories were issued because water quality monitoring showed that bacteria levels exceeded health and safety standards (one percentage point higher than last year)."

Of course, the vast majority of the good burghers of the United States have not been privy to the information published by the *Los Angeles Times* and the NRDC. For example, if you happen to depend on papers such as the *Tallahassee Democrat,* you remain largely ignorant of just about anything environmental as such papers do not want to make its public uncomfortable with news of our imperfect regulatory and political system when it is involved in environmental matters.

And thus we are faced with the gulf between actual environmental risks and the public's erroneous perception of such risks. The differences between the results of detailed scientific facts and what is reported to the public have been outlined in this book. Chapter 1 listed the factors that were necessary for successful restoration in an ideal world. Of course, our world is anything but ideal, and the current state of the aquatic environment is a reflection of the careless way that human beings treat their natural environment.

The question remains: What are the factors that contribute to the discrepancies between reality and the art of restoration of aquatic systems?

13.2 Scientific Research

The development of a comprehensive scientific database in an important aquatic ecosystem is fundamental to the development of planning and management programs to protect the resource. Various programs have been successful (such as the Apalachicola effort) and some have not. However, in the overwhelming majority of situations, adequate scientific data are not available — either to evaluate the system for management purposes or to restore lost resources. Most databases are composed of a series of unrelated studies over varying time periods that cannot be used in an organized fashion to define the current state of the system in question. This lack of information supports unrestricted growth efforts and camouflages the loss of natural resources. The needed research depends on long-term, adequate funding with continuous and relevant scientific efforts that are based on system-specific questions. Ecosystem-level scientific research is fundamental to the restoration process.

The early successes of the Apalachicola management initiatives were due to various, unrelated factors. The close association between local and state officials and university scientists was responsible, in large part, for the coordinated effort to protect the Apalachicola resources. During the 1970s, the seafood industry was strong, and oyster representatives and associated elected officials at various levels contributed to the development of a far-reaching management plan. The basin was largely undeveloped with a small human population. The symbolic importance of the Apalachicola oyster industry served as a rallying point for various actions that eventually led to the preservation of intrinsic wetlands values.

However, by 1984, the various elements of management and research in the Apalachicola system were opposed by development interests that viewed the bay research as a threat to the urbanization of areas such as St. George Island. There was a shift from the open generation and public use of objective scientific data to political and bureaucratic control of information for the advantage of powerful economic interests. These interests were successful in the cancellation of federal (Florida Sea Grant) research funds and the matching money generated at the local level. At the same time, there was a trend toward the strict control of information by local and regional news media, which has increased in recent years. The Apalachicola Research Reserve has never had adequate funding for a research program to evaluate the state of the bay. The present condition of the Apalachicola system remains unknown. The information vacuum created and sustained by developmental interests and maintained by regional news media has contributed to a real threat to the Apalachicola resources as the area undergoes unregulated land development.

The loss of research funding for the Apalachicola system was not an isolated instance. Based on objections by the Leon County Commission to making the lake database available to the public, research funds were withdrawn from the FSU Center, which had carried out the studies that showed the deterioration of the regional lake resource. Publicity associated with the failure of the Florida Department of Environmental Protection led to a loss of state funding for the FSU group that exists to this day. The office of the governor of Florida actually established a "clearinghouse" on the FSU campus, the purpose of which was to review federal grant proposals. By the early 1990s, the FSU state and federal funding was seriously reduced, and the only outlet for prolonged, ecosystem-level funding was based on private organizations and outside foundations. Control of information generated by regional, state, and federal granting agencies is dominated by politics in Florida, and this does not seem to be an isolated instance of such control.

Scientific research can have a major impact on the management of an important resource. A combined descriptive and experimental approach is effective only when such a program anticipates resource questions that have not been asked (Livingston, 1983a) and is of sufficient scope to address systemwide problems (Livingston, 2002). Some form of popularization of the research results is also needed so that informed laypeople can understand the scientific issues. However, even if the research is correctly carried out and the information is delivered in an appropriate format, there is still no real guarantee of resource protection unless there is a political will to translate the scientific data into an effective management plan as well as an ethical basis for the implementation of such a plan. According to the experience in other alluvial systems such as the Chesapeake, the establishment of an adequate scientific database does not necessarily lead to a successful restoration program.

The gulf between scientific research and political action in the United States is currently at an all-time high. Recently, more than 4000 scientists, including 48 Nobel Prize winners, signed a statement that opposed the Bush administration's misuse of scientific information. The policy of sacking scientists on advisory boards whose opinions do not reflect the politics of the moment has continued with a vengeance. The rewriting of EPA reports with elimination of sections that do not fit the political take on issues such as global warming represents a blatant disregard for scientific facts on important environmental issues. The governance of scientific processes by narrow interpretations of ethical and even religious beliefs is not representative of the scientific method. What started as a purge of scientific advisors in the EPA in the 1980s has been transformed into the wholesale misrepresentation of scientific findings in the present administration. A broad range of the news media has tacitly accepted the political transformation of the role of science in the determination of environmental issues, and the subject of the role of science in the resolution of restoration processes remains a non-issue in terms of public discourse.

13.3 Regulation and Enforcement

In March 1983, the opening shot for political control of the U.S. EPA was fired by the Reagan administration.

> SCIENTISTS WHO DREW CONSERVATIVE IRE DISMISSED BY EPA: "The Environmental Protection Agency has removed more than 50 scientists from its technical advisory boards after conservative groups provided lists characterizing dozens of scientists as 'horrible,' 'real activist,' or 'bleeding heart liberal.' ...According to the list, EPA's research division was full of 'invidious environmental activists.'"
>
> —H. Kurtz, *Washington Post*, March 3, 2003

> EPA'S LITTLE LIST OF WHO WON'T BE MISSED: "An ever-widening congressional probe of the Environmental Protection Agency expanded Wednesday with an inquiry into whether a 'hit list' was used to remove dozens of top scientists from their roles as EPA advisors. ... 'This so-called hit list simply outlines a policy throughout the Reagan administration to replace scientists

with representatives of the industry that each representative is supposed to regulate.' (Dr. Samuel S.) Epstein said."

—J. Coates and D. Collin, *Chicago Tribune*, March 4, 1983

LOCAL SCIENTIST IS ON EPA 'HIT LIST': "For one Tallahassee biologist, recent scandals at the Environmental Protection Agency have taken a surprising personal twist... His name surfaced recently on a controversial 'hit list' purportedly targeting liberal scientists for dismissal from the agency's Science Advisory Board. ... Livingston's was the only name from North Florida. The list describes him as a 'fair scientist, competent' but with 'bad policy.'"

—Karen Olsen, *Tallahassee Democrat*, March 2003

The political control of the EPA extended to top administrators and bureaucrats who were viewed as environmental advocates or those sympathetic to environmental causes. Elimination of scientists who do not adhere to a conservative political line continue to this day. Political control exists at all levels of the regulatory realm — local, state, and federal. The news media no longer provide protection for those who speak out concerning the destruction of public resources.

The regulation of sewage treatment in the United States represents one example of the growing lack of regulation in the United States. The release of untreated or improperly treated sewage wastes present certain problems related to water quality. Sewage is composed of a complex mixture of inorganic and organic compounds, along with various forms of bacteria, protozoans, and viruses (Mearns et al., 1991; Goudie, 1994). Although the constituents of sewage may vary from area to area, such effluents have certain general characteristics. Human sewage usually contains high concentrations of nutrients (nitrogen and phosphorus compounds), organic carbon, particulates (turbidity and suspended solids), heavy metals (lead, mercury, cadmium, etc.), petroleum compounds (polynuclear aromatic hydrocarbons or PAHs), polychlorinated biphenyls (PCBs), organic contaminants such as pesticides, and microorganisms (bacteria, viruses). The presence of industrial sources and the relationship of sewage collection to storm water disposal can affect the forms of toxic substances that may be present in sewage discharges. The technology exists for complete treatment of sewage wastes. However, the more complete the treatment, the higher the costs. The increasing use of package plants, small installations run by individual developers, has added to the problems associated with sewage discharges.

There is considerable scientific evidence that receiving systems are stressed due to multiple effects from spills of raw and inadequately treated sewage (Hutchinson, 1973; Goudie, 1994; Mearns et al., 1991). Many of the metals and organic contaminants associated with raw sewage are carried by fine particulates that eventually end up in the sediments. Because of the high partition coefficients of certain organic contaminants and some metal complexes, there can be rapid bioconcentration of these components by benthic organisms. Trophic interactions can lead to food web biomagnification of toxic components of the sewage. In this way, many of the persistent toxic components of sewage can accumulate in important marine populations in concentrations that are thousands of times the original concentrations in the water and sediments of the receiving areas. The exact fate of these contaminants depends on the persistence of the parent compounds and their by-products, which can be measured over years and even decades. The cumulative effects of heavy metal complexes, pesticides, PCBs, and PAHs associated with sewage discharges on receiving aquatic systems can be extensive.

There is evidence that nutrient imbalances due to sewage spills have also led to increased blooms of noxious algae that secrete toxic agents (Murphy et al., 1976). Documented alteration of phytoplankton and SAV by sewage effluents is often followed by changes in water and sediment quality (hypoxia, increased concentrations of ammonia, increased turbidity, and reduced light penetration). The combined effects of these habitat changes lead to decreased production of invertebrates and fishes, reduced trophic diversity, further concentration of organic carbon in the sediments with associated increases in BOD and sediment oxygen demand (SOD), and enhanced near-bottom hypoxia and anoxia. Hypereutrophication is now considered a major problem in a broad range of freshwater and estuarine systems in the United States today due to the release of nutrients in sewage spills (Livingston, 2000, 2002). A classic example of such impacts is related to sewage releases in Perdido Bay (see Chapter 5).

The avoidance of up-front infrastructure costs such as sewage treatment is one of the basic tenets of local governments. In Leon County, Florida, the problem with sewage treatment is not atypical of many communities in the United States. Long-term water and sediment quality data indicated that a package sewage treatment plant (STP) on southeastern Lake Lafayette (Figure 2.16) drains into the lake through groundwater connections (Livingston, 1995b). High chloride concentrations in receiving areas of the lake were traced to the package plant. High concentrations of nutrients (including ammonia) and certain other toxic agents were found in the vicinity of the lake adjacent to the STP. Despite years of comprehensive water quality data, the Florida Department of Environmental Protection took no regulatory action.

Sewage spills occur frequently throughout Leon County. Some of the spills are known to enter Lake Jackson through Megginnis Arm (Figure 2.14). The effects of these spills remain largely undocumented, but the cumulative effects on the receiving systems could be significant. A sewage spill occurred on October 3, 1993, at a lift station facility located in the headwaters of the Indian Mounds drainage into Lake Jackson. A power failure, and shutdown of a pumping station, allowed approximately 5712 gallons of raw sewage to escape into an adjacent ditch over a 1-hour and 42-minute period. A clean-up crew treated the affected area with HTH chlorine, pine oil, and lime. The Leon County Lakes Project collected water and biological samples. There was an extensive kill of aquatic organisms in the Indian Mound Creek that was associated with the dispersal of granulated chlorine in the upper creek. There was little publicity associated with such spills. The lack of regulation of sewage treatment plants extends to monitoring requirements by local and state regulatory agencies. Orthophosphate data are not required, and nutrient loading is not even recognized as an important indicator of sewage problems by the FDEP.

The problems associated with the regulation of toxic substances such as mercury and dioxin were outlined in Chapter 11. These regulatory lapses should not be limited to the agencies involved. The political control of regulatory agencies is the chief source of these problems; it is the politicians at the local, state, and federal levels who are responsible for the current lack of environmental regulation in the United States.

13.4 Public Education

Public education has been considered an important part of the restoration effort. This includes active participation of citizens in the monitoring efforts. Development of associated educational activities includes involvement of students from secondary school systems, undergraduate and graduate programs of regional colleges and universities, and public education via the usual media outlets. Analysis, modeling, and publication of project data include the development of symposia and other forms of meetings with publication of project findings that the general public can understand. The impact of public educational programs remains uneven, however, with mixed results depending on the project involved. Unfortunately, the American public remains largely indifferent to and ignorant of even the most basic environmental concepts and events. The rise of the NIMBY movement does not fill the knowledge gap in the public arena.

13.5 Legal Action

The closest we can come to a level playing field in the process of restoration remains in the legal process, especially at the federal level. There are still powerful environmental laws that exist to protect aquatic systems. A court-ordered study to provide the objective basis for restoration can work under the right circumstances. One drawback to this alternative is that there is often great expense in time and money. Many cases drag on for years. A successful prosecution of the remediation also depends on objective scientific testimony. However, progressive legal action is still operable under the right circumstances. There are various examples of such success: the Hudson River and PCBs, phosphorus pollution of the Florida Everglades, some of the problems with the Great Lakes, and examples given in Chapter 11. There are many approaches that can lead to successful remediation that do not involve court action, but the most effective way to restore a given aquatic system still lies in the legal system.

13.6 News Media

There is still a cadre of environmental writers who continue to write objectively on environmental subjects. However, these reporters do not set policy for the news media. They do not compose headlines for their stories. They do not write editorials, and they certainly do not have much to say concerning the order or significance of environmental reporting. Despite the fact that modern journalism has a synergistic relationship with the American public, it remains that much of what is reported about environmental affairs follows a formula that is not designed to tell the truth. The usual environmental coverage is meant to titillate and create controversy. The sale of news often dominates the environmental agenda.

The comparison between the scientific facts and what is actually reported to the public, as outlined above, does not represent a monolithic conspiracy to subvert fact into self-serving fiction. However, there are many ways to fill an article with facts without revealing the truth:

1. *Omission.* The single most effective way to misrepresent what is going on is to simply ignore the facts of a given issue. If it isn't reported, it didn't happen. The creative omission of scientific facts, wrapped in assumptions of the ignorance of the American public, is a favorite ploy of many journalists. Why bother the public with boring scientific facts when a simplified version of a given issue can be presented without necessarily falsifying the information?
2. *Sensationalism.* The pumping up of the "facts," usually in the form of misrepresenting or unduly emphasizing the threat to human health, not only undermines public understanding of a given issue, but also leads to a misdirection of the real causes and effects of a given environmental problem.
3. *Creation of controversy.* One of the most common tricks of the media is to create interest by pitting one side vs. the other, usually in the form of unenlightened "environmentalists" vs. the polluter in question. If there ever was a formula for

misrepresentation of the facts of a given issue, this is it. Once again, this approach to environmental "journalism" ignores the scientific facts in favor of creating an interesting diversion that titillates rather than informs.

4. *Presentation of two sides of a one-sided issue.* When there is enough scientific information to come to a reasonable conclusion concerning a given environmental issue, the press often will continue to present both sides as equally valid. This supposedly maintains the claim of objectivity when, in fact, it is a subtle way of misrepresentation of the facts. Global warming is one example of such a ruse. Although there is overwhelming scientific evidence that global warming is happening and that human activities are exacerbating such warming, major elements of the media still treat this issue as debatable.

5. *Deferred representation of the facts.* Often, an environmental story will be presented on page 1 with blaring headlines that are often misleading. When it turns out that there are errors in the story, the corrections of such errors are published at the bottom of page 38.

6. *Publication of outright lies.* One of the most disturbing new trends is the dissemination of false information that, when repeated over and over again, becomes accepted fact. Somewhere along the way, the news media have discovered that, thanks to a short attention span together with public indifference to environmental matters, they can get away with almost anything in the way of misrepresentation of the facts. This used to be the primary *modus operendi* of the tabloids and right-wing media outlets, but is now embraced by major elements of the mainstream news media.

7. *Misleading headlines.* This is the simplest way to mislead the public. In part, this method assumes that the reader will not necessarily understand the story. This assumption is not altogether without substance.

8. *Re-ordering of the facts.* Leading the story with irrelevant and/or outlandish "facts" and anecdotes is one way to catch the attention of the reader. Scientific findings that are relegated to the end of a long story often dispute such "facts," but the emphasis on the anecdotal often prevails.

9. *Ignorance.* There is considerable evidence that many environmental writers have little scientific training, and are not capable of understanding scientific issues, much less reporting on them.

10. *Reduced investigative reporting.* There is a dearth of true investigative reporting concerning environmental issues. False and misleading statements by politicians and their bureaucratic hacks are rarely checked for accuracy, and when such statements are contradicted by scientific information, they are presented as a valid alternative to the science. In many instances, the scientific facts are simply ignored in favor of the political explanation.

11. *The Pollyanna effect.* In some cases, a mild but reasonable-sounding explanation is made concerning an environmental problem that is counter to the scientific data even though it sounds "reasonable" to the unenlightened public ear. This form of media chicanery is particularly damaging because it sounds reasonable when it actually is designed to conceal important information.

12. *The media as a special interest.* There is little effort in media circles to present conflicts of interest in the reportage of an environmental issue. With the increasing control of media outlets by large corporations, this problem will continue to grow.

Examples of the above list of falsification of environmental news by the media have been given in the comparison of scientific information with what was actually reported. The north Florida experience is simply a microcosm of the world at large. There is little doubt that the gulf between what the public perceives as important environmental issues and what constitutes the actual threats to the environment is the product of the constellation of factual misrepresentation by media outlets that range from tabloids and right-wing organizations to the major TV networks and big-city papers.

13.7   The Ecology of Restoration

The assumption that merely eliminating the loading of a given pollutant will lead to reversion of the aquatic system to its previous (pristine and/or productive) state is without scientific verification, and is probably not valid. The long-term studies in the Perdido system (Livingston, 2000, 2002) indicate that bay recovery as a response to restoration activities (i.e., reduced nutrient loading) was not the mirror image of impact history, and changes in nannoplankton populations and the increased activity of some bloom species indicated complex biological interactions during the recovery phase. After bloom species were established in Perdido Bay due to nutrient loading from the pulp mill, it took less nutrient loading to stimulate a bloom than when the bay was not infested by such bloom species. Altered food webs due to such blooms did not return to their previous state with cessation of the blooms after reductions in nutrient loading. Biological processes are nonlinear, and reflect adaptation by populations and corresponding responses of trophic organization that are not predictable by linear models of recovery. These uncertainties require a different approach to anticipated responses of aquatic systems to restoration activities.

13.8   Economic/Political Considerations

The most important single factor in most environmental matters is the economic consideration. American politics are controlled by money. The economical aspects of a given situation also dominate the entire restoration process. There is a common theme that runs through the history of restoration in aquatic systems in the United States. During early phases of economic development of a given area, there is a clear suppression of environmental considerations. This includes interference with scientific efforts to understand the aquatic system in question. In most cases, the process of economic development ends in debilitation of the natural resources. Often, the combined impacts of industrialization, urbanization, and agricultural production remain undetermined. In more notable cases, such as the Chesapeake system or the Florida Everglades, there is the "discovery" of the damage to the system in question at some time after the actual development has been completed. At that time, there is then a much-publicized effort, usually at public expense, to restore the damaged ecosystem. The very same politicians and their representatives in the field (i.e., the U.S. Army Corps of Engineers) are then put in charge of, and take the credit for, the restoration effort. The scientific effort is carefully controlled. Dissenters are either disparaged or disciplined, usually in the form of withholding funds from research interests or job replacement or outright job elimination for state and federal workers. Much of this is choreographed by major elements of the news media, with cheerleading by the now-domesticated organized environmental groups. Everybody is happy with the possible exception of those who are economically dependent on the lost resources. This often involves Native Americans and/or fishing interests who rank at the bottom of the economic/sociological organization of American society.

The reason that there has been a general decline in aquatic habitats can be found in the cultural values of American society. Contrary to the popular image created by the news media, the natural environment does not rank high in the American hierarchy of values. This attitude has been in place in America for centuries. Those who stand in the way of "progress" are eliminated in one way or another. The overall "optimism" engendered in the pseudo-populist media forces has maintained the popularity of restoration, together with the glossy reports of those whose job it is to carry out the rehabilitation process. The truth is, while there have been successful efforts to restore various aquatic systems, the list of failures keeps growing. Some of the major aquatic areas of concern include the Mississippi River and associated parts of the Gulf of Mexico, the Colorado River system, the Chesapeake system, the Kissimmee–Okeechobee–Florida Everglades–Florida Bay–Florida Keys–coral reef system, and countless lakes, rivers, and estuaries that remain in various stages of disrepair.

Despite the growing list of failures, future restoration efforts should not be counted out. There will always be ways to remediate at least parts of the damaged aquatic systems that can then yield useful productivity while maintaining habitat for the growing list of endangered species. Issues such as global warming are so important that failure to remediate in a timely fashion will result in disaster, both economically and environmentally. However, if same patterns are followed for global warming as those outlined above (i.e., do the damage first, then remediate if possible), the results to future generations will be catastrophic. The same pitfalls in the restoration process that have been outlined in this book remain: rationalization, denial, greed, ignorance, indifference, political chicanery, and media duplicity. These pitfalls constitute the new constellation of seven deadly sins that will have an increasingly adverse effect on the quality of life for future generations unless human beings become more concerned with environmental issues. This will take a major revision of a culture that has thus far fallen short of providing the political will to live with nature rather than against it.

section VI

Appendices

Appendix I

Field/Laboratory Research Outlines and Methods Used for Studies by the Florida State University Research Group (1970–2004)

The Leon County Lakes Project started during the winter of 1988, and has been carried out continuously to the present time. This project is somewhat unique in that most of the research has been carried out without any formal financial support. It started inadvertently as a class exercise, and has continued to the present time with the voluntary support of undergraduate students, graduate students, staff, and scientists of the Department of Biological Science, the Center for Aquatic Research and Resource Management of Florida State University, and Environmental Planning and Analysis, Inc. (Tallahassee, Florida). The Leon County Commission provided a 6-year grant (the only formal funding by a public agency) from the summer of 1991 through the fall of 1997. This funding was administrated through Mr. Helge Swanson and his staff, which included Ms. Karen Kebart, Mr. Thomas Ballantine, and Mr. Tom Greene. Funding for a program designed to teach aquatic ecology to Native Americans was funded by the Elizabeth Ordway-Dunn Foundation and the Millstone Institute for Preservation. There were many people who contributed to the research effort over the past 16 years. It is impossible to acknowledge all of the hundreds of undergraduate students, graduate students, technicians, and staff personnel who took part in the long-term lakes program.

A. North Florida Lake Systems

1. Introduction

The overall database includes a 16-year study of Lake Jackson and a series of lakes in Leon County. Analyses include background monitoring for various physical/chemical parameters, including field (physical/chemical) data, nutrients, productivity indices, sediment chemistry, field descriptive and biological (experimental) data concerning benthic infaunal macroinvertebrates, and field analysis and field/laboratory experiments concerning fish condition and distribution. We initiated a field experimental program to determine the relationship of storm water runoff, the distribution of submerged aquatic vegetation in the receiving lakes, and the various habitat variables that are affected by the resulting hypereutrophication.

2. Project Outline

The following is an outline of the work that has been covered in the long-term lake analyses:

- Background physical/chemical field monitoring
- Nutrient analyses: water
- Sediment nutrient analyses
- Heavy metal and PAH concentrations in sediments
- Infaunal macroinvertebrate distribution
- Experiments concerning toxicity of contaminated sediments from Megginnis Arm on infaunal macroinvertebrate assemblages in Lake Jackson
- Fish distribution and population analysis of incidence of disease factors in Lake Jackson
- Stormwater event analyses: water quality and effects on various lake systems
- Distribution of submerged aquatic vegetation (SAV) in space and time: aerial analysis, qualitative population occurrence, quantitative analyses
- Chlorophyll *a* distribution and relationship to nutrient loading, nutrient distribution, and SAV analyses
- Water/sediment/animal analyses for metals: dynamics of transfer of metals into lake food webs
- Experimental analysis of the interactions of nutrient/organic carbon inputs with microphyte/macrophyte distribution and water quality indices

The major forms of controlling factors such as depth, light distribution, nutrient input and cycling, sediment and water oxygen demand, nutrient limitation and productivity of microphytes and macrophytes, and the effects of plant growth on important habitat conditions have been reviewed extensively with reference to the ongoing experimental program.

3. Methods and Materials

Water quality and biological parameters currently under investigation are given below:

a. Lake Jackson

Temperature
Conductivity
Dissolved oxygen (DO)
pH
Color
Turbidity
Secchi depths/depths
Nutrients (N, P, C compounds of various types)
Chlorophyll *a*
Metals and PAH distribution in water, sediments, and animals
Light distribution
Standing crop by species of SAV
Gross/net phytoplankton and SAV productivity
Respiration of phytoplankton/SAV
Sediment oxygen demand (SOD)
Nutrient limitation of SAV
Infaunal macroinvertebrates

Epifaunal macroinvertebrates
Fishes

b. Lake Hall

Temperature
Conductivity
Dissolved oxygen (DO)
pH
Color
Turbidity
Secchi depths/depths
Nutrients (N, P, C compounds of various types)
Chlorophyll *a*
Metals and PAH distribution in water, sediments, and animals
Infaunal macroinvertebrates

c. Lake Ella

Temperature
Conductivity
Dissolved oxygen (DO)
pH
Color
Turbidity
Secchi depths/depths
Nutrients (N, P, C compounds of various types)
Chlorophyll *a*
Metals and PAH distribution in water, sediments, and animals
Infaunal macroinvertebrates

d. Lake Lafayette

Temperature
Conductivity
Dissolved oxygen (DO)
pH
Color
Turbidity
Secchi depths/depths
Nutrients (N, P, C compounds of various types)
Chlorophyll *a*
Metals and PAH distribution in water, sediments, and animals
Infaunal macroinvertebrates

e. Lake McBride

Temperature
Conductivity
Dissolved oxygen (DO)
pH
Color
Turbidity

Secchi depths/depths
Nutrients (N, P, C compounds of various types)
Chlorophyll *a*
Metals and PAH distribution in water, sediments, and animals
Infaunal macroinvertebrates

f. No-Name Pond

Temperature
Conductivity
Dissolved oxygen (DO)
pH
Color
Turbidity
Secchi depths/depths
Nutrients (N, P, C compounds of various types)
Chlorophyll *a*
Metals and PAH distribution in water, sediments, and animals
Infaunal macroinvertebrates

g. Lake Munson

Temperature
Conductivity
Dissolved oxygen (DO)
pH
Color
Turbidity
Secchi depths/depths
Nutrients (N, P, C compounds of various types)
Chlorophyll *a*
Metals and PAH distribution in water, sediments, and animals
Infaunal macroinvertebrates

h. Lake Talquin

Temperature
Conductivity
Dissolved oxygen (DO)
pH
Color
Turbidity
Secchi depths/depths
Nutrients (N, P, C compounds of various types)
Chlorophyll *a*
Metals and PAH distribution in water, sediments, and animals
Infaunal macroinvertebrates

Nutrient components and associated factors include the following: NO_3, + NO_2, NH_3, (DIN), TN, (ON), PO_4, TP, (OP), DOC, POC, TC, SiO_2, chlorophyll *a*, and TSS (seston). Data concerning the temporal and spatial distribution of these factors will be used in the experimental phase of the Lakes Project. Actual field investigations may vary and can be adjusted to avoid duplication or to target specific research questions.

We have used a series of GPS (Global Positioning System) analyses that allow determinations of latitude and longitude for each fixed station in the various study areas. This system represents a network of satellites with bearings taken from three satellites at any given time. Within a set of specific conditions, these readings are thought to be highly accurate. The stations used for the various studies are shown in Figures 2.1 through 2.3.

B. River–Estuarine and Coastal Systems: Gulf of Mexico

The research effort in the northern Gulf of Mexico is based on written, peer-reviewed protocols for all field and laboratory operations. Water quality and biological analyses have been taken by personnel from CARRMA (Florida State University, Tallahassee, Florida) and Environmental Planning & Analysis, Inc. (Tallahassee, Florida), with additional work carried out at a series of laboratories in various countries. Water quality methods and analyses and specific biological methods have been continuously certified through the Quality Assurance Section of the Florida Department of Environmental Protection (Comprehensive QAP #940128 and QAP #920101). Various methods that have been used to take field information and run laboratory analyses have been published in the reviewed literature. This includes the following: Flemer et al., 1997; Livingston, 1975a,b, 1976a,b, 1980a, 1982a,b, 1984a,b, 1985a, 1987c, 1988a,b,c, 1992a,b, 1997a; Livingston et al., 1974, 1976a,b, 1997, 1998a,b.

Protocol for Comparative Field Analyses

In addition to using the background (biological) data for hypothesis development in terms of experimental initiatives, the various biological indices (which include phytoplankton, zooplankton, infaunal macroinvertebrates, epifaunal macroinvertebrates, and fishes) and the trophic equivalents (transformations based on extensive data concerning feeding processes and natural history phenomena) will be used to evaluate the present status of the Perdido system. This will be done by comparing the first 2 years of data with similar data from a series of other river–estuarine systems. Some of these systems are in relatively natural condition and others have been affected by human activities, including the release of pulp mill effluents. It is not feasible to determine the status of a given system without using the comparative approach. Some of the systems that will be used for this analysis include the Econfina and Fenholloway estuaries (Apalachee Bay), the Apalachicola River estuary, the Choctawhatchee River estuary, the Escatawpa–Pascagoula system, the Mobile River estuary, and the Winyah Bay system.

Our research group has carried out a series of studies concerning the above systems over the past 35 years and we will use such data as part of the comparative analysis of the Perdido River estuary. These studies have been carried out with submerged aquatic vegetation, infaunal and epifaunal macroinvertebrates, and fishes. Phytoplankton and zooplankton collections are based on methods that are commonly used in the literature. We have quantified the various other biological sampling techniques so that we can estimate the relative effectiveness of such samplers and so that we can use a standard method for comparative purposes. In most cases, multiple pseudo-replicates (e.g., not true replicates in a statistical sense) were taken for the purpose of estimating sampler precision (accountability of variation within the capability of the given sampler). Quantification was based on an evaluation based on the rarefaction analysis of cumulative biological indices as a function of multiple sub-samples. The asymptotic relationship of species accumulation curves was used to determine the sampling effort for a given sampler.

The results of these preliminary studies are given in a series of articles published in peer-review journals. A representative list is given below:

Livingston, R.J. 1976a. Diurnal and seasonal fluctuations of estuarine organisms in a north Florida estuary: Sampling strategy, community structure, and species diversity. *Est. Coastal Mar. Sci.*, 4: 373–400.

Livingston, R.J., R.S. Lloyd, and M.S. Zimmerman. 1976a. Determination of adequate sample size for collections of benthic macrophytes in polluted and unpolluted coastal areas. *Bull. Mar. Sci.*, 26: 569–575.

Stoner, A.W., H.S. Greening, J.D. Ryan, and R.J. Livingston. 1982. Comparisons of macrobenthos collected with cores and suction sampler in vegetated and unvegetated marine habitats. *Estuaries*, 6: 76–82.

In this way, the various methods that have been used to collect organisms in the various study areas will allow a comparative analysis with other systems that is both consistent and standardized according to the demonstrated effectiveness of the various sampling techniques.

The following outline will provide the working order of the comparative analyses:

- Determination of station comparability based on physical habitat conditions (e.g., salinity, surface and bottom, in areas deeper than 1 m).
- Determination, wherever possible, of comparable water years in terms of freshwater runoff in the respective systems.
- A comparison of the age structure and size of important fish and invertebrate populations can provide insight concerning the ecological status of the Perdido system. We will focus on any indications of life history disruption or dysfunction when compared with systems that are known to be relatively free of human impact. This will include an analysis of habitat-associated trends in the various systems that could be associated with the possible absence of important groups of organisms. The evaluation of seasonally smoothed data will concentrate on the following indices of the above biological groups:
  - Growth rates by population
  - Numbers and biomass (Total biota and top dominants)
  - Species richness
  - Species diversity and evenness
  - Trophic units
  - Guild associations

We will look for comparative trends, to include similarities and differences due to natural and anthropogenous stress.

Sediment Analyses

Sediments were analyzed for particle size distribution and organic composition according to methods described by Mahoney and Livingston (1982) as defined by Galehouse (1971).

a. Particle Size Analysis

Sediment samples were taken with coring devices (Plexiglas: 7.6 cm diam., 45 cm² cross-section). Analyses were taken on 10-cm samples for regular percent organics and particle size analysis.

- Unpreserved sediments were divided into coarse (>62 μm) and silt-clay (<62 μm) fractions.
- The coarse fraction was analyzed by wet-sieving using 1/2 phi unit intervals.
- The silt-clay fraction was analyzed using a pipette method in 1/2 phi unit intervals from 4 to 6 phi, and in 1-phi intervals for the finer fractions (6 to 10 phi).
- Statistical analyses were carried out using phi units and sediment designations as outlined in the Wentworth scale.
- Statistical treatments of the granulometric results followed the method of moments for calculation of mean grain size (in phi units, with 1 mm = 0.5 phi), skewness (a measure of non-normality of distribution), and kurtosis (a measure of the spread of the distribution). Sorting coefficients represented the measure of grain size as follows: well sorted = 0.50; moderately well sorted = 0.71; moderately sorted = 1.00; poorly sorted = 2.00.

b. Percent Organics

After placing the sediments in a drying oven set at 104°C for a minimum of 12 hr, the sediments were ashed in a muffle furnace for 1 hr or more at temperatures approximating 550°C. The percent organic fraction of the sediments was calculated as a percent of the dry weight.

Sediment analysis is a two-part process; the first part consists of forming a data file and the second part is the actual granulometric analysis. The first part can be accomplished by either running the EZseds program to enter and reformat raw data or by exporting previously entered sediment data from one of the EP&A 4th Dimension Data Management System (DMS) sediment data files. EZseds is a FORTRAN program that prepares raw sediment granulometric data for analysis with the MacGranulo program. MacGranulo is a FORTRAN application that performs sediment grain-size analysis on the data in the input file. Results include summary statistics such as mean grain size (as determined by three different methods) and percent organics, particle-size distribution according to the Wentworth classification, and a frequency distribution plot.

c. Mercury Analyses

Sediment samples were collected at a series of sampling sites in the system. Three sub-samples were taken at each established station. Sediment samples were placed on ice. The samples were shipped to Enviro-Test Laboratories in Canada. Prior to shipping, all samples were logged into chain-of-custody forms and a sample logbook. Samples were examined to ensure that accurate labels had been affixed and that all lids were tight. Prior to closing each cooler containing samples, the chain-of-custody form was completed and placed in a Ziplock® bag that was then placed inside the cooler. The forms were placed in plastic bags. The originator retained a copy of each chain-of-custody form. As a general procedure, samples were taken in order from supposedly clean areas to possible polluted areas. No sampling was carried out in the rain.

Quality control samples (blanks, etc.) were labeled as above. After collection, identification, and preservation, the sample was maintained under chain-of-custody procedures discussed below. A blank consisted of an empty sampling container. For each group of samples (inorganic mercury, methyl mercury), one blank was used for each lot number of jars.

Samples of sediments were taken with corers (the top 2 cm were used for mercury analysis; top 10 cm for sediment particle size and percent organics). A PVC corer was used for the mercury samples. Chemical analyses were run on sediments taken from each of the sampling stations. Three replicate sediment samples were taken at each station for total mercury and methyl mercury analysis. Each of the three replicates was composed of three core sub-samples that were cut into squares (all edges "shaved") to eliminate the outer parts of the core sample. The "shaving" of the sample was carried out with an acid (5 to 10% nitric acid) washed plastic knife. Samples were placed in acid-washed, 1-pint, wide necked, Teflon-lined, screw-top glass jars. Nitric acid (10% pesticide grade) was used to wash the jars. Samples were immediately preserved in the field with concentrated nitric acid to pH 2. Care was taken to mix the sediment and check for the proper pH. Each bottle was labeled as indicated above. Acid preservation for each sample was checked with pH paper to make sure that the pH 2 target is maintained. Samples were then placed on ice and bubble wrapped.

Sampling for toxic agents in sediments in other studies followed similar procedures. All sediment samples were taken with corers (the top 2 cm). A stainless steel corer was used for the organic toxicant samples and a PVC corer was used for the metal samples.

Physicochemical Sampling and Analysis

Designated, fixed stations in a given system were sampled (surface and bottom) for water quality factors. Vertical collections of water temperature, salinity, conductivity, dissolved oxygen, depth, Secchi depth, and pH were taken at the surface (0.1 m) and at 0.5-m intervals below the surface to about 0.5 m above the bottom. Surface and bottom water samples were taken with Niscon bottles for further chemical analyses in the laboratory. All sampling was carried out on a synoptic basis with all samples in a given system taken within one tidal cycle.

A list of variables is given below.

- *Physicochemical Factors:*
  Temperature
  Salinity/conductivity
  Dissolved oxygen (DO)
  pH
  Secchi depth
  Depth
  Light transmission
  True color, NCASI colorimetric method
  True Color, Spectroscopic Method
  Turbidity, nephelometric method
  Total dissolved solids
  Total suspended solids
  Dissolved oxygen, YSI DO meter
  Particulate organic carbon
  Dissolved organic carbon
  Total inorganic carbon
  Total organic carbon
  Total carbon
  Chlorophyll *a*
  Chlorophyll *b*
  Chlorophyll *c*

- *Chemical analyses (loading):*
  - Nitrogen nitrate and nitrite
  - Ammonia nitrogen

- Organic nitrogen and organic phosphate
  - Total organic nitrogen and total organic phosphate
  - Dissolved organic nitrogen and dissolved organic phosphate
  - Particulate organic nitrogen and particulate organic phosphate
- Total nitrogen
- Orthophosphate
- Total dissolved phosphorus
- Total phosphorus
- Dissolved reactive silicate
- Inorganic carbon
- Organic carbon
  - Dissolved organic carbon
  - Particulate organic carbon
  - Total organic carbon
- Total carbon

Physical/chemical data and water samples for chemical analysis were taken according to methods defined by the U.S. Environmental Protection Agency (1983). Temperature and salinity were measured with YSI Model 33 S-C-T meters calibrated in the laboratory with commercial standards. DO was taken with YSI Model 57 oxygen meters calibrated by the azide modified Winkler technique. The pH was measured with an Orion Model 250A pH meter equipped with a calomel electrode. Aquatic light readings were taken with a Li-Cor LI-1800UW underwater spectroradiometer, which measures the spectral composition of photon flux density at 1 to 2-nm intervals from 300 to 850 nm.

A Beckman DU-64 spectrophotometer was used to analyze true color (APHA, 1989). Both ratio and nephelometric turbidity analyses were carried out. A ratio turbidimeter and a Hach model 2100A turbidimeter were used for turbidity analyses. Quantification of dissolved inorganic nitrogen (nitrite, nitrate) and dissolved inorganic phosphorus (orthophosphate) followed methods outlined by Parsons et al. (1984). Ammonium was measured with an ion electrode (U.S. EPA 1983, Method 350.3) after raising the pH to 11. This method had some essential features (e.g., minimal interference from waters highly stained with humic materials and paper mill effluents) that required special attention to the results. Consequently, the level of detection was relatively high (e.g., 2.0 µm ammonium-N) but adequate for this study. Particulate organic nitrogen (U.S. EPA 1983, Method 351.4) and particulate organic phosphorus (U.S. EPA 1983, Method 364.4) were collected on Gelman glass fiber filters combusted at 500°C, oxidized to inorganic fractions and organic nitrogen measured as ammonia and organic phosphorus measured as particulate organic phosphorus. Total dissolved solids (APHA Method 209-B), total suspended solids (APHA Method 209-C), dissolved organic carbon (APHA Method 415.1), chlorophyll *a* (APHA Method 1002-G), and biochemical oxygen demand (APHA Method 405.1) were carried out according to the methods outlined by the American Public Health Association (APHA, 1989). Particulate organic carbon was analyzed according to Parsons et al. (1984) (Method 3.1). Nutrient analyses were carried out according to APHA (1989).

Nutrient Loading Models

The U.S. Geological Survey (Tallahassee, Florida; M. Franklin, personal communication) provided flow data for the Perdido River system, Bayou Marcus Creek, and Elevenmile Creek. Nutrient sampling was carried out in the various rivers leading into the bay (Figure 5.2). Monthly nutrient concentrations from these rivers were used in the models. Overland runoff from unmonitored areas was estimated using models devised by

A. Niedoroda (personal communication). Loading models for the 11-year database were run at monthly, quarterly, and annual intervals.

The nutrient loading models were based on a ratio estimator developed by Dolan et al. (1981). This model corrects for the biases due to sparse temporal sampling. It uses auto- and cross-covariance values of flows and nutrients for correction of the loading calculations. Nutrient data for the models were usually taken monthly, whereas river data were taken daily. Program modifications were made by A. Niedoroda and G. Han (personal communication) wherein (1) due to discontinued river monitoring at Bayou Marcus Creek and the Blackwater River in September 1991, river flows for these areas were estimated from regressions (log/log, Elevenmile Creek and Bayou Marcus Creek; Styx and Blackwater Rivers); (2) Stations 9 and 21 were sampled monthly from March 1993 to October 1994 and flows were then estimated by the following, St. 09--(2.52 *St. 05) + (1.01 * St. 04) + (1.10 * St. 07), St. 21--(2.64*[St. 13-32]) + 32. These data are considered questionable at best and should not be used when forming any serious conclusions.

Dr. Alan Niedoroda and Mr. Gregory Han of Woodward-Clyde designed the loading estimator program. Data were grouped according to methods described in previous reports. Model applications were carried out according to a formula given by Dolan et al. (1981).

$$\overline{\mu}_y = \mu_x \frac{m_y}{m_x} \left(\frac{1 + \frac{1}{n} \frac{S_{xy}}{m_x m_y}}{1 + \frac{1}{n} \frac{S_{x^2}}{m_{x^2}}} \right)$$

where

μ_y = Estimated Load
μ_x = mean daily flow for the year
m_y = mean daily loading for the days on which concentrations were determined
m_x = mean daily flow for the days on which concentrations were determined

$$S_{xy} = \frac{1}{(n-1)} \sum_{i=1}^{n} x_i y_i - n m_x m_y$$

$$S_{x^2} = \frac{1}{(n-1)} \sum_{i=1}^{n} x_{i^2} - n m_{x^2}$$

n = number of days on which concentrations were determined
x_i = individual measured flows
y_i = daily loading for each day on which concentrations were determined

This formulation corrects for biases introduced by sparse (temporal) sampling by using auto and cross covariance values of the flows and nutrients to correct the loading calculations for variability in the river flow which influences the nutrient concentrations. This applies where flow data are usually taken regularly (usually daily), whereas nutrient concentrations are usually taken at monthly or quarterly intervals.

Light Determinations

Light penetration depths were taken using standard Secchi disks. Field light transmission data were taken with a Li-Cor LI-1800UW underwater spectroradiometer. The underwater light field was characterized by incident radiant flux per unit surface area as quanta m^{-2} s^{-1}. Samples included three to five scans (replicates) taken at 1 to 2-nm intervals that were averaged for each reading. Flux measurements were taken for photosynthetically active radiation (400 to 700 nm; PAR) and individual wavelengths. For any series of collections, field samples were taken during relatively calm conditions between the hours of 1000 and 1400 hr. Multiple air light readings were taken to correct for short-term radiation variability during light measurements.

Field/Laboratory Methods

All sampling efforts were carried out in a similar fashion with respect to sample collection, transfer, and analysis. For all types of organisms, outside corroboration has been carried out continuously by recognized taxonomists. Laboratory specimen libraries have been established and continuously maintained. Chain-of-custody forms are filled out with each transfer with the usual procedure as follows:

- Samples are delivered (chain-of-custody form signed). A copy of the chain-of-custody form is given to the person transferring samples; the original is kept with samples. Samples are then stored in a secure area.
- Samples are processed (if required), and date and time are recorded (ID# kept with each sample).
- Organisms are picked (if required), and the date and time are recorded (ID# kept with sample).
- Preparation of specimens are made for identification (ID# kept with sample).
- Data are recorded in proper form for reporting to customer.
- Data, samples, and reference collection are transferred to client (chain-of-custody form signed). A copy of original is requested.

ID labels are kept on the outside of the sample container as well as inside the sample container with the organisms. This information includes project, sample location and date, and is made with indelible ink on a suitable label material outside and on a paper label inside (preferably white rag paper).

1. Phytoplankton and Zooplankton

Net phytoplankton samples were taken with two 25-μm nets (bongo configuration) in duplicate runs for periods of 1 to 2 min. Repetitive (three) 1-L whole water phytoplankton samples were taken at the surface. Phytoplankton samples were immediately fixed in Lugol's solution in its acid version (Lovegrove, 1960). Samples were analyzed by methods described by Prasad et al. (1990) and Prasad and Fryxell (1991).

Zooplankton were taken in various ways. Multiple tows were made with 202-μm nets run over varying time periods that depended on zooplankton density. Sample volumes were recorded so that a quantitative estimate could be made of zooplankton numbers. Another approach, used under certain conditions, involved pumping water (a volume that was determined in the field) through 202-μm nets. Zooplankton preservation was made using 10% formalin. All counts were made to species.

2. Periphyton

A set of slides was anchored at given stations 2 to 2.5 cm below the surface. After 2 to 3 weeks, we randomly took five slides and placed them in labeled jars filled with distilled water. Jars were transported on ice to the lab where they were scraped on both sides and preserved with 5% formalin for a composite sample that will be stored in screw-cap vials. Each sample was then diluted with distilled water, depending on the density of algae, thoroughly shaken, and three aliquots of 0.05 or 0.1 mL were placed on slides, covered, counted, and identified.

3. Benthic Macrophytes

Quantitative seagrass sampling used divers to take eight (randomly distributed) quadrat samples ($1/4$-m^2) according to methods described by Livingston et al. (1976a,b). Samples were identified to species. Dry weight determinations were to be made for macrophytes (rooted and epiphytic) by species for above and below-ground dry weight biomass. Samples were heated in the oven at about 105°C until there was no further weight loss (at least 12 hr). Samples were then weighed to the nearest hundredth of a gram.

A map of the offshore seagrasses was developed from just west of the Aucilla River mouth to a point near the Spring Warrior Creek entrance (Figure 3.3). A set of 166 natural-color aerial photographs was taken (AeroMap U.S. Inc., personal communication) along four flight lines between 9:06 and 9:50 AM E.S.T. on November 15, 1992. Ground-truthing was carried out as a series of diver-transects at 320-m intervals from land to 5 to 7 km offshore: data included visual observations of seagrass species and density, habitat distribution, and bottom type. Seagrass density was estimated as percent cover (bare, no observed macrophytes; very sparse, less than 10% coverage; sparse, 10 to 40% coverage; moderate, 40 to 70% coverage; and dense, 70 to 100% coverage). The aerial photography was transferred to a computerized seagrass basemap at a scale of 1:24,000. Overlay maps of the transect data were printed at the same scale. The center point of every photograph in the series was then positioned and interpreted using the diver transect data. The seagrass signature (density, distribution) was identified. Density classification was accomplished by visually comparing the photograph to examples of various density classes from an enlarged Crown Density Scale. The final basemap was then photographically reduced in scale and scanned into a computer system. A detailed analysis was carried out concerning the inshore areas of the SAV survey (within 3 km of shore). The survey region was divided into a series of 17 zones along the coast that were 2.5 km wide and 3 km offshore with the river mouths of the Aucilla, Econfina, and Fenholloway Rivers positioned in the center of individual zones. Absolute and relative areas of the different grass densities and other habitat classes, by zone, were then calculated and used for statistical analysis, along with water/sediment quality analyses at stations within each zone.

4. Field Collections of Fishes and Invertebrates

All field biological collections were standardized according to statistical analyses carried out early in the program (Hooks et al., 1975; Livingston, 1976a,b, 1987a,b; Greening and Livingston, 1982; Livingston et al., 1976a,b; Meeter and Livingston, 1978; Stoner et al., 1982). Infaunal macroinvertebrates were taken using coring devices (7.6 cm diameter, 10 cm depth). The number of cores taken was determined using large initial samples (40 cores) and a species accumulation analysis based on rarefaction of cumulative biological indices (Livingston et al., 1976a,b). Based on this analysis, multiple (10 to 12) core samples were usually taken randomly at each station, with the assured representation of at least 80% of the species taken in the initial sampling. All infaunal samples were preserved in 10% buffered formalin in the field, sieved through 500-μm screens and identified to species wherever possible.

Where roots, etc. prevented coring, suction dredges were used as prescribed with multiple sweeps of specific time (number and time to be determined during first field trip). Animals were preserved in 10% buffered formalin and stored in 100% denatured ethanol.

Epibenthic fishes and invertebrates were collected with 5-m otter trawls (1.9-cm mesh wing and body, 0.6-cm mesh liner) towed at speeds of about 3.5 to 4 km.hr^{-1} for 2 min, resulting in a sampling area of about 600 m² per tow. Repetitive samples (two or seven) were taken at each site; sampling adequacy of such nets had been determined by Livingston (1976). All collections were made during the same lunar and tidal phase at the first quarter moon. Samples were taken within 2 hr of high tide, with day and night trawls separated by one tidal cycle. Larger fishes (>300 mm SL) were taken with 100-m nylon trammel nets and stomachs were removed in the field and injected with a 10% buffered formalin solution.

All organisms were preserved in 10% buffered formalin, sorted and identified to species, counted, and measured (standard length for fishes; total length for penaeid shrimp, carapace width for crabs). Representative samples of fishes and invertebrates were dried and weighed, and regressions were run so that data from the biological collections could be converted into dry and ash-free dry mass. The numbers of organisms were then converted to biomass m⁻². All biological data (infauna, epibenthic macroinvertebrates, fishes) were expressed as numbers m⁻² mo⁻¹ or biomass m⁻² mo⁻¹.

5. Trophic Determinations

Gut analyses were carried out in the same manner throughout the 30-year field analysis period. Fishes and invertebrates were placed in 5, 10, or 20-mm size classes (depending on size), and food items were taken from the stomachs of up to 25 animals in a given size class with pooling by sampling date and station. Determinations similar to those described above were carried out and it was determined that at least 15 animals were needed for a given size class × station × sampling period. Stomach contents were removed and preserved in 70% isopropanol and a dilute solution of Rose Bengal stain. In larger animals, we pumped the stomachs out and released the subject. The content analysis was carried out through gravimetric sieve fractionization (Carr and Adams, 1972). Contents were washed through a series of six sieves (2.0 to 0.75-mm mesh), and the frequency of occurrence of each food type was recorded for each sieve fraction. As the food items were comparable in size, the relative proportion of each food type was measured directly by counting. Dry weight and ash-free dry weight were determined for each food type by size class. In general, mutually exclusive categories were used for the food types that included both plant and animal remains. Stomach contents by food type by size class were then calculated and related back to station/time designations. Based on the long-term stomach content data for each size class, the fishes and invertebrates were then reorganized first into their trophic ontogenetic units using cluster analyses (Czekanowsli [Bray-Curtis] or C-Lamba similarity measure; flexible grouping cluster strategy [β = –0.25). The basis for this analysis is given by Sheridan (1978) and Livingston (1984).

All biological data (as biomass m⁻² mo⁻¹ of the infauna, epibenthic macroinvertebrates and fishes) were transformed from species-specific data into a new data matrix based on trophic organization as a function of ontogenetic feeding stages of the species found in Perdido Bay over the multi-year sampling program. Ontogenetic feeding units were determined from a series of detailed stomach content analyses carried out with the various epibenthic invertebrates and fishes in the near-shore Gulf region (Sheridan, 1978, 1979; Laughlin, 1979; Sheridan and Livingston, 1979, 1983; Livingston, 1980, 1982a,b, 1984b, unpublished data; Stoner and Livingston, 1980, 1984; Laughlin and Livingston, 1982; Stoner, 1982; Clements and Livingston, 1983, 1984; Leber, 1983, 1985). Based on the long-term stomach content data for each size class, the fishes and invertebrates were reorganized first into their trophic ontogenetic units using cluster analyses. Infaunal macroinvertebrates were also organized by feeding preference based on a review of the scientific literature (Livingston, unpublished data). The field data in Perdido Bay were then reordered into trophic levels so that temporal changes in the overall trophic organization of this system could be determined over the 8-year study period. We assumed that feeding habits (at this level of detail) did not change over the period of observation; this assumption is based on previous analyses of species-specific fish feeding habits that remained stable in a given area over a 6- to 7-year period (Livingston, 1980).

The long-term field data were used to establish a database based on the total ash-free dry biomass.m^{-2}.mo^{-1} as a function of the individual feeding units of the infauna, epibenthic macroinvertebrates, and fishes. Data from the various stations in Perdido Bay were used (g m^{-2}) for all statistical analyses. The data were summed across all taxonomic lines and translated into the various trophic levels that included herbivores (feeding on phytoplankton and benthic algae), omnivores (feeding on detritus and various combinations of plant and animal matter), primary carnivores (feeding on herbivores and detritivorous animals), secondary carnivores (feeding on primary carnivores and omnivores), and tertiary carnivores (feeding on primary and secondary carnivores and omnivores). All data are given as ash-free dry mass m^{-2} mo^{-1} or as percent ash-free dry mass m^{-2} mo^{-1}. In this way, the long-term database of the collections of infauna, epibenthic macroinvertebrates, and fishes was reorganized into a quantitative and detailed trophic matrix based not solely on species (Livingston, 1988a,b,c) but on the complex ontogenetic feeding stages of the various organisms. A detailed review of this process and the resulting feeding categories is given by Livingston et al. (1977).

6. Oyster Studies

Oyster samples (multiple) were taken with full head tongs (16-tooth head; 4.5-m handles) at each station on a monthly basis in Apalachicola Bay from March 1985 through October 1986. The use of tongs was standardized with respect to opening widths and sampling effort. To quantify the sampling effort, a series of 30 standardized, random tong samples was taken at the Big Bayou and Cat Point reefs in February 1985. The cumulative size frequency distribution (in 10-mm increments) was determined and plotted for each sampling site. The number of samples necessary for a specified level of quantification was determined according to a method described by Livingston et al. (1976a,b). The method allowed determination of the number of sub-samples necessary to achieve specific levels of size class accumulation when compared to the results of 30 sub-samples. Seven sub-samples accounted for 80.8% (Big Bayou) and 87.5% (Cat Point) of the sampling variability for the total sample. Accordingly, this number of sub-samples (located beyond the asymptote for size class accumulation) was considered to constitute a representative sample taken in each of the test regions. Numbers of oysters per tong were recorded and converted to numbers.m^{-2}.

All oysters were measured to the nearest millimeter in the field according to the greatest distance from beak to lip using linear calipers. A total of 140 oysters taken from four stations (n = 35 at each site) were used to determine the relationship of shell length and weight of oyster meat. Four separate length/weight equations were developed to account for known differential growth characteristics in different regions of the bay:

$$\ln(\text{AFDW}) = 2.505 \times \ln(\text{LEN}) - 10.980 \text{ (Cat Point, } r^2 = 0.83)$$

$$\ln(\text{AFDW}) = 2.303 \times \ln(\text{LEN}) - 10.306 \text{ (East Hole, } r^2 = 0.84)$$

$$\ln(\text{AFDW}) = 2.202 \times \ln(\text{LEN}) - 9.125 \text{ (Paradise, } r^2 = 0.90)$$

$$\ln(\text{AFDW}) = 2.465 \times \ln(\text{LEN}) - 10.190 \text{ (Scorpion, } r^2 = 0.86)$$

where AFDW is the ash-free dry weight of oyster meat and LEN is oyster shell length in millimeters (mm). *F*-tests (α = 0.05) disclosed that equations from both bars in the eastern bay (Cat Point and East Hole) were significantly different from equations developed for bars in the western bay (Paradise and Scorpion). Although the differences were not significant within the respective regions, east and west, the site-specific equations were used for the transformations of the length data into ash-free dry weights. This allowed the most comprehensive and accurate use of the data for such transformations. All tong data (in terms of numerical abundance and ash-free dry weight) were calculated on a unit m^{-2} basis. These data were then transformed to estimates of total numbers and biomass on each bar based on the estimated size of the bars. Areas were established from computer simulations of oyster-producing areas (Livingston, unpublished data) based on historic records, interviews with oystermen and state environmental agencies, and our past and ongoing field studies (Livingston, 1984).

The tong data (oyster density and length frequency) were standardized by a quantitative comparison with data derived from a series of multiple (0.25 m^{-2}) quadrats taken on the same sampling sites (Cat Point Bar, East Hole, Paradise) over the same period of study by other researchers. Scattergrams of the respective density and length frequency databases were log-transformed to approximate the best fit for a normalized distribution. Statistical comparisons were made of the monthly data by station. Independence tests were used to examine each sample for autocorrelation at a number of lags. For each sample, autocorrelation is computed for the lesser of 24 or $n/4$ lags, where n is the number of observations. The program then computes the Q statistic of Ljung and Box (1978) for an overall test of the autocorrelations as a group, where the null hypothesis was that correlation at each lag was equal to zero. The parametric F-test for comparison of variances and t-test for comparison of means were used to test the hypothesis that the data came from normally distributed populations. Means testing was carried out with both parametric and non-parametric tests. For independent, random samples from normally distributed populations, the parametric t-test was used to compare the sample means. Where there was serial dependence of data from a given pair of stations, it was removed by differencing the observations and calculating and plotting the autocorrelations of the differences. If the differences were not serially correlated, we applied the Wilcoxon sign-rank test on the differences to compare the two sets of numbers (P = 0.05).

Spatfall accumulation was analyzed from a subset of 12 stations. Spat baskets, constructed of plastic-coated wire (25 × 25 × 25 cm; 2.5-cm mesh), were filled with about 20 sun-bleached oyster shells and placed at each site. Bricks were placed at the bottom of each basket so that the oysters remained off the bottom so as to lessen problems with sedimentation. Samples were retrieved, and new sets of oyster shells were set out at 2- to 3-week intervals. One spat basket was used at each station. Seven shells were randomly chosen from each basket for analysis. Spat counts were made from the inner surface of each valve; this standardization was based on test results that indicated less variance of such adherence on inner surfaces than on outer surfaces.

Oyster data were grouped in two ways prior to analysis: (1) bay-wide totals and (2) eastern vs. western reefs. In the latter grouping strategy, selected reefs from the eastern bay (Sweet Goodson, Cat Point, East Hole, Porter's Bar) were compared to selected reefs from the western bay (Paradise, Pickalene, Scorpion). These reefs were chosen because of their relative commercial importance to overall oyster production in the bay. Intervention models were used to analyze the effects of the two hurricanes on the monthly total oyster numbers (in thousands) and the monthly average shell length (in mm) of oysters in the bay (Box and Jenkins, 1976; Pankratz, 1991). Three major interventions occurred in the Apalachicola Bay during the study period: Hurricane Elena (combined with the cessation of commercial oystering after September 1985), Hurricane Kate, and the resumption of commercial oystering in May 1986. The models were used to test the possibility that the interventions were significantly associated with level changes in the monthly total oyster numbers {$N(t)$} and the monthly average shell lengths {$S(t)$}.

Three indicator variables corresponding to the interventions are defined as

$$X_1(t) = \begin{cases} 0, \ t < September \quad 1985 \\ \hline 1, \ t \ge September \quad 1985 \end{cases}$$

$$X_2(t) = \begin{cases} 0, \ t < December \quad 1985 \\ \hline 1, \ t \ge December \quad 1985 \end{cases}$$

$$X_3(t) = \begin{cases} 0, \ t < May \quad 1986 \\ \hline 1, \ t \ge May \quad 1986 \end{cases}$$

Using B as the backward shift operator such that $BX(t) = X(t-1)$, the relationship between the monthly total oyster number $\{N(t)\}$ and the three interventions can be described by the following general intervention model:

$$N(t) = \beta_0 + v_1(B)X_1(t) + v_2(B)X_2(t) + v_3(B)X_3(t) + \xi(t) , \qquad (A.1)$$

where $v_1(B)$, $v_2(B)$, and $v_3(B)$ are polynomials with the typical form;

$$v(B) = \omega_0 - \omega_1 B - \ldots - \omega_R B^R$$

where $\{\omega_0, \omega_1, \ldots, \omega_h\}$ are parameters to be estimated.

In model (A.1), β_0 is a constant and $\xi(t)$ is a noise term which is often modeled as a stationary autoregressive moving average (ARMA; p, q) process where

$$\xi(t) - \phi_1 \xi(t-1) \ldots - \phi_p \xi(t-p) = \varepsilon(t) - \theta_1 \varepsilon(t-1) - \ldots - \theta_q \varepsilon(t-q),$$

where p is the order of the autoregression (AR) term, and q is the order of the moving average (MA) term. The $\varepsilon(t)$'s values are assumed to be independent and normally distributed with mean zero and variance σ^2.

Model (A.1) extends the traditional linear regression models in two directions. First, the monthly total oyster number series $\{N(t)\}$ may react to an intervention with a time lag. For example, the term $v_1(B)X_1(t) = \omega_0 X_1(t) - \omega_1 X_1(t) - \omega_1 X_1(t-1) - \ldots - \omega_h X_1(t-h)$ in model (A.1) represents that the impact of the first intervention on N(t) is distributed across several time periods. Second, instead of assuming that the errors $\xi(t)$ are independently distributed, ARMA(p,q) models are used for $\xi(t)$ which incorporate possible serial correlations in the response series $\{N(t)\}$. Model (A.1) was fitted using the Linear Transfer Function Identification Method proposed by Pankratz (1991, Chapter 5). The relationship between the monthly average shell length series {S(*t*)} and the three interventions was modeled in a similar fashion.

In addition to the overall oyster data, intervention models were fitted to eastern and western sections of the bay using numbers.m^{-2} and average shell length (in mm) of oysters from stations in these areas. The four series were denoted by EN(*t*) and WN(*t*) for eastern and western densities and by ES(*t*) and WS(*t*) for eastern and western shell lengths, respectively. Intervention models were fitted to the four series separately.

7. Modeling

Physical processes in Apalachicola Bay were simulated using a time-dependent, three-dimensional model of the Blumberg–Mellor family (Blumberg and Mellor, 1980, 1987). This model solves a system of coupled differential (prognostic) equations for free surface elevation, two components of the horizontal velocity, temperature, salinity, turbulence energy, and turbulence macroscale. The spatial integration is explicit in the horizontal and implicit in the vertical. For its forcing, the model admits a comprehensive database comprised of time-dependent temperature, salinity, surface heat and humidity fluxes and wind stress, tides, and residual signals, when available. In the horizontal, the model uses a body-fitted curvilinear orthogonal coordinate system that allows one to better adhere to the convoluted coastline. In the vertical, a sigma-coordinate system converts the free surface and seabed into coordinate surfaces, and allows for better resolution of the surface and near-bottom boundary layers. The sophisticated second-order turbulence closure model of Mellor and Yamada (1982), as modified by Galperin et al. (1988), describes the vertical mixing.

The model was calibrated and verified using hydrographic data collected at 0.5-hr intervals from 23 instruments located at sites throughout the bay during the 6-month period from June to November 1993 (Wu and Jones, 1992; Huang and Jones, 1997). Measured river inflows, surface wind stress, free surface elevations, and temperature and salinity signals at the open boundaries were utilized as time-dependent boundary conditions. The open boundaries were comprised of five single-point and multi-point grid locations through which Apalachicola Bay connects to the Gulf directly. Freshwater runoff into the bay was drawn from four rivers: Apalachicola River and its tributaries, Whiskey George Creek, Cash Creek, and the Carrabelle River. Flow values measured by the U.S. Geological Survey's Sumatra gage were used to develop the input from the main stem of the Apalachicola River and its tributaries using a one-dimensional hydrologic model (DYNHYD; Ambrose and Barnwell, 1989). Local rainfall data were used as input to hydrologic models (SWMM; Huber and Dickinson, 1988) for the Whiskey George Creek, Cash Creek, and Carrabelle River as well as the ungaged portion of the Apalachicola River downstream of the Sumatra gage. Wind forcing was obtained from a continuous record collected at a weather station at the St. George Island causeway. For modeling the 1985–1986 period used in the present study, real-time temperature and salinity signals at the open boundaries were unavailable; composite climatological forcing functions were developed based upon Fourier representation of the mean seasonal data in a strategy similar to that of Galperin and Mellor (1990a). Velocity estimates were not calibrated against field observations as currents are more sensitive to precise location. However, the salinity calibration efforts effected an indirect calibration of velocity because salinity is a direct result of velocity-induced advection and diffusivity transport processes.

For this study, model runs were made for the entire years 1985 and 1986, in which hourly values were calculated for 715 cells in the Apalachicola Bay system. The model output included elevation and surface and bottom values for salinity, temperature, and components of current velocity (horizontal and vertical). Additional variables were derived from the raw model output and included averages, standard deviations, maximum and minimum values, and residual current velocities and direction. Such variables were computed for specific time periods as described below.

8. Statistical Analyses and GIS Mapping

Dependent variables for all statistical models included larval densities in 1985 and 1986, spatfall from closed sides in 1985 and 1986, average total oysters.m^{-2}, new growth (young of the year cohort: October 1985–April 1986), average growth of whole oyster bars (October 1985–April 1986), old growth (measured oysters in baskets), and mortality (1985). Two sets of independent variables were developed. The first set of variables was taken from the monthly field database and included surface color, surface turbidity, surface oxygen anomaly, bottom temperature, bottom salinity, bottom DO, Secchi depth, water depth, and salinity stratification. We also used average river flow and total rainfall (East Bay Tower, Florida Division of Forestry) over the 10-day period preceding sampling at each station. The second set of independent variables comprised data derived from the hydrodynamic model and included both surface and bottom values for the average and standard deviation of salinity and temperature, average and residual (resultant) current velocity, and minimum and maximum values for salinity and temperature. Data were calculated for the hydrodynamic model cells that included the geographic location of our stations. Data were time-averaged over the 30-day period preceding the biological collection at each station. Dependent and independent variables were transformed for normality. Associations between the oyster data and the various habitat characteristics were examined using stepwise multiple linear regression. Regression models for the biological variables were developed separately for the monthly field-observed data and the hydrodynamic model output so that a comparison could be made concerning statistical models based on traditional monthly sampling information and statistical models based on continuous data derived from hydrodynamic simulations. A P-value of 0.01 was chosen as the cutoff for including explanatory variables into a final model. Residuals were checked routinely for normality.

We used the best of both groups of statistical models to select a variable (mortality) for testing the application of the hydrodynamic model in the projection of oyster population response under alternative regimes of river flow. Experimental oyster mortality data and the output from the hydrodynamic model were used to develop a seasonal (time-averaged) statistical model. We used a stepwise linear regression method with average monthly percent mortality at 13 stations during May to August 1985 as the dependent variable. As predictors, we used averaged hydrodynamic model output values over the same time period. We used the following surface and bottom water predictors: average salinity, standard deviation of the salinities, maximum and minimum salinities, and average water velocity. We also used average surface elevation and salinity stratification. Thus, the resulting statistical model expressed mortality solely in terms of hydrodynamic model output parameters. We then substituted into the regression equation the hydrodynamic model values from all the hydrodynamic model cells to calculate predicted mortality over the whole bay for the May to August period of 1985.

We entered the resulting mortality estimates into an ArcView® Geographic Information System (GIS) database in which latitude and longitude coordinates were input for each of the hydrodynamic model cells. Then we used the GIS software to link each mortality estimate with a geographic location to generate a map surface depicting the spatial distribution of predicted mortality within the bay. An inverse squared-distance weighting method was used for interpolation of the surface. The interpolation was performed for a grid of cells 100 m on a side and incorporated data values from eight nearest neighbors. We applied the same statistical model (developed with 1985 data) to averaged hydrodynamic model output values for the May to August period of 1986 and repeated the GIS mapping process to obtain a prediction surface for potential mortality during 1986 (Christensen et al., 1998).

Experimental Methodology

1. Nutrient Limitation Experiments

We constructed 18-L, all-glass microcosms (width: 20, length: 21.5, and depth: 48 cm) using nontoxic sealant. A wooden rack assembly held microcosms at the surface of large, temperature-controlled reservoirs. Bay ambient water temperatures were maintained within ±2°C through the use of heating and cooling coils placed in the thermal reservoir. Large volumes of bay water were gently pumped (Ruhl Model 1500 rubber diaphragm pump) from approximately 0.5 m depth at each station into polyethylene containers by first passing through a set of nested Nitex plankton nets held in a stand-pipe system to equilibrate water pressure (i.e., a 202-μm net placed inside a 64-m net; 0.5 × 2.0 m) and to remove larger zooplankton.

Sampling was conducted during late afternoon and the samples were transported to an experimental "microcosm tank facility" near Tallahassee, Florida, during early evening to reduce heating from day-time peak temperatures. Experiments began the next day. Because a continuous supply of estuarine phytoplankton was logistically infeasible, we used a modified static renewal approach with daily removal of 10% of the volume of each microcosm. A supply of the field sample for each station was maintained as renewal water. This water was filtered through a 25-μm Nitex® net and held at 4°C in the dark. Thus, early experimental artifacts such as production of allelopathic substances would be less likely in such a system compared to a static experimental design. Before use, all collecting vessels, culture tanks, and Tygon air lines were acid washed (10% HCl), rinsed twice with fresh water, rinsed once with deionized water, and air dried. Microcosms were sealed with nontoxic silicone sealant. Air was delivered to each microcosm to maintain mixing via Tygon tubing, plastic control valves, and "air stones" from an oil-free aquaculture air pump.

Water collected from a station the previous day was placed in a fiberglass mixing tank during early morning to minimize light shock and was gently mixed with a rubber paddle to ensure homogeneous filling requirements. Water was poured sequentially in small volumes into glass microcosms to ensure homogeneous filling requirements. Reference samples were taken from the mixing tank for chlorophyll a, dissolved inorganic phosphate (PO_4), ammonium (NH_3), nitrite (NO_2), and nitrate (NO_2), particulate organic nitrogen (PON), and phosphorus (POP) to characterize initial nutrient conditions and standing stock of phytoplankton. Chlorophyll a was collected on Gelman A/E glass fiber filters and extracted with 90% acetone buffered with MgCO, and measured according to the method of Parsons et al. (1984). Quantification of dissolved inorganic nitrogen (i.e., NO_2 and NO_3) and phosphorus (PO_4) followed Parsons et al. (1984). Ammonium was measured as NH_3 with an ion electrode (U.S. EPA 1983, Method 350.3) after converting NH_4 to NH_3 by raising the pH to 11. This method has some essential features (e.g., minimal interference from waters highly stained with humic materials and paper mill effluents). However, the level of detection typically was relatively high (e.g., 2.0 μM NH_3-N) but adequate for this study. Particulate organic nitrogen (U.S. EPA, 1983, Method 351.4) and POP (U.S. EPA, 1983, Method 364.4) were collected on Gelman glass fiber filters, combusted at 500°C, oxidized to inorganic fractions, and organic nitrogen measured as NH_3 and organic phosphorus measured as PO_4.

Single samples were collected daily at approximately 0800 hr from each microcosm for chlorophyll analysis to measure changes in phytoplankton biomass. Nutrient enrichment experiments included the following treatments triplicated for each station: three control tanks, three phosphorus-enriched tanks at 10 μM PO_4 above ambient, three nitro-gen-enriched tanks at 50 μM NH_3 above ambient NH_3, and three combined $NH_3 + PO_4$ (referred to as N+P) above ambient as described for single additions. We used NH_3 enrichments because of interest in potential increased riverine transport into the bay. Although aware of possible suppression of NO_3 uptake by NH_3, this process was judged relatively unimportant in bioassays lasting over 5 to 12 d (D'Elia et al., 1986) and at high NO_3 concentrations (Pennock, 1987). Dortch (1990) concluded from a comprehensive lit-erature review that NH_3 suppression of NO_3 uptake under field conditions is quite variable and often undetectable. Nutrient additions were made at time-0 of the experiment. On day 2 and daily thereafter, 18 L of water were collected from each microcosm for chemical and biological analyses. This volume was replaced by an equal volume of renewal water adjusted to ambient temperature before introduction into the microcosms. Hydrographic variables (e.g., surface water temperature [°C] and salinity [8b] [YSI model 33 conductivity meter]) and 30-cm all-white Secchi disk readings (m) were collected generally within 1 week of samples for nutrient enrichment experiments.

A randomized block design was used with statistical significance of ($P < 0.05$) within and among stations. The four treatments in each row were randomly selected so that each of the three station-specific experiments had 12 treatments (randomized four treatments by three replicates for three stations). Treatments were oriented in a north/south direction to minimize shading from changes in the annual sun angle and to allow detection of a possible blocking effect. A repeated-measures, randomized analysis of variance (ANOVA) was run for each experiment. ANOVA assumptions were checked using residual boxplots, the Lilliefors test on the standard residuals, and the Bartletts test for equality of variances. A log transformation was necessary to achieve equality of variances. The two-way inter-action p-values for treatment by day and for block by day were calculated. To determine how treatments differed by day, a randomized block analysis of variance was run for each station, month, and day. As above, a log transformation was necessary to satisfy assump-tions. The Scheffe's S-test was used as a post-hoc treatment.

2. Diver-Deployed (in situ) Benthic Lander
a. Introduction

We have analyzed the relatively broad literature on this subject and the method chosen reflects our interpretation of results from the studies of others in addition to our own experience with SOD analyses in our South Carolina work. Replicate samples were used as an indication of experimental error. *In situ* tests have the most potential for error. The laboratory approach, while suffering from the problem of not being carried out in the field, has proven to be the most reliable in terms of variability of results. We followed techniques that have a record of being the most reliable on the basis of intersystem comparisons.

We worked with *in situ* samplers developed and constructed at Texas A&M University (Gilbert Rowe). The diver-deployed Benthic Lander used in the SOD measurements was designed and built by a research team at Texas A&M University under the direction of Rowe. A description of the instrument is given in an article given by Rowe and Boland ("Benthic Oxygen Demand and Nutrient Regeneration on the Continental Shelf of Loui-siana and Texas," abstract for *LUMCON Meeting, 1991*).

b. Instrument Specifications

Duplicate samples are taken after the instrument is grounded in sediments at a given station. DO probes (one for each of the duplicate units) are attached to YSI DO meters, which are monitored at 2- to 5-min intervals at the surface. The battery-operated machine has dual stirrers that maintain a regular rate of stirring in each 7-L (vol.) chamber. Each chamber covers 0.09 m² of sediment. Syringe ports are available for removal of water within each chamber for chemical analysis. The Lander is housed in a basic 3/4-in. aluminum frame to which is attached a battery housing with two power Sonic batteries. Ballast weights are used to keep the device in the sediments. Total negative weight is − 28.23 kg and the total positive lift (largely from the 13-in. instrument spheres) is 13 kg.

c. Operation of the Lander

1. Laboratory Preparations

- The YSI meters should be prepared. This includes installation of probe membranes, redlining the instrument at salinity 0, adjusting the temperature to ambient for calibration, calibrating probes for ambient temperature, setting meter scales to desired DO range, checking calibration with 5 to 10% sodium sulfite solution (0 ppm oxygen). The temperature should be reset to the expected bottom temperature (based on previous experience). Salinity should also be set in a similar fashion. Glass and gaskets should be cleaned with a solvent (toluene).
- Make sure that the silicon oil fills the impeller housing.
- Charge batteries to +13 volts and place in housing.
- Check O-ring on battery and use light treatment of silicone grease on the battery chamber O-rings. Close chamber.
- Check for chamber volume by measuring distance from the bottom of the rim to the flange (40 mm = 7.0 L).
- Grease and clean the pins for attachment to the DO batteries (large pin is +). Attach battery wires and start/check stirring assemblies.
- Secure device in frame and set/secure for travel.
- 2.Boat Operations (Deployment)
- Turn on stirrers by connecting the battery leads. Calibrate the DO meters and place probes in housing.
- Check syringes for the removal of water.
- Slide apparatus off boat and lower gently with marker buoy attached to frame.
- Divers should remove the chamber unit and secure the bungee cords. One diver on each side should lift the apparatus out and away from the frame. Then the divers should flip the chamber unit over to release all air bubbles. Make sure that stirrers are operating.
- Look for flat area to place chambers and set chambers down gently on sediment, pressing down until the gray flange is sitting on the sediment surface. Again, check the stirrers. DO chamber wires should be bundled together.
- Take water samples (200 mL) with syringes. Filter samples through a 0.2-μm filter.
- Take DO and temperature readings at 5-min intervals, recording data on the sample sheet. Such sheets should be part of the overall data sheet (see BIFS field data sheet summary).
- After a significant drop in DO has been recorded (1 to 3 hr), divers should take a final water sample with the syringes (same as above). The chamber should be picked up and turned upside down to check for impeller movement (sediment may interfere with such a mechanism). The chamber unit should be placed in lander housing and attached with bungee cords. Dive lifts should be inflated and divers should escort apparatus to the surface. The apparatus should be carefully placed on the boat and the battery leads detached.

3. Determination of Oxygen Consumption Rates

$$\text{Rate R} = \frac{\left(\text{Final } O_2, \ \text{mL/L} - \text{Initial } O_2, \ \text{mL/L}\right)\left(\text{Volume of chamber} = 7 \ \text{L}\right)}{\left(\text{Area of chamber, } 0.091 \ \text{m}^2\right)\left(\text{Total time, hr-}T_{final} - T_{initial}\right)}$$

Conversion of mg L^{-1} to mL L^{-1}: 1 mg = 0.7 mL

3. Laboratory (in vivo) SOD Determinations

Four replicate samples of undisturbed core samples were taken with the SOD corer apparatus at each of the three stations. Sediment filled the corer to at least half the height of the PVC coring tube. Water filled the remainder of the corer. Great care was taken so that there was minimal disturbance to the sample. Detachable cores were capped and placed in racks. The samples were sealed with PVC end caps and placed in an upright position in the rack in a Gott cooler. The seawater bath in the Gott cooler was maintained at the level of water in the cores to lessen pressure on the seals. The samples were taken immediately to the laboratory and equilibrated for 10 hr with gentle aeration without disturbing the sediment. Laboratory water temperature was controlled at field levels.

We followed the batch method of analysis. The laboratory SOD measurements were made on undisturbed sediments in the original core tubes. Water circulation was maintained with an internal mixing device. Two replicates from each station set were poisoned with HCN; the remaining two samples were left without any additions or changes. DO probes were pre- and post-calibrated using the micro-Winkler technique. Temperature and depletion of dissolved oxygen were monitored (without stirring) every 30 min for 4 to 6 hr or until 1.5 to 3.0 mg L^{-1} DO depletion was obtained. If the initial DO readings in the cores were below 4.0 mg L^{-1}, the core water was aerated without sediment disturbance in order to have an initial DO reading of at least 4.0 mg L^{-1}.

DO depletion curves were constructed with the cumulative DO depletion data plotted against time. The slope was calculated from the linear portion of the DO depletion curve using the first-degree regression function: the results are to be represented in units of gO_2 m^{-2} h^{-1}.

4. Primary Productivity (in situ Light/Dark Bottle Study)

Primary productivity is the rate at which inorganic carbon is converted to organic carbon via chlorophyll-bearing organisms such as phytoplankton. The plankton respiration term was an important part of evaluating the oxygen uptake mechanism in the bay. Therefore, we evaluated the phytoplankton net and gross photosynthesis using the classical light and dark bottle oxygen technique. Some unpublished data using the ^{14}C uptake method were obtained by personnel of the Environmental Research Laboratory, Gulf Breeze, for Perdido Bay. Data taken using this method were comparable to our own.

Presently, there is no available method of measuring primary productivity that is free of artifacts. Various authors have recently emphasized the importance of production processes occurring at strong physical gradients (Legendre and Demers, 1985) and an emerging area of emphasis has concentrated on ecohydrodynamics. The approach in this study examined bay primary productivity in terms of vertical stratification and its relationship to the euphotic zone, the longitudinal salinity gradient in the bays, and temporal patterns over the warm season when hypoxic conditions become prevalent. The scaling of measurements was made mostly at the meso- to macro-scale; fine-grain scaling of observations was based on the present design. Integral water column rates of photosynthesis allowed an estimate of system production minus the relatively small contribution from submerged aquatic vegetation and a somewhat more important contribution of phytomicrobenthos.

All procedures involving bottle cleaning, checking for supersaturation, the handling of water samples, and calculations are given in the above citation. The oxygen method will be used, whereby clear (light) and darkened (dark) bottles will be filled with water samples and suspended at regular depth intervals for appropriate intervals. The duration of exposure will depend on the rate of net photosynthesis and respiration with a need to avoid bubble formation in light bottles or hypoxia in dark bottles except at depths where hypoxia is present. We will start with sunrise-to-sunset deployment, with possible adjustment to 0900 to 1500 hr or some equal distribution of time around noon (CST). Any given bubble development (>0.3 mm diameter) at a single depth will not result in a change in deployment schedule. Bubble development at different depths will be reason to change such deployment based on an initial screening by periodically checking the bottle strings.

a. Endpoints
Changes in the concentration of DO will be measured with a YSI Model 58 DO meter equipped with a BOD probe. A magnetic stirrer and stirring bar will be required (a 12-volt system will be used). The specifications of this DO unit is ±0.01 mg^{-1} with an accuracy of ±0.03 mg^{-1}. These specifications are minimally satisfactory for the warm season in Perdido Bay. More sensitive measurement equipment is not seen to be seaworthy in a small research vessel. If necessary, a more sensitive system will be available at a shore-based system, which will require transport of samples to this laboratory setup. The present method will allow a spatial coverage that is important to the project.

b. Methods
Two numbered light bottles and two dark bottles will be used for each depth at each station for a schematic of the experimental setup. Non-brass ("marine") wire will be used to attach a strong (plastic or stainless steel) snap-catch to the neck of each 300-mL BOD bottle. Wheaton "400" brand borosilicate (high-purity) glass bottles are to be used. Plastic bottle caps will be used to protect the stoppers. The bottles are to be acid-cleaned (with warm 10% HCl) and rinsed with distilled water prior to use. Just before filling each bottle, the container will be rinsed with the water to be tested. Phosphorus-containing cleaning agents will not be used. The experimental setup has been constructed so that the supporting lines or racks do not shade the suspended bottles.

c. Calculations
The increase in oxygen concentration in the light bottle during incubation is a measure of net production (somewhat less than gross production due to respiration). The loss of oxygen in the dark bottle is an estimate of total plankton respiration.

Net photosynthesis = Light bottle DO − Initial DO
Respiration = Initial DO − Dark bottle DO
Gross photosynthesis = Light bottle DO − Dark bottle DO

Results from duplicates are to be averaged. The calculations of the gross and net production for each incubation depth should be plotted.

$$\text{mg carbon fixed m}^{-2} = \text{mg oxygen released}/1 \times 12/32 \times 1000 \times K$$

where K = photosynthetic quotient ranging from 1 to 2, depending on the N supply. One mole of oxygen (32 g) is released for each mole of carbon (12 g) fixed. The productivity of a vertical column of water 1 m square is determined by plotting productivity for each exposure depth and graphically integrating the area under the curve.

Perdido System: Ecological Study of the Perdido Bay Drainage System (1988–1991)

Project Personnel for the Perdido Study

Arceneaux, Mr. David (Project management)

> Champion International Corporation
> 375 Muscogee Road
> P.O. Box 87
> Cantonment, FL 32533-0087

Birkholz, Dr. Detlef A. (Toxicology, Residue analyses)

> Environmental toxicologist
> Enviro-Test Laboratories
> 9936 67th Avenue
> Edmonton, Alberta T6E OP5
> Canada

Brush, Dr. Grace S. (Long-core analysis, Long-term changes in estuaries)

> Department of Geography and Environmental Engineering
> The Johns Hopkins University
> 313 Ames Hall
> Baltimore, MD 21218

Cifuentes, Dr. Luis A. (Nutrient studies, Isotope analyses)

> Department of Oceanography
> Texas A&M University
> Texas A&M Research Foundation
> Box 3578
> College Station, TX 77843

Coffin, Dr. Richard A. (Nutrient studies, Isotope analyses)

> U.S. Environmental Protection Agency
> Environmental Research Laboratory
> Sabine Island
> Gulf Breeze, FL 32561-5299

Davis, Dr. William P. (Biology of fishes)

> U.S. Environmental Protection Agency
> Environmental Research Laboratory
> Sabine Island
> Gulf Breeze, FL 32561-5299

Epler, Dr. John H. (Aquatic insects)

> Rt. 3, Box 5485
> Crawfordville, FL 32327

Flemer, Dr. David A. (Nutrients and productivity in aquatic systems)

U.S. Environmental Protection Agency
Environmental Research Laboratory
Sabine Island
Gulf Breeze, FL 32561-5299

Franklin, Dr. Marvin. (Riverine hydrology)

U.S. Geological Survey
Tallahassee, FL 32306

Gallagher, Dr. Tom (Hydrological modeling)

Hydroqual, Inc.
1 Lethbridge Plaza,
Mahwah, NJ 07430

Gilbert, Dr. Carter R. (Fish systematics)

Curator in Fishes
Florida Museum of Natural History
Department of Natural Sciences
Museum Road, University of Florida
Gainesville, FL 32611-2035

Isphording, Dr. Wayne C. (Marine geology)

Tierra Consulting
P.O. Box 2243
Mobile, AL 36688

Kalke, Dr. Richard D. (Estuarine zooplankton)

Marine Science Institute
The University of Texas
P.O. Box 1267
Port Aransas, TX 78373-1267

Karsteter, Mr. William R. (Aquatic macroinvertebrates)

Environmental Planning & Analysis, Inc.
933 1/2 West Tharpe Street
Tallahassee, FL 32303

Klein, Dr. C. John III (Aquatic engineering, Estuarine modeling)

Klein Engineering & Planning Consultants
3011 North Branch Lane
Baltimore, MD 21234

Koenig, Dr. Christopher C. (Biology of fishes)

Environmental Planning & Analysis, Inc.
933 1/2 West Tharpe Street
Tallahassee, FL 32303

Lane, Dr. Jacqueline M. (Study plan and protocol review)

>   Friends of Perdido Bay
>   10738 Lilian Highway
>   Pensacola, FL 32406

Livingston, Dr. Robert J. (Aquatic ecology)

>   Environmental Planning & Analysis, Inc.
>   933 1/2 West Tharpe Street
>   Tallahassee, FL 32303

Niedoroda, Dr. Alan W. (Oceanographic modeling)

>   Environmental Science & Engineering, Inc.
>   P.O. Box 1703
>   Gainesville, FL 32602-1703

Prasad, Dr. Alshinthala K. (Aquatic algae)

>   Environmental Planning & Analysis, Inc.
>   933 1/2 West Tharpe Street
>   Tallahassee, FL 32303

Price, Ms. Janet H. (Project management)

>   Champion International Corporation
>   375 Muscogee Road
>   Cantonment, FL 32533-0087

Ray, Dr. Gary L. (Aquatic macroinvertebrates)

>   Environmental Planning & Analysis, Inc.
>   933 1/2 West Tharpe Street
>   Tallahassee, FL 32303

Woodsum, Mr. Glenn C. (Computer programming, Data management)

>   Environmental Planning & Analysis, Inc.
>   933 1/2 West Tharpe Street
>   Tallahassee, FL 32303

Project Outline

>   *Overall Scope:* To address questions regarding conditions in Elevenmile Creek
>   and Perdido Bay and the impact of Champion's Pensacola Mill on those con-
>   ditions. To provide Champion with background scientific information for iden-
>   tification of additional treatment needs and an objective evaluation of
>   alternatives for mitigation of environmental impact.

I. Introduction

The Perdido Study has been developed with an interdisciplinary team to answer specific questions concerning the impact of Champion International Corporation's Pensacola Mill on Elevenmile Creek and the Perdido Bay system. An ecosystem approach has been used because both natural (background) conditions and other forms of human impact must be determined if mill impact is to be evaluated. Field descriptive data are used in conjunction with experimental and modeling efforts to form a thorough analysis of the Perdido drainage system. Maps of the area with the location of physicochemical and biological sampling stations are given in Figure 5.1.

II. Historical Review

A. Background

To provide appropriate background and guidance for the proposed studies, a review was made of historic trends in the Perdido drainage system. This review included remembrances of long-term residents in the region, an annotated bibliography of scientific literature, a description of ongoing studies, an analysis of the historical trends of the Perdido Bay system during this century, and a synopsis of existing data with attention to climatological trends, water quality characteristics, biological organization, and changes in the Pensacola Mill treatment facilities. From the integration of these analyses, a series of research questions was developed.

B. Research Questions

Upon review of the historical data, a list of research questions was compiled. From this list, a concise statement of the research questions to be answered by this study was developed. The potential problems associated with the mill effluent include the following: color/transparency impacts, organic/inorganic nutrient loading with consequent impacts on the DO regime in Elevenmile Creek and Perdido Bay, spatial/temporal response of the biological components of the Perdido system to changes in DO, potential for toxic impacts, and the overall sediment/water interactions (nutrients, organic matter, biological components) that determine the ecological relationships and productivity of the Perdido river-estuarine system.

C. Research Needs

Based on data available from the historical overview and the research questions, a detailed listing of research needs was developed:

- Distribution of sediments (as characterized by particle size, percent organics, etc.) and toxic agents (such as metals and chlorinated organics) in the bay.
- Seasonal changes of the vertical and horizontal distribution of physicochemical components in basic water quality factors in the river–bay system. This should include the interaction of salinity/temperature changes, hypoxic/anoxic conditions, and the relationship of nutrients to such processes, with particular attention to land use and upland discharges.
- Two annual cycles of the spatial distribution of phytoplankton, zooplankton, infaunal macroinvertebrates, epifaunal macroinvertebrates, and fishes in the Perdido River–Bay system that will provide an evaluation of existing conditions in the receiving areas as well as the needed descriptive (background) data for the associated experimental program.

- Experimental studies including *in situ* tests and microcosm analyses that are designed to examine specific hypotheses related to sub-lethal biological responses to the mill effluent. This should incorporate the ecological aspects of population and community distribution at various levels of biological organization (plant and animal) relative to gradients of specific water quality conditions found in the Perdido system.
- Development of an ecological model that relates the various field and laboratory components to the research questions. Such questions address the spatial/temporal relationships of the natural system, with an emphasis on the potential impact of the mill on important ecological processes. These studies should be designed to include the interactive factors of hydrological conditions, mass flows of nutrients and organic substances, the color/transparency issue, the relationship of the pycnocline to hypoxic conditions in the bay (with attention to the effects of wind and tides), and the relationship of past and present biological conditions of the Perdido system to the various aspects of habitat distribution and quality. The model will also incorporate the historical data through trend analyses and time series tests. The final report will integrate these issues with specific answers to the research questions.

D. Sediment/Metals Analysis

During the investigation into existing data, it was discovered that there was a complete data set from 1983 of sediment characteristics in Perdido Bay done by Dr. Wayne Isphording of the University of South Alabama. These data were very useful in establishing the location of sampling stations and in making an early determination of the concentration of metals in the Perdido Bay system. In addition, by duplicating the original study, a comparison could be made of present conditions with those noted in 1983.

The final report contains x-ray sedigraph (270 mesh screen) analyses of Perdido Bay sediment samples with analyses that will include detailed contour maps of the following variables: organic carbon; carbonate carbon; particle size (phi units); sediment types (sand, silt, clay; sandy shell gravel, shell gravel, sandy, shelly silt, silty sand, silty shelly sand, sandy silt, silty sand – Shepard's classification); depth; pH; eH; DO; temperature; salinity; and metals (Cu, Zn, Ag, Ti, Co, Fe, Cr, Va, Pb, Hg, Cd, Ba, Al). The report also provides station-specific ternary diagrams of sand-silt-clay percentages for the Perdido Bay system.

E. Descriptive Monitoring Program

During the early development, a review of the past database and then current studies concerning the Perdido Bay system indicated that very little background biological data had been taken or were proposed to be taken. There was relatively little comprehensive descriptive and process-oriented information. If the basic functions of the river–bay system were to be established and impact analyses were to be successfully carried out, there would be a need for a combined field/laboratory program that included a monitoring program and an integrated laboratory/field experimental program.

The descriptive monitoring program was carried out on a monthly, quarterly, and annual basis (depending on the factor) for 2 years with a starting time of October 1988. River and bay stations were determined after a review of existing data and analysis of the Isphording sediment data and preliminary water quality analyses. A series of preliminary field trips during the fall of 1988 provided information in support of habitat stratification, the design of sampling methods, and verification of appropriate sampling stations. In the bay, a well-developed biological program was coupled with an analysis of the vertical distribution of salinity, temperature, DO, and other variables considered important in evaluating the vertical structure of the physicochemical system. Biological components sampled included phytoplankton, zooplankton, infaunal and epibenthic macroinvertebrates and fishes. All biological samples were analyzed to the species level. The biological components will be transformed into various levels of analysis: population indices (numbers, biomass); community factors (relative dominance, species richness, species diversity, evenness); trophic organization; and guild associations.

In cooperation with Dr. J. Lane, Dr. D. Flemer, and Dr. W. Davis, through funding from the U.S. Environmental Protection Agency (EPA), permanent stations were established on the bay to monitor wind strength and direction, local rainfall, and water levels. These data are now being organized by Drs. Flemer and Davis. Wind, visual wave observations, and water flow measurements were also made during routine monitoring operations.

The descriptive monitoring program was designed to serve as the background for a series of experimental studies testing hypotheses regarding mill effluent impact. At the end of the initial 2-year field sampling period, the data were reviewed to determine how the background sampling program could be reduced and still be maintained to follow important trends in the Perdido system. The various physical, chemical, and biological variables were clustered by station so that representative sampling sites could be established. This reduced monitoring program was designed to allow for continuing appraisal of specific water/sediment factors and important biological components of the freshwater and estuarine portions of the Perdido drainage system.

F. Stratigraphic Analysis of Perdido Bay Sediments

This part of the Perdido project will explore the potential for using the stratigraphic record of bay sediments to reconstruct the historical sequence of submerged macrophytes, centric and pennate diatoms, sedimentation rates, and nutrient influxes in Perdido Bay. Long core analysis will be made using carbon-14 dating, and through identification of historically dated deforestation, recognized by an increase in herbaceous plants and decreases in arboreal pollen in the sediments. In addition, cesium-137 dating will be used to identify the 1957 and 1963 horizons, which relate to the time of atomic bomb testing. Changes in total organic carbon (TOC), total organic nitrogen (TON), and the degree of pyritization of iron will be used to evaluate the possible existence of hypoxia and/or anoxia at the time of sediment deposition.

III. Detailed Analysis of Elevenmile Creek

A. Rationale

The work on Elevenmile Creek will proceed along two tracks, the first being an in-depth evaluation of the biological relationships of the system and the second being studies in support of the QUAL-2E model refinement. All components of this investigation will be compared to observations made in reference streams, which are relatively free of the effects of human activity. These reference streams include Holliger Creek, Styx River, and Pond Creek. The location of these reference streams and the location of the sampling stations on these streams are indicated on Figure 5.2.

B. Biological Relationships

With regard to biological relationships, sampling and analysis of microalgae, infaunal and epifaunal invertebrates, and fishes will be used to determine the organization of this system. Biological indices will be calculated to aid in this evaluation. Such factors will then be related to gradients of DO and other physicochemical and hydrological aspects of Elevenmile Creek. This information will be related to the nutrient/oxygen consumption pathways and to the results of the experimental leaf pack transfer program.

The use of Teflon leaf packs was developed in the examination of the impact of toxic substances on streams. As part of the monitoring program, leaf packs will be used to determine the spatial/temporal aspects of the biota of Elevenmile Creek and the reference streams. Five leaf packs will be placed at each station, tethered at each site by lines that are anchored to stream banks. Each leaf pack will consist of 20 3.0-cm² Teflon "leaves," which will be placed in an 8.0-mm mesh bag. The area of the Teflon "leaves" will approximate 1000 cm². The leaf packs will be retrieved after 3 to 4 weeks in the field and prepared according to the protocol. The animals will be preserved in 10% buffered formalin, counted, and identified to species.

A series of transfer experiments will be carried out between the reference streams and Elevenmile Creek. Leaf packs inoculated in the reference streams will be moved to Elevenmile Creek and vice versa. Adequate controls will be initiated so that the results will give indications of the potential influence of the mill effluent on the creek biota at various locations along the Elevenmile Creek system. Results from the leaf pack experiments can be compared to the field collections along the subject streams to gain insight into the potential effects of the mill on Elevenmile Creek biological diversity. The use of leaf packs will also allow a comparison of this sampling method with the current use of Hester–Dendy samplers. Leaf-packs will also be used in microcosm experiments to evaluate the presence of toxic substances.

C. QUAL-2E Model Refinement

As part of a waste-load allocation in support of an NPDES permit renewal, the EPA, in cooperation with Champion and the Florida Department of Environmental Regulation, developed a QUAL-2E model of Elevenmile Creek. Data developed for the model included a reaeration study, dye studies, and several field sampling surveys. This study will look at refinements of that model.

The studies in support of the QUAL-2E model refinement will include: (1) detailed hydraulic survey; (2) synoptic sampling; and (3) evaluation of monthly and synoptic water quality data. The detailed hydraulic survey will establish the hydraulic characteristics of the Elevenmile Creek system at a level adequate to support calibration, verification, and operation of the water quality numerical model. These hydraulic parameters will be obtained from a one-time-only field measurement program. Channel cross-sections will be obtained at approximately the following intervals on the main channel: (1) 1/4-mile intervals in the upper 3 miles of the channel; (2) 1/2-mile intervals from 3 miles to 8 miles from the head; (3) 1-mile intervals from 8 miles to the outlet. Cross-sections will also be obtained in the lower portions of the main tributaries. The precise location of all cross-sections will be selected in the field by a hydraulic engineer to obtain representative sections. The cross-sections will extend from bank to bank to provide information for flows up to bankfull and elevations will be referenced to mean sea level. Water level elevations will be taken to provide a water surface profile that will be used to calculate channel roughness coefficients. A report will be prepared that will present the data and define the hydraulic characteristics of Elevenmile Creek. Hydraulic analyses will be completed to evaluate flow velocities and other characteristics for flows up to bankfull.

Synoptic water quality sampling and hydrological measurements will be conducted along Elevenmile Creek and its tributaries. This sampling effort will measure water quality and hydrological parameters identical to those described in the descriptive monitoring program. In addition, at least two samplings (e.g., early morning and late afternoon) will be conducted and a greater number of sampling points will be included than in previous surveys. This study will provide the greater temporal and spatial resolution that is required to refine QUAL-2E.

The synoptic studies will be conducted at least three times (twice during the warm season and once during the winter) with Champion, Environmental Science & Engineering (ES&E), and Environmental Planning & Analysis (EP&A) personnel participating in the field investigation. ES&E will assist Champion in the collection of water quality samples and will conduct instantaneous stream discharge measurements at least once during the day on Elevenmile Creek and its tributaries. These streamflow measurements, in conjunction with color measurements, will be used to estimate the contribution of tributaries and groundwater base flow to Elevenmile Creek. The contribution determined by color will follow methods originally employed by Champion and will be used to support (not supplant) streamflow measurements. Sediment characteristics and infaunal macroinvertebrates will be analyzed by EP&A.

The water quality data collected during the monthly water quality sampling on Elevenmile Creek by Champion (described in Section V below) and the synoptic surveys will be used as described in Section VII to refine the QUAL-2E model and in the biological evaluation described above.

IV. Effects of Color

A. Rationale

Watercolor is an important factor in the ecological processes of aquatic systems. Color can alter both the quality and quantity of light transmission in rivers and estuaries; such changes can affect the distribution of SAV and the nature of phytoplankton productivity (qualitatively and quantitatively). There are complex interactions of watercolor with nutrient relationships as they pertain to the overall trophic organization of a given system. This includes changes in the limiting factors and altered nutrient utilization by microalgae and SAV. Thus, any analysis of the impact of color components of the mill effluent should be associated with the basic analysis of nutrient relationships in the bay and the overall trophic organization of the Perdido River–Bay system.

Because the color question is closely associated with the nutrient question, the proposed research on color in regard to trophic organization will be carried out in conjunction with the nutrient loading and nutrient limitation programs described in Section V. Station locations will be carefully chosen to make optimal use of the descriptive data, nutrient limitation work, and the proposed experimental work. Color analyses will demand the differentiation of changes in the spectral composition of light as it is used by the plants in question through the use of a spectroradiometer.

B. Seagrass Survey

The first part of this analysis will be a survey of the SAV and microalgae of the Elevenmile Creek–Perdido Bay system, establishment of a suitable reference stream or streams for a comparison with Elevenmile Creek, and a determination of existing habitat features of both systems. The second stage of the color evaluation will be a series of experiments designed to evaluate the problem of possible adverse effects of watercolor (either from the mill or as background from the natural streams) on the productivity of the Perdido system. This analysis will examine the reduction of light (as measured by watercolor, Secchi disk readings, turbidity, etc.), and the qualitative alteration of wavelength distribution as factors in the possible changes of biological components such as the biomass and productivity of SAV and microalgae in the Perdido system. Such analysis will take into account spatial distributions and seasonal changes of variables affecting light transmission as documented in the 2-year descriptive database.

C. Seagrass Experiments

Submerged aquatic vegetation (in the upper bay, this would apply to *Vallisneria americana*) will be addressed in a series of experiments that examine the possible influence of light distribution (and the changes of transmitted light characteristics due to the influence of color) on the distribution and productivity of SAV in the bay. The results of the historical studies will establish the necessary background for this research question. The visible light spectrum from 400 to 700 nm will be the subject of this study. Studies of past and current distribution of SAV (to be carried out by the U.S. Fish and Wildlife Service) and habitat distribution (thermo-halocline changes, nutrient distributions) in the bay will serve as the basis for the experimental portion of the program. Field experiments will include transplantation of SAV in various parts of the bay. Microcosm experiments will include the use of mill effluents to determine if the changes in the color spectrum (changes in individual wavelengths of light needed by the seagrass species) and/or components due to mill discharges have an influence on the productivity and well-being of SAV species in the bay. We will then analyze the descriptive and experimental data to determine if changes in the color regime due to the discharge of mill effluents are having an effect on the distribution of SAV in related portions of the Perdido estuary.

V. Perdido Bay Dissolved Oxygen Studies

A. Rationale

The hypoxic conditions at depth in Perdido Bay could be due to one or more factors that include the following: (1) overloading of organics in an estuary that has too little turnover in the water budget to handle the particulate and/or dissolved organic matter; (2) enhanced phytoplankton production leading to a disjunct food web (too little zooplankton predation to handle the organic matter produced by the phytoplankton; such zooplankton response may be due to overproduction and/or phytoplankton shifts to inedible species); this disjunct food web situation, in turn, leads to benthic (bottom) loading of organics from overproduction of the phytoplankton, high BOD, and anoxia or hypoxia at depth with thermohaline stratification of the water column; and (3) a combination of seasonal events that lead to the nutrient/organic loading that simply overloads the capability of the bay to process the organic matter deposited on the bottom.

The DO regime of Perdido Bay can be characterized by examining the interactions of processes known to control the DO balance in estuarine systems. Such processes include:

- Stratification, hydrography, and water circulation patterns
- Basin-wide nutrient loading
- Nutrient limitation
- Sediment characteristics (particle size, percent organics, nutrients)
- SOD
- Trophic status

By characterizing the interrelationships between these processes and the DO regime, judgements can be made regarding:

- Whether the bay hypoxia is related to hypereutrophication
- The nature of the causative factors
- The role of the Pensacola Mill in contributing to hypoxic conditions

B. Hydrographic Characterization

The hydrography and circulation of Perdido Bay will be studied at a level of detail adequate to support the assessment of the nutrient and pollutant loading from all sources. This will generally consist of: (1) monthly monitoring of physical, chemical, and biological parameters at bay stations by EP&A personnel as described elsewhere in this plan; (2) diurnal measurement programs to describe parameters over a tidal cycle during characteristic periods of river flow and bay mixing/stratification; (3) specific diurnal measurement programs to define event-scale variations of physical parameters; and (4) a screening analysis combining theory with field measurements to specify the basic physical characteristics of the estuary. Perdido Bay is well suited for applying a screening analysis because of its size, depth, and geometry.

In summary, the hydrographic characterization study will include (1) definition of the physical boundaries of Perdido Bay; (2) establishment of the physical classification of the estuary; (3) evaluation of the flushing time of nutrient and pollutant loads; (4) evaluation of the slowly varying patterns of mixing/stratification; (5) evaluation of event-scale processes that have major impacts on the hydrography and water quality of the bay; and (6) evaluations of the conditions in the lower portions of the Perdido River and Elevenmile Creek.

C. Basin-Wide Nutrient Loading

The aim of the Nutrient Loading Study is to characterize the loading of Perdido Bay from the surrounding watersheds by (1) establishing the composition and relative magnitude of the loads from each major drainage basin; (2) determining the average and extreme loads for different times during annual cycles; (3) correlating river runoff levels with rainfall events, nutrient loads, and other pollutant loads; (4) estimating the relative loads from point, non-point, and residential sources at a screening level; and (5) establishing the interannual variability and long-term trends in the load inputs to the bay.

The approach consists of a combination of a screening analysis, field sampling, laboratory analysis, hydrological modeling, statistical analysis, and assimilation of previous studies and measurements. The major data sources for this study are the descriptive monitoring program outlined in Section II.E, a monthly and storm event sampling program by Champion, and data from the "Perdido Bay Interstate Project," which was a joint project between Alabama Department of Environmental Management (ADEM) and FDER. To verify data results, a chemistry intercalibration exercise was undertaken with ADEM and FDER.

D. Nutrient Limitation

This part of the project is designed to analyze the relation of nutrient limitation to phytoplankton growth in Perdido Bay, which has been characterized by hypoxic conditions and occasional anoxia in various areas in recent years. The basic thesis for this work is that any evaluation of potential sources of excess nutrients depends, in part, on the relative importance of nutrients to phytoplankton production in Perdido Bay. An increase in limiting nutrient(s) could lead to the exacerbation of the existing DO problem.

A series of monthly nutrient (N and/or P) limitation experiments will be carried out for 12 months. After completion of the first year of nutrient limitation work, a new avenue of research that will concentrate on other factors such as silica and on specific food web aspects of the nutrient limitation question will be undertaken. As part of this research, whole bay water treatments will be used to allow for the analysis of the effects of the experimental setup on whole water samples (net phytoplankton, zooplankton). Such experiments will aid in the evaluation of the effects of changes in zooplankton numbers on the utilization of phytoplankton blooms induced by addition of various factors such as N/P.

E. Sediment Characteristics

Data concerning the distribution of bottom sediment types and size gradations will be obtained from the sediment study described in Section II.D. These will be used in an analysis of the frequency at which bottom sediments are re-entrained by the action of storms and flood discharges. These same data will be combined with field measurements of current patterns (from the descriptive monitoring program) to evaluate potential areas of long-term sediment accumulation or erosion.

All other data needed for this analysis are available from existing sources and other parts of the study program. Wind data will be used to establish the wave climatology at key locations of the bay through standard depth- and fetch-limited wave prediction algorithms. Depths will be developed through our own hydrological analyses. River discharges from the Nutrient Loading Study and the Elevenmile Creek Study will be combined with circulation data from the Hydrographic Characterization Study to estimate currents corresponding to the wave conditions at the key locations. These values will be used to compute the fluid shear stresses acting on the bottom sediments, which, in turn, will be used to determine how often sediment erosion threshold conditions are exceeded. Depending on the sediment grain size distribution and porosity, the depth of bottom sediment erosion for different levels of storm events can also be computed. The bottom currents for different types of events will be estimated from patterns measured in the field monitoring studies. These will be used with the results of the previous calculations to compute sedimentation and erosion patterns.

In addition to some routine work on the structural components (particle size frequency, nutrient composition, percent organics), some nutrient transfer experiments will be run in Perdido Bay to evaluate the storage/release capacity of the bay sediments as part of the estuarine productivity question. These experiments will be run in conjunction with the continued nutrient loading analyses. These studies will be integrated with models run by ES&E to determine how often the sediments are stirred up and what effect depth-related wind events have on nutrient transfer rates.

F. Sediment Oxygen Demand (SOD)

During the third year of study, a series of SOD analyses will be made. These studies will be combined with sediment–nutrient determinations. Based on such work, specific experiments will be designed during the third year to determine the nature of nutrient transport between the sediments and the overlying water column. This will include a determination of the effects of seasonal changes in temperature, salinity, DO, and the thermohaline structure of the water column on nutrient relationships of the estuary. The results of such tests will aid in the interpretation of the descriptive nutrient data and will be used to assess the possible role of the mill in the perceived changes in the DO regime of the Perdido system.

G. Isotope Analysis

An examination will be made of the contribution of the Pensacola Mill to the dissolved organic carbon (DOC) and nitrogen pools in the bay using stable isotope analysis by Drs. R. Coffin and L. Cifuentes. Carbon stable isotope analyses of dissolved organic and sediment carbon will be used to determine the contribution of mill effluents to the DOC pools in the estuary. Carbon stable isotope analyses of bacteria, coupled with short-term oxygen consumption experiments and ultimate BOD determinations, will be used to examine the effects of mill discharges on DO concentrations in the bay. In addition, nitrogen stable isotope analysis will be used to evaluate the contribution of mill effluents to the nitrogen pools in the estuary. The experimental data will be combined with field evaluations to determine the potential effects of the Pensacola Mill on the DO regime of the bay and to evaluate whether or not the observed hypoxic conditions have affected background estuarine productivity.

H. Trophic Status

This part of the project will be carried out as a joint effort with the U.S. EPA (Gulf Breeze Environmental Research Laboratory) and EP&A personnel. EP&A will provide the taxonomic and ecological analysis of phytoplankton samples taken monthly by the EPA scientists.

Phytoplankton and zooplankton will be sampled at various bay stations (to be determined in conjunction with the other studies) at monthly intervals over the 2-year period of study. The monthly phytoplankton collections by EPA scientists will be processed as part of this program. The seasonal trends of phytoplankton and zooplankton will be analyzed as part of the ecological background in the evaluation of the current status of the bay.

In addition, the SOD and sediment nutrient concentrations will be evaluated to determine the possible role of sediments in the nutrient limitation of phytoplankton in the bay. The relationship of phytoplankton production to nutrient loading will be evaluated to understand the trophic status of Perdido Bay. These studies are related to the nutrient studies as well as the color issue because ambient rates of production may also be affected by the levels and types of color present in the water. Specifically, the nutrient limitation experiments will determine which nutrients are limiting to phytoplankton in Perdido Bay. An increase in specific nutrients due to human activities could enhance the phytoplankton productivity, which, in turn, could cause hyper-eutrophication due to overproduction, and under-utilization of phytoplankton biomass by higher trophic levels. This, in turn, would result in a loss of oxygen due to either the chemical or biochemical oxygen demand of sediments that were enriched with the excess organic matter.

VI. Field Evaluation of Toxic Agents

There are several questions relating to the possible effects of toxicants that are released by the Pensacola Mill into the Perdido system. This aspect of the study will evaluate chlorinated compounds, conductivity and un-ionized ammonia in Elevenmile Creek, and other nonspecific toxic agents.

To address the issues of sedimentation of chlorinated compounds, a set of cores will be taken on established Elevenmile Creek and Pond Creek stations and the 24 bay stations and analyzed for chlorinated organic compounds. Such data analysis will be carried out as a follow-up to the sediment metal analyses. A total of 50 to 60 cores will be taken for this preliminary analysis of the chlorinated organic compounds. The top 2 cm will be used for the analysis.

The chemical analyses will be made by Enviro-Test Laboratories (Edmonton, Alberta, Canada; Dr. Detlef A. Birkholz). A screening analysis will be run on EOCl (solvent-extracted organochlorine compounds). If relatively high levels of EOCl are found in a given sample, further tests (quantitative, with replicates) will be run to characterize the qualitative composition of the chlorinated compounds that are suspected to be part of pulp mill discharges.

Field toxic effect experiments will be carried out along Elevenmile Creek and in Perdido Bay. An *in situ* 30-day test has been successfully developed where indigenous species are placed in containers along gradients of freshwater runoff. A series of tests will be run whereby live animals will be taken in collections along the Perdido system; these animals will be placed in experimental chambers at sites along the river and bay (stations to be determined by historic surveys). Indigenous (cryptic) benthic species of fishes and invertebrates will be used for these tests. Live animals will be placed in test chambers (one organism per chamber).

The test chambers are constructed of sections of PVC pipe covered tightly at both ends with perforated plastic plugs. Test containers will be filled by immersing the plugged ends in a water-filled trough; recently collected animals will be randomly placed into each chamber and the end will be plugged. The chambers will then be lowered on a line so that five replicates are held at or near the bottom by a brick and the other five chambers are held near the surface by a float. Water quality conditions (DO, temperature, conductivity/salinity) will be noted in each chamber at deployment and retrieval. Test organisms will be measured before and after the 30-day exposure along existing gradients of various water quality factors.

These field experiments will evaluate the conditions that are possibly affecting the organisms in the Perdido drainage system. As such, they are designed as screening tests to determine if there are mill-related toxic agents that are affecting the distribution of invertebrates and fishes. The leaf-pack transfer experiments described in Section III.B, and the *in situ* experiments described above will be used to set up a series of fresh and estuarine microcosms. These microcosms can be established with different concentrations of mill effluent, as well as different concentrations of specific components of mill effluent. Inoculated leaf packs will then be transferred to these microcosms to determine the effects of specific components of mill effluents. Using these tests, hypotheses regarding toxic effects can be tested and specific effluent limits can be established in regard to acute and chronic toxicity.

VII. Comparison of the Perdido System with Other River–Estuarine Systems

The current status of the various important elements of the Perdido system (e.g., population densities, community indices, and trophic organization) will be evaluated by detailed comparisons with other systems along the northern Gulf coast. Previously generated databases will be used for a rigorous comparison of various river–estuarine systems (natural and affected by various anthropogenous activities, including the release of pulp mill effluents). Because all such data have been taken by identical (quantitative) sampling techniques as those used in the Perdido system, the comparative approach used should be very effective in determining the present status of the Perdido system.

VIII. Predictive Modeling

A. Rationale

Various forms of modeling can be used to integrate and interpret the data obtained from the multiple projects that make up the Perdido Basin Drainage Study. In addition to computer models, statistical models that include time series analyses, forecasting, and intervention analysis can be used to better understand the complex system being measured.

The Klein models will be used to evaluate state relationships of major physical/chemical relationships in Perdido Bay. Data (tides, wind, rainfall) from the citizen monitoring group (Friends of Perdido Bay) will be used for this analysis and for the other modeling efforts. This screening model framework will provide a preliminary assessment of the hydrodynamic and water quality behavior of Perdido Bay, with identification of those processes that control circulation and its effect on important water quality factors.

Calibrated water quality models can provide a means of evaluating the effect of wastewater discharges on water quality in Elevenmile Creek, the lower portion of the Perdido River, and in Perdido Bay. Separate models will be used to represent the nontidal and tidal portions of the study area. These models will be calibrated with water quality data that reflect existing point source inputs, but they can also be used to assess the potential impact of proposed discharges or the effect of changes in existing discharge quality or quantity.

B. Hydrology

1. Elevenmile Creek

There are several standard techniques for estimating low flow characteristics at ungaged locations. The choice of technique depends on the available information and the level of effort required. These techniques include the use of one or more nearby long-term stream-flow records and drainage area ration, correlation, regional regression, and hydrologic simulation using long-term rainfall records.

To estimate the low flows, flow duration data and low flow frequency data will be developed. There are approximately 4 years of continuous streamflow data for Elevenmile Creek at the U.S. Highway 90 bridge. The flows at the gage location include Champion's effluent in addition to other discharges and watershed runoff. The Elevenmile Creek streamflow data will need to be adjusted for Champion's effluent discharge prior to use in estimating natural streamflows. Champion's continuous discharge records will be used for this adjustment. A characterization of Elevenmile Creek natural (without Champion's process water discharges) low flows, including 7-day, 10-year flow (7Q10), can then be calculated. Because of the short length of record available for Elevenmile Creek, the adjusted flow record will need to be extended using records from one or more suitable (geographically close and physiologically similar) watersheds. Potential primary stations include Pine Barren Creek, Pond Creek, Brushy Creek, and the Perdido River near Bar-rineau Park. The estimates will be evaluated by comparing the climatic conditions during the 4-year gaging period with long-term climatic conditions and the selected long-term streamflow records.

2. Perdido Bay

A large amount of data has been collected from Perdido Bay and the rivers that feed into it. These data indicate that strong vertical density stratification often develops between the fresher and warmer surface water layer and the bottom water of the bay. Low DO concentrations in the bottom water are often found along with the strongly stratified conditions. The critical conditions are associated with the occurrence of low river input to the bay during the warm summer months.

The large database will be used to establish the modeling scenarios that will be used to assess the impact of Champion's discharge on the water quality of the bay during summer low flow conditions. Hydrographic and water quality cross-sections along the thalweg of the bay will be plotted for 24 months of data collected by EP&A and 36 months collected by the EPA. These will be used to distinguish periods where near-critical conditions exist. Time series of water quality parameters will be plotted covering these 60 measurement intervals for key upper- and lower-bay stations to assess the occurrence and duration of steady-state conditions of the bay systems. River inputs during the selected periods of near-critical conditions will be computed from the river gages adjusted for ungaged portions of the overall drainage basin. The 7Q10 river input discharge will be evaluated. The measured steady-state period(s) with near-critical conditions that occur when river inputs were close to the 7Q10 will be designated as "worst-case" candidates for water quality modeling.

C. Water Quality Modeling

1. Elevenmile Creek

As discussed in Section III, the QUAL-2E model was calibrated with data collected from Elevenmile Creek by the EPA. The data collected as part of the study described in Section III will be used to further refine the QUAL-2E model calibration. The model segmentation of Elevenmile Creek begins just upstream of Champion's wastewater discharge and extends downstream to approximately 2.3 miles from the mouth where the tidal influence of Perdido Bay begins. This model serves two purposes: (1) it provides a tool for evaluating the impact of wastewater discharges from Cantonment's publicly owned treatment works (POTW) and Champion on water quality in the creek, and (2) it also provides boundary conditions to the model of Perdido Bay. In this way, the effect that point source discharges to the creek have on water quality in the Bay can be properly evaluated.

2. Perdido Bay

Salinity and DO concentrations measured in Perdido Bay and the lower portions of Elevenmile Creek and the Perdido River indicate that vertical stratification occurs from time to time. Lateral gradients, however, do not appear to be significant and, therefore, a two-dimensional segmentation of the bay is appropriate. The model will include the lower 2.3 miles of Elevenmile Creek, the tidal portion of the Perdido River, and will extend through Perdido Bay to Inerarity Point.

Water circulation in the model will be determined by the Pritchard technique (*ASCE, JHD*, Vol. 95(11), January 1969), which is based on a balance of ocean-derived salinity. This analysis determines vertical flows and dispersion, as well as longitudinal flows that are in the seaward direction in the surface layer and in the landward direction in the bottom layer. In the upper portion of Perdido Bay, where Elevenmile Creek and the Perdido River enter, lateral gradients in salinity occasionally are observed. This technique, however, will still calculate the appropriate average transport across this section of the bay. This approximation seems reasonable in contrast to the alternative, which is a three-dimensional, time-variable hydrodynamic model.

Because a two-dimensional segmentation is required to properly represent the vertical gradients in water quality, the QUAL-2E model used in Elevenmile Creek cannot be used for the bay. The model that has been selected is the U.S. EPA's WASP3 program. The kinetic structure of WASP3 is consistent with WASP4, but "bugs" appear to remain in the WASP4 program and therefore, WASP4 will not be used. HydroQual has used WASP3 previously. For this project, salinity will be added as a state variable so that DO saturation can be properly calculated as a function of both temperature and salinity.

The WASP3 model will be calibrated with water quality data collected within the period from 1988 to 1991. As described in the hydrology section, data collected during this period will be reviewed to identify critical conditions with respect to DO conditions in Perdido Bay. Fortunately, during this period, low flows in the Perdido River reached critical levels, providing data in the bay that reflect minimum levels of dilution. Even with low freshwater inflow, the hydraulic residence time in the bay is short enough to make steady-state analyses appropriate, if the freshwater inputs are reasonably steady. The WASP3 model can also be used in a time-variable mode by assigning initial conditions based on observed data and then comparing calculated concentrations with water quality data from subsequent sampling surveys. It is planned to use two or three periods to calibrate the model. These periods will include the critical period for DO, as well as periods that represent different environmental conditions. Calibrating the model with data from varying environmental conditions will provide confidence that the factors affecting DO in Perdido Bay have been properly represented.

The extensive water quality database for Perdido Bay developed as part of this study allows comparison of water quality conditions during the critical periods chosen for calibration with conditions at other times during this period. This database can also provide a means for assessing the temporal variability of water quality data and the duration of periods that can be analyzed as steady state.

The Perdido Bay WASP3 model will quantitatively evaluate the impact of existing or proposed point source discharges on DO concentrations in Perdido Bay and the lower portions of Elevenmile Creek and the Perdido River. The effect of changing the magnitude or location of discharges can directly and easily be investigated in projections. Each of the models used in this work will be provided to FDER for their review and use.

IX. Timetable for Completion of Project: 1988–2004

The timing of the Perdido program will follow a predetermined series of field and laboratory objectives that have been designed to answer the questions that came out of the historic survey. The following chart represents the chronological course of the Perdido project:

```
Year 1988            1989      1990                1991      1992          etc. to 2004
Month JJASONDJFMAMJJASONDJFMAMJJASONDJFMAMJJASONDJFMAMJJ
Historic XXXXX
Program dev. XXXXXXXXXXXXXXXXXXXXXXXX
Monitoring  XXXXXXXXXXXXXXXXXXXXXXXXXx x x x x x x x x x x x
Experimental xxxxxxxxxxxxxxxxxxxxxxxxxxxx¹XXXXXXXXXXXXXXXXXXXX
  In situ tests  XXXXXXXXXXXXXXXXXXXXX      XXXXXXX
  FW transfer tests XXXXXXXXXXXXXXXXXXXXX
  Nutrient limitation      XXXXXXXXXXXXXXXXXXX
  Light expermtsXXXXXXXXXXXXXXXXXXXXXXXXXXXXXXXXXXXXXXXXX
  Submerged aquatic vegetation   XXXXXXXXXXXXXXXXXXXXXXX
  Sediment/nutrient exchange/SOD      XXXXXXXXX
  Isotope expermts......................................    XXXXXXXXXXX
  Toxics analysis                    XXXXXXXXXX
  Microcosm expermts            XXXXXXXXXXXXXXXXXXXX
Nutrient loadingXXXXXXXXXXXXXXXXXXXXXXXXXXXXXXXXXXXXXXXXX
Elevenmile Creek intensive analyses   X   X   X
Stratigraphic study            XXXXXXXX
Hydrographic study             XXXXXXXX
Vertical field analyses/modeling data       XXXXXXXX
Analytical/modeling         XXXXXXXXXXXXXXXXXXXXXXXX
Final integration of various data sets      XXXXXXXXXXXXXXX
Reports                            InterimX      FinalX
```

Long-Term Monitoring (1988–2004)

System	Met.	PC	LC	POL	PHYTOPL	SAV	ZOOPL	INFAUNA	INV	FISHES	FW	NL	NR
Perdido Bay°	50***	16**	16**	3	12.5*	3*	3*	16*	16**	16**	16**	16**	1*

Met. = River flow, rainfall

PC = Salinity, conductivity, temperature, DO, oxygen anomaly, pH, depth, Secchi   * = monthly

LC = NH_3, NO_2, NO_3, TIN, PON, DON, TON, TN, PO_4, TDP, TIP, POP, DOP, TOP, TP   ** = mixed (monthly, quarterly)

DOC, POC, TOC, IC, TIC, TC, BOD, Silicate, TSS, TDS, DIM, DOM, POM, PIM   *** = daily, monthly, yearly

System	Met.	PC	LC	POL	PHYTOPL	SAV	ZOOPL	INFAUNA	INV	FISHES	FW	NL	NR

NCASI color, turbidity, chlorophyll *a*, *b*, *c*, sulfide

POL = water/sediment pollutants (pesticides, metals, PAH)

PHYTOPL = whole water and net phytoplankton

SAV = submerged aquatic vegetation

ZOOPL = net zooplankton

INFAUNA = infaunal macroinvertebrates taken with cores and/or ponars

INV = invertebrates taken with seines in freshwater areas and around the edges of the bay and with otter trawls in the bay

FISHES = fishes taken with seines in freshwater areas and around the edges of the bay and with otter trawls in the bay

FW = food web transformations

NL = nutrient loading

NR = nutrient limitation experiments

° = Long-term data: 10/88-9/03

nd = no data.

Stations sampled:

Monthly

System	Met.	PC	LC	POL	PHYTOPL	SAV	ZOOPL	INFAUNA	INV	FISHES	FW	NL	NR
Metered chemistry, lab chemistry (surface, bottom)	2	4	5	7	11	13	17	18	R3				
	22	23	25	26	29	31	33	37	40				
ww phytoplankton (surface, bottom)	18	22	23	24a	25	26	29	31	37				
	40	41	42	42a	42b	42c							
Infauna (cores)	18	22	23	25	26	29	31	33	37	40			

Quarterly

System	Met.	PC	LC	POL	PHYTOPL	SAV	ZOOPL	INFAUNA	INV	FISHES	FW	NL	NR
Metered chemistry, lab chemistry (surface, bottom)	2	4	5	7	9	11	13	17	18	R3			
	21	22	23	25	26	29	31	33	37	40			
	41	42	43	44	45	46	47	48	42a	42b			
	42c	SC1	SC2										
Infauna (cores)	2	4	5	7	11	13	17	18	R3				
Freshwater fishes	2	4	5	7	11	13	17	18	R3				
Bay invertebrates	18	22	23	25	26		31	33	37	40			
Bay fishes	18	22	23	25	26		31	33	37	40			

Comparison of Econfina and Fenholloway Systems: Gulf Coast

I. Methods Used for Data Collection in the Long-Term Apalachee Bay Research Program

The research effort in the northern Gulf of Mexico is based on written, peer-reviewed protocols for all field and laboratory operations. Water quality and biological analyses have been taken by personnel from the Center for Aquatic Research and Resource Management (CARRMA, Florida State University, Tallahassee, Florida) and Environmental Planning & Analysis, Inc. (Tallahassee, Florida), with additional work carried out at a series of laboratories in various countries. Water quality methods and analyses and specific biological methods have been continuously certified through the Quality Assurance Section of the Florida Department of Environmental Protection (Comprehensive QAP #940128 and QAP #920101). Various methods that have been used to take field information and run laboratory analyses have been published in the reviewed literature. This includes the following: Flemer et al. (1997); Livingston (1975a,b, 1976a,b, 1980b, 1982a,b, 1984a,b, 1985a, 1987c, 1988a,b,c, 1992a,b, 1997a); and Livingston et al.), 1974, 1976a,b, 1997, 1998a,b).

A. Sediment Analyses

Sediments were analyzed for particle size distribution and organic composition according to methods described by Mahoney and Livingston (1982) as defined by Galehouse (1971).

1. Particle Size Analysis

Sediment samples were taken with coring devices (Plexiglas: 7.6 cm d., 45 cm² cross-section). Analyses were taken on 10-cm samples for regular percent organics and particle size analysis.

- Un-preserved sediments were divided into coarse (>62 μm) and a silt-clay (<62 μm) fractions.
- The coarse fraction was analyzed by wet-sieving using 1/2 phi unit intervals.
- The silt-clay fraction was analyzed using a pipette method in 1/2 phi unit intervals from 4 to 6 phi, and in 1-phi intervals for the finer fractions (6 to 10 phi).
- Statistical analyses were carried out using phi units and sediment designations as outlined in the Wentworth Scale.
- Statistical treatments of the granulometric results followed the method of moments for calculation of mean grain size (in phi units, with 1 mm = 0.5 phi), skewness (a measure of non-normality of distribution), and kurtosis (a measure of the spread of the distribution). Sorting coefficients represented the measure of grain size as follows: well sorted = 0.50; moderately well sorted = 0.71; moderately sorted = 1.00; poorly sorted = 2.00.

2. Percent Organics

After placing the sediments in a drying oven set at 104°C for a minimum of 12 hr, the sediments were ashed in a muffle furnace for 1 hr or more at temperatures approximating 550°C. The percent organic fraction of the sediments was calculated as a percent of the dry weight.

Sediment analysis is a two-part process where the first consists of forming a data file and the second is the actual granulometric analysis. The first part can be accomplished by either running the EZseds program to enter and reformat raw data or by exporting previously entered sediment data from one of the EP&A 4th Dimension Data Management System (DMS) sediment data files. EZseds is a FORTRAN program that prepares raw sediment granulometric data for analysis with the MacGranulo program. MacGranulo is a FORTRAN application that performs sediment grain-size analysis on the data in the input file. Results include summary statistics such as mean grain size (as determined by three different methods) and percent organics, particle-size distribution according to the Wentworth classification, and a frequency distribution plot.

3. Sediment Toxic Substances Analyses

Sediment samples were collected at a series of sampling sites in the system. Three sub-samples were taken at each established station. Sediment samples were placed on ice. The samples were shipped to Enviro-Test Laboratories in Canada. Prior to shipping, all samples were logged into chain-of-custody forms and a sample logbook. Samples were examined to ensure that accurate labels had been affixed and that all lids were tight. Prior to closing each cooler containing samples, the chain-of-custody form was completed and placed in a Ziplock® bag that was then placed inside the cooler. The forms were placed in plastic bags. The originator retained a copy of each chain-of-custody form. As a general procedure, samples were taken in order from supposedly clean areas to possible polluted areas. No sampling was carried out in the rain.

Quality control samples (blanks, etc.) were labeled as above. After collection, identification, and preservation, the sample was maintained under chain-of-custody procedures discussed below. A blank consisted of an empty sampling container. For each group of samples (inorganic mercury, methyl mercury), one blank was used for each lot number of jars.

Samples of sediments were taken with corers (the top 2 cm were used for mercury analysis, and top 10 cm for sediment particle size and percent organics). A PVC corer was used for the mercury samples. Chemical analyses were run on sediments taken from each of the sampling stations. Three replicate sediment samples were taken at each station for total mercury and methyl mercury analysis. Each of the three replicates was composed of three core sub-samples that were cut into squares (all edges "shaved") to eliminate the outer parts of the core sample. The "shaving" of the sample was carried out with an acid (5 to 10% nitric acid) washed plastic knife. Samples were placed in acid-washed, 1-pint, wide-necked, Teflon-lined, screw-top glass jars. Nitric acid (10% pesticide grade) was used to wash the jars. Samples were immediately preserved in the field with concentrated nitric acid to pH 2. Care was taken to mix the sediment and check for the proper pH. Each bottle was labeled as indicated above. Acid preservation for each sample was checked with pH paper to make sure that the pH 2 target was maintained. Samples were then placed on ice and bubble wrapped.

Sampling for toxic agents in sediments in other studies followed similar procedures. All sediment samples were taken with corers (the top 2 cm). A stainless steel corer was used for the organic toxicant samples and a PVC corer was used for the metal samples.

B. Chemical-Specific Operations

1. Metals

Field Operations. All samples were taken with corers (the top 2 cm were used for analysis). Field sediment samples were taken at each sampling station using an acid-washed PVC corer. Three cores were taken for each sample. Samples were placed in acid-washed, 1-pint, wide-necked, Teflon-lined, screw-top glass jars. Nitric acid (10% pesticide grade) was used to wash the jars. Samples were immediately preserved in the field with concentrated nitric acid to pH 2. Care was taken to mix the sediment and check for the proper pH. One sample × 6 metals (zinc, lead, mercury, copper, nickel, and aluminum) × 20 stations represent a total of 120 metal analyses. Each bottle was labeled as indicated above. Acid preservation for each sample should be checked with pH paper to make sure that the pH 2 target is maintained. Samples are then placed on ice and bubble wrapped.

2. EOCl

Field Operations. Field sediment samples (3) were taken at all stations using a metal corer. Samples were placed in acetone/hexane-washed aluminum foil and then put in 1-pint, wide-necked, Teflon-lined screw-top glass jars. Each bottle was labeled as indicated above. Samples were placed on dry ice in the field. Care was taken to pack the glass jars in bubble wrap for the trip to Canada. Each jar was wrapped separately and taped individually. Three samples × 20 stations represents a total of 60 EOCl samples.

C. Laboratory Operations

Samples were prepared and shipped to Canada for analysis. Samples were sent to:

Dr. Detlef A. Birkholz, Executive Vice President
Enviro-Test Laboratories
9935 67 Avenue
Edmonton, Alberta, Canada
T6E OP5
Phone: 780-413-5227

D. Biological Sampling

All field biological collections were standardized according to statistical analyses carried out early in the program (Hooks et al., 1975; Livingston, 1976a,b, 1987a,b,c; Greening and Livingston, 1982; Livingston et al., 1976a,b; Meeter and Livingston, 1978; Stoner et al., 1982). Infaunal macroinvertebrates were taken using coring devices (7.6-cm diameter, 10-cm depth). The number of cores to be taken was determined using large initial samples (40 cores) and a species accumulation analysis based on rarefaction of cumulative biological indices (Livingston et al., 1976a,b). Based on this analysis, multiple (10 to 12) core samples were usually taken randomly at each station, with the assured representation of at least 80% of the species taken in the initial sampling. All infaunal samples were preserved in 10% buffered formalin in the field, sieved through 500-μm screens, and identified to species wherever possible.

Where roots, etc. prevented coring, suction dredges were used as prescribed with multiple sweeps of specific time (number and time determined during first field trip). Animals were preserved in 10% buffered formalin and stored in 100% denatured ethanol.

Epibenthic fishes and invertebrates were collected with 5-m otter trawls (1.9-cm mesh wing and body, 0.6-cm mesh liner) towed at speeds of about 3.5 to 4 km.hr^{-1} for 2 min, resulting in a sampling area of about 600 m^2 per tow. Repetitive samples (2 or 7) were taken at each site; sampling adequacy of such nets had been determined by Livingston (1976a,b). All collections were made during the same lunar and tidal phase at the first quarter moon. Samples were taken within 2 hr of high tide, with day and night trawls separated by one tidal cycle. Larger fishes (>300 mm SL) were taken with 100-m nylon trammel nets and stomachs were removed in the field and injected with a 10% buffered formalin solution.

All organisms were preserved in 10% buffered formalin, sorted and identified to species, counted and measured (standard length for fishes, total length for penaeid shrimp, carapace width for crabs). Representative samples of fishes and invertebrates were dried and weighed, and regressions were run so that data from the biological collections could be converted into dry and ash-free dry mass. The numbers of organisms were then converted to biomass.m^{-2}. All biological data (infauna, epibenthic macroinvertebrates, fishes) were expressed as numbers.m^{-2}.mo^{-1} or biomass.m^{-2}.mo^{-1}.

E. Trophic Determinations

Gut analyses were carried out in the same manner throughout the 30-year field analysis period. Fishes and invertebrates were placed in 5-, 10-, or 20-mm size classes (depending on size), and food items were taken from the stomachs of up to 25 animals in a given size class with pooling by sampling date and station. Determinations similar to those described above were carried out and it was determined that at least 15 animals were needed for a given size class × station × sampling period. Stomach contents were removed and preserved in 70% isopropanol and a dilute solution of Rose Bengal stain. In larger animals, we pumped the stomachs out and released the subject. The content analysis was carried out through gravimetric sieve fractionization (Carr and Adams, 1972). Contents were washed through a series of six sieves (2.0 to 0.75-mm mesh), and the frequency of occurrence of each food type was recorded for each sieve fraction. As the food items were comparable in size, the relative proportion of each food type was measured directly by counting. Dry weight and ash-free dry weight were determined for each food type by size class. In general, mutually exclusive categories were used for the food types that included both plant and animal remains. Stomach contents by food type by size class were then calculated and related back to station/time designations. Based on the long-term stomach content data for each size class, the fishes and invertebrates were then reorganized first into their trophic ontogenetic units using cluster analyses (Czekanowsli [Bray–Curtis] or C-Lamba similarity measure; Flexble Grouping cluster strategy [$\beta = -0.25$]). The basis for this analysis is given by Sheridan (1978) and Livingston (1984a).

All biological data (as biomass m^{-2} mo^{-1} of the infauna, epibenthic macroinvertebrates and fishes) were transformed from species-specific data into a new data matrix based on trophic organization as a function of ontogenetic feeding stages of the species found in Perdido Bay over the multi-year sampling program. Ontogenetic feeding units were determined from a series of detailed stomach content analyses carried out with the various epibenthic invertebrates and fishes in the near-shore Gulf region (Sheridan, 1978, 1979; Laughlin, 1979; Sheridan and Livingston, 1979, 1983; Livingston, 1980b, 1982a,b, 1984b, unpublished data; Stoner and Livingston, 1980, 1984; Laughlin and Livingston, 1982; Stoner, 1982; Clements and Livingston, 1983, 1984; Leber, 1983, 1985). Based on the long-term stomach content data for each size class, the fishes and invertebrates were reorganized first into their trophic ontogenetic units using cluster analyses. Infaunal macroinvertebrates were also organized by feeding preference based on a review of the scientific literature (Livingston, unpublished data). The field data in Perdido Bay were then reordered into trophic levels so that temporal changes in the overall trophic organization of this system could be determined over the 8-year study period. We assumed that feeding habits (at this level of detail) did not change over the period of observation; this assumption is based on previous analyses of species-specific fish feeding habits that remained stable in a given area over a 6 to 7-year period (Livingston, 1980b).

The long-term field data were used to establish a database based on the total ash-free dry biomass m^{-2} mo^{-1} as a function of the individual feeding units of the infauna, epibenthic macroinvertebrates and fishes. Data from the various stations in Perdido Bay were used (g m^{-2}) for all statistical analyses. The data were summed across all taxonomic lines and translated into the various trophic levels that included herbivores (feeding on phytoplankton and benthic algae), omnivores (feeding on detritus and various combinations of plant and animal matter), primary carnivores (feeding on herbivores and detritivorous animals), secondary carnivores (feeding on primary carnivores and omnivores), and tertiary carnivores (feeding on primary and secondary carnivores and omnivores). All data are given as ash-free dry mass m^{-2} mo^{-1} or as percent ash-free dry mass m^{-2} mo^{-1}. In this way, the long-term database of the collections of infauna, epibenthic macroinvertebrates and fishes was reorganized into a quantitative and detailed trophic matrix based not solely on species (Livingston, 1988a,b,c), but on the complex ontogenetic feeding stages of the various organisms. A detailed review of this process and the resulting feeding categories is given by Livingston et al. (1977).

F. Suction Dredge Data

1. Quantification of Sampling Program

The suction dredge data represent a highly quantified sampling effort involving the capture of aquatic invertebrates across a broad spectrum of riverine and estuarine habitats. We quantified various biological sampling techniques so that we could estimate the relative effectiveness of such samplers. This quantification also allowed the use of a particular sampling method for comparative purposes (through time and between ecosystems).

For the quantification process, we used three suction dredge samples for each station. For each sample, the suction dredge was run for a given period of time. Each sample was broken down into multiple sub-samples or pseudo-replicates (i.e., not true replicates in a statistical sense) by spreading the sample on a tray that was subdivided into a number of equal parts. This approach was used to determine the number of sub-samples that could be analyzed as a representative subset of the original sample. Quantification was based on the rarefaction analysis of cumulative biological indices (in this case, new species and species richness) as a function of multiple sub-samples. The asymptotic relationship of species accumulation curves was used to determine the sampling effort for a given sampler. The basis for these preliminary studies has been given in a series of articles that were published in peer-reviewed journals (Livingston, 1976a,b; Livingston et al., 1976a,b; Stoner et al., 1982). In this way, the following analysis allowed a consistent and standardized approach to the quantification of the suction dredge collections and an evaluation of the demonstrated effectiveness of this sampling technique.

A total of three sub-samples from each sample would provide a representative comparison of the two systems based on the cumulative numbers of new species taken. Just over 60% of the overall (total) species in the full sample would be taken in the Econfina River and just under 60% of these species would be taken in the Fenholloway with a total of nine sub-samples. There was no clear asymptotic relationship in the accumulation of new species in either system but the high numbers of new species taken leveled out by the third sub-sample in of each of the three suction dredge samples, This accumulation in terms of mean numbers of species also reflected a leveling by the ninth sub-sample.

II. Field Protocol: Econfina/Fenholloway Project — Final

1. SAV Mapping

Starting in April, we will start the field SAV mapping with both inshore boats (unless one boat is needed for other field sampling). We will operate 7 days a week, weather permitting, until the mapping is completed. Transect priority is in the area of the seven permanent stations in each system; then the rest of the study area will be surveyed. Standard mapping protocol is to be observed.

B. Field Monitoring

a. Water Quality: (Chain-of-Custody Forms Filled Out). A synoptic survey will be made with three teams in one system at the same time (river, estuary, Gulf) with samples collected and taken back to Tallahassee during the same day. Surface and bottom samples taken in water >1 m; mid-water samples at <1m.

Stations:

F01D, F01F, F03, F03B, F02, E00D, E01, E01A, E00F
F04, F05, F05A, F06, E03, E04, E05, E06
F9, F11, F12, F10, F16, F14, F15 and two unimpacted stations
E7, E8, E9, E10, E11, E12, E13 and two depth complementary stations

Factors:

- Dissolved oxygen (DO)
- Specific conductance (freshwater only)
- Ammonia
- Light transmission (color, turb., Secchi)
- Nutrients (NO_2, NO_3, o-P, TP, TN)
- Total/dissolved/particulate, organic carbon
- Total dissolved/suspended solids
- Temperature
- pH
- CI, SO_4 (freshwater only)
- BOD, COD
- Alkalinity (freshwater only), hardness (freshwater only)
- Salinity (saltwater only)
- Silica

b. Biology: (Chain-of-Custody Forms Filled Out). Take only metered physicochemical data and Secchi disk readings at biological stations.

1. Econfina and Fenholloway Rivers

Stations: F01D, F01F, F03, F02, F03B, F02, E00D, E01, E01A, E00F

Factors:

- *Periphyton:* Anchor each setup with proper floats, 2.5 cm below surface. Mark apparatus as to system, station, date. After 2 to 3 weeks, randomly take five slides and place in labeled jars filled with distilled water. Jars will be transported on ice to lab where slides will be scraped on both sides and preserved with 5% formalin = composite which will be stored in screw-cap vials. Each sample will be diluted with distilled water, depending on density of algae, thoroughly shaken, three aliquots of 0.05 ml or 0.1 ml placed on slide, covered, counted, and identified.
- *Benthic invertebrates:* Suction dredge to be used as prescribed with multiple sweeps of specific time (number and time to be determined during first field trip). Preserved in 10% buffered formalin, stored in 100% denatured ethanol.

- *Fishes:* Multiple seine tows taken until no new species are taken. For larger fishes, identify sex, measure (standard length), examine for disease, and release. Smaller fishes, 10% buffered formalin in glass jars with plastic lids. Store in 50% isopropyl alcohol.

2. Econfina and Fenholloway Estuaries

Stations: F04, F05, F05A, F06, E03, E04, E05, E06

Factors:

1. *Phytoplankton:* Taken with 4 Rep. 2×2 Net (28-μm bongo, 15.2-cm plankton net hauls). Two duplicate runs of two net setups. Preserve with 5% formalin and place in labeled glass bottles. Also take three 1-L samples, 5% formalin.
2. *Infaunal and epibenthic macroinvertebrates:* Take 12 cores and place in glass jars, 10% buffered formalin. Also take suction dredge samples (see above).
3. *Epifaunal invertebrates and fishes:* Seven 2-min otter trawl tows and multiple traps. Five leaf packs. Check fishes for external pathology and condition factor.

3. Econfina and Fenholloway Gulf Areas

Stations: F9, F11, F12, F10, F16, F14, F15 and two unimpacted stations; E7, E8, E9, E10, E11, E12, E13 and two depth complementary stations

Factors:

- *Phytoplankton:* Same as above for estuaries.
- *Submerged aquatic vegetation (SAV):* Eight 0.25-m square sample, randomly sampled, whole plants taken, placed wet in labeled plastic bags. Separated, identified to species, weighed (dry weight).
- *Epiphytes:* Select entire shoots (undamaged), including sheaths of *Thalassia* and *Syringodium* (ten shoots of each species at each site). Place wet in labeled plastic tubes and place on ice.
- *Epibenthic macroinvertebrates and fishes:* Seven 2-minute otter trawl tows and multiple traps. Five leaf packs. Check fishes for external pathology and condition factor.

Analysis of Pensacola River–Bay System: 1997

I. Development of the Study Plan

The Pensacola Bay system is composed of a series of rivers (Escambia River, Blackwater River, Yellow River; Figure 6.1) that drain into a major estuarine system (Escambia Bay, Pensacola Bay, Blackwater Bay, East Bay). It is possible that the Cantonment mill will change its discharge site to the Escambia River.

A. Study Components

The basic purpose of this study is to evaluate the assimilative capacity of the Pensacola Bay system for loading from the Pensacola mill into the Escambia River.

There are three basic approaches that will be integrated to make this evaluation: field analyses of basic habitat variables, determination of the biological condition of the receiving system, and nutrient/productivity models. This includes the following:

- Distribution of sediment quality (as characterized by particle size, percent organics, etc.) in the study area. The sediment quality will be directly related to the biological (infaunal) characteristics of the system.
- Seasonal changes of the vertical and horizontal distribution of physicochemical components in the Pensacola River–Bay system. This includes the interaction of salinity/temperature changes, hypoxic–anoxic conditions, and the relationship of nutrients to such processes with particular attention to land use and upland discharges.
- One annual cycle of the spatial distribution of phytoplankton, infaunal macroinvertebrates, epifaunal macroinvertebrates, and fishes in the Pensacola Bay system; this information will provide an evaluation of existing conditions in the receiving areas as well as the needed descriptive (background) for the evaluation of the assimilative capacity of the system to receive mill effluents.
- Translation of the biological data into a trophic organization that can be compared to other such data in a series of Gulf systems; this will allow a relatively rapid evaluation of the current biological condition of the Pensacola Bay system that can be determined within the relatively restricted time limits of the study.
5. Development of an ecological model that relates the various field and laboratory components to the basic question. The study will address the spatial/temporal relationships of the natural system with an emphasis on the potential impact of the mill on important ecological processes. These studies will be designed to include the interactive factors of hydrological conditions, mass flows of nutrients and organic substances, the resulting changes in phytoplankton constituents and productivity, the relationship of the pycnocline to hypoxic conditions in the bay (with attention to the effects of wind and tides), and the relationship of past and present biological conditions of the Pensacola system to the various aspects of habitat distribution and quality.

The habitat data (water/sediment quality) will be related to the biological conditions of the receiving areas and the nutrient analyses will be used to determine nutrient loading relative to the productivity conditions of the bay. These data will also be combined with various experimental data (Sediment Oxygen Demand [SOD], water column productivity) for use in the models that will be developed for the final evaluations of assimilative capacity of the receiving system. Seasonal variation of response will be determined.

B. Rationale for the Study

This study is designed to evaluate whether or not there is a fatal flaw in the proposal to discharge mill effluents into the Escambia River. The rationale for the proposed study plan is as follows.

Physicochemical data will be used to determine the current state of the Pensacola system relative to habitat status, nutrient loading, state of eutrophication, and distribution of productivity in receiving areas. These data will also be used in the modeling effort. Sediment quality data will be used to evaluate the possible storage of nutrients and the qualitative aspects of the distribution of infaunal macroinvertebrates in the receiving area. The phytoplankton data will be used to define the trophic status of the Pensacola system;

such data are crucial in the definition of the nutrient dynamics and productivity in the area. The infaunal and epifaunal macroinvertebrates and fishes will be transformed into the trophic equivalents for a rapid and relatively accurate method of determining the biological status of the proposed receiving area. The resulting trophic indices will be compared to those of other bay systems for such an evaluation. The light data, SOD information, and light/dark (productivity) data will be used in the modeling effort. The data generated will allow a review of the present ecological status of the receiving system in addition to evaluating the assimilative capacity for the loading of mill effluents into the Escambia River. The time constraints for answering the relatively complex question of assimilative capacity are considerable; a 12-month database is minimal for such a study.

C. Study Areas

Program development and station placement were determined by a comprehensive analysis of water quality factors taken on a set of preliminary sampling stations in May 1997. Based on this information, a diving survey was made in June 1997 to stratify the habitats in the Pensacola system for the development of a Study Plan. Physicochemical data were again taken, and the exact positions of the sampling stations for the proposed study were finalized (Figure 6.1).

The preliminary May and June 1997 surveys were carried out during a period of relatively heavy rainfall. Consequently, the study areas were characterized by water quality conditions that reflected high river runoff (low salinity, high color, etc.). The results of the preliminary sampling effort (May 1997) indicated that the Escambia River was characterized by high color and turbidity, relatively high dissolved oxygen, low chlorophyll, low BOD, moderate inorganic nitrogen concentrations, and low orthophosphate concentrations. The relatively shallow upper Escambia Bay (above the I-10 bridge) was characterized by low salinity, moderately high DO (no salinity stratification), low Secchi depths, relatively high chlorophyll concentrations (Stations T2C, T3A, T3C), moderately high BOD and ammonia, and very low orthophosphate. The data indicate that upper Escambia Bay is nutrient enriched, with orthophosphate as the primary limiting factor and the possibility of moderately developed phytoplankton blooms in some parts of the bay. Lower Escambia Bay showed signs of hypoxia at depth in the deeper portions of the bay that were characterized by salinity stratification. Chlorophyll levels were moderate to low in this part of the bay, with relatively low concentrations of the various nitrogen and phosphorus compounds. Pensacola Bay was characterized by higher salinities, no real hypoxia at depth due to salinity stratification, low chlorophyll concentrations, and, with the exception of Stations T7C and T10C, low ammonia concentrations. Orthophosphates were nondetectable throughout Pensacola Bay. Turbidity and color were low in this area, with increasing Secchi depths along the N–S axis. Santa Rosa Sound tended to follow this trend, with high salinity, relatively clear water, and low chlorophyll and nutrient concentrations. Overall, there were north–south trends of water quality that were dependent on point and nonpoint sources, river flow, and the physiography of the bay receiving areas.

The June 1996 diving survey showed similar distributions of the water column water quality variables as the previous survey. Upper Escambia Bay was characterized by low light transmission (especially from the bottom to about 1 m above the bottom), and sediments were generally characterized by high silt–clay concentrations. Deeper areas down the middle of lower Escambia Bay were characterized by hypoxia at depth (where there was salinity stratification), increasing water clarity, and relatively sandy sediments in areas <2 m deep (along the east and west sides of the bay). The lower bay was also characterized by sediments with fine particles. A new series of stations was established in the Blackwater drainage system; here, water quality was very good with relatively high light penetration and high DO in both river and bay areas. This system was considered a very good reference area for the Escambia system as there was little development in the Blackwater system with no heavy industry and well-developed wetlands that were not adversely affected by human development. There was a little evidence of point or non-point sources of pollution in this system.

Based on the diving survey, we adjusted various station positions, dropped some stations, and added others. During the diving survey, six stations (T6E, T6F, T9A, T10A, T11A, and T11B) from the previous survey were dropped. We added eight stations (2A [the proposed mill discharge site], BW1, BW2, BW3, BW4, BW5. BW6, and BW7). The addition of the Blackwater system was considered an essential analysis of a cognate (unpolluted) system that will allow a comparison with the more affected Escambia system. The highest concentrations of stations were located in the Escambia River–upper Escambia Bay as the primary receiving areas for the mill effluents. Transects along the lower Escambia Bay were considered important in secondary effects of point and non-point runoff and with the Blackwater system (East Bay) as the unpolluted reference system for the survey.

D. Timetable

The timetable includes an 18-month survey starting in May 1997.

1. Proposed Timetable

The timetable for the study included an 18-month survey starting in May 1997.

Year	1997								1998									
Month	M	J	J	A	S	O	N	D	J	F	M	A	M	J	J	A	S	O
Program development	X	X																
Water quality monitoring	X	X	X	X	X	X	X	X	X	X	X	X	X	X	X	X	X	X
Sediment analysis					X													
SOD/productivity				X	X	X	X											
Phytoplankton analysis	X	X	X	X	X	X	X	X	X	X	X	X	X	X	X	X	X	X
Infaunal macroinvertebrates	X		X		X		X		X		X							
Epifaunal macroinvertebrates	X		X		X		X		X		X							
Fishes			X				X			X			X			X		
FII Trophic determination											X							
Analytical/modeling	X	X	X	X	X	X	X	X	X	X	X	X	X	X	X			

II. Methods and Materials

A. Field Analyses

1. Habitat Stratification and Study Development

Based on the May 1997 water quality data, an evaluation was made of the Pensacola Bay system. Habitat stratification was carried out at each of the preliminary stations. Stations were moved according to such analysis. In addition to water quality data, a qualitative sediment analysis was made along with a diving survey of the various stations. Based on the field analysis, station locations were reestablished and stations were either eliminated or added as needed. The new station complement was determined based on the May 1997 and June 1997 surveys.

2. Physicochemical Analysis
The following forms of data will be taken during each monthly field survey.

a. Water Chemistry.   Depth profiles (surface or 0.1 m and at 0.5 intervals to the bottom):

Dissolved oxygen (DO)
Conductivity
Salinity
pH
Secchi depth
Temperature

b. Laboratory Chemistry.   Color

Turbidity
Chlorophyll *a*
BOD5
Nitrogen series: NH_3, NO_2, NO_3, TON, DON, TN
Phosphorus series: PO_4, POP, DOP, TP
Carbon series: DOC, POC, TOC

For all data analyses, we will follow protocols outlined in our Comprehensive Quality Assurance Plan filed with the Florida Department of Environmental Protection.

3. Sediment Analysis
Sediment samples (top 2 cm) will be taken on the biological stations with a PVC corer (7.6 cm diameter). Sediment samples will be analyzed for particle size distribution and organic composition according to methods described by Mahoney and Livingston (1982) as defined by Galehouse (1971). Sediment samples will analyzed for the following parameters:

- Grain size analysis
  - Phi unit distribution
- Percent organics
- Physical characteristics
  - Total organic carbon (TOC)
- Nutrients
  - Total nitrogen (TN)
  - Total phosphorus (TP)

4. Biological Analysis
Biological samples (infauna, epibenthic macroinvertebrates, fishes) will be taken quarterly at fixed stations in the respective systems. These stations were based on habitat stratification. Infaunal macroinvertebrates will be taken at the various river and bay biological stations. Infaunal macroinvertebrates will be taken using coring devices (7.6 cm diameter, 10 cm depth). Multiple (10) core samples will be taken randomly at each station, with the assured representation of at least 80% of the species taken in the initial sampling. All infaunal samples will be preserved in 10% buffered formalin in the field, sieved through 500-µm screens, and identified to species wherever possible. Epibenthic fishes and invertebrates will be collected at the bay stations with 5-m otter trawls (1.9-cm mesh wing and body, 0.6-cm mesh liner) towed at speeds of about 3.5 to 4 km hr^{-1} for 2 min, resulting in a sampling area of about 600 m^2 per tow. Repetitive samples (7) will be taken at each site.

All organisms will be preserved in 10% buffered formalin, sorted and identified to species, counted, and measured (standard length for fishes; total length for penaeid shrimp, carapace width for crabs). The numbers of organisms will then be converted to biomass.m^{-2} based on previously determined count/weight regressions.

Ontogenetic feeding units for fishes and invertebrates will be determined based on a series of detailed stomach content analyses carried out with the various epibenthic invertebrates and fishes in a series of Gulf estuaries (Sheridan, 1978, 1979; Laughlin, 1979; Sheridan and Livingston, 1979, 1983a,b; Livingston, 1980a, 1982, 1984a, unpublished data; Stoner and Livingston, 1980, 1984; Laughlin and Livingston, 1982; Stoner, 1982; Clements and Livingston, 1983, 1984; Leber, 1983, 1985). Based on the long-term stomach content data for each size class, the fishes and invertebrates will be reorganized first into their trophic ontogenetic units using cluster analyses. Infaunal macroinvertebrates will also be organized by feeding preference based on a review of the scientific literature (Livingston, unpublished data). The field data will then be reordered into trophic levels so that monthly changes in the overall trophic organization of the Pensacola system can be determined over the study period. The assumption that feeding habits (at this level of detail) will not change over the period of observation is based on previous analyses of species-specific fish feeding habits that remained stable in a given area over a 6- to 7-year period (Livingston 1980b). All biological data (as biomass.m^{-2}.mo^{-1} of the infauna, epibenthic macroinvertebrates, and fishes) will be transformed from species-specific data into a new data matrix based on trophic organization as a function of ontogenetic feeding stages of the species found in various Gulf estuaries.

The long-term field data from the Pensacola system will be used to establish a database based on the total ash-free dry biomass m^{-2} mo^{-1} as a function of the individual feeding units of the infauna, epibenthic macroinvertebrates, and fishes. Data from the various stations in the Pensacola Bay system will be used (gm^{-2}) for all statistical analyses. The data will be summed across all taxonomic lines and translated into the various trophic levels that include herbivores (feeding on phytoplankton and benthic algae), omnivores (feeding on detritus and various combinations of plant and animal matter), primary carnivores (feeding on herbivores and detritivorous animals), secondary carnivores (feeding on primary carnivores and omnivores), and tertiary carnivores (feeding on primary and secondary carnivores and omnivores) (Livingston et al., 1997). Data will be given as ash-free dry mass m^{-2} mo^{-1} or as % ash-free dry mass m^{-2} mo^{-1}.

In this way, the database of the collections of infauna, epibenthic macroinvertebrates, and fishes will be reorganized into a quantitative and detailed trophic matrix based not solely on species (Livingston, 1988a,b,c), but on the complex ontogenetic feeding stages of the various organisms in the Pensacola system. This will allow a relatively rapid yet comprehensive evaluation of the biological status of this system relative to other such estuaries in the northern Gulf of Mexico.

5. Statistical Analyses

Scattergrams of the field data will be examined and either logarithmic or square root transformations will be made, where necessary, to approximate the best fit for a normalized distribution. These transformations will be used in all statistical tests of significance. Statistical comparisons will be made of the monthly field physicochemical and the quarterly biological data by station or groups of stations. Independence tests will be used for preliminary analyses to examine each sample for autocorrelation at a number of lags. The program will compute the Q statistic of Ljung and Box (1978) for an overall test of the autocorrelations as a group, where the null hypothesis is that correlation at each lag is equal to zero. The parametric F-test for comparison of variances and t-test for comparison of means will be used to test the hypothesis that the data came from normally distributed populations. Means testing will be carried out with both parametric and nonparametric tests. For independent, random samples from normally distributed populations, the parametric *t*-test will be used to compare the sample means. Where there is serial dependence of data from a given pair of stations, it will be removed by differencing the observations and calculating and plotting the autocorrelations of the differences. If the differences are not serially correlated, we will apply the Wilcoxon sign-rank test on the differences to compare the two sets of numbers (P = 0.05). A two-sample spreadsheet will be developed that includes a Chi-square test, which compares the shapes of the distributions for the two samples.

Principal Component Analysis (PCA) and associated correlation matrices will be determined using monthly data and the SAS™ statistical software package. A PCA will be carried out as a preliminary review of the data using the river information, sediment data, and water quality variables. The PCA will be used to reduce the physicochemical variables into a smaller set of linear combinations that can account for most of the total variation of the original set. For the physicochemical variables in this study, the series will be stationary after appropriate transformations; thus, the sample correlation matrix of the variables will be a good estimate of the population correlation matrix. Therefore, the standard PCA can be carried out based on the sample correlation matrix.

A matrix of the water and sediment quality data associated with the sampling stations in each system will be prepared. The data will be grouped by month or quarter, depending on the variables. Values for the dependent variables (total biomass m^{-2}, herbivore biomass m^{-2}, omnivore biomass m^2, primary carnivore biomass m^{-2}, secondary carnivore biomass m^{-2}, tertiary carnivore.m^{-2}) will then be paired with the water/sediment quality data (independent variables) taken for stations within each sector of the study area. Unless otherwise defined, these statistics will be run using SAS™, Systat™, and SuperAnova™. Data analyses will be run on the independent data sets. Significant principal components will then be used to run a series of regression models with the biological factors as dependent variables. Residuals will be tested for independence using serial correlation (time series) analyses and the Wald–Wolfowitz (Wald and Wolfowitz, 1940) runs test. A Chi-square test will be run to evaluate normality. The purpose of this evaluation will be to establish the current ecological status of Escambia Bay relative to the reference system.

6. Sediment Oxygen Demand (SOD)

SOD will be taken on ten stations during the first quarterly sampling and on five stations during the succeeding three quarters. SOD data will be used in the modeling process.

7. Light Readings, Phytoplankton Collections, and Light/Dark Bottle Tests

The light/phytoplankton/productivity work will be carried out at three stations in Escambia Bay. Aquatic light readings will be taken with a Li-Cor LI-1800UW underwater spectroradiometer that measures the spectral composition of photon flux density at 1- to 2-nm intervals from 300 to 850 nm. Three to five scans (replicates) will be taken and averaged for each reading. Spectral radiometric light readings will be taken only on cloudless days, with winds less than 10 to 12 knots and within 2.5 hr of solar noon. By dividing each aqueous irradiance reading by the corresponding atmospheric irradiance, the spectral radiometric readings will be "normalized." Spectral irradiance will be given as units of energy (micromoles photons) nm^{-1} m^{-2}. The K_d (downwelling) attenuation values will be calculated with reference to different wavelengths; the K_d (PAR) is the attenuation coefficient of the photosynthetically active radiation (400 to 700 nm). The lower limit of the euphotic zone is considered the PAR at 1% of the surface light (expressed as depth).

Phytoplankton samples will be taken with whole water samples at designated stations with 1-L bottles at the surface. Three 1-L water samples, treated as separate replicates, will be collected from each station and will be immediately fixed in Lugol's solution in its acid version (Lovegrove, 1960) to give the sample a weak brown color. Fixed samples will be transported back to the laboratory for processing. Examination of whole water samples will provide a better estimate of nannoplankton abundance than net plankton abundance because small cells are more abundant, and hence more likely to be observed in settled water samples. The unconcentrated samples will be settled and concentrated by decanting or by pipeting to 100 or 200 mL, depending on the density of the samples. Phytoplankton cells will be counted with a standard research microscope (see Semina in Sournia, 1978) with phase contrast illumination at 400X magnification. Three aliquots of 0.05 mL each will be transferred separately using a disposable pipette onto a glass slide and covered with a 24 × 30 mm coverglass. All phytoplankton cells will be counted in each aliquot and identified to species or to the lowest taxonomic category possible (genus/order/class). The location (coordinates on the stage of the microscope) of specimens difficult to identify during the counting procedure will be noted; these will be examined immediately (after the counting was over) with oil-immersion objectives. Diatoms, dinoflagellates and cryptophytes whose orientation is unsuitable for taxonomic determination will be examined by slightly touching the coverglass with a preparation needle to make such cells turn around. Abundance will be expressed in number of cells per liter. Permanent slides of diatoms will be prepared using standard procedures of acid-cleaning described in Prasad et al. (1990). Material for scanning electron microscopy will be prepared as described by Prasad and Fryxell (1991). After preparation, electron micrographs will be taken on a Polaroid 4 × 5 Land film type 55/positive–negative using the JEOL-840 scanning electron microscope at Florida State University, operating at an accelerating voltage of 10 or 20 KV.

Presently, there is no available method of measuring primary productivity that is free of artifacts. Recent literature continues to elucidate problems with methods, assumptions, and experimental designs in the measurement of primary productivity (Nixon, 1988a,b). Various authors have recently emphasized the importance of production processes occurring at strong physical gradients (Legendre and Demers, 1985) and an emerging area of emphasis is concentrating on ecohydrodynamics. The productivity data will be used in the modeling process.

III. Comparison of Nassau and Amelia Systems: Atlantic Coast

A. Field Methods and Materials

Preliminary analyses for water quality factors, together with light transmission data (spectroradiometric determinations), were used to delineate the distribution of mill effluents so that isopleths of important variables could be determined and associated gradients verified through factor-specific spatial differentiation. Stations were determined that defined the distribution of mill effluents in the receiving area. Matching stations, chosen for comparability of habitat characteristics (temperature, salinity), were established in the Nassau River estuary as reference sites for comparative analyses (Figure 7.1). Data were taken monthly over two 12-month sampling periods (1994–1995; 1997–1998). Microcosm and mesocosm experiments with pulp mill effluents and ammonia were carried out during the 1997–1998 sampling period under conditions approximating those observed in the field.

1. Water Quality and Light Transmission

Detailed descriptions of methods for the collection of physicochemical field data are given by Flemer et al. (1997); Livingston (1979, 1982a,b, 2000); and Livingston et al. (1997, 1998a,b, 2000). Field data (temperature, salinity, conductivity, DO, pH) were taken with Datasonde

4 multiprobes (AMJ, Inc.). The DO anomaly was calculated from the field measurements as the difference between the measured DO and the oxygen solubility at the observed temperature and salinity (Weiss, 1970). Chemical analyses were based on protocols of the American Public Health Association (APHA, 1989); these included chlorophyll *a* (APHA Method 1002-G), Biochemical Oxygen Demand (APHA Method 405.1), true (NCASI) color (colorimeter, Pt-Co Units), and nutrients (ammonia, nitrite, nitrate, particulate and dissolved organic nitrogen, total nitrogen, orthophosphate, particulate and dissolved organic phosphorus, total phosphorus). Ammonium was measured as ammonia with an ion electrode (U.S. EPA 1983, Method 350.3). This method has limited sensitivity to interference from humic materials and paper mill effluents (Flemer et al., 1997). Particulate organic carbon was analyzed according to methods by Parsons et al. (1984). Turbidity was determined with a ratio turbidimeter. Light penetration depths were taken using standard Secchi disks. Field light transmission data were taken with a Li-Cor LI-1800UW underwater spectroradiometer. The underwater light field was characterized by incident radiant flux per unit surface area as quanta $m^{-2} s^{-1}$. Samples included three to five scans (replicates) taken at 1 to 2-nm intervals that were averaged for each reading. Flux measurements were taken for photosynthetically active radiation (400–700 nm; PAR) and individual wavelengths. For any series of collections, field samples were taken during relatively calm conditions between the hours of 1000 and 1400 h. Multiple air light readings were taken to correct for short-term radiation variability during light measurements.

2. Biological Sampling

Net phytoplankton samples were taken with two 25-μm nets (bongo configuration) in duplicate runs for periods of 1 to 2 min. Repetitive (three) 1-L whole water phytoplankton samples were taken at the surface. Phytoplankton samples were immediately fixed in Lugol's solution in its acid version (Lovegrove, 1960). Samples were analyzed by methods described by Prasad et al. (1990) and Prasad and Fryxell (1991). Zooplankton were taken with two 202-μm nets (bongo configuration) in duplicate runs. Samples were preserved with 10% formalin. Plankton identifications were made to species (Prasad and Livingston, 1995).

Methods used for the comparison of monthly data (water quality, biological factors) were developed to determine significant differences between matching Amelia and Nassau sites (polluted and unpolluted) over the 12-month study periods (Livingston et al., 1998a,b; Livingston, 2000). For independent, random samples from normally distributed populations, the parametric *t*-test was used to compare the sample means. For cases where one or both of the data sets violated the assumption of normality, a data transformation was made to bring the data into normality. Tests were also developed to compare two serially correlated populations of numbers taken at subject stations by calculating differences of the observations and plotting the autocorrelations (months) of the differences. If differences were not serially correlated, we applied the Wilcoxon sign-rank test to compare (0.05 confidence level) the two sets of numbers.

Field data were analyzed using a PCA as a preliminary review of the water quality variables (Livingston et al., 1998a,b). The PCA was used to reduce the physicochemical variables into a smaller set of linear combinations that could account for most of the total variation of the original set. Significant principal components were then applied to regression models with phytoplankton and zooplankton abundance and species richness as dependent variables. Residuals were tested for independence using serial correlation (time series) analyses and the Wald–Wolfowitz (Wald and Wolfowitz, 1940) runs test. A Chi-square test was run to evaluate normality. Statistics were run using SAS™, Systat™, and SuperAnova™.

B. Experimental Methods

A combination of background field monitoring, controlled laboratory experiments using microcosms of *Skeletonema costatum* (Grev.) Cleve and field mesocosm experiments (multispecies) was used to evaluate the effects of pulp mill effluents and ammonia on plankton assemblages in the Amelia River estuary. Measured solutions of ammonia were used to evaluate the effects of ammonia by itself relative to the effects of ammonia as part of the whole mill effluent. Target concentrations for the ammonia experiments were based on known field concentrations in polluted areas of the Amelia system. Concentrations of pulp mill effluents were determined by field color analyses at ambient conditions at station R1 in the Amelia system. We carried out one microcosm test (6/29/98 to 7/4/98) using lab-cultured *Skeletonema* with measured injections of ammonia, and two tests (7/17/98 to 7/22/98; 8/28/98 to 9/1/98) with pulp mill effluents with ammonia concentrations approximating those in the field. We performed six larger-volume mesocosm tests in the field with natural phytoplankton assemblages taken from the reference Nassau area. Two tests (8/19/97 to 8/21/97; 10/27/97 to 10/29/97) were run with measured injections of ammonia and four tests (5/20/98 to 5/22/98; 6/24/98 to 6/26/98; 8/4/98 to 8/6/98; 9/23/98 to 9/25/98) were carried out with pulp mill effluents that were added to basal mixtures to approximate ammonia concentrations determined in the field. For all tests, ammonia dosages were tested daily and ammonia was added where necessary to maintain target concentrations.

Laboratory microcosm tests were established using 18 1000-mL Erlenmeyer flasks in a randomized block array. The basal mixture (700 mL) was offshore water enriched with nitrate, orthophosphate, and silicon dioxide. Each flask was inoculated with the lab-cultured test species (*Skeletonema costatum*). After addition of ammonia solution or mill effluent, ammonia concentrations (five treatments and a control) were measured with an Orion ammonia-sensitive electrode (Flemer et al., 1997). Experiments were run at 22°C ± 1°C. Growlight® lights were used; light levels were checked with a spectroradiometer for treatment comparability. Experimental light levels were comparable to those in taken in the field (PAR, 0.5 m: 275-300 µEinsteins.m^{-2} sec^{-1}). Day lengths of 10:14 hr (light:dark) were used for the experiments. Water quality collections were taken daily for each test. Chlorophyll *a* concentrations (indicative of *Skeletonema* abundance) were analyzed using a Wetlabs fluorometer. Test results were determined for days 1, 3, and 5 of the test period.

Field mesocosms were established in the Amelia River estuary with water and natural phytoplankton assemblages taken from the Nassau system. Zooplankton were removed by passing water through a 64-µm plankton net. A standard mesocosm, run as a closed system, was a 20-L clear, polypropylene cubitainer fitted with closure adapters that allowed acceptance of the Hydrolab datasonde for monitoring purposes. Mesocosms were suspended 0.1 m below the water surface in the Amelia system in areas distant from mill effects. Five treatments (using ammonia or mill effluent) were established with a control. Three replicates per treatment were randomly distributed in a meshed frame for containment. Through experimentation, we established an effluent/ammonia spiking routine at 1-day intervals. Mesocosms were monitored individually for ammonia, color, chlorophyll *a*, DO, and pH on a daily basis. Ammonia concentrations were determined to ascertain the concentration of inoculants. Color was assayed to determine the concentration of the mill effluent. Chlorophyll *a* was taken as an index of phytoplankton growth. The maximum duration of the tests was 2 to 3 days, as determined by a series of preliminary tests.

A one-way ANOVA model was used to analyze the microcosm results. Six treatments, with the first as the control (no added ammonia), were arranged in a randomized (six treatments × three replicates) experiment. The variable of interest was chlorophyll *a* (representative of numbers or biomass of *Skelatonema costatum*). The same experiment was repeated during three time periods: June 29 to July 4 as Experiment 1; July 17 to July 22 as Experiment 2; and August 28 to September 1 as Experiment 3. Multiple comparison tests were performed on the experimental results. Based on the recommendation by Kirk (1995), *post-hoc* contrasts were tested by Tukey's HSD (honestly significant difference) test, Bonferroni's test, and Scheffe's S test. The SAS statistical software was used for the analysis. Statistical assumptions were tested using residual box plots. We used scattergrams of the residuals vs. the Fitted Y Dependent and scattergrams of cell means vs. the standard deviations. In addition, interactive bar charts were constructed showing cell means with standard deviation error bars.

Appendix II

Statistical Analyses Used in the Long-Term Studies of Aquatic Systems (1971–2004)

I. Difference Testing

A series of statistical methods was used to analyze the long-term data. Scattergrams of the long-term field data were examined and either logarithmic or square root transformations were made, where necessary, to approximate the best fit for a normalized distribution. These transformations were used in all statistical tests of significance.

To examine spatial differences in mean values of various field characteristics, the nonparametric Wilcoxin Signed Rank test for paired differences (i.e., applied to the differences between paired observations through time at the sites being compared) was used. The Wilcoxin test (Wilcoxin, 1949) is well suited for these types of field data comparisons because the differencing used tends to remove much of the serial correlation in the data sets, and the test does not require the normality assumption, which is often difficult to meet with field data. In all such comparisons, the null hypothesis ($\mu_1 = \mu_2$) was tested against the alternative ($\mu_1 \neq \mu_2$) using a 2-tailed test with $\alpha = 0.05$. Variances of the field characteristics were tested for significant differences using a modified Levene's statistic (Levene, 1960) when the underlying populations were non-normal. Both W_{10} and W_{50} statistics were examined using the recommendations of Brown and Forsythe (1974). Nonparametric testing, while avoiding problems with normality, is still subject to the assumptions of independence. Dependence was examined in both the Wilcoxon and Levene tests with an overall Q statistic (Ljung and Box, 1978).

We used a number of basic statistical tests regarding comparisons of two or more data sets. These included the following:

A. Independence Tests

Random, independent observations are integral to the comparison of two samples. The analysis can check to see if there are certain forms of numeric dependence in the data; if there is dependence due to problems of design or sampling methodology, for example, it cannot be detected. The independence tests used for preliminary analyses examine each sample for autocorrelation at a number of lags. For each sample, autocorrelation is computed for the lesser of 24 or $n/4$ lags, where n is the number of observations. The program then computes the Q statistic of Ljung and Box (1978) for an overall test of the autocorrelations as a group, where the null hypothesis is that correlation at each lag is equal to zero. This statistic is approximately Chi-square distributed under the null hypothesis with $(K - m)$ degrees of freedom, where K is the number of lags examined and m is the number of estimated parameters (1 if an AR1 filter has been applied to the data, 0 otherwise). The p-value of the Chi-square test is checked and if it remains less than 0.05, the null hypothesis is rejected.

The output sheet also contains plots of the autocorrelation functions for the two samples. These graphs are barcharts showing the values of autocorrelation coefficients at all computed lags, and are overlain with dashed lines at the critical value (value for which any individual correlation may be considered significant) for each series. Critical values are equal to $2/\sqrt{n}$, where n is the number of observations.

B. Normality tests

The parametric F-test for comparison of variances and t-test for comparison of means assume that the data come from normally distributed populations. In support of those tests, the two-sample spreadsheet provides a (Chi-square) goodness-of-fit test that compares distributions of the samples to normal distributions having the same means and variances as the samples. Because the number of observations is generally fairly small, the program is designed to divide the data into seven equally sized intervals (frequency classes) for comparison with expected (normally distributed) frequencies. The program also examines the expected frequencies for the leftmost and rightmost classes and expands (doubles) the width of these classes if the initial expected frequency is less than one. These procedures both help to minimize the problem of having extremely low expected frequencies in the denominator of the Chi-square computation. However, very low sample sizes and sample distributions with extreme departures from normality can still interfere with the test. For this reason, the program prints a warning when sample size is below 25. The Chi-square test should not be considered reliable for such small sample sizes.

The spreadsheet contains a graph of the frequency distribution (with a normal curve overlain for visual comparison purposes) for each of the two samples. Additionally, the sheet contains calculated values for skewness, kurtosis, and Geary's g statistic, which is another measure of kurtosis and which may be compared with the value 0.7979 (g value for a normal distribution). All of these values can be used by investigators to assist in their consideration of the normality issue.

C. Variance Testing

For independent, random samples from normally distributed populations, the F-test can be used to compare the variances of two samples. F is computed as the ratio of the variances of the samples:

$$F = \frac{s_1^2}{s_2^2} ,$$

where the null hypothesis is

$$H_0 : \frac{\sigma_1^2}{\sigma_2^2} = 1 ,$$

and where s_1^2 is the larger of the two sample variances. The program prints the value of F, the associated p-value from an F-table with $n_1 - 1$ numerator degrees of freedom and

$n_2 - 1$ denominator degrees of freedom. It also prints a decision to reject or not to reject the null hypothesis (using an alpha level of 0.05), and a warning if it appears that one or more test assumptions have been violated. If one or both of the data series violate the independence assumption due to serial dependence at lag 1, one can use the AR(1) filter option in the two-sample menu. Running the filter will prompt the program to use a modified formula for the calculation of F, which computes F in the presence of lag 1 dependence.

For those cases where the samples do not satisfy the assumption of normality, two modifications to Levene's statistic (Levene, 1960) for variance comparisons are included in the spreadsheet. The statistics are W_{10} and W_{50}, which are both robust to departures from normality in the underlying populations (Brown and Forsythe, 1974). They use robust measures of central location (W_{10} uses the mean of the middle 80% of the observations; W_{50} uses the median observation) and thereby decrease or eliminate the influence of extreme values in testing variance. Brown and Forsythe (1974) recommend using W_{50} for testing the variances of asymmetric distributions and using W_{10} for testing long-tailed distributions. Both formulas involve the generation of a new data series z_{ij}, which also must be free of serial dependence. If a warning message is printed with the robust tests, it applies to the new series, not the original series, and it is possible to have dependence in the z_{ij} series even if the original data were independent. Dependence is examined with an overall Q statistic (Ljung and Box, 1978) and Chi-square test developed in a fashion similar to that described above in the independence testing section. The output of the variance tests should not be used if a transformation (log, square root, arcsin, normalization) has been applied to the data. Also, performing a sequential average tends to "smooth" the data (i.e., variation is lowered in both data sets) and may have an effect on the variance test.

D. Means Testing

For independent, random samples from normally distributed populations, the parametric *t*-test can be used to compare the sample means. For unpaired samples, the spreadsheet provides either a *t*-test with equal variance or a *t*-test with unequal variance (the results of the variance test above are used as the criterion for deciding which *t*-test to output). For paired samples, the sheet provides the results of a paired *t*-test. For all tests, the output contains the computed value of the *t* statistic, a probability value (two-sided test, $\alpha = 0.05$) from the *t* distribution, a reject or do not reject decision regarding the null hypothesis, and a warning if it appears that one or more test assumptions have been violated. If one or both of the data series violate the independence assumption due to serial dependence at lag 1, one can use the AR(1) filter option from the two-sample menu. Running the filter will prompt the program to use a modified formula for the calculation of *t*, which computes *t* in the presence of lag 1 dependence.

For cases where one or both of the data sets violate the assumption of normality, a data transformation can be made to bring the data into normality (although variance testing is affected), or use one of the nonparametric tests that have been included in the spreadsheet. The tests are the rank sum test for unpaired comparisons (Wilcoxin, 1949) and the signed rank test for paired differences (Wilcoxin, 1949). Both tests are subject to the assumption of random independent samples. For the rank sum test, both of the data series should be independent; while for the signed rank test, only the series of differences needs to be independent. This can make the signed rank test a valuable tool for two-sample testing because, in practice, serial correlation often disappears when dependent data sets are differenced. Wilcoxin test results should not be used if an AR1 filter has been applied to the data.

Output for the Wilcoxon tests includes a description of the rejection region, the computed value for the test statistic T for small samples and z for larger samples ($n > 25$ for signed rank test, $n > 10$ for rank sum), a reject or do not reject decision regarding the null hypothesis (with a probability value based on a two-tailed test and alpha level of 0.05), and a warning if the independence assumption has been violated. Output for the signed rank Wilcoxon test also includes for the differenced data series an overall Q statistic and Chi-square probability value, along with a graph of the autocorrelation function.

E. Distributional Shape Testing

The two-sample spreadsheet also includes a Chi-square test, which compares the shapes of the distributions for the two samples. Although the samples may individually exhibit dependence, the assumption is made that the two samples are independent of each other. There are no assumptions regarding the form of the underlying distributions. To examine shape only, the program adjusts the two data sets by subtracting the respective mean value from each observation such that both series then have a common mean value of zero. Then the overall range of data in the two series is divided into the lesser of 9 or $n/4$ classes and a frequency distribution is produced for each sample. The frequency distributions are compared with a Chi-square test.

Output for the shape test includes the computed values for the Chi-square statistic and its associated probability value, a reject or do not reject decision regarding the null hypothesis (based on a one-tailed test and alpha level of 0.01), and a warning if the sample size is too small for a reliable test ($n < 25$) or if the independence assumption has been violated.

F. Principal Components Analysis and Regressions

PCA and associated correlation matrices were determined using monthly data and the SAS™ statistical software package. A PCA was carried out as a preliminary review of the data using river flow information, sediment data, and water quality variables. The PCA was used to reduce the physicochemical variables into a smaller set of linear combinations that could account for most of the total variation of the original set. For the physicochemical variables in this study, the series were stationary after appropriate transformations; thus, the sample correlation matrix of the variables was a good estimate of the population correlation matrix. Therefore, the standard PCA could be carried out based on the sample correlation matrix.

A matrix of the water and sediment quality data associated with the sampling stations in the Perdido system was prepared. The data were then grouped by station by year. Values for the dependent variables (total biomass.m^{-2}, herbivore biomass.m^{-2}, omnivore biomass.m^{-2}, primary carnivore biomass.m^{-2}, secondary carnivore biomass.m^{-2}, tertiary carnivore.m^2) were then paired with the water/sediment quality data (independent variables) taken for stations within each sector. Unless otherwise defined, these statistics were run using SAS™, Systat™, and SuperAnova™. Data analyses were run on the three independent data sets. Significant principal components were then used to run a series of regression models with the biological factors as dependent variables. Residuals were tested for independence using serial correlation (time series) analyses and the Wald–Wolfowitz (Wald and Wolfowitz, 1940) runs test. A Chi-square test was run to evaluate normality.

References

Adams, D.A., J.S. O'Connor, and S.B. Weisberg. 1996. Sediment Quality in the NY/NJ Harbor System. An investigation under the Regional Environmental Monitoring and Assessment Program (R-EMAP). Draft Final Report.

Adams, D.A., J.S. O'Connor, and S.B. Weisberg. 1998. Sediment quality of the NY/NJ Harbor System. Final Report. EPA/902-R-98-001.

Admiraal, W. 1977. Tolerance of estuarine benthic diatoms to high concentrations of ammonia, nitrite ion, nitrate ion, and orthophosphate. *Mar. Biol.* 43: 307–313.

Admiraal, W. and H. Peltier. 1980. Distribution of diatom species on an estuarine mud flat and experimental analysis of the selective effect of stress. *J. Exp. Mar. Biol. Ecol.* 46: 157–175.

Alam, P. 1988. Lake Jackson Conservation Plan, prepared for the Leon County Board of County Commissioners by the Department of Public Works Environmental Management. 14 pp.

Ambrose, R.B. Jr. and T.O. Barnwell. 1989. Environmental software at the U.S. Environmental Protection Agency's Center for Exposure Assessment Modeling. *Environmental Software* 4, 76–93.

American Public Health Association (APHA). 1989. *Standard Methods for the Examination of Water and Wastewater, 17th ed.* Amer. Pub. Health Assoc., Washington, D.C. 1193 pp.

Anderson, D.M. 1996. Control and mitigation of red tides. Project Start (Solutions to Avoid Red Tide), Sarasota, FL. 59 pp.

Anderson, D.A. and D.J. Garrison. 1997. The ecology and oceanography of harmful algal blooms. *Limnol. Oceanogr.* 42, pp. 1–1305.

Armstrong, N.E. 1982. Responses of Texas estuaries to freshwater inflows. In V.S. Kennedy (Ed.), *Estuarine Comparisons.* Academic Press, New York, NY, pp. 103–120.

Babcock, S. 1976, Lake Jackson Studies, Annual Report for Research Project, Florida Game and Freshwater Fish Commission. 120 pp.

Bacchus, S.T. 2002. The "Ostrich" component of the multiple stressor model: undermining South Florida. In J.W. Porter and K.G. Porter (Eds.), *The Everglades, Florida Bay and Coral Reefs of the Florida Keys: An Ecosystem Sourcebook.* CRC Press, Boca Raton, FL. pp. 677–748.

Baird, D. and R.E. Ulanowicz. 1989. The seasonal dynamics of the Chesapeake Bay ecosystem. *Ecol. Monogr.* 59: 329–364.

Belton, T.J., R. Hazen, B.E. Ruppel, K. Lockwood, R. Mueller, E. Stevenson, and J.J. Post. 1985. A Study of Dioxin (2,3,7,8-tetrachlorodibenzo-*p*-dioxin) Contamination in Select Finfish, Crustaceans, and Sediments of New Jersey Waterways. Unpublished report. Office of Science and Research, New Jersey Department of Environmental Protection, CN-409, Trenton, NJ.

Bevis, T.H. 1995. The Response of Fish Assemblages to Stormwater Discharge into a North Florida Solution Lake. M.S. thesis, Department of Biological Sciences and Center for Aquatic Research and Resource Management, Florida State University, Tallahassee, FL.

Birkholz, D. A. 1999. Chemical Analysis Report. Enviro-Test Laboratories. Edmonton, Alberta, Canada.

Blanchet, R.H. 1979. The distribution and abundance of ichthyoplankton in the Apalachicola Bay, Florida area. M.S. thesis, Florida State University, Tallahassee, FL.

Blumberg, A.F. and G.L. Mellor. 1980. A coastal ocean numerical model. In J. Sundermann and K.P. Holz (Eds.), *Mathematical Modeling of Estuarine Physics, Proceedings of the International Symposium,* Hamburg, 24–26 August 1978. Springer-Verlag, Berlin. pp. 203–214.

Blumberg, A.F. and G.L. Mellor. 1987. A description of a three-dimensional coastal ocean circulation model. In N.S. Heaps (Ed.), *Three-Dimensional Coastal Ocean Models.* American Geophysical Union, Washington, D.C., pp. 1–16.

Bobbie, R.J., S.J. Morrison, and D.C. White. 1978. Effects of substrate biodegradability on the mass and activity of the associated estuarine microbiota. *Appl. Envir. Microbiol.* 35: 179–184.

Boesch, D.F., N.E. Armstrong, C.F. D'Elia, N.G. Maynard, H.W. Paerl, and S.L. Williams. 1993. Deterioration of the Florida Bay Ecosystem: An Evaluation of the Scientific Evidence. National Fish and Wildlife Foundation, National Park Service, South Florida Water Management District. Unpublished report.

Bolgiano, R.W. 1980. Mercury contamination of the floodplains of the South and South Fork Shenandoah Rivers. Basic Data Bulletin 47. State Water Control Board, Richmond, VA.

Bolgiano, R.W. 1981. Mercury contamination of the floodplains of the South and South Fork Shenandoah Rivers. Basic Data Bulletin 48. State Water Control Board, Richmond, VA.

Bonin, D.J., M.R. Droop, S.Y. Maestrini, and M.C. Bonin. 1986. Physiological features of six microalgae to be used as indicators of seawater quality. *Crypt. Algol.* 7: 23–83.

Bopp, R.F., M.L. Gross, H. Tong, H.J. Simpson, S.J. Monson, B.L. Deck, and F.C. Moser. 1991. A major incident of dioxin contamination: sediments of New Jersey estuaries. *Environ. Sci. Toxicol.* 25: 951–956.

Bopp R.F., S.N. Chillrud, E.L. Shuster, and H.J. Simpson. 1998. Trends in chlorinated hydrocarbon levels in Hudson River basin sediments. *Envir. Health Perspec.* 106(Suppl. 4), pp. 1075–1081.

Bortone, S. 1991. Sea Grass Mapping of Perdido Bay. Unpublished report.

Boschen, C.J. 1996. The feeding ecology of the dominant fishes that occur in a eutrophicated north Florida solution lake. M.S. thesis, Department of Biological Sciences and Center for Aquatic Research and Resource Management, Florida State University, Tallahassee, FL.

Box, G.E.P. and G.M. Jenkins. 1976. Time Series Analysis: Forecasting and Control. Holden-Day, Inc. San Francisco, CA.

Boyer, J.N., J.W. Fourqurean, and R.D. Jones. 1999. Seasonal and long-term trends in the water quality of Florida Bay. In J.W. Fourqurean, M.B. Robblee, and L.A. Deegan (Eds.). Florida Bay: A dynamic subtropical estuary. *Estuaries* 22: 417–430.

Boyer, J.N. and R.D. Jones. 2002. A view from the bridge: external and internal forces affecting the ambient water quality of the Florida Keys National Marine Sanctuary (FKNMS). In J.W. Porter and K.G. Porter (Eds.), *The Everglades, Florida Bay and Coral Reefs of the Florida Keys: An Ecosystem Sourcebook.* CRC Press, Boca Raton, FL. pp. 609–628.

Boynton, W.R. 1997. Estuarine ecosystem issues of the Chesapeake Bay. In R.D. Simpson and N.J.L. Christensen, Jr. (Eds.), *Ecosystem Function and Human Activities.* Chapman and Hall Publishers, New York. pp. 71–93.

Boynton, W.R., J.H. Garber, R. Summers, and W.M. Kemp. 1995. Inputs, transformations, and transport of nitrogen and phosphorus in Chesapeake Bay and selected tributaries. *Estuaries* 18: 285–314.

Boynton, W.R., W.M. Kemp, and C.W. Keefe. 1982. A comparative analysis of nutrients and other factors influencing estuarine phytoplankton production. In V.S. Kennedy (Ed.), *Estuarine Comparisons.* Academic Press, New York, New York. pp. 69–90.

Brady, K. 1982. Larval Fish Distribution in Apalachee Bay. M.S. thesis, Florida State University, Tallahassee, FL.

Brand, L.E. 2000. An Evaluation of the Scientific Basis for "Restoring" Florida Bay by Increasing Freshwater Runoff from the Everglades. Reef Relief. Unpublished report.

Brand, L.E. 2002. The transport of terrestrial nutrients to South Florida coastal waters. 2002. In J.W. Porter and K.G. Porter (Eds.), *The Everglades, Florida Bay and Coral Reefs of the Florida Keys: An Ecosystem Sourcebook.* CRC Press, Boca Raton, FL. pp. 361–413.

Brank, C.W.C. and P.A. Senna. 1994. Factors influencing the development of *Cylindrospermopsis raciborskii* and *Microcystis aeruginosa* in the Paranoa Reservoir, Brasilia, Brazil. *Algological Studies* 75: 85–96.

Breitburg, D.L. 1990. Near-shore hypoxia in the Chesapeake Bay: patterns and relationships among physical factors. *Est. Coastal Shelf Sci.* 30: 593–609.

Bricker, S.B., C.G. Clement, D.E. Pirhalla, S.P. Orlando, and D.R.G. Farrow. 1999. National Estuarine Eutrophication Assessment; Effects of Nutrient Enrichment in the Nation's Estuaries. National Oceanic and Atmospheric Administration, National Ocean Service, Special Projects Office and the National Centers for Coastal Ocean Science, Silver Springs, MD. 71 pp.

Brouard, D., C. Demers, R. Lalumiere, R. Schetagne, and R. Verdon. 1990. Evolution of Mercury Levels in Fish of the La Grande Hydroelectric Complex, Quebec. Summary Report, Hydro Quebec (Montreal) and Schooner Inc. (Quebec), P.Q. 97 pp.

Brown, M.B. and A.B. Forsythe. 1974. Robust tests for the equality of variances. *J. Am. Statis. Assoc.* 69: 364–367.

Brush, G.S. 1991. Long-Term Trends in Perdido Bay: A Stratigraphic Study. Unpublished Report, Florida Department of Environmental Regulation.

Burgess, N.M., D.C. Evers, J.D. Kaplan, M. Duggan, and J.J. Kerekes. 1998. Ecological impacts of mercury. In *Mercury in the Atlantic: A Progress Report*. Regional Science Coordinating Committee, Environment Canada.

Byrne, C.J. 1980. The geochemical cycling of hydrocarbons in Lake Jackson, Florida. Ph.D. dissertation, Department of Oceanography, Florida State University, Tallahassee, FL.

Cairns, D.J. 1981. Detrital Production and Nutrient Release in a Southeastern Flood-Plain Forest. M.S. thesis, Florida State University, Tallahassee, FL.

Camp Dresser and McKee, Inc. 1998. Site Investigation Report; Holtrachem Manufacturing Site, Orrington, Maine. Unpublished report, Cambridge, MA.

Campbell, K.R., C.J. Ford, and D.A. Levine. 1998. Mercury distribution in Poplar Creek, Oak Ridge, Tennessee, USA. *Envir. Toxicol. Chem.* 17: 1191–1198.

Carr, W.E.S. and C.A. Adams. 1972. Food habits of juvenile marine fishes: evidence of the cleaning habit in the leatherjacket, *Oligoplites saurus*, and the spottail pinfish, *Diplodus holbrooki. Fish. Bull.* 70: 1111–1120.

Cassidy, R.O. 1992. An Interim Waer Quality Evacuation of Seven Tallahassee Lakes. Unubl. Report. Growth Management Dept., City of Tallahassee, Tallahassee, Florida.

Causy, B.D. 2002. The role of the Florida Keys National Marine Sanctuary in the South Florida ecosystem restoration initiative. In J.W. Porter and K.G. Porter (Eds.), *The Everglades, Florida Bay and Coral Reefs of the Florida Keys: An Ecosystem Sourcebook*. CRC Press, Boca Raton, FL. pp. 883–894.

Chaky, D.A. 2003. Polychlorinated Biphenyls, Polychlorinated Dibenzo-*p*-dioxins, and Furans in the New York Metropolitan Area: Interpreting Atmospheric Deposition and Sediment Chronologies. Ph.D. dissertation, Rensselaer Polytechinic Institute, Troy, NY.

Chang, E.H. 1988. Distribution, abundance and size composition of phytoplankton off Westland, New Zealand, February 1982. *N.Z. J. Mar. Freshwater Res.* 22: 345–367.

Chang, F.J., C. Anderson, and N.C. Boustead, 1990. First record of *Heterosigma* (Raphidophyceae) bloom with associated mortality of cage-reared salmon in Big Glory Bay, New Zealand. *N.Z. J. Mar. Freshwater Res.* 24: 461–469.

Chanton, J. and F.G. Lewis. 1999. Plankton and dissolved inorganic carbon isotopic composition in a river-dominated estuary: Apalachicola Bay, Florida. *Estuaries* 22: 575–583.

Chanton, J. and F.G. Lewis. 2002. Examination of coupling between primary and secondary production in a river-dominated estuary: Apalachicola Bay, Florida, U.S.A. *Limnol. Oceanogr.* 47: 683–697.

Chemical Land Holdings. 2000. Passaic River Tissue Data Summary Table Dioxin/Furan. Draft report (unpublished).

Chesapeake Bay Foundation. 1999. Shenandoah Waters, Toxic Contaminated Fish. Unpublished report.

Chesapeake Bay Program. 2003. Website information.

Christensen, J.D., M.E. Monaco, R.J. Livingston, G. Woodsum, T.A. Battista, C.J. Klein, B. Galperin, and W. Huang. 1998. Potential Impacts of Freshwater Inflow on Apalachicola Bay, Florida Oyster (*Crassostrea virginica*) Populations: Coupling Hydrologic and Biological Models. NOAA/NOS Strategic Environmental Assessments Division Report. Silver Spring, MD. 58 pp.

Clement, B. and G. Merlin. 1995. The contributions of ammonia and alkalinity to landfill leachate toxicity to duckweed. *Sci. Total Environ.* 170: 71–79.

Clements, W.H. and R.J. Livingston. 1983. Overlap and pollution-induced variability in the feeding habits of filefish (Pisces: Monacanthidae) from Apalachee Bay, Florida. *Copeia* (1983): 331–338.

Clements, W.H. and R.J. Livingston. 1984. Prey selectivity of the fringed filefish *Monacanthus ciliatus* (Pisces: Monacanthidae): role of prey accessibility. *Mar. Ecol. Prog. Ser.* 16: 291–295.

Clewell, A.F. 1977. Geobotany of the Apalachicola River region. In R.J. Livingston (Ed), *Proceedings of the Conference on the Apalachicola Drainage System*. Florida Department of Natural Resources, Marine Resources Publication 26. St. Petersburg, FL. pp. 6–15.

Codd, G.A., C. Edwards, K.A. Beattle, L.A. Lawton, D.L. Campbell, and S.G. Bell. 1995. Toxins from cyanobacteria (bluegreen algae). In W. Wiessner, E. Schnept, and R.C. Starr (Eds.). *Algae, Environment and Human Affairs*. Biopress Ltd., Bristol, England. Pp. 1–17.

Collard, S.B. 1991a. The Pensacola Bay System: Biological Trends and Current Status. Water Resources Special Report 91-3. Havana: Northwest Florida Water Management District.

Collard, S.B. 1991b. Management Options for the Pensacola Bay System: The Potential Value of Seagrass Transplanting and Oyster Bed Refurbishment Programs. Water Resources Special Report 91-4. Havana: Northwest Florida Water Management District.

Crawford, D.W., N.L. Bonnevie, C.A. Gillis, and R.J. Wenning. 1994. Historical changes in the ecological health of the Newark Bay Estuary, New Jersey. *Ecotoxicol. Environ. Saf.* 29: 276–303.

Crawford, D.W., N.L. Bonnevie, C.A. Gillis, and R.J. Wenning. 1995. Sources of pollution and sediment contamination in Newark Bay, New Jersey. *Ecotoxicol. Environ. Saf.* 30: 85–100.

Cross, R.D. and D.L. Williams. 1981. *Proceedings of the National Symposium on Freshwater Inflow to Estuaries*. U. S. Fish and Wildlife Service, Office of Biological Services. FWS/OBS-81/04.

Dauer, D.M., R.W. Ewing, G.H. Tourtellotte, W.T. Harlan, J.W. Sourbeer, and H.R. Baker. 1982. Predation, resource limitation and the structure of lower Chesapeake Bay. *Int. Rev. Gesamt. Hydrobiol.* 67: 477–489.

Dauer, D.M., A.J. Rodi, and J.A. Ranasinghe. 1992. Effects of low dissolved oxygen events on the macrobenthos of the lower Chesapeake Bay. *Estuaries* 15: 384–391.

Dauer, D.M., J.A. Ranasinghe, and S.B. Weisberg, 2000. Relationships between benthic community condition, water quality, sediment quality, nutrient loads, and land use patterns in Chesapeake Bay. *Estuaries* 23: 80–96.

Davis, S.M. and J.C. Ogden. 1994. *Everglades: The Ecosystem and Its Restoration*. St. Lucie Press, Delray Beach, FL.

Davis, W.P. and S.A. Bortone. 1992. Effects of kraft mill effluents on the sexuality of fishes: an environmental early warning? In T. Colborn and C. Clements (Eds.), *Chemically Induced Alterations in Sexual and Functional Development: The Wildlife/Human Connections*. Princeton Scientific Publishing, Princeton, NJ. pp. 113–127.

Davis, W.P., M.R. Davis, and D.A. Flemer. 1999. Observations on the regrowth of subaquatic vegetation following transplantation: a potential method to assess environmental health of coastal habitats. In S. Bortone (Ed.), *Seagrasses; Monitoring, Ecology, Physiology, and Management*. CRC Press, Boca Raton, FL. pp. 231–238.

De Jong, V.N. 1995. The Ems Estuary, The Netherlands. In A.J. McComb (Ed.), *Eutrophic Shallow Estuaries and Lagoons*. CRC Press, Boca Raton, FL. pp. 81–108.

De Jong, V.N. and W. van Raaphorst, 1995. Eutrophication of the Dutch Wadden Sea (Western Europe), an estuarine area controlled by the River Rhine. In A.J. McComb (Ed.), *Eutrophic Shallow Estuaries and Lagoons*. CRC Press, Boca Raton, FL. pp. 129–150.

Deegan, L.A., J.W. Day, Jr., J.G. Gosselink, A. Yanez-Arancibia, G. Soberon Chavez, and P. Sanchez-Gil. 1986. Relationships among physical characteristics, vegetation distribution, and fisheries yield in Gulf of Mexico estuaries. In D.A. Wolfe (Ed), *Estuarine Variability*. Academic Press, New York. pp. 83–100.

Deevey, E.S. Jr. 1988. Estimation of downward leakage from Florida lakes. *Limnol. Oceanogr.* 33(6, part 1), 1308–1320.

D'Elia, C.F., W.R. Boynton, and J.G. Sanders. 2003. A watershed perspective on nutrient enrichment, science and policy in the Patuxent River, Maryland: 1960–2000. *Estuaries* 26: 171–185.

Denison, J.G., S.H. Olsen, and E.G. Hoffman. 1977. Baseline receiving environment study prior to secondary treatment by the Fernandina Division of ITT Rayonier Incorp. Unpublished report. ITT Rayonier, Inc. Shelton, Washington..

Dennison, W.C., R.J. Orth, K.A. Moore, J.C. Stevenson, V. Carter, S. Kollar, P.W. Bergstrom, and R.A. Batuik. 1993. Assessing water quality with submersed aquatic vegetation. *Bioscience* 43: 86–94.

Dolan, D.M, A.K. Yui, and R.D. Geist. 1981. Evaluation of river load estimation methods for total phosphorus. *Great Lakes Res.* 7: 207–214.

Dortch, Q. 1990. The interaction between ammonium and nitrate uptake in phytoplankton. *Mar. Ecol. Prog. Ser.* 61: 183–201.

Downing, K.M. and J.C. Merkens. 1955. The influence of dissolved oxygen concentrations on the toxicity of un-ionized ammonia to rainbow trout (*Salmo gairdnerii* Richardson). *Ann. App. Biol.* 43: 243–246.

Duarte, C.M. 1995. Submerged aquatic vegetation in relation to different nutrient regimes. *Ophelia* 41: 87–112.

Dugan, P.J. and R.J. Livingston. 1982. Long-term variation in macroinvertebrate communities in Apalachee Bay, Florida. *Est. Coast. Shelf Sci.* 14: 391–403.

Duncan, J.L. 1977. Short-Term Effects of Storm Water Runoff on the Epibenthic Community of a North Florida Estuary (Apalachicola, Florida). M.S. thesis, Florida State University, Tallahassee, FL.

Edmiston, H.L. 1979. The zooplankton of the Apalachicola Bay system. M.S. thesis, Florida State University, Tallahassee, FL.

Ehrlich, R., R.J. Wening, G.W. Johnson, S.H. Su, and D.J. Paustenbach. 1994. A mixing model for polychlorinated dibenzo-*p*-dioxins and dibenzofurans in surface sediments from Newark Bay, New Jersey using polytopic vector analysis. *Arch. Environ. Contam. Toxicol.* 27: 486–500.

Ekino, S., T. Ninomiya, and M. Susa. 2004. Letter to the editor, More on mercury content in fish. *Science* 303: 764.

Elder, J.F. and D.J. Cairns. 1982. Production and Decomposition of Forest Litter Fall on the Apalachicola River Floodplain, Florida. U.S. Geological Survey Water-Supply Paper 2196.

Ernst, H.R. 2003. *Chesapeake Bay Blues.* Rowman & Littlefield Publishers, Inc., Oxford, U.K. 203 pp.

Estabrook, R.H. 1973. Phytoplankton ecology and hydrography of Apalachicola Bay. M.S. thesis, Florida State University, Tallahassee, FL.

Estevez, E.D., L.K. Dixon, and M.F. Flannery. 1991. West-coastal rivers of peninsular Florida. In R.J. Livingston (Ed.), *The Rivers of Florida*. Springer-Verlag, New York. pp. 187– 222.

Federle, T.W., M.A. Hullar, R.J. Livingston, D.A. Meeter, and D.C. White. 1983a. Spatial distribution of biochemical parameters indicating biomass and community composition of microbial assemblies in estuarine mud flat sediments. *Appl. Envir. Microbiol.* 45: 58–63.

Federle, T.W., R.J. Livingston, D.A. Meeter, and D.C. White. 1983b. Modifications of estuarine sedimentary microbiota by exclusion of epibenthic predators. *J. Exp. Mar. Biol. Ecol.* 73: 81–94.

Federle, T.W., R.J. Livingston, L.E. Wolfe, and D.C. White. 1986. A quantitative comparison of microbial community structure of estuarine sediments for microcosms in the field. *Can. J. Microbiol.* 32: 319–325.

Finley B.L., K.R. Trowbridge, S. Burton, D.M. Proctor, J.M. Panko, and D.J. Paustenbach. 1997. Preliminary assessment of PCB risks to human and ecological health in the lower Passaic River. *J. Toxicol. Environ. Health* 52(2): 95–118.

Fisher, T.R., L.W. Harding, Jr., D.W. Stanley, and L.G. Ward. 1988. Phytoplankton, nutrients, and turbidity in the Chesapeake, Delaware, and Hudson estuaries. *Est. Coastal Shelf Sci.* 27: 61–88.

Fisher, T.R., E.R. Peele, J.W. Ammerman, and L.W. Harding, Jr. 1992. Nutrient limitation of phytoplankton in Chesapeake Bay. *Mar. Ecol. Prog. Ser.* 82: 51–63.

Flemer, D.A., R.J. Livingston, and S.E. McGlynn. 1997. Phytoplankton nutrient limitation: seasonal growth responses of sub-temperate estuarine phytoplankton to nitrogen and phosphorus — an outdoor microcosm experiment. *Estuaries* 21: 145–159.

Fletcher, R. 1990. Impact of Cultural Eutrophication on Aquatic Invertebrates in Lake Jackson (Tallahassee, Florida). Final report, Honors and Scholars Program, Department of Biological Sciences and Center for Aquatic Research and Resource Management, Florida State University, Tallahassee, FL.

Florida Department of Environmental Protection. 1996. Biological Assessment of Air Products and Chemicals, Inc., Tallahassee, FL.

Florida Department of Environmental Protection. 1997. Biological Sterling Fibers, Inc. Tallahassee, FL.

Florida Department of Environmental Protection (FDEP). No date. About the Everglades Ecosystem. Accessed September 11, 2000, at http://www.dep.state.fl.us/everglades/About.htm.

Florida Department of Environmental Regulation. 1991. Biological Assessment of ITT Rayonier, Inc. of Nassau County NPDES # Fl0000701, Tallahassee, FL.

Fourqurean, J.W, I.R.D. Jones, and J.C. Zieman. 1993. Processes influencing water column nutrient characteristics and phosphorus limitation of phytoplankton biomass in Florida Bay, FL, USA: Inferences from spatial distributions. *Est. Coastal Shelf Sci.* 36: 295–314.

Fourqurean, J.W. and M.B. Robblee. 1999. Florida Bay: a history of recent ecological changes. *Estuaries* 22: 345–357.

Fourqurean, J.W., M.B. Robblee, and L.A. Deegan. 1999. Florida Bay: a dynamic subtropical estuary. *Estuaries* 22.

Funicelli, N.A. 1984. Assessing and managing effects of reduced freshwater inflow to two Texas estuaries. In V.S. Kennedy (Ed.), *The Estuary as a Filter.* Academic Press, New York. pp. 435–446.

Galehouse, J.S. 1971. Sedimentary analysis. In R.E. Carver (Ed.), *Procedures in Sedimentary Petrology.* Wiley-Interscience, New York. pp. 69–94.

Galperin, B. and G.L. Mellor, 1990a. A time-dependent, three-dimensional model of the Delaware Bay and River. 1: Description of the model and tidal analysis. *Est. Coastal Shelf Sci.* 31: 231–253.

Galperin, B., L.H. Kantha, S. Hassid, and A. Rosati. 1988. A quasi-equilibrium turbulent energy model for geophysical flows. *J. Atmospher. Sci.* 45: 55–62.

Gameson, A.L.H., N.J. Barrett, and J.S. Strawbridge. 1973. The aerobic Thames estuary. In S.H. Jenkins (Ed.), *Advances in Water Pollution Research, Sixth Information Congress,* Jerusalem. Pergamon Press, New York. pp. 843–850.

Gilbert, P.M., R. Magnien, M.W. Lomas, J. Alexander, C. Fan, E. Haramoto, M. Trice, and T.M. Kana. 2001. Harmful algal blooms in the Chesapeake and coastal bays of Maryland, USA: comparison of 1997, 1998, and 1999 events. *Estuaries* 24: 875–883.

Gillis, C.A., N.L. Bonnevie, and J. Wenning. 1993. Mercury contamination in the Newark Bay Estuary. *Ecotoxicol. Environ. Saf.* 25: 214–226.

Gillis, C.A., N.L. Bonnevie, S.H. Su, J.G. Ducy, S.L. Huntley, and J. Wenning. 1995. DDT, DDD, and DDE contamination of sediment in the Newark Bay Estuary, New Jersey. *Arch. Environ. Contam. Toxicol.* 28: 85–92.

Gorham, P.R. 1964. Toxic algae. In D.F. Jackson (Ed.), *Algae and Man.* Plenum Press, New York. pp. 307–336.

Gossett, R.W., D.A. Brown, and D.R. Young. 1983. Predicting the bioaccumulation of organic compounds in marine organisms using octanol/water partition coefficients. *Mar. Pollut. Bull.* 14: 387–392.

Goudie, A. 1994. *The Human Impact on the Natural Environment, fourth edition.* The MIT Press, Cambridge, MA.

Greening, H.S. 1980. Seasonal and diel variations in the structure of macroinvertebrate communities: Apalachee Bay, Florida. M.S. thesis, Florida State University, Tallahassee, FL.

Greening, H.S. and R.J. Livingston. 1982. Diel variations in the structure of epibenthic macroinvertebrate communities of seagrass beds (Apalachee Bay, Florida). *Mar. Ecol. Prog. Ser.* 7: 147–156.

Gunster, D.G., N.L. Bonnevie, C.A. Gillis, and R.J. Wenning. 1993. Assessment of chemical loadings to Newark Bay, New Jersey from petroleum and hazardous chemical accidents occurring from 1986 to 1991. *Ecotox. Environ. Saf.* 25: 202–213.

Hallegraeff, G.M. 1995. Harmful algal blooms: a global overview. In G.M. Hallegreaff, D.M. Anderson, and A.D. Cembella (Eds.), *Manual on Harmful Marine Microalgae.* IOC Manuals and Guides No. 33. UNESCO. pp. 1–22.

Hallegraeff, G.M., D.M. Anderson, and A.D. Cembella. 1995. *Manual on Harmful Marine Microalgae.* IOC Manuals and Guides No. 33. UNESCO.

Hallegraeff, G.M., D.M. Anderson, and A.D. Cembella. 2003. *Manual on Harmful Marine Microalgae.* UNESCO Publishing, Landais, France, 793 pp.

Hand, J., J. Col, and L. Lord. 1996. 1996 Florida Water Quality Assessment, 305(b) Technical Appendix. Florida Department of Environmental Protection, Tallahassee.

Hara, Y. and M. Chihara. 1987. Morphology, ultrastructure, and taxonomy of the Raphidophycean alga *Heterosigma akashiwo*. *Bot. Mag. Tokyo* 100: 151–163.

Harlow, V.H. and R.W. Alden. 1997. Analysis of organomercury from fish collected in conjunction with the Shenandoah River mercury monitoring: risk analysis of organomercury fish contamination in the South and South Fork Shenandoah Rivers. Applied Marine Research Laboratory, Old Dominion University. Unpublished report for Chesapeake Bay Foundation.

Harris, H.H., I.J. Pickering, and G.N. George. 2003. The chemical form of mercury in fish. *Science (Brevia)* 1203.

Harris, H.H., I.J. Pickering, and G.N. George. 2004. Letter to the editor, More on mercury content in fish. *Science* 303: 764.

Harriss, R.C. and R.R. Turner. 1974. Job Completion Report: Lake Jackson Investigations, Job 5: Nutrients, Water Quality, and Phytoplankton Productivity, Job 7: Data Analysis and Reporting. Florida Game and Fresh Water Fish Commission. 231 pp.

Harvey, R. and K. Havens. 1999. Lake Okeechobee Action Plan. Lake Okeechobee Issue Team for the South Florida Ecosystem Restoration Working Group.

Hayes, P.F. and R.W. Menzel. 1981. The reproductive cycle of early setting *Crassostrea virginica* (Gmelin) in the northern Gulf of Mexico, and its implications for population recruitment. *Biol. Bull.* 160: 80–88.

Hayward, R.S. and F.J. Margraf. 1987. Eutrophication effects on prey size and food available to yellow perch in Lake Erie. *Trans. Am. Fish. Soc.* 116(2): 210–223.

Heard, W.H. 1977. Freshwater mollusca of the Apalachicola drainage. In R.J. Livingston (Ed.), *Proceedings of the Conference on the Apalachicola Drainage System.* Florida Department of Natural Resources, Marine Resources Publication 26. St. Petersburg, FL. pp. 20–21.

Hecky, R.E. and P. Kilham. 1988. Nutrient limitation of phytoplankton in freshwater and marine environments. *Limnol. Oceanogr.* 33: 796–822.

Hein, M., M.F. Pedersen, and K. Sand-Jensen. 1995. Size-dependent nitrogen uptake in micro- and macroalgae. *Mar. Ecol. Prog. Ser.* 118: 247–253.

Hodgkiss, I.J. and W.S Yim. 1995. A case study of Tolo Harbour, Hong Kong. In A.J. McComb,(Ed.), *Eutrophic Shallow Estuaries and Lagoons.* CRC Press, Boca Raton, FL. pp. 41–58.

Holland, A.F., N.K. Mountford, and J.A. Mihursky. 1977. Temporal variation in upper bay mesohaline benthic communities. 1. The 9-m mud habitat. *Ches. Sci.* 18: 370–378.

Holland, A.F., N.K. Mountford, M.H. Hiefel, K.R. Kaunmeyer, and J.A. Mihursky. 1980. Influence of predation on infaunal abundance in upper Chesapeake Bay, USA. *Mar. Biol.* 57: 221–235.

Hooks, T.A., K.L. Heck, and R.J. Livingston. 1975. An inshore marine invertebrate community: structure and habitat associations in the N.E. Gulf of Mexico. *Bull. Mar. Sci.* 26: 99–109.

Houlihan, J. and R. Wiles. 2001. Brain Food: What Women Should Know about Mercury Contamination of Fish. Environmental Working Group, Washington, D.C.

Howarth, R.W. 1988. Nutrient limitation of net primary production in marine ecosystems. *Annu. Rev. Ecol. Syst.* 19: 89–110.

Howarth, R.W. and R. Marino. 1998. A mechanistic approach to understanding why so many estuaries and brackish waters are nitrogen limited. In T. Hellstrom (Ed.), *Effects of Nitrogen in the Aquatic Environment.* KVA Report 1, Royal Swedish Academy of Sciences, Stockholm.

Howarth, R.W., H.S. Jensen, R. Marino, and H. Postma. 1995. Transport to and processing of phosphorus in near-shore and oceanic waters. In H. Tiessen (Ed.), *Phosphorus in the Global Environment.* Wiley, New York.

Howarth, R.W., D.M. Anderson, T.M. Church, H. Greening, C.S. Hopkinson, W.C. Huber, N. Marcus, R.J. Nainman, K. Segerson, A.N. Sharpley, and W.J. Wiseman. 2000. Clean Coastal Waters: Understanding and Reducing the Effects of Nutrient Pollution. Ocean Studies Board and Water Science and Technology Board, National Academy Press, Washington, D.C. 391 pp.

Huang, W. and W.K. Jones. 1997. Three-Dimensional Modeling of Circulation and Salinity for the Low River Flow Season in Apalachicola Bay, Florida. Water Resources Special Report 97-10. Northwest Florida Water Management District, Havana, FL.

Huber, W.C. and R.E. Dickinson. 1988. Storm Water Management Model, Version 4, User's Manual. Environmental Research Laboratory, Office of Research and Development. U.S. Environmental Protection Agency, Athens, GA.

Hudson, R.J.M. and C.W. Shade. 2003. Letter to the editor, More on mercury content in fish. *Science* 303: 763.

Hudson, T.J. and D.R. Wiggins. 1996. Comprehensive Shellfish Harvesting Area Survey of Pensacola Bay System Escambia and Santa Rosa Counties, Florida. Florida Department of Environmental Protection, Tallahassee.

Hulburt, E.M. and N. Corwin. 1970. Relation of the phytoplankton to turbulence and nutrient renewal in Casco Bay, Maine. *J. Fish. Res. Bd. Can.* 27: 2081–2090.

Hulburt, E.M. and J. Rodman. 1963. Distribution of phytoplankton species with respect to salinity between the coast of southern New England and Bermuda. *Limnol. Oceanogr.* 8: 263–269.

Huntley, S.L., N.L. Bonnevie, R.J. Wenning, and H. Bedbury. 1993. Distribution of polycyclic aromatic hydrocarbons (PAHS) in three northern New Jersey waterways. *Bull. Environ. Contam. Toxicol.* 51: 865–872.

Huntley, S.L., R.J. Wenning, D.J. Paustenbach, A.S. Wong, and W.J. Luksemburg. 1994. Potential sources of polychlorinated dibenzothiophenes in the Passaic River, New Jersey. *Chemosphere* 29: 257–272.

Huntley, S.L., N.L. Bonnevie, and R.J. Wenning. 1995. Polycyclic aromatic hydrocarbon and petroleum hydrocarbon contamination in sediment from the Newark Bay Estuary, New Jersey. *Arch. Environ. Contam. Toxicol.* 28: 93–107.

Hutchinson, G. E. 1951. *A Treatise on Linmology. Vol. 1: Geography, Physics, and Chemistry.* John Wiley & Sons, New York. pp. 102–104.

Hutchinson, G.E. 1973. Eutrophication. *Am. Sci.* 61: 269–279.

Imai, I., M. Yamaguchi, and M. Watanabe. 1997. Ecophysiology, life cycle, and bloom dynamics of *Chattonella* in the Seto inland sea, Japan. In D.M. Anderson, A.D. Cembella, and G.M. Hallagraeff (Eds.), *Physiological Ecology of Harmful Algal Blooms.* Springer-Verlag, Berlin. pp. 93–112.

Ingle, R.M. 1951. Spawning and setting of oysters in relation to seasonal environmental changes. *Bull. Mar. Sci. Gulf Caribbean* 1: 111–135.

Ingle, R.M. and C.E. Dawson. 1952. Growth of the American oyster *Crassostrea virginica* (Gmelin) in Florida waters. *Bull. Mar. Sci. Gulf Caribbean* 2: 393–404.

Ingle, R.M. and C.E. Dawson. 1953. A Survey of Apalachicola Bay. Technical Series No. 10. State of Florida Board of Conservation. 38 pp.

Isphording, W.C. and R.J. Livingston. 1989. Report on synoptic analyses of water and sediment quality in the Perdido Drainage System. Florida Department of Environmental Regulation., Tallahassee, FL. Unpublished report.

Iverson, R.L. and H.F. Bittaker. 1986. Seagrass distribution and abundance in eastern Gulf of Mexico coastal waters. *Est. Coastal Shelf Sci.* 22: 577–602.

Iverson, R.L., W. Landing, B. Mortazawi, and J. Fulmer. 1997. Nutrient transport and primary productivity in the Apalachicola River and Bay. In F.G. Lewis (Ed.), *Apalachicola River and Bay Freshwater Needs Assessment.* Report to the ACF/ACT Comprehensive Study. Northwest Florida Water Management District, Havana, FL. Unpublished report.

Jackson, J.B.C., M.X. Kirby, W.H. Wolfgang, H. Berger, K.A. Bjorndal, L.W. Botsford, B.J. Bourque, R.H. Bradbury, R. Cooke, J. Erlandson, J. Estes, T.P. Hughes, S. Kidwell, C.B. Lange, H.S.Lenihan, J.M. Pandolfi, C.H. Peterson, R.S. Steneck, M.J. Tegner, and R.R. Warner. 2001. Historical overfishing and the recent collapse of coastal ecosystems. *Science* 293: 629–638.

Jaworski, N.A. 1981. Sources of nutrients and the scale of eutrophication problems in estuaries. In B.J. Neilson and L.E. Cronin (Eds.), *Estuaries and Nutrients.* Humana Press, Clifton, NJ. pp. 81–100.

Jaworski, N.A., D.W. Lear, Jr., and O. Villa. 1972. Nutrient management in the Potomac estuary. In G.E. Likens (Ed.), *Nutrients and Eutrophication. Am. Soc. Limnol. Oceanogr. Spec. Symp.,* Vol. 1. Lawrence, KS. pp. 246–272.

Jenkins, E.W. 2003. Environmental education and the public understanding of sciences. *Front. Ecol. Environ.* 1: 437–443.

Johnson, J.S. 1987. A Multivariante Analysis of Storm Water Runoff in the Tallahassee, Florida Area. M.S. Thesis. Department of Geography, Florida State University, Tallahassee, Florida.

Johnson, P.S. and J. NcN. Sieburth. 1979. Chroococcoid cyanobacteria in the sea: a ubiquitous and diverse photographic biomass. *Limnol. Oceanogr.* 24: 928–935.

Jones, K.C., J.A. Stratford, K.S. Waterhouse, and N.B. Vogt. 1989. Organic contaminants in Welsh soils: polynuclear aromatic hydrocarbons. *Environ. Sci. Technol.* 23: 540–550.

Jordan, T.E., D.E. Weller, and D.L. Correll. 2003. Sources of nutrient inputs to the Patuxent River Estuary. *Estuaries* 26: 226–243.

Keister, J.E., E.D. Houde, and D.L. Breitburg. 2000. Effects of bottom-layer hypoxia on abundance and depth distributions of organisms in Patuxent River, Chesapeake Bay. *Mar. Ecol. Pro. Ser.* 205: 43–59.

Keller, B.D. and A. Itkin. 2002. Shoreline nutrients and chlorophyll *a* in the Florida Keys, 1994–1997: a preliminary analysis. In J.W. Porter and K.G. Porter (Eds.), *The Everglades, Florida Bay and Coral Reefs of the Florida Keys: An Ecosystem Sourcebook.* CRC Press, Boca Raton, FL. pp. 649–658.

Kemp, W.M. and W.R. Boynton. 1992. Benthic-pelagic interactions: nutrient and oxygen dynamics. In D.E. Smith, M. Leffler, and G. Mackiernan (Eds.), *Oxygen Dynamics in the Chesapeake Bay. A Synthesis of Recent Research.* Maryland Sea Grant College, College Park, MD.

Kennish, M.J. 1997. *Estuarine and Marine Pollution.* CRC Press, Boca Raton, FL. 524 pp.

Kennish, M., R.J. Livingston, D. Raffaelli, and K. Reise. 2003. The future of estuaries. Review paper for the *International Conference on Environmental Future* (Zurich, Switzerland).

King, R.J. and B.R. Hodgson. 1995. Tuggerah Lakes system, New South Wales, Australia. 1995. In A.J. McComb (Ed.), *Eutrophic Shallow Estuaries and Lagoons.* CRC Press, Boca Raton, FL. pp. 19–30.

Kirk, D.E. 1999. Analysis of Data Concerning Size Distribution of Largemouth Bass in Lake Jackson, Florida. Unpublished report.

Kirk, J.T.O. 1981. Estimation of the scattering coefficient of natural waters using underwater irradiance measurements. *Aust. J. Mar. Freshwater Res.* 32: 533–539.

Kirk, R.E. 1995. *Experimental Design, Procedure for the Behavioral Sciences.* Brooks/Cole Publishing Company, New York.

Koenig, C.C., R.J. Livingston, and C.R. Cripe. 1976. Blue crab mortality: interaction of temperature and DDT residues. *Arch. Environ. Contam. Toxicol.* 4: 119–128.

Komarek, J. 1991. A review of water-bloom forming *Microcystis* species, with regard to populations from Japan. *Algological Studies* 64: 115–127.

Kórner, S., S.K. Das, S. Veenstra, and J.E. Vermaat. 2001. The effect of pH variation at the ammonium-ammonia equilibrium in wastewater and its toxicity to *Lemna gibba. Aquat. Bot.* 71, 71–78.

Kushlan, J.A. 1991. The Everglades. In R.J. Livingston (Ed.), *The Rivers of Florida.* Springer-Verlag, New York. pp. 121–142.

Lapointe, B.E. 1997. Nutrient thresholds for bottom-up control of macroalgal blooms on coral reefs in Jamaica and southeast Florida. *Limnol. Oceanogr.* 42: 1583–1586.

Lapointe, B.E. 1999. Simultaneous top-down and bottom-up forces control macroalgal blooms on coral reefs. (Reply to the comment by Hughes et al.). *Limnol. Oceanogr.* 44: 1586–1592.

Lapointe, B.E. and P.J. Barile. 1999. Seagrass die-off in Florida Bay: an alternative interpretation. *Estuaries* 22: 460–470.

Lapointe, B.E., W.R. Matzie, and P.J. Barile. 2001. Biotic phase-shifts in Florida Bay and bank reef communities of the Florida Keys: linkages with historical freshwater flows and nitrogen loading from the Everglades runoff. In J.W. Porter and K.G. Porter (Eds.), *The Florida Everglades, Florida Bay, and Coral Reefs of the Florida Keys. An Ecosystem Sourcebook.* CRC Press, Boca Raton, FL.

Lapointe, B.E. and W.R. Matzie. 2002. Biotic phase-shifts in Florida Bay and fore reef communities of the Florida Keys: linkages with historical freshwater flows and nitrogen loading from Everglades runoff. In J.W. Porter and K.G. Porter (Eds.), *The Everglades, Florida Bay and Coral Reefs of the Florida Keys: An Ecosystem Sourcebook.* CRC Press, Boca Raton, FL. pp. 629–648.

Lapointe, B.E. and P.J. Barile. 2004. Comment on J.C. Zieman, J.W. Fourqurean, and T.A. Frankovich. 1999. Seagrass die-off in Florida Bay: long-term trends in abundance and growth of turtlegrass, *Thalassia testudinum. Estuaries* 22: 460–470. *Estuaries* 27: 157–164.

Lassus, P. and J.P. Berthome. 1988. Status of 1987 Algal Blooms in IFREMER. ICES/annex III C.: 5–13.

Laughlin, R.A. 1979. Trophic ecology and population distribution of the blue crab, *Callinectes sapidus* Rathbun in the Apalachicola estuary (North Florida, U.S.A.). Doctoral dissertation, Florida State University, Tallahassee, FL.

Laughlin, R.A. and R.J. Livingston. 1982. Environmental and trophic determinants of the spatial/temporal distribution of the brief squid (*Lolliguncula brevis*) in the Apalachicola Estuary (North Florida, USA). *Bull. Mar. Sci.* 32: 489–497.

Lawler, Matusky & Skelly Engineers. 1982. Engineering Feasibility Study of Rehabilitating the South River and South Fork Shenandoah River. Unpublished report for Du Pont de Nemours and Co.

Lawler, Matusky & Skelly Engineers. 1989. (Update of) Engineering Feasibility Study of Rehabilitating the South River and South Fork Shenandoah River. Unpublished report for Du Pont de Nemours and Co.

Leach, J.H., M.G. Johnson, J.R.M. Kelso, J. Hartmann, W. Numann, and B. Entz. 1977. Responses of percid fishes and their habitats to eutrophication. *J. Fish. Res. Board Can.* 34: 1964–1971.

Leber, K.M. 1983. Feeding Ecology of Decapod Crustaceans and the Influence of Vegetation on Foraging Success in a Subtropical Seagrass Meadow. Ph.D. dissertation, Florida State University, Tallahassee, FL.

Leber, K.M. 1985. The influence of predatory decapods, refuge, and microhabitat selection on seagrass communities. *Ecology* 66: 1951–1964.

Lee, T.N., E. Williams, E. Johns, D. Wilson, and N.P. Smith. 2002. Transport processes linking South Florida coastal ecosystems. In J.W. Porter and K.G. Porter (Eds.), *The Everglades, Florida Bay and Coral Reefs of the Florida Keys: An Ecosystem Sourcebook.* CRC Press, Boca Raton, FL. pp. 309–342.

Legendre, L. and S. Demers. 1985. Auxiliary energy, ergoclines, and aquatic biological production. *Nat. Can.* 112: 5–14.

Leitman, H.M., J.E. Sohm, and M.A. Franklin. 1982. Wetland Hydrology and Tree Distribution of the Apalachicola River Floodplain, Florida. U.S. Geological Survey Report 82. U.S. Government Printing Office, Washington, D.C. 92 pp.

Leitman, H.M., J.E. Sohm, and M.A. Franklin. 1983. Wetland Hydrology and Tree Distribution of the Apalachicola River Flood Plain, Florida. U.S. Geological Survey Water-Supply Paper 2196-A. 52 pp.

Leitman, S.F. 2003a. An Evaluation of Evapo-Precipitation Loses from Impoundments in the ACF Basin. Unpublished report.

Leitman, S.F. 2003b. Review of Cumulative Monthly Deficits during Three Major Drought Events. Unpublished report.

Leitman, S.F. 2003c. Overview of Consumptive Demands in the Apalachicola-Chattahoochee-Flint Drainage Basin. Prepared for The Nature Conservancy, Florida Office.

Leitman, S.F. 2003d. A Review of Reservoir Management in the ACF Basin. Prepared for The Nature Conservancy, Florida Office.

Leitman, S.F., L. Ager, and C. Mesing. 1991. The Apalachicola experience: environmental effects of physical modifications to a river for navigational purposes. In R.J. Livingston (Ed.), *The Rivers of Florida.* Springer-Verlag, New York. pp. 223–246.

Lenihan, H.S. and C.H. Peterson. 1998. How habitat degradation through fishery disturbance enhances impacts of hypoxia on oyster reefs. *Ecol. Appl.* 8: 128–140.

Levene, H. 1960. Robust tests for equality of variances. In I. Olkin (Ed.), *Contributions to Probability and Statistics.* Stanford University Press, Palo Alto, CA. pp. 278–292.

Lewis, F.G. III, and R.J. Livingston. 1977. Avoidance of bleached kraft pulp mill effluent by pinfish (*Lagodon rhomboides*) and gulf killifish (*Fundulus grandis*). *J. Fish. Res. Bd. Can.* 34: 568–570.

Lewis. F.G. III and A.W. Stoner. 1983. Distribution of macro-fauna within seagrass beds: an explanation for patterns of abundance. *Bull. Mar. Sci.* 33: 296–304.

Light, H.M., M.R. Darst, and J.W. Grubbs. 1998. Aquatic habitats in relation to river flow in the Apalachicola River floodplain, Florida. U.S. Geological Survey Professional Paper 1594, pp. 1–59.

Livingston, R.J. 1973. Analysis of Mulat-Mulatto Bayou. Unpublished report. Florida Department of Transportation.

Livingston, R.J. 1975a. Impact of kraft pulp-mill effluents on estuarine and coastal fishes in Apalachee Bay, Florida, USA. *Mar. Biol.* 32: 19–48.

Livingston, R.J. 1975b. Resource management and estuarine function with application to the Apalachicola drainage system. *Estuarine Pollution Control and Assessment* 1: 3–17.

Livingston, R.J. 1975c. Long-term fluctuations of epibenthic fish and invertebrate populations in Apalachicola Bay, Florida. *U.S. Fish Bull.* 74: 311–321.

Livingston, R.J. 1976a. Diurnal and seasonal fluctuations of organisms in a north Florida estuary. *Est. Coastal Mar. Sci.* 4: 373–400.

Livingston, R.J. 1976b. Environmental considerations and the management of barrier islands: St. George Island and the Apalachicola Bay system. In *Barrier Islands and Beaches. Technical Proceedings of the 1976 Barrier Islands Workshop.* Annapolis, MD. pp. 86–102.

Livingston, R.J. 1977. The Apalachicola dilemma: wetlands development and management initiatives. In *National Wetlands Protection Symposium* Environmental Law Institute and the Fish and Wildlife Service. Washington, D.C. pp. 163–177.

Livingston, R.J. 1979. Multiple factor interactions and stress in coastal systems: a review of experimental approaches and field implications. In F.J. Vernberg (Ed.), *Marine Pollution: Functional Responses.* Academic Press, New York. pp. 389–413.

Livingston, R.J. 1980a. Ontogenetic trophic relationships and stress in a coastal sea grass system in Florida. In V.S. Peterson (Ed.), *Estuarine Perspectives.* Academic Press, New York. pp. 423–435.

Livingston, R.J. 1980b. The Apalachicola experiment: research and management. *Oceanus* 23: 14–28.

Livingston, R.J. 1981. Man's impact on the distribution and abundance of Sciaenid fishes. *Sixth Annual Marine Recreational Fisheries Symposium; Sciaenids: Territorial Demersal Resources.* National Marine Fisheries Service. Houston, TX. pp. 189–196.

Livingston, R.J. 1982a. Trophic organization in a coastal seagrass system. *Mar. Ecol. Prog. Ser.* 7: 1–12.

Livingston, R.J. 1982b. Between the idea and the reality: an essay on the problems involved in applying scientific data to research management problems. In A. Donovan and A.L. Berge (Eds.), *Working Papers in Science and Technology Studies.* Virginia Polytechnic Institution, Blacksburg, VA. pp. 31–59.

Livingston, R.J. 1983a. Resource atlas of the Apalachicola estuary. Florida Sea Grant College Publication. Gainesville, FL.

Livingston, R.J. 1983b. Identification and Analysis of Sources of Pollution in the Apalachicola River and Bay System. Final Report, Florida Department of Natural Resources, Tallahassee, FL.

Livingston, R.J. 1984a. Trophic response of fishes to habitat variability in coastal seagrass systems. *Ecology* 65: 1258–1275.

Livingston, R.J. 1984b. The Ecology of the Apalachicola Bay system: An Estuarine Profile. U.S. Fish and Wildlife Service FWS/PBS 82/05, Gainesville, FL. 148 pp.

Livingston, R.J. 1984c. River-derived input of detritus into the Apalachicola estuary. In R.D. Cross and D.L. Williams (Eds.), *Proceedings of the National Symposium on Freshwater Inflow to Estuaries.* Fish and Wildlife Service, Washington, D.C. pp. 320–332.

Livingston, R.J. 1984d. Long-Term Effects of Dredging and Open-Water Disposal on the Apalachicola Bay System. Final Report, National Oceanic and Atmospheric Administration, Washington, D.C.

Livingston, R.J. 1985a. The relationship of physical factors and biological response in coastal seagrass meadows. Proceedings of a Seagrass Symposium, Estuarine Research Foundation. *Estuaries* 7: 377–390.

Livingston, R.J. 1985b. Aquatic field monitoring and meaningful measures of stress. In H.H. White (Ed.), *Concepts in Marine Pollution Measurements.* National Oceanic and Atmospheric Administration, Washington, D.C. pp. 681–692.

Livingston, R.J. 1985c. Application of scientific research to resource management: case history, the Apalachicola Bay system. In N.L. Chao and W. Kirby-Smith (Eds.), *Proceedings of the International Symposium on Utilization of Coastal Ecosystems: Planning, Pollution, and Productivity.* Fundacao Universidad di Rio Grande. Rio Grande, Brazil, pp. 103–125.

Livingston, R.J. 1986a. The Choctawhatchee River-Bay System. Final Report. Northwest Florida Water Management District, Havana, FL. Unpublished report.

Livingston, R.J. 1986b. Analysis of Field Data Concerning Old Pass Lagoon (Choctawhatchee Bay, Florida: September, 1985– February, 1986). Final Report. Northwest Florida Water Management District, Havana, FL. Unpublished Report.

Livingston, R.J. 1987a. Historic trends of human impacts on seagrass meadows in Florida (invited paper). *Symposium Proceedings: Subtropical-Tropical Seagrasses of the Southeastern U.S. Florida Mar. Res. Publ.* 42: 139–151.

Livingston, R.J. 1987b. Field sampling in estuaries: the relationship of scale to variability. *Estuaries* 10: 194–207.

Livingston, R.J., 1987c. Distribution of Toxic Agents and Biological Response of Infaunal Macroinvertebrates in the Choctawhatchee Bay System. Final Report, Office of Coastal Management, Florida Department of Environmental Regulation, Tallahassee, FL. Unpublished report.

Livingston, R.J., 1988a. The Ecology of Lake Jackson: An Interim Report. Department of Biological Sciences and Center for Aquatic Research and Resource Management, Florida State University, Special Publication Series No. 88-01, Tallahassee, FL.

Livingston, R.J. 1988b. Inadequacy of species-level designations for ecological studies of coastal migratory fishes. *Env. Biol. Fishes* 22: 225–234.

Livingston, R.J. 1988c. Projected Changes in Estuarine Conditions Based on Models of Long-Term Atmospheric Alteration. Report CR-814608-01-0, U.S. Environmental Protection Agency, Washington, D.C.

Livingston, R.J. 1989a. Lake McBride Report. Unpublished report.

Livingston, R.J. 1989b. The Ecology of the Choctawhatchee River System. Final Report. Northwest Florida Water Management District, Havana, FL. Unpublished report.

Livingston, R.J. 1990. Inshore marine habitats. In R.L Myers and J.J. Ewel (Eds.), *Ecosystems of Florida*. University of Central Florida Press, Orlando, FL. pp. 549–573.

Livingston, R.J. 1991a. *The Rivers of Florida*. Springer-Verlag, New York. 229 pp.

Livingston, R.J. 1991b. Medium sized rivers: gulf coastal plain. In C.T. Hackney (Ed.), *Biodiversity of the Southeastern U.S.* Ecological Society of America. pp. 351–385.

Livingston, R.J. 1991c. Historical relationships between research and resource management in the Apalachicola River-estuary. *Ecol. Applic.* 1: 361–382.

Livingston, R.J. 1992a. The Ecology of Lake McBride. Unpublished report. Leon County Commission, Tallahassee, FL

Livingston, R.J. 1992b. Ecological Study of the Perdido Drainage System. Florida Department of Environmental Regulation. Tallahassee, FL. Unpublished report.

Livingston, R.J. 1993a. The Ecology of the Lakes of Leon County. Unpublished report. Leon County Commission, Tallahassee, FL.

Livingston, R.J. 1993b. River-Gulf Study. Florida Department of Environmental Regulation. Tallahassee, FL. Unpublished report.

Livingston, R.J. 1993c. Estuarine Wetlands. In J. Berry and M. Dennison (Eds.), *Wetlands*. Noyes Publications. pp. 128–153.

Livingston, R.J. 1994. Ecological Study of Jack's Branch and Soldier Creek. Florida Department of Environmental Regulation. Tallahassee, FL. Unpublished report.

Livingston, R.J. 1995a. Volumes I and II. The Ecology of the Lakes of Leon County. Leon County Commission, Tallahassee, FL. Unpublished report.

Livingston, R.J. 1995b. Effects of sewage (untreated and treated) on Lakes Jackson and Lafayette. Leon County Commission, Tallahassee, FL. Unpublished report.

Livingston, R.J. 1995c. Review of Leon Lakes Project and Proposal. Leon County Commission, Tallahassee, FL. Unpublished report.

Livingston, R.J. 1995d. Nutrient Analysis of Upper Perdido Bay. Florida Department of Environmental Regulation. Tallahassee, FL. Unpublished report.

Livingston, R.J. 1996a. The Ecology of the Lakes of Leon County. Lake Jackson and Lake Lafayette: Bluegreen Algae Blooms. Leon County Commission, Tallahassee, FL. Unpublished report.

Livingston, R.J. 1996b. Ecological Study of the Amelia and Nassau River-Estuaries. Florida Department of Environmental Protection, Tallahassee, FL.

Livingston, R.J. 1997a. Effects of Urban Development on the Lake McBride Drainage System (Millstone Creek-NoName Pond-Upper Lafayette Basin). Leon County Commission. Unpublished report.

Livingston, R.J., 1997b. Effects of urban development on the lakes of Leon County (1988–1997). Leon County Commission, Tallahassee, FL. Unpublished report.

Livingston, R.J. 1997c. Eutrophication in estuaries and coastal systems: relationships of physical alterations, salinity stratification, and hypoxia. In F.J. Vernberg, W.B. Vernberg, and T. Siewicki (Eds.), *Sustainable Development in the Southeastern Coastal Zone.* University of South Carolina Press, Columbus, SC. pp. 285–318.

Livingston, R.J. 1997d. Update on the Ecological Status of Perdido Bay. Florida Department of Environmental Regulation. Tallahassee, FL. Unpublished report.

Livingston, R.J. 1997e. Trophic response of estuarine fishes to long-term changes of river runoff. *Bull Mar. Sci.* 60: 984–1004.

Livingston, R.J. 1997f. Final Analyses of Perdido Data Base: The Soldier Creek System. Florida. Florida Department of Environmental Regulation. Tallahassee, FL. Unpublished report.

Livingston, R.J., 1998a. Effects of urban development on the lakes of Leon County (1988-1997). Addendum. Leon County Commission. Tallahassee, Florida. Unpublished report.

Livingston, R.J. 1998b. Perdido Bay Analysis: 10/88– 9/98. Florida Department of Environmental Regulation. Tallahassee, FL. Unpublished report.

Livingston, R.J. 1999a. Effects of urban development on the lakes of Leon County (1988–1999). Leon County Commission, Tallahassee, FL. Unpublished report.

Livingston, R.J., 1999b. Holding Pond Ecology: Application to Protection of Lake McBride System. Unpublished report.

Livingston, R.J. 1999c. Pensacola Bay System Environmental Study: Ecology and Trophic Organization. Florida Department of Environmental Regulation. Tallahassee, FL. Unpublished report.

Livingston, R.J., 2000. Eutrophication Processes in Coastal Systems: Origin and Succession of Plankton Blooms and Effects on Secondary Production. CRC Press, Boca Raton, FL. 327 pp.

Livingston, R.J. 2001a. Ecological Study of the Amelia and Nassau River-Estuaries: Florida Department of Environmental Regulation. Tallahassee, FL. Unpublished report.

Livingston, R.J., 2001b. Mercury Distribution in Sediments and Mussels in the Penobscot River–Estuary. Unpublished report for the Natural Resources Defense Council, New York, NY.

Livingston, R.J., 2002. *Trophic Organization in Coastal Systems.* CRC Press, Boca Raton, FL.

Livingston, R.J. 2003. Effects of Sewage Disposal (Bayou Marcus Creek Facility), Agricultural Runoff and Urban Development on the Perdido Bay System. Florida Department of Environmental Regulation. Tallahassee, Florida. Unpublished report.

Livingston, R.J. 2004a. Final Report: Effects of Pulp Mill Effluents on the Fenholloway River–Bay System. Florida Department of Environmental Regulation. Tallahassee, FL. Unpublished report.

Livingston, R.J. 2004b. Current Status of Perdido Bay. Florida Department of Environmental Regulation. Tallahassee, Florida. Unpublished report.

Livingston, R.J., R.L. Iverson, R.H. Estabrook, V.E. Keys, and J. Taylor, Jr. 1974. Major features of the Apalachicola Bay system: physiography, biota, and resource management. *Flor. Sci.* 4: 245–271.

Livingston, R.J., R.S. Lloyd, and M.S. Zimmerman. 1976a. Determination of adequate sample size for collections of benthic macrophytes in polluted and unpolluted coastal areas. *Bull. Mar. Sci.* 26: 569–575.

Livingston, R.J., G.J. Kobylinski, F.G. Lewis III, and P.F. Sheridan. 1976b. Long-term fluctuations of epibenthic fish and invertebrate populations in Apalachicola Bay, Florida. *Fish. Bull.* 74: 311–321.

Livingston, R.J. and E.A. Joyce, Jr. 1977. *Proceedings of the Conference on the Apalachicola Drainage System,* April 23–24, 1976, Gainesville, Florida. Florida Department of Natural Resources. Florida Marine Research Publications 26. St. Petersburg, FL.

Livingston, R.J., P.S. Sheridan, B.G. McLane, F.G. Lewis III, and G.G. Kobylinski. 1977. The biota of the Apalachicola bay system: functional relationships. In R.J. Livingston (Ed.), *Proceedings of the Conference on the Apalachicola Drainage System.* Florida Department of Natural Resources, Marine Resources Publication 26. St. Petersburg, FL. pp. 75–100.

Livingston, R.J. and O.L. Loucks. 1978. Productivity, trophic interactions, and food-web relationships in wetlands and associated systems. In P.E. Greeson, J.R. Clark, and J.E. Clark (Eds.), *Wetland Functions and Values: the State of Our Understanding*. American Water Resources Association. Lake Buena Vista, FL. pp. 101–119.

Livingston, R.J., N.P. Thompson, and D.A. Meeter. 1978. Long-term variation of organochlorine residues and assemblages of epibenthic organisms in a shallow north Florida (USA) estuary. *Mar. Biol.* 46: 355–372.

Livingston, R.J. and J.L. Duncan. 1979. Climatological control of a north Florida coastal system and impact due to upland forestry management. In R.J. Livingston (Ed.), *Ecological Processes in Coastal and Marine Systems*. Plenum Press, New York. pp. 339–382.

Livingston, R.J. and G.L. Ray. 1989. A Simplified and Rapid Method for Assessing the Biological Disturbances Resulting from Stormwater and Marina Discharges in Estuaries. Final Report, The Florida Institute of Government and The Florida Department of Environmental Regulation, Tallahassee, FL. Unpublished report.

Livingston, R.J. and H. Swanson. 1993. Project Overview: The Ecology of the Lakes of Leon County. Unpublished report to the Leon County Board of County Commissioners. Florida State University and The Department of Growth and Environmental Management of Leon County, FL.

Livingston, R.J. and S.E. McGlynn. 1994. Polynucleated Aromatic Hydrocarbons in Leon County Lakes. Unpublished report.

Livingston, R.J., S.E. McGlynn, X. Niu, and G.C. Woodsum. 1996. Effects of Fluridone on Lake Jackson. Unpublished report.

Livingston, R.J., X. Niu, F.G. Lewis, and G.C. Woodsum. 1997. Freshwater input to a Gulf estuary: Long-term control of trophic organization. *Ecol. Appl.* 7: 277–299.

Livingston, R.J., S.E. McGlynn, and X. Niu. 1998a. Factors controlling seagrass growth in a gulf coastal system: water and sediment quality and light. *Aquat. Bot.* 60: 135–159.

Livingston, R.J., A.W. Niedoroda, T.W. Gallagher, and A. Thurman. 1998b. Environmental Studies of Perdido Bay. Florida Department of Environmental Regulation. Tallahassee, FL. Unpublished report.

Livingston, R.J., R.L. Howell, X. Niu, F.G. Lewis, and G.C. Woodsum, 1999. Recovery of oyster reefs (*Crassostrea virginica*) in a gulf estuary following disturbance by two hurricanes. *Bull. Mar. Sci. Gulf Caribbean* 64: 75–94.

Livingston, R.J., F.G. Lewis III, G.C. Woodsum, X. Niu, R.L. Howell IV, G.L. Ray, J.D. Christensen, M.E. Monaco, T.A. Battista, C.J Klein, B. Galperin, and W. Huang. 2000. Coupling of physical and biological models: response of oyster population dynamics to fresh water input in a shallow Gulf estuary. *Est. Coast. Shelf Sci.* 50: 655–672.

Livingston, R.J. and G.C. Woodsum. 2001. Analysis of Data Concerning the Florida Everglades Ecosystem. Unpublished report, Natural Resources Defense Council.

Livingston, R.J., A.K. Prasad, X. Niu, and S.E. McGlynn. 2002. Effects of ammonia in pulp mill effluents on estuarine phytoplankton assemblages: field descriptive and experimental results. *Aquat. Bot.* 74: 343–367.

Livingston, R.J., S. Leitman, G.C. Woodsum, B. Galperin, P. Homann, D. Christensen, and M.E. Monaco. 2003. Relationships of River Flow and Productivity of the Apalachicola River-Bay System. National Oceanic and Atmospheric Administration. Unpublished report.

Ljung, R. and J. Box. 1978. On a measure of lack of fit in time series models. *Biometricka* 65: 297–303.

Locarnini, S.J.P. and B.J. Presley. 1996. Mercury concentrations in benthic organisms from a contaminated estuary. *Mar. Env. Res.* 41: 225–239.

Lockwood, J.L., M.C. Ross, and J.P. Sah. 2003. Smoke on the water: the interplay of fire and water flow on Everglades restoration. *Ecol. Environ.* 1: 462–468.

Loesch, H. 1960. Sporadic mass shoreward migrations of demersal fish and crustaceans in Mobile Bay, Alabama. *Ecology* 41: 292–298.

Lohman, R., E. Nelson, S.J. Eisenreich, and K.C. Jones. 2000. Evidence for dynamic air-water exchange of PCDD/Fs: a study in the Raritan Bay/Hudson River Estuary. *Environ. Sci. Technol.* 34: (15) 3086–3093.

Long, E.R., G.M. Sloane, R.S. Carr, K.J. Scott, G.B. Thursby, E. Crecelius, C. Peven, and H.L. Windom. 1997. Magnitude and Extent of Sediment Toxicity in Four Bays of the Florida Panhandle: Pensacola, Choctawhatchee, St. Andrew, and Apalachicola. NOAA Technical Memorandum NOS ORCA 117, Silver Spring, MD.

Lovegrove, T. 1960. An improved form of sedimentation apparatus for use with an inverted microscope. *J. Cons. Chem.* 25: 279–284.

Maestrini, S.Y., D.J. Bonin, and M.R. Droop. 1984a. Phytoplankton as indicators of seawater quality: bioassay approach and protocols. In L.E. Shubert (Ed.), *Algae as Ecological Indicators.* Academic Press, New York. pp. 71–131.

Maestrini, S.Y., M.R. Droop, and D.J. Bonin. 1984b. Test algae as indicators of sea water quality: prospects. In L. E. Shubert (ed.), *Algae as Ecological Indicators.* Academic Press, New York. pp. 132–188.

Mahoney, B.M.S. 1982. Seasonal fluctuations of benthic macrofauna in the Apalachicola estuary, Florida. The role of predation and larval availability. Ph.D. dissertation. Florida State University. Tallahassee, FL.

Mahoney, B.M.S. and R.J. Livingston. 1982. Seasonal fluctuations of benthic macrofauna in the Apalachicola estuary, Florida, USA. *Mar. Biol.* 69: 207–213.

Main, K.L. 1983. Behavioral Response of Acaridean Shrimp to a Predatory Fish. Ph.D. dissertation, Department of Biological Science, Florida State University, Tallahassee, FL.

Maine Department of Environmental Protection. 1998. Mercury in Maine. Land and Water Resources Council, 1997 Annual Report, Appendix A.

Malone, T.C. 1977. Environmental regulation of phytoplankton productivity in the lower Hudson estuary. *Est. Coastal Mar. Sci.* 13: 157–172.

Malone, T.C., L.H. Crocker, S.E. Pike, and B.A. Wendler. 1988. Influences of river flow on the dynamics of phytoplankton in a partially stratified estuary. *Mar. Ecol. Prog. Ser.* 32, 149–160.

Malone, T.C., D.J. Conley, T.R. Fisher, P.M. Gilbert, L.W. Harding, and K.G. Sellner. 1996. Scales of nutrient-limited phytoplankton productivity in Chesapeake Bay. *Estuaries* 19: 371–385.

Marcomini, A., A. Sfriso, B. Pavoni, and A.A. Orio. 1995. Eutrophication of the Lagoon of Venice: Nutrient loads and exchanges. In A.J. McComb (Ed.), *Eutrophic Shallow Estuaries and Lagoons.* CRC Press, Boca Raton, FL. pp. 59–80.

Marshall, H.G. 1982a. Meso-scale distribution patterns for diatoms over the northeastern continental shelf of the United States. *7th Diatom-Symposium.* pp. 393–400.

Marshall, H.G. 1982b. Phytoplankton distribution along the eastern coast of the USA. IV. Shelf waters between Cape Lookout, North Carolina, and Cape Canaveral, Florida. *Proceedings of the Biological Society of Washington.* 95: 99–113.

Marshall, H.G. 1984. Phytoplankton distribution along the eastern coast of the USA. V. Seasonal density and cell volume patterns for the northeastern continental shelf. *J. Plankton Res.* 6: 169–193.

Marshall, H.G. 1988. Distribution and concentration patterns of ubiquitous diatoms for the northeastern continental shelf of the United States. *Proc. 9th Int. Diatom Symp. 1986.* Otto Koeltz, Stuttgart. pp. 75–85.

Marshall, H.G. and J.A. Ranasinghe. 1989. Phytoplankton distribution along the eastern coast of the U.S.A. VII. Mean cell concentrations and standing crop. *Cont. Shelf Res.* 9: 153–164.

Martin, B.J. 1980. Effects of Petroleum Compounds on Estuarine Fishes. U.S. Environmental Protection Agency report. EPA-600/3-80-019.

Mason, D. R. 1977. Lake Jackson Water Quality Monitoring Study (November 1976–December 1977). Department of Science and Engineering, Florida Institute of Technology, Melbourne, FL. Unpublished report.

Mattraw, H.C. and J.F. Elder. 1982. Nutrient and Detritus Transport in the Apalachicola River, Florida. 1984. U.S. Geological Survey Water-Supply Paper 2196-C.

McCain, B.B., S. Chan, M.M. Krahn, D.W. Brown, M.S. Myers, J.T. Landahl, S. Pierce, R.C. Clark, and U. Varanasi. 1992. Chemical contamination and associated fish diseases in San Diego Bay. *Environ. Sci. Technol.* 26, 725–733.

McComb, A.J. 1995. *Eutrophic Shallow Estuaries and Lagoons.* CRC Press, Boca Raton, FL. 240 pp.

McComb, A.J. and R.J. Lukatelich, 1995. The Peel-Harvey estuarine system, Western Australia. In A.J. McComb (Ed.), *Eutrophic Shallow Estuaries and Lagoons.* CRC Press, Boca Raton, FL, pp. 5–18.

McCormick, P.V., S. Newman, S. Miao, D.E. Gawlik, and D. Marley. 2002. Effects of anthropogenic phosphorus inputs on the Everglades. In J.W. Porter and K.G. Porter (Eds.), *The Everglades, Florida Bay and Coral Reefs of the Florida Keys: An Ecosystem Sourcebook.* CRC Press, Boca Raton, FL, 83–126.

McLane, B.G. 1980. An Investigation of the Infauna of East Bay-Apalachicola Bay. M.S. thesis. Florida State University, Tallahassee, FL.

McPherson, B.F. and K.M. Hammett. 1991. Tidal rivers of Florida. In R.J. Livingston (Ed.), *The Rivers of Florida.* Springer-Verlag, New York. pp. 31– 46.

Means, D.B. 1977. Aspects of the significance to terrestrial vertebrates of the Apalachicola River drainage basin, Florida. In R. J. Livingston (Ed.), *Proceedings of the Conference on the Apalachicola Drainage System.* Florida Department of Natural Resources, Marine Resources Publication 26. St. Petersburg, FL. pp. 37–67.

Mearns, A.J., M. Matta, G. Shigenaka, D. MacDonald, M. Buchman, H. Harris, J. Golas, and G. Lauenstein. 1991. Contaminant Trends in the Southern California Bight: Inventory an Assessment. NOAA Tech. Mem. NOS ORCA 62, Seattle, WA.

Meeter, D.A. and R.J. Livingston. 1978. Statistical methods applied to a four-year multivariate study of a Florida estuarine system. In J. Cairns, Jr., K. Dickson, and R.J. Livingston (Eds.), *Biological Data in Water Pollution Assessment: Quantitative and Statistical Analyses.* American Society for Testing and Materials, Special Technical Publication 652: pp. 53–67.

Meeter, D.A., R.J. Livingston, and G.C. Woodsum. 1979. Short and long-term hydrological cycles of the Apalachicola drainage system with application to Gulf coastal populations. In R.J. Livingston (Ed.), *Ecological Processes in Coastal and Marine Systems.* Plenum Press, New York. pp. 315–338.

Mellor, G.L. and T. Yamada. 1982. Development of a turbulence closure model for geophysical fluid problems. *Rev. Geophys. Space Phys.* 20: 851–875.

Menzel, R.W. 1955a. Effects of two parasites on the growth of oysters. *Proc. Natl. Shellfish Assoc.* 45: 184–186.

Menzel, R.W. 1955b. The growth of oysters parasitized by the fungus *Dermocystidium marinum* and by the trematode *Bucephalus cuculus. J. Parasit.* 41: 333–342.

Menzel, R.W. and F.E. Nichy. 1958. Studies of the distribution and feeding habits of some oyster predators in Alligator Harbor, Florida. *Bull. Mar. Sci. Gulf Car.* 8: 125–145.

Menzel, R.W., N.C. Hulings, and R.R. Hathaway. 1957. Causes of depletion of oysters in St. Vincent Bar, Apalachicola Bay, Florida. *Proc. Natl. Shellfish Assoc.* 48: 66– 71.

Menzel, R.W., N.C. Hulings, and R.R. Hathaway. 1966. Oyster abundance in Apalachicola Bay, Florida in relation to biotic associations influenced by salinity and other factors. *Gulf Res. Rep.* 2: 73–96.

Menzie, C.A., B.B. Potocki, and J. Santodonato. 1992. Exposure to carcinogenic PAH's in the environment. *Environ. Sci. Technol.* 26: 1278–1284.

Messing, A.W., A.M. Dombrowski, and V.H. Harlow. 1997. Analysis of Organomercury from Fish Collected in Conjunction with the Shenandoah River Mercury Monitoring : 1992, 1994, and 1996 Collections. Applied Marine Research Laboratory, Old Dominion University. Unpublished report.

Mitsch, W. J. and J.G. Gosselink. *Wetlands.* John Wiley & Sons, New York. 1993.

Monselise, B.E. and D. Kost, 1993. Different ammonium uptake, metabolism, and detoxification efficiencies in two Lemnaceae. *Planta* 189: 167–173.

Morgan, E. (Coordinator). 1998. Mercury in Maine. Land & Water Resources Council, 1997 Annual Report, Appendix A. Maine Department of Environmental Protection.

Morrison, S.J., J.D. King, R.J. Bobbie, R.E. Bechtold, and D.C. White. 1977. Evidence of microfloral succession on allochthonous plant litter in Apalachicola Bay, Florida, USA. *Mar. Biol.* 41: 229–240.

Mortazavi, B., R.L. Iverson, W.M. Landing, and W. Huang. 2000a. Phosphorus budget of Apalachicola Bay: a river-dominated estuary in the northeastern Gulf of Mexico. *Mar. Ecol. Prog. Ser.* 198: 33–42.

Mortazavi, B., R.L. Iverson, W.M. Landing, F.G. Lewis, and W. Huang. 2000b. Control of phytoplankton production and biomass in a river-dominated estuary: Apalachicola Bay, Florida USA. *Mar. Ecol. Prog. Ser.* 198: 19–31.

Mortazavi, B., R.L. Iverson, and W. Huang. 2000c. Dissolved organic nitrogen and nitrate in Apalachicola Bay, Florida: spatial distributions and monthly budgets. *Mar. Ecol. Prog. Ser.* 214: 79–91.

Moshiri, G.A. 1976. Interrelationships between Certain Microorganisms and Some Aspects of Sediment-Water Nutrient Exchange in Two Bayou Estuaries. Phases I and II. University of Florida Water Resources Research Center Res. Ctr. Publ. No. 37. 45 pp.

Moshiri, G.A. 1978. Certain Mechanisms Affecting Water Column-to-Sediment Phosphate Exchange in a Bayou Estuary. *J. Water Pollut. Cont. Fed.* 50: 392–394.

Moshiri, G.A. 1981. Study of Selected Water Quality Parameters in Bayou Texar. Final Report on Contract No. DACW01-80-0252, University of West Florida, Pensacola, FL.

Moshiri, G.A. and W.G. Crumpton. 1978. Some aspects of redox trends in the bottom muds of a mesotrophic bayou estuary. *Hydrobiologia* 57: 155–158.

Moshiri, G.A., W.G. Crumpton, and D.A. Blaylock. 1978. Algal metabolites and fish kills in a bayou estuary: an alternative explanation to the low dissolved oxygen controversy. *J. Water Pollut. Cont. Fed.* 50: 2043–2046.

Moshiri, G.A., W.G. Crumpton, and N.G. Aumen. 1979. Dissolved glucose in a bayou estuary, possible sources and utilization by bacteria. *Hydrobiologia* 62: 71–74.

Moshiri, G.A., N.G. Aumen, and W.G. Swann III. 1980. Water Quality Studies in Santa Rosa Sound, Pensacola, Florida. U.S. Environmental Protection Agency, Gulf Breeze Environmental Research Lab. Pensacola, FL.

Moshiri, G.H., N.G. Aumen, and W.G. Crumpton. 1987. Reversal of the eutrophication process: a case study. In B.G. Neilson and L.E. Cronin (Eds.), *Estuaries and Nutrients*. Humana Press, Clifton, NJ. pp. 370–390.

Mower, B., J. Sowles, and S. Ladner. 1997. Evaluation of Sediments Contaminated with Mercury Near Holtrachem. Unpublished report. Maine Department of Environmental Protection.

Murphy, T.P., D.R.S. Lean, and C. Nalewajko. 1976. Blue-green algae: their excretion of iron-selective chelators enables them to dominate other algae. *Science* 192: 900–902.

Myers, V.B. 1977. Nutrient Limitation of Phytoplankton Productivity in north Florida Coastal Systems: Technical Considerations; Spatial Patterns; and Wind Mixing Effects. Ph.D. dissertation. Florida State University. Tallahassee, FL.

Myers, V.B. and R.J. Iverson. 1977. Aspects of nutrient limitation of phytoplankton productivity in the Apalachicola Bay system. In R.J. Livingston (Ed.), *Proceedings of the Conference on the Apalachicola Drainage System*. Florida Department of Natural Resources, Marine Resources Publication 26. St. Petersburg, FL. pp. 68–74.

Myers, V.B. and R.J. Iverson. 1981. Phosphorus and nitrogen limited phytoplankton productivity in Northeastern Gulf of Mexico coastal estuaries. In B.J. Nielson and L.E. Cronin (Eds.), *Estuaries and Nutrients*. Humana Press, Clifton, NJ, pp. 569–582.

Nakazima, M. 1965. Studies on the source of shellfish poison in Lake Hamana. III. Poisonous effects on shellfishes feeding on *Prorocentrum* sp. *Bull. Jap. Soc. Sci. Fish.* 31: 281–285.

National Oceanic and Atmospheric Administration (NOAA). 1995. Magnitude and Extent of Sediment Toxicity in the Hudson-Raritan Estuary. National Ocean Service, NOAA Technical Memorandum NOS ORCA 88.

National Oceanic and Atmospheric Administration (NOAA). 2003. Watershed Database and Mapping Projects/Newark Bay. National Ocean Service, NOAA.

New Jersey Department of Environmental Protection. 2000. Restoration of the Passaic River Conceptual Proposal. Maritime Resources New Jersey Department of Transportation. Revised March 17, 2000.

New Jersey Department of Environmental Protection. 2002. Estimate of Cancer Risk to Consumers of Crabs Caught in the Area of the Diamond Alkali Site and Other Areas of the Newark Bay Complex from 2,3,7,8-TCDD and 2,3,7,8-TCDD Equivalents. Unpublished report.

Newell, R.I.E. 1988. Ecological changes in Chesapeake Bay: are they the result of over-harvesting the American oyster, *Crassostrea virginica*? Understanding the estuary: Advances in Chesapeake Bay Research. Proceedings of a conference. March 29–31, 1988. Baltimore, MD. Chesapeake Research Consortium Publication 129, CBP/TRS 24/88/. pp. 536–546.

Nixon, S.W. 1988a. Comparative ecology of freshwater and marine systems. *Limnol. Oceanogr.* 33: 1–1025.

Nixon, S.W. 1988b. Physical energy inputs and the comparative ecology of lake and marine ecosystems. *Limnol. Oceanogr.* 33: 1005–1025.

Northwest Florida Water Management District. 1992. Characterization of Karst Development in Leon County, Florida for the Delineation of Wellhead Protection Areas.

Northwest Florida Water Management District. 1994a. Preliminary nonpoint source loading rate analysis of the Pensacola Bay system. Havana: Northwest Florida Water Management District.

Northwest Florida Water Management District. 1994b. District Water Management Plan. Program Development Series 96-2. Havana: Northwest Florida Water Management District.

Odum, W.E., J.S. Fishes, and J.C. Pickral. 1979. Factors controlling the flux of particulate organic carbon from estuarine wetlands. In R.J. Livingston (Ed.), *Ecological Processes in Coastal and Marine Systems.* Plenum Press, New York. pp. 69–82.

Odum, W.E., C.C. McIvor, and T.G. Smith. 1982. The Florida Mangrove Zone: a Community Profile. U.S. Fish and Wildlife Service, Office of Biological Services, Washington, D.C., FWS/OBS-82/24. 144 pp.

Olinger, L.W., R.G. Rogers, P.L. Force, T.L. Todd, B.L. Mullings, F.T. Bisterfeld, and L.A. Wise II. 1975. Environmental and Recovery Studies of Escambia Bay and the Pensacola Bay System, Florida. Washington: U.S. Environmental Protection Agency.

O'Shea, M.L. and T.M. Brosnon. 2000. Trends in indicators of eutrophication in western Long Island Sound and the Hudson-Raritan Estuary. *Estuaries* 23: 877–901.

Pankratz, A. 1991. *Forecasting with Dynamic Regression Models.* John Wiley & Sons, New York.

Parks, J.W., A. Lutz, and J.A. Sutton. 1989. Water column methylmercury in the Wabigoon/English River–Lake system: factors controlling concentrations, speciation, and net production. *Can. J. Fish. Aquat. Sci.* 46: 2184–2202.

Parsons, T.R., Y. Maita, and C.M. Lalli. 1984. *A Manual of Chemical and Biological Methods for Seawater Analysis.* Pergamon Press, New York. 173 pp.

Patrick, R.M., 1967. Diatom communities in estuaries. *Estuaries* 83: 311–315.

Patrick, R.M. 1971. Diatom communities. *Am. Microscop. Soc. Symposium. Monogr.* 3: 151164.

Patrick, R.M. and C.W. Reimer. 1966. The Diatoms of the United States. Vol. 1. Monograph No. 13, Monographs of the Academy of Natural Sciences of Philadelphia, 688 pp.

Pearce, J.W. 1977. Florida's environmentally endangered land acquisition program and the Apalachicola River system. In R.J. Livingston (Ed.), *Proceedings of the Conference on the Apalachicola Drainage System.* Florida Department of Natural Resources, Marine Resources Publication 26. St. Petersburg, FL. pp. 141–145.

Pearson, T.H. 1980. Marine pollution effects of pulp and paper industry wastes. *Helgol. Wiss. Meer.* 33: 340–365.

Pearson, T.H. and R. Rosenberg. 1978. Macrobenthic succession in relation to organic enrichment and pollution of the marine environment. *Oceanogr. Mar. Biol. A. Rev.* 16: 229–311.

Pennak, R.W. 1978. Freshwater Invertebrates of the United States. John Wiley & Sons, New York.

Pennock, J.R. 1987. Temporal and spatial variability in plankton, ammonium, and nitrate uptake in the Delaware River. *Est. Coastal Shelf Sci.* 24: 841–857.

Pennock, J.R. and J.H. Sharp. 1994. Temporal alteration between light- and nutrient limitation of phytoplankton production in a coastal plain estuary. *Mar. Ecol. Prog. Ser.* 111: 275–288.

Pensacola Bay System Technical Symposium. 1997. Pensacola, FL.

Peterson, B.J. and R.W. Howarth. 1987. Sulfur, carbon, and nitrogen isotopes used to trace organic matter flow in the salt-marsh estuaries of Sapelo Island, Georgia. *Limnol. Oceanogr.* 32: 1195–1213.

Peterson, C.H. 1979. Predation, competitive exclusion, and diversity in the soft-sediment benthic communities of estuaries and lagoons. In R.J. Livingston (Ed.), *Ecological Processes in Coastal and Marine Systems.* Plenum Press, New York. pp. 233–264.

Peterson, B.J. and R.W. Howarth. 1987. Sulfur, carbon, and nitrogen isotopes used to trace organic matter flow in the salt-marsh estuaries of Sapelo Island, Georgia. *Limnol. Oceanogr.* 32: 1195–1213.

Philips, E.J., M. Cichra, F.J. Aldridge, and J. Jembeck. 2000. Light availability and variations in phytoplankton standing crops in a nutrient-rich blackwater river. *Limnol. Oceanogr.* 45: 916–929.

Porter, J.W. and K.G. Porter. 2002. *The Everglades, Florida Bay and Coral Reefs of the Florida Keys: an Ecosystem Sourcebook.* CRC Press, Boca Raton, FL. 1000 pp.

Porter, J.W., V. Kosmynin, K.L. Patterson, K.G. Porter, W.C. Jaap, J.L. Wheaton, K. Hackett, M. Lybolt, C.P. Tsokos, G. Yanev, D.M. Marcinek, J. Dotten, D. Eaken, M. Patterson, O.W. Meier, M. Brill, and P. Dustan. 2002. Detection of coral reef change by the Florida Keys Coral Reef monitoring project. 2002. In J.W. Porter and K.G. Porter (Eds.), *The Everglades, Florida Bay and Coral Reefs of the Florida Keys: An Ecosystem Sourcebook.* CRC Press, Boca Raton, FL, pp. 749–769.

Potts, M. 1980. Blue-green algae (Cyanophyta) in marine coastal environments of the Sinai Peninsula; distribution, zonation, stratification and taxonomic diversity. *Phycologia* 19: 60–73.

Prasad, A.K.S.K. and G.A. Fryxell. 1991. Habit, frustule morphology and distribution of the Antarctic marine benthic diatom *Entopyla australis* var. *gigantea* (Greville) Fricke (Entopylaceae). *Br. Phyc. J.* 26: 101–122.

Prasad, A.K.S.K., J.A. Nienow, and R.J. Livingston, 1990. The genus *Cyclotella* (Bacillariophyta) in Choctawhatchee Bay, Florida, with special reference to *C. striata* and *C. choctawhatcheeana* sp. nov. *Phycologia* 29: 418–436.

Premla, V.E. and M.U. Rao. 1977. Distribution and seasonal abundance of *Oscillatoria nigroviridis* Thwaites ex. Gomant in the waters of Visakhapatnam Harbour. *Ind. J. Mar. Sci.* 3: 79–91.

Prescott, G.W. 1962. *Algae of the Western Great Lakes.* Wm. C. Brown Company, Dubuque, IA.

Prescott, G.W. 1980. *How to Know the Aquatic Plants.* W. C. Brower Company, Dubuque, IA.

Purcell, B.H. 1977. The ecology of the epibenthic fauna associated with *Vallisneria americana* beds in a north Florida estuary. M.S. thesis, Florida State University. Tallahassee, FL.

Rabalais, N.N., R.E. Turner, D. Justic, Q. Dortch, W.J. Wiseman, Jr., B.K. Sen Grupta. 1996. Nutrient changes in the Mississippi River and system responses on the adjacent continental shelf. *Estuaries* 19: 386–407.

Rabalais, N.N., R.E. Turner, D. Justic, Q. Dortch, and W.J. Wiseman, Jr. 1999. Characterization of Hypoxia: Topic 1 Report for the Integrated Assessment of Hypoxia in the Gulf of Mexico. NOAA Coastal Ocean Program, Silver Springs, MD. 167 pp.

Rada, R.G., J.E. Findley, and J.G. Wiener. 1986. Environmental fate of mercury discharged into the upper Wisconsin River. *Water, Air Soil Pollut.* 29: 57–76.

Rainville, R.P., B.J. Copeland, and W.T. McKean. 1975, Toxicity of Kraft mill wastes to an estuarine phytoplankton. *J. Wat. Pol. Cont. Fed.* 47: 487–503.

Reardon, J. 1999. Ecology of Phytoplankton Communities in Lake Jackson. M.S. thesis, Department of Biological Science, Florida State University, Tallahassee, FL.

Reddy, P.M. and V. Venkateswarlu. 1986. Ecology of algae in the paper mill effluents and their impact on the River Tungabhadra. *J. Environ. Biol.* 7: 215–223.

Reeve, D.W. and P.F. Earl. 1988. Chlorinated organic compounds in bleached chemical pulp and pulp-mill effluents. 1. The potential for effluent regulation. Canadian Pulp & Paper Association Branch Paper for Pacific Coast-Western Branches Joint Conference.

Regnell, O. and G. Ewald. 1997. Factors controlling temporal variation in methyl mercury levels in sediment and water in a seasonally stratified lake. *Limnol. Oceanogr.* 42: 1784–1795.

Reich, C.D., E.A. Shinn, T.D. Hickey, and A.B. Tihansky. 2002. Tidal and meteorological influence on shallow marine groundwater flow in the upper Florida Keys. In J.W. Porter and K.G. Porter (Eds.), *The Everglades, Florida Bay and Coral Reefs of the Florida Keys: An Ecosystem Sourcebook.* CRC Press, Boca Raton, FL. pp. 659–676.

Richey, J.E., R.C. Wissmar, A.H. Devol, G.E. Likens, J.S. Eaton, W.E. Odum, N.M. Johnson, O.L. Loucks, R.T. Prentke, and P.H. Rich. 1978. Carbon flow in four lake ecosystems: a structural approach. *Science* 202, 1183–1186.

Robertson, B. 1982. Guardian of Apalachicola Bay. *Oceans* 5:65–67.

Rosenberg, D.M., R.A. Bodaly, R.E. Hecky, and R.W. Newbury. 1987. The environmental assessment of hydroelectric impoundments and diversions in Canada. In M.C. Healy and R.R. Wallace (Eds.), Canadian Aquatic Resources. *Canadian Bulletin of Fisheries and Aquatic Science* 215: 71–104.

Round, F.E. 1981. *The Ecology of Algae.* Cambridge University Press, Cambridge and London. Sournia, A. (Ed.). Phytoplankton Manual. Monographs on Oceanic Methodology. UNESCO Paris. 653 pp.

Rudnick, D.T., Z. Chen, D.L. Childers, J.N. Boyer, and T.D. Fontaine III. 1999. Phosphorus and nitrogen inputs to Florida Bay. The importance of the Everglades watershed. In J.W. Fourqurean, M.B. Robblee, and L.A. Deegan (Eds.), Florida Bay: A dynamic subtropical estuary. *Estuaries* 22: 398–416.

Ryther, J.H. and W.M. Dunstan. 1971. Nitrogen, phosphorus, and eutrophication in the coastal marine environment. *Science* 171: 1008–1013.

Santos, S.H. and S.A. Bloom. 1980. Stability in an annually defaunated estuarine soft-bottom community. *Oecologia* 46: 290–294.

Santos, S.L. and J.L. Simon. 1980a. Marine soft-bottom community-establishment following annual defaunation: larval or adult recruitment? *Mar. Ecol.* 2: 235–241.

Santos, S.L. and J.L. Simon. 1980b. Response of soft-bottom benthos to annual catastrophic disturbance in a south Florida estuary. *Mar. Ecol.* 3: 347–355.

Schmidt-Gengenbach, J. 1991. A Study of the Effects of Pollutants on the Benthic Macroinvertebrates in Lake Jackson, Florida: A Descriptive and Experimental Approach. M.S. thesis, Department of Biological Sciences and Center for Aquatic Research and Resource Management, Florida State University, Tallahassee, FL.

Seal, T.L., F.D. Calder, G.M. Sloane, S.J. Schropp, and H.L. Windom. 1994. Florida Coastal Sediment Contaminants Atlas: A Summary of Coastal Sediment Quality Surveys. Florida Department of Environmental Protection, Tallahassee, FL.

Seliger, H.H. and J.A. Boggs. 1988. Long term pattern of anoxia in the Chesapeake Bay. In M.P. Lynch and E.C. Krome (Eds.), *Understanding the Estuary: Advances in Chesapeake Bay Research.* Chesapeake Research Consortium, Solomons, MD. pp. 570–583.

Sellner, K.G., S.E. Shumway, M.W. Luckenbach, and T.L. Cucci. 1995. The effects of dinoflagellate blooms on the oyster *Crassostrea virginica* in Chesapeake Bay. In P. Lassus, G. Arzul, E. Erard, P. Gentien, and C. Marcalliou (Eds.), *Harmful Marine Algal Blooms.* Lavoisier, Intercept Ltd. pp. 505–511.

Sheridan, P.F. 1978. Trophic Relationships of Dominant Fishes in the Apalachicola Bay System (Florida). Dissertation, Florida State University. Tallahassee, FL.

Sheridan, P.F. 1979. Trophic resource utilization by three species of sciaenid fishes in a northwest Florida estuary. *Northeast Gulf Sci.* 3: 1–15.

Sheridan, P.F. and R.J. Livingston. 1979. Cyclic trophic relationships of fishes in an unpolluted, river-dominated estuary in North Florida. In R.J. Livingston (Ed.), *Ecological Processes in Coastal and Marine Systems.* Plenum Press, New York. pp. 143–161.

Sheridan, P.F. and R.J. Livingston. 1983. Abundance and seasonality of infauna and epifauna inhabiting a *Halodule wrightii* meadow in Apalachicola Bay, Florida. *Estuaries* 6: 407–419.

Shimada, M., T.H. Murakami, T. Imahayashi, H.S. Ozaki, T. Toyashima, and T. Okaichi. 1983. Effects of sea bloom, *Chattonella antiqua*, on grill primary lamellae of the young yellowtail, *Seriola quinqueradiata. Acta Histochem. Cytochem.* 16: 232–244.

Shoplock, B. Ecology of Zooplankton in Lake Jackson 1999. M.S. thesis, Department of Biological Science, Florida State University, Tallahassee, FL.

Sittig, M.H. 1985. *Handbook of Toxic and Hazardous Chemicals and Carcinogens.* Noyes Publications, Park Ridge, NJ.

Sklar, F., C. McVoy, R. VanZee, D.E. Gawlik, K. Tarboton, D. Rudnick, and S. Miao. 2002. The effects of altered hydrology on the ecology of the Everglades. In J.W. Porter and K.G. Porter (Eds.), *The Everglades, Florida Bay and Coral Reefs of the Florida Keys: An Ecosystem Sourcebook.* CRC Press, Boca Raton, FL. pp. 39–82.

Skulberg, O.M., B. Underdal, and H. Utkilen. 1994. Toxic waterblooms with cyanophytes in Norway — current knowledge. *Algological Studies* 75: 279–289.

Smayda, T.J. 1980. Phytoplankton species succession. In I. Morris (Ed.), *The Physiological Ecology of Phytoplankton.* University of California Press. Berkeley, CA. pp. 493–570.

Smayda T.J. 1989. Primary production and the global epidemic of phytoplankton blooms in the sea: a linkage? In E.M. Cosper, J. Carpenter, and V.M. Bricelj (Eds.), *Novel Phytoplankton Blooms: Causes and Impacts of Recurrent Brown Tides and Other Unusual Blooms.* Springer-Verlag, Berlin.

Smayda, T.J. 1990. Novel and nuisance phytoplankton blooms in the sea. Evidence for a global epidemic. In E. Granneli, B. Sundstrom, R. Edler, and D.M. Anderson (Eds.), *Toxic Marine Phytoplankton: Proceedings of the Fourth International Conference.* Elsevier, New York. pp. 29–40.

Smith, G.A., J.S. Nickels, W.M. Davis, R.F. Martz, R.H. Findlay, and D.C. White. 1982. Perturbations in the biomass, metabolic activity, and community structure of the estuarine detrital microbiota: resource partitioning in amphipod grazing. *J. Exp. Mar. Biol. Ecol.* 64: 125–143.

Smith, N.P. and P.A. Pitts. 2002. Regional-scale and long-term transport patterns in the Florida Keys. In J.W. Porter and K.G. Porter (Eds.), *The Everglades, Florida Bay and Coral Reefs of the Florida Keys: An Ecosystem Sourcebook.* CRC Press, Boca Raton, FL. pp. 343–360.

Smith, S.V. 1981. Response of Kaneohe Bay, Hawaii to relaxation of sewage stress. In B.G. Neilson and L.E. Cronin (Eds.), *Estuaries and Nutrients.* Humana Press, Clifton, NJ. pp. 391–410.

Sournia, A. 1978. *Phytoplankton Manual.* Monographs on Oceanic Methodology 6. UNESCO, Paris. 337 pp.

Sournia, A., M.J. Chretiennot-Dinet, and M. Ricard. 1991. Marine phytoplankton: how many species in the world ocean? *J. Plankton Res.* 13: 1093–1099.

South Florida Water Management District. 1994. An Update of the Surface Water Improvement and Management Plan for Biscayne Bay. Draft Final Report, West Palm Beach, FL.

South Florida Water Management District. 1999. Everglades Interim Report. West Palm Beach, FL.

South Florida Water Management District. 2000. 2000 Everglades Consolidated Report. West Palm Beach, FL.

South Florida Water Management District. 2001a. 2001 Everglades Consolidated Report. West Palm Beach, FL.

South Florida Water Management District. 2001b. Kissimmee River Restoration Project: Current Events. Accessed February 7, 2001, at http://www.sfwmd.gov/org/erd/krr/events/3_krrce.html.

Sowles, J.W. 1997a. Memo to Stacey Ladner. Subject: Interpretation of Hg in Penobscot Sediments. Maine Department of Environmental Protection. Augusta, ME.

Sowles, J.W. 1997b. Water resources survey — HoltraChem (Part II). September 29, 1997 memorandum to Stacey Ladner, Maine Department of Environmental Protection. Augusta, ME.

Sowles, J.W. 1999. Mercury contamination in the Penobscot River estuary at HoltraChem Manufacturing Company — An evaluation of monitoring data and interpretation of toxic potential and ecological implications. April 23, 1999 data summary report. Maine Department of Environmental Protection. Augusta, ME.

Squires, L.E. and N.A. Sinnu. 1982. Seasonal changes in the diatom flora in the estuary of the Damour River, Lebanon. *7th Diatom Symposium.* pp. 359–372.

Stankelis, R.M., M.D. Naylor, and W.R. Boynton. 2003. Submerged aquatic vegetation in the mesohaline region of the Patuxent estuary: past, present and future status. *Estuaries* 26: 186–195.

Steinman. A.D., K.E. Havens, H.J. Carrick, and R. VanZee. 2002. The past, present, and future hydrology and ecology of Lake Okeechobee and its watersheds. In J.W. Porter and K.G. Porter (Eds.), *The Everglades, Florida Bay and Coral Reefs of the Florida Keys: An Ecosystem Sourcebook.* CRC Press, Boca Raton, FL. pp. 19–37.

Stern, A.H. 2004. Letter to the editor, More on mercury content in fish. *Science* 303: 763.

Stewart, W.D.P. 1974. *Alga Physiology and Biochemistry.* Blackwell Scientific Publications, London. 989 pp.

Stockner, J.G. and D.D. Cliff. 1976. Effects of pulp mill effluent on phytoplankton production in coastal marine waters of British Columbia. *J. Fish. Res. Bd. Can.* 33: 2433–2442.

Stockner, J.G. and D.D. Costella. 1976. Marine phytoplankton growth in high concentrations of pulp mill effluent. *J. Fish. Res. Bd. Can.* 33: 2758–2765.

Stoner, A.W. 1976. Growth and food conversion efficiency of pin-fish (*Lagodon rhomboides*) exposed to sublethal concentrations of bleached kraft mill effluents. M.S. thesis, Florida State University. Tallahassee, FL.

Stoner, A.W. 1979a. The macrobenthos of seagrass meadows in Apalachee Bay, Florida, and the feeding ecology of Lagodon rhomboides (Pisces: Sparidae). Ph.D. dissertation, Florida State University, Tallahassee, FL.

Stoner, A.W. 1979b. Species-specific predation on amphipod Crustacea by the pinfish, Lagodon rhomboides: mediation by macrophyte standing *Crop. Mar. Biol.* 55: 201.

Stoner, A.W. 1980a. The role of seagrass biomass in the organization of benthic macrophyte assemblages. *Bull. Mar. Sci.* 30: 537–551.

Stoner, A.W. 1980b. Abundance, reproductive seasonality and habitat preferences of amphipod crustaceans in seagrass meadows of Apalachee Bay, Florida. *Contr. Mar. Sci.* 23: 63–77.

Stoner, A.W. 1980c. Feeding ecology of the Lagodon rhomboides (Pisces: Sparidae): variation and functional responses. *Fish. Bull.* 78: 337–352.

Stoner, A.W. 1982. The influence of benthic macrophytes on the foraging behavior of pinfish *Lagodon rhomboides* (Linnaeus). *J. Exp. Mar. Biol. Ecol.* 58: 271–284.

Stoner, A.W. and R.J. Livingston. 1980. Distributional ecology and food habits of the banded blenny *Paraclinus fasciatus* (Clinidae): a resident in a mobile habitat. *Mar. Biol.* 56: 239–246.

Stoner, A.W. and R.J. Livingston. 1984. Ontogenetic patterns in diet and feeding morphology in sympatric sparid fishes from seagrass meadows. *Copeia* (1984): 174–187.

Stoner, A.W., H.S. Greening, J.D. Ryan, and R.J. Livingston. 1982. Comparison of macrobenthos collected with cores and suction dredge. *Estuaries* 6: 76–82.

Su, S.H., L.C. Pearlman, J.A. Rothrock, T.J. Iannuzzi, and B.L. Finley. 2002. Potential long-term ecological impacts caused by disturbance of contaminated sediments: a case study. *Environ. Manage.* 29: 234–249.

Sullivan, M.J. 1978. Diatom community structure: taxonomical and statistical analysis of a Mississippi salt marsh. *J. Phycol.* 14: 468–475.

Swanson, J.R. 1991. Quantification of Environmental Impact Factors from Various Physiographic Regions of Leon County Using a Delphi Consensus Technique. M.S. thesis, Department of Geography, Florida State University, Tallahassee, FL.

Tanner, W.F. 1960. Florida Coastal Classification. *Trans. Gulf Coast Assoc. Geol. Soc.* 10: 259–266.

Thornton, J.A., H. Beekman, G. Boddington, R. Dick, W.R. Harding, M. Lief, I.R. Morrison, and A.J.R. Quick. 1995a. The ecology and management of Zandvlei (Cape Province, South Africa), an enriched shallow African estuary. In A.J. McComb (Ed.), *Eutrophic Shallow Estuaries and Lagoons.* CRC Press, Boca Raton, FL. pp. 109–128.

Thorpe, P., R. Bartel, P. Ryan, K. Albertson, T. Pratt, and D. Cairns. 1997. The Pensacola Bay System Surface Water Improvement and Management Plan. Northwest Florida Water Management District, Havana, FL. Unpublished report.

Todd, G. 1980. Annual study of the mercury contamination of the fish and sediments in the South, South Fork, Shenandoah, and Shenandoah Rivers. Virginia Water Control Board. Unpublished report.

Tomas, C.R., B. Bendis, and D.K. Johns. 1999. Role of nutrients in regulating plankton blooms in Florida Bay. In H. Kumpf, K. Steidinger, and K. Sherman (Eds.), *The Gulf of Mexico: Large Marine Ecosystem.* Blackwell Science, New York. pp. 323–337.

Toner, W. 1975. Oysters and the good 'ol boys. *Planning* 41: 10–15.

Tougas, J.I. and J.W. Porter. 2002. Differential coral recruitment patterns in the Florida Keys. In: J.W. Porter and K.G. Porter (Eds.), *The Everglades, Florida Bay and Coral Reefs of the Florida Keys: An Ecosystem Sourcebook.* CRC Press, Boca Raton, FL. pp. 289–811.

Tuovila, B.J., T.H. Johegen, P.A. LaRock, J.B. Outland, D.H. Esry, and M. Franklin. 1987. An evaluation of the Lake Jackson filter system and artificial marsh on nutrient and particulate removal from stormwater runoff. In K.R. Reddy and W.H. Smith (Eds.), *Aquatic Plants for Water Treatment and Resource Recovery.* Magnolia Publ. pp. 271–278.

Turner, R.E., W.W. Schroeder, and W.J. Wiseman, Jr. 1987. The role of stratification in the deoxygenation of Mobile Bay and adjacent shelf bottom waters. *Estuaries* 10: 13–20.

U.S. Army Corps of Engineers. 1976. Statement of Findings: Perdido Pass Channel (maintenance dredging), Baldwin County, AL.

U.S. Environmental Protection Agency. 1971. Conference in the Matter of Pollution of the Interstate Waters of the Escambia River Basin (Alabama-Florida) and the Intrastate Portions of the Escambia Basin within the State of Florida. Pensacola, FL.

U.S. Environmental Protection Agency. 1976. Quality Criteria for Water. Office of Water and Hazardous Materials, Washington, D.C.

U.S. Environmental Protection Agency. 1983. Methods for chemical analysis of water and wastes. EPA-600/4-79-020. Environmental Monitoring and Support Laboratory, Office of Research and Development, United States Environmental Protection Agency, Cincinnati, OH.

U.S. Environmental Protection Agency. 1989. Ambient water quality criteria for ammonia (saltwater). Office of Water Regulations and Standards, Criteria and Standards Division, Washington, D.C. (EPA 440/S88004).

U.S. Environmental Protection Agency. 1990. Analysis of the section 301(h) secondary treatment variance application by Los Angeles County Sanitation Districts for Joint Water Pollution Control Plant. Region 9, Water Management Division.

U.S. Environmental Protection Agency. 1993. Water-quality protection program for the Florida Keys National Marine Sanctuary. Phase II report. U. S. Environmental Protection Agency, Washington, D.C.

U.S. Environmental Protection Agency. 1997a. Mercury Report to Congress. An Assessment of Exposure to Mercury in the United States. Office of Air Quality Planning and Standards, and Office of Research and Development, EPA452/R-97-006.

U.S. Environmental Protection Agency. 2000. Fenholloway Nutrient Study, Perry FL. U.S. Environmental Protection Agency. Region 4, Athens, GA.

U.S. Environmental Protection Agency. 2003. River Corridor and Wetland Restoration. Washington D.C.

U.S. Fish and Wildlife Service. 1990. Aerial Photography of the Seagrass Beds of Perdido Bay. Unpublished report, Pensacola, Florida.

Valiela, L., J. McClelland, J. Hauxwell, P.J. Behr, D. Hersh, and K. Foreman. 1997. Macroalgal blooms in shallow estuaries: controls and ecophysiological and ecosystem consequences. *Limnol. Oceanogr.* 42, 1105–1118.

Van Es, F.B., M.A. Van Arkel, L.A. Bowman, and H.G.J. Schroder. 1980. Influence of organic pollution on bacterial, macrobenthic and meiobenthic populations in intertidal flats of the Dollard. *Neth. J. Sea Res.* 14: 288–304.

Van Raalte, C.D., I. Valeila, and J.M. Teal. 1976. The effect of fertilization on the species composition of salt marsh diatoms. *Water Res.* 10, 1–4.

Wagner, J.R. 1984. Hydrogeologic assessment of the October 1982 draining of Lake Jackson, Leon County, Florida. Unpublished report. Northwest Florida Water Management District, Wat. Res. Special Rep. 84-1.

Wald, A. and J. Wolfowitz. 1940. On a test whether two samples are from the same population. *Ann. Math. Statistics* 11: 147–162.

Wallin, J.M., M.D. Hattersley, D.F. Ludwig, and T.J. Iannuzzi. 2002. Historical assessment of the impacts of chemical contamination in sediments on benthic invertebrates in the tidal Passaic River, New Jersey. *Human & Ecological Risk Assessment* 8: 1156–1177.

Wanielista, M.P. 1976. Nonpoint Source Effects (with appendices). Unpublished report #ESEI -76-2. Florida Department of Environmental Regulation.

Wanielista, M.P. and Y.A. Yousef. 1985. Stormwater Management: An Update. Unpublished report. University of Central Florida. Environmental Systems Engineering Institute, Publication #85-1.

Wanielista, M.P., H.H. Harper, and S. Kuo. 1984. Bottom sediments: Megginnis Arm, Lake Jackson, Tallahassee, Florida. Unpublished report. College of Engineering, University of Central Florida.

Waterbury, J., S. Watson, R. Guillard, and L. Brand. 1979. Widespread occurrence of a unicellular, marine, planktonic, cyanobacterium. *Nature* 277: 293–294.

Weiss, R.F. 1970. Helium isotope effect in solution in water and sea water. *Science* 168, 247–248.

Wenning, R.J., M.A. Harris, B. Finley, D.J. Paustenbach, and H. Bedbury. 1993. Application of pattern recognition techniques to evaluate polychlorinated dibenzo-*p*-dioxin and dibenzofuran distributions in surficial sediments from the lower Passaic River and Newark Bay. *Ecotoxicol. Environ. Saf.* 25: 103–125.

White, D.C. 1983. Analysis of microorganisms in terms of quantity and activity in natural environments. Microbes in their natural environments. *Soc. Gen. Microbiol. Symp.* 34: 37–66.

White, D.C., R.J. Bobbie, S.J. Morrison, D.K. Oesterhof, C.W. Taylor, and D.A. Meeter. 1977. Determination of microbial activity of lipid biosynthesis. *Limnol. Oceanogr.* 22: 1089–1099.

White, D.C., R.J. Livingston, R.J. Bobbie, and J.S. Nickels. 1979a. Effects of surface composition, water column chemistry, and time of exposure on the composition of the detrital microflora and associated macrofauna in Apalachicola Bay, Florida. In R.J. Livingston (Ed.), *Ecological Processes in Coastal and Marine Systems*. Plenum Press, New York. pp. 53–67.

White, D.C., W.M. Davis, J.S. Nickels, J.D. King, and R.J. Bobbie. 1979b. Determination of the sedimentary microbial biomass by extractable lipid phosphate. *Oecologia* 40: 51–62.

Whitfield, M. 1974. The hydrolysis of ammonium ions in sea water — a theoretical study. *J. Mar. Biol. Assoc. U.K.* 54, 565–580.

Wieland, R. 2003. Loss of credibility could undermine Chesapeake cleanup effort. *Bay Journal* 13: comment 2.

Wiener, J.G. and P.J. Shields. 2000. Mercury in the Sudbury River (Massachusetts, U.S.A.): pollution history and a synthesis of recent research. *Can. J. Fish. Aquat. Sci.* 57: 1053–1061.

Wikfors, G.H. and R.M. Smolowitz. 1993. Detrimental effects of a *Prorocentrum* isolate upon hard clams and bay scallops in laboratory feeding studies. In T.J. Smayda and Y. Shimizu (Eds.), *Toxic Phytoplankton Blooms in the Sea. Proceedings of the Fifth International Conference on Toxic Marine Phytoplankton.* Elsevier, New York. pp. 447–452.

Wilcoxin, F. 1949. *Some rapid approximate statistical procedures.* American Cyanamid Company, Stamford, CN.

Wild, S.R., S.P. Mcgrath, and K.C. Jones. 1990a. Organic contaminant in an agricultural soil with a known history of sewage sludge amendments: polynuclear aromatic hydrocarbons. *Environ. Sci. Technol.* 24, 1706–1711.

Wild, S.R., S.P. Mcgrath, and K.C. Jones. 1990b. The polynuclear aromatic hydrocarbon (PAH) content of archived sewage sludges. *Chemosphere* 20(6), 703–716.

Woelke, C.E. 1961. Pacific oyster *Crassostrea gigas* mortalities with notes on common oyster predators in Washington waters. *Proc. Nat. Shellfish. Assoc.* 50: 53–66.

Wolfe, D.A., E.R. Long, and G.B. Thursby. 1996. Sediment toxicity in the Hudson–Raritan Estuary: distribution and correlations with chemical contamination. *Estuaries* 19: 901–912.

Wolfe, M.F., S. Schwarzbach, and R.A. Sulaiman. 1998. Effects of mercury on wildlife: a comprehensive review. *Env. Toxicol. Chem.* 17: 146–160.

Wright, R.T. and B.J. Nebel. 2002. *Environmental Science: Toward a Sustainable Future.* Prentice Hall, Upper Saddle River, NJ.

Wu, T.S. and W.K. Jones. 1992. Preliminary circulation simulations in Apalachicola Bay. In M.L. Spaulding, K. Bedford, A. Blumberg, R. Cheng, and C. Swanson (Eds.), *Estuarine and Coastal Modeling. Proceedings of the 2nd International Conference.* American Society of Civil Engineering, New York. pp. 344–356.

Yerger, R.W. 1977. Fishes of the Apalachicola River. In R.J. Livingston (Ed.), *Proceedings of the Conference on the Apalachicola Drainage System.* Florida Department of Natural Resources, Marine Resources Publication 26. St. Petersburg, FL. pp. 22–33.

Young, D.L. and R.T. Barber. 1973. Effects of waste dumping in New York Bight on the growth of natural populations of phytoplankton. *Environ. Pollut.* 5: 237–252.

Zeeman, C. 2001. Memo to S. Ladner regarding Media Protection Goals for Contaminants of Concern Released by the HoltraChem Manufacturing Facility, Orrington, Maine. Maine Department of Environmental Protection.

Zieman, J.C., J.W. Fourqurean, and T.A. Frankovich. 1999. Seagrass die-off in Florida Bay: long-term trends in abundance and growth of turtle grass, *Thalassia testudinum. Estuaries* 22: 460–470.

Zieman, J.C., J.W. Fourqurean, and T.A. Frankovich. 2004. Reply to B.E. Lapointe and P.J. Barile (2004), Comment on J.C. Zieman, J.W. Fourqurean, and T.A. Frankovich. 1999. Seagrass die-off in Florida Bay: long-term trends in abundance and growth of turtle grass, *Thalassia testudinum. Estuaries* 22: 460–470. *Estuaries* 27: 165–178.

Zillioux, E.J., D.B. Porcella, and J.M. Benoit. 1993. Mercury cycling and effects in freshwater wetland ecosystems. *Env. Toxicol. Chem.* 12: 2245–2264.

Zimmerman, M.S. and R.J. Livingston. 1976a. The effects of kraft mill effluents on benthic macrophyte assemblages in a shallow bay system (Apalachee Bay, North Florida, U.S.A.). *Mar. Biol.* 34: 297–312.

Zimmerman, M.S. and R.J. Livingston. 1976b. Seasonality and physico-chemical ranges of benthic macrophytes from a north Florida estuary (Apalachee Bay). *Cont. Mar. Sci., University of Texas* 20: 34–45.

Zimmerman, M.S. and R.J. Livingston. 1979. Dominance in benthic macrophyte assemblages from a north Florida estuary (Apalachee Bay). *Bull. Mar. Sci.* 29: 27–40.

Zongwei, C., V.M.S. Ramanujam, and M.L. Gross. 1994a. Levels of polychlorodibenzo-*p*-dioxins and dibenzofurans in crab tissues from the Newark/Raritan Bay System. *Environ. Sci. Technol.* 28: 1528–1534.

Zongwei, C., V.M.S. Ramanujam, and M.L. Gross. 1994b. Mass-profile monitoring in trace analysis: identification of polychlorodibenzothiophenes in crab tissues collected from the Newark/Raritan Bay system. *Environ. Sci. Technol.* 28: 1535–1538.

Index